CAMBRIDGE STUDIES IN
ADVANCED MATHEMATICS : 40

Explicit Brauer Induction

Already published

1 W.M.L. Holcombe *Algebraic automata theory*
2 K. Petersen *Ergodic theory*
3 P.T. Johnstone *Stone spaces*
4 W.H. Schikhof *Ultrametric calculus*
5 J.-P. Kahane *Some random series of functions, 2nd edition*
6 H. Cohn *Introduction to the construction of class fields*
7 J. Lambek & P.J. Scott *Introduction to higher-order categorical logic*
8 H. Matsumura *Commutative ring theory*
9 C.B. Thomas *Characteristic classes and the cohomology of finite groups*
10 M. Aschbacher *Finite group theory*
11 J.L. Alperin *Local representation theory*
12 P. Koosis *The logarithmic integral I*
13 A. Pietsch *Eigenvalues and s-numbers*
14 S.J. Patterson *An introduction to the theory of the Riemann zeta-function*
15 H.J. Baues *Algebraic homotopy*
16 V.S. Varadarajan *Introduction to harmonic analysis on semisimple Lie groups*
17 W. Dicks & M. Dunwoody *Groups acting on graphs*
18 L.J. Corwin & F.P. Greenleaf *Representations of nilpotent Lie groups and their applications*
19 R. Fritsch & R. Piccinini *Cellular structures in topology*
20 H Klingen *Introductory lectures on Siegel modular forms*
22 M.J. Collins *Representations and characters of finite groups*
24 H. Kunita *Stochastic flows and stochastic differential equations*
25 P. Wojtaszczyk *Banach spaces for analysts*
26 J.E. Gilbert & M.A.M. Murray *Clifford algebras and Dirac operators in harmonic analysis*
27 A. Frohlich & M.J. Taylor *Algebraic number theory*
28 K. Goebel & W.A. Kirk *Topics in metric fixed point theory*
29 J.F. Humphreys *Reflection groups and Coxeter groups*
30 D.J. Benson *Representations and cohomology I*
31 D.J. Benson *Representations and cohomology II*
32 C. Allday & V. Puppe *Cohomological methods in transformation groups*
33 C. Soulé et al *Lectures on Arakelov geometry*
34 A. Ambrosetti & G. Prodi *A primer of nonlinear analysis*
35 J. Palis & F. Takens *Hyperbolicity, stability and chaos at homoclinic bifurcations*
37 Y. Meyer *Wavelets and operators*
38 C. Weibel *An introduction to homological algebra*
39 W. Bruns & J. Herzog *Cohen–Macaulay rings*

EXPLICIT BRAUER INDUCTION
with applications to algebra and number theory

Victor P. Snaith
Britton Professor of Mathematics
McMaster University

Published by the Press Syndicate of the University of Cambridge
The Pitt Building, Trumpington Street, Cambridge CB2 1RP
40 West 20th Street, New York, NY 10011-4211, USA
10 Stamford Road, Melbourne, Victoria 3166, Australia

First published 1994

Printed in Great Britain at the University Press, Cambridge

Library of Congress cataloguing in publication data available

A catalogue record for this book is available from the British Library

ISBN 0 521 46015 8 hardback

Contents

Preface

'Funny that you should ask. As it happens I have a complete mathematical vision of power. You could call it a programme. Some pieces are in place, some theorems are proved, others await proof and revelation. It is based on the theory of schemes.'

'Of finite type?' asked Zhilin, with rapt curiosity.

'Any type, my friend, any type at all.'

'Schemes, eh.' Eli nodded his approval.'Good. Good.'

'Schemes are a little like the crania of topology with little light bulbs of algebra, called sheaves, stuck all over them.'

'Sheaves, very good. Very agriculdural.'

from *The Yukiad* (Snaith, 1990c)

This volume began as a one-term advanced graduate course in algebra which I gave at the beginning of 1990 at McMaster University. As originally conceived my plan was to give a brief introduction to the representation theory of finite groups in characteristic zero. This sketch was to have been succeeded by an outline of the topological construction of my original Explicit Brauer Induction formula (Snaith, 1988b, 1989b) followed by a description of the behaviour of Explicit Brauer Induction with respect to Adams operations 4.1.6 as originally proved in theorem 2.33 of Snaith (1989a). Equipped with 4.1.6 the course was then to have concluded with a discussion of class-groups of group-rings and a proof of M.J. Taylor's conjecture concerning determinantal congruences 4.3.10 (see also the stronger congruences of 4.3.37).

However, in 1989, I learnt of the work of Robert Boltje (1989, 1990) which axiomatised Explicit Brauer Induction formulae and, entirely algebraically, found a different formula. This second formula was easier to use than my original one — being a homomorphism rather than a derivation (see 2.3.28) — and its purely algebraic derivation was far better suited for exposition in a graduate course on algebra! As it happens, the formulae of Snaith (1988b) and Boltje (1990) are related by an equation which is to be found in 2.5.11. Indeed, for *p*-groups this relation may be used to derive either formula from the other (see 2.5.16).

By the time I had concluded the course I had collected quite a number of applications of Explicit Brauer Induction, several of which had figured vaguely in my original motivation but had taken sufficiently long to develop that they could not be included in Snaith (1989b). At that point it seemed to me advantageous to have an algebraic treatment of Explicit Brauer Induction, together with a selection of typical applications for the benefit of those who did not like the topological proofs of Snaith (1989b) which were apt, on occasion, to resort to stable homotopy theory to derive algebraic results!

Having described the origins of this book, let me move on to sketch its contents. Each chapter has a more detailed introduction to which the reader is referred for a more complete account. In general, the topics concern examples of 'applied representation theory' in which algebraic objects (class-groups, for example) are studied by means of finite-dimensional, complex representations of finite groups. Under these circumstances Explicit Brauer Induction enables one to use Brauer's induction theorem either *constructively* (see 6.3.6, for example) or with greater control (see 4.3.37). As a result, almost all the main results presented here are either new (for example, 4.3.37, 6.3.6 and 6.3.20) or are proved by a new method (for example, 4.5.39, 4.6.3 and 7.3.56).

Chapter 1 quickly covers the basic material concerning the finite-dimensional representations of a finite group. We specialise almost immediately to the case of complex representations, emphasising the properties of induced representations.

Chapter 2 begins with an account of Brauer's canonical version of Artin's induction theorem. Brauer's induction theorem, in its classical existential (non-canonical) form, is proved topologically by an extension of Snaith's (1988b) method. The formalism of Explicit Brauer Induction is recapitulated from Snaith (1989a,b) and Boltje's axioms are stated and shown to yield an Explicit Brauer Induction homomorphism *with rational coefficients.* The difficult part of the proof is to show that integrality of the Explicit Brauer Induction map, a_G, which is accomplished by Boltje's (1990) argument. However, for completeness, I have included a description of the topological construction of a_G which is due to Peter Symonds (1991) and which provides those who have topological tastes with a more conceptual, alternative point of view. As I have remarked previously, natural Explicit Brauer Induction formulae are not unique, since Snaith's (1988b) and Boltje's (1990) formulae are generally different. The chapter concludes by deriving a third Explicit Brauer Induction homomorphism, d_G, which is due to Robert Boltje. The map, d_G, has rational coefficients, is natural and commutes with induction. This remarkable homomorphism was discovered during the course of the joint work, which appears in Chapter 6, by R. Boltje, G-M. Cram and myself on conductors in the non-separable residue field case.

Chapter 3 studies the Explicit Brauer Induction formula when applied to an irreducible representation of the group of invertible, 2×2 matrices with entries in the field with q elements. Each of the irreducible representations of GL_2F_q is constructed and the leading terms of the Explicit Brauer Induction formulae are evaluated. In addition, it is shown how this 'leading term data' enables one to construct the Shintani correspondence between Frobenius-invariant, irreducible representations of one general linear group and the irreducible representations of the subgroup of Frobenius-fixed matrices. Our construction, which makes no mention of character-values or the Shintani norm map, offers an appealingly intrinsic approach to the 'base-change' which indicates just how useful Explicit Brauer Induction might prove to be if it were possible to extend the technique to the case of admissible representations of the general linear group of a local field (Gérardin & Labesse, 1979).

Chapter 4 introduces the Adams operations in the complex representation ring of a finite group (to be precise, in this special case these operations were originally due to Burnside). It is shown how the Explicit Brauer Induction formula enables one to write its image under any Adams operation as a linear combination of monomial representations. Next the Hom-description is given for the class-group of finitely generated, projective modules over the integral group-ring of a finite group. This description presents the class-group as a quotient of idèle-valued functions on the representation ring of the group modulo global-valued functions and other special functions which are called determinants. Martin Taylor (1978) conjectured that the Adams operations, when applied to determinantal functions, would satisfy some congruences modulo the residue degree. This determinantal congruence conjecture is proved, strengthened and reformulated to give some new homomorphisms whereby to detect class-groups. The chapter concludes with new proofs of two results concerning determinants. It is shown that the determinantal subgroup satisfies Galois descent in tamely ramified extensions, which was one of the three mains steps in M.J. Taylor's proof of Fröhlich's conjecture (Taylor, 1981). It is also shown that Adams operations preserve the determinantal subgroup, which was originally proved in a different manner by Ph. Cassou-Noguès and M.J. Taylor (Taylor, 1984).

Chapter 5 deals with six topics which are united merely by the fact that they are related to the class-group of a group-ring. In the previous chapter new maps out of the class-group were constructed by means of determinantal congruences. This chapter commences with the construction of *restricted determinants*, which are new maps into the class-group. The construction of the restricted determinant homomorphisms is later refined to yield a technique for the detection of the class-group of a maximal order in the rational group-ring of a finite group. In Chapter 7 this detection technique, which uses some new types of Hom-groups, is applied to give a new proof of David Holland's theorem, which states that the Fröhlich–Chinburg conjecture is true in the class-group of a maximal order. This chapter also contains a calculation of the class-group of the integral group-ring of the quaternion group of order eight. Every text on class-groups has its own version of this calculation; mine is accomplished by means of a little homological algebra and a new type of reduced norm invariant. Two sections deal with the subgroup, called the Swan subgroup, which is generated in the class-group by the projective modules which were so elegantly constructed by R.G. Swan (1960). By topological techniques using groups actions on spheres new families of relations between Swan modules are derived. These relations, together with the algebra of determinantal congruences, are used to calculate the Swan subgroup of some types of nilpotent groups. These calculations prompt us to venture a conjecture as to the identity of the Swan subgroup of any nilpotent group. The final section illustrates the non-triviality of the class defined in the class-group by the roots of unity which reappear in the material on real cyclotomic Galois module structure at the end of Chapter 7.

Chapter 6 deals with the problem, raised in Serre (1960) and Kato (1989), of constructing a Swan conductor for Galois representations of complete, discrete valuation fields whose residue fields are inseparable. The classical theory of the Artin and Swan conductors is recalled, together with Kato's definition of a Swan conductor for a one-dimensional representation. The Explicit Brauer Induction formula is used to define the required Swan conductor in general, and many properties are derived for this new conductor, which coincides with the classical Swan conductor when the residue field extension is separable. Examples are given to show that no such generalisation of the Swan conductor can possess the classical property of 'inductivity in dimension zero'. However, using the rational-valued Explicit Brauer Induction homomorphism, d_G, of Chapter 2 we show that our Swan conductor is 'inductive in dimension zero' if it is so on all p-subgroups of a particularly simple type.

Chapter 7 treats the Galois module structure of the unit group in a ring of algebraic integers. The main conjecture in this topic is due to A. Fröhlich and T. Chinburg and concerns the identity in the class-group of the group-ring of a Galois group of certain invariants which are constructed from the theory of class formations. In the tamely ramified case one of these invariants is simply the projective module furnished by the additive group of the ring of integers in the tame Galois extension of number fields. In general, one studies the global Chinburg invariant and attempts to equate it with an analytically defined class due to A. Fröhlich and Ph. Cassou-Noguès. The chapter begins with the construction of the local Chinburg invariants. These are described explicitly by means of the theory of central simple algebras and from this description a new proof is given of the result, due to T. Chinburg (1985), that the local invariants vanish in the tamely ramified case. The global Chinburg invariant is defined and the main conjecture is stated. The chapter contains a new proof, using the method of Chapter 5, of D. Holland's (1992) result, which states that the Fröhlich–Chinburg conjecture is true in the class-group of a maximal order. The chapter closes with some new families of (cyclotomic) extensions for which the Fröhlich–Chinburg conjecture is true in the class-group.

Each chapter is endowed with a selection of exercises, which vary from relatively straightforward problems, such as the completions of omitted proofs, to research problems suggested by the topics treated in the text.

As mentioned above, parts of this book have served as the basis for an advanced graduate course in algebra. In fact the material lends itself to such a purpose in a number of ways and the reader may find the following suggestions helpful in designing such courses. Twice I have given a course which covered Chapters 1 and 2 and then concluded by applying the results on Adams operations in Chapter 4 to derive the determinantal congruences. Such a course was particularly successful, since it assumes only a minimal algebraic background and culminates in a proof of M.J. Taylor's conjecture concerning determinantal congruences — all this being accomplished within 40 hours of lectures. A one-term introductory course on representation theory can be made from Chapter 1, and Chapter 3 and can be given some additional mystique by assuming the existence of the Explicit Brauer Induction homomorphism, a_G, in order to analyse the Shintani correspondence in the manner of Chapter 3. Finally, several two-term courses may be designed by combining the contents of the first two chapters with any one of Chapter 4 (and possibly Chapter 5), Chapter 6 or Chapter 7. The last three alternatives are sufficiently advanced that, in the course of the related exercises, they introduce the student to a number of interesting research problems.

In the course of writing this book I have been helped by conversations, correspondence and comments from many mathematicians. In particular, I am very grateful to Robert Boltje for his improvements on my original Explicit Brauer Induction formula and for his help in our joint work with Georg-Martin Cram, which forms Chapter 6. Conversations with Greg Hill, Charles Curtis and my student, Brian McCudden, gave rise to the treatment of Shintani descent which appears in Chapter 2. I am also grateful to Ted Chinburg, Ali Fröhlich, David Holland and Martin Taylor for their interest in and advice concerning Galois module structure. In addition, it was Martin Taylor who started me thinking about the Swan subgroup of a nilpotent group and about the problem of Terry Wall, concerning the Swan subgroup of a product, which appears in Chapter 5.

Victor Snaith,
McMaster University,
September 1993

1
Representations

Introduction

In this chapter we recall the most elementary facts about the finite-dimensional representations of a finite group over a field. In our applications we shall mainly be interested in representations which are realised over some subfield of the complex numbers.

Therefore, in Section 1, we introduce unitary, orthogonal and symplectic representations in preparation for our topological treatment of Brauer's Induction Theorem 2.1.20.

In Section 2 we specialise to the case of complex representations and recapitulate in brief the theory of the character (or trace) of a representation and the orthogonality relations which are satisfied by characters of irreducible representations. Induced representations are introduced and their standard adjointness properties (e.g. Frobenius reciprocity, the Double Coset formula) are derived.

Section 3 consists of exercises of an introductory nature.

1.1 Basic definitions

Let G be a finite group. Let K be a field and let V be a finite-dimensional vector space over K. Let $GL(V)$ denote the group of K-linear automorphisms of V. A homomorphism

1.1.1
$$\rho : G \longrightarrow GL(V)$$

gives rise to an action of G on V by means of K-linear automorphisms. Explicitly, if $g \in G$ and $v \in V$ the action is given by

1.1.2
$$g \cdot v = \rho(g)(v).$$

A *finite-dimensional K-representation* of G is the K-isomorphism class of such an action. That is, the representation given by a G-action on V_1 is equivalent to the representation given by a G-action on V_2 if and only if there is a K-linear isomorphism, $\beta : V_1 \xrightarrow{\cong} V_2$, such that $\beta(g \cdot v_1) = g \cdot \beta(v_1)$ for all $g \in G$, $v_1 \in V_1$. In terms of homomorphisms, ρ, of 1.1.1 two homomorphisms

1.1.3 $$\rho_1, \rho_2 : G \longrightarrow GL(V)$$

give rise to the same representation if and only if there exists $B \in GL(V)$ such that $B\rho_1(g)B^{-1} = \rho_2(g)$ for all $g \in G$. Very often we will choose an isomorphism between V and K^n, where $dim_K(V) = n$ is the dimension of V. In that case a representation will become a conjugacy class of homomorphisms of the form

1.1.4 $$\rho : G \longrightarrow GL_n(K) = GL(K^n).$$

Example 1.1.5 (i) *One-dimensional representations* Let $K^* = K - 0 \cong GL_1(K)$ denote the multiplicative group of non-zero elements of K. A homomorphism, $\rho : G \longrightarrow K^*$, gives rise to a unique one-dimensional representation, since K^* is abelian.

(ii) *Permutation representations* Let G act on a finite set, X. Form the vector space, V, whose K-basis consists of the elements of X. Therefore G acts on V by permuting the basis vectors. In terms of a homomorphism into $GL_n(K)$, where $n = \#(X)$, we obtain a homomorphism, $\rho : G \longrightarrow GL_n(K)$, in which $\rho(g)$ has only one non-zero entry in each row or column. If we order the basis $\{x_i \in X; 1 \leq i \leq n\}$ then the (i, j)th entry in $\rho(g)$ is 1 if $g(x_j) = x_i$ and zero otherwise. Such a matrix is called a *permutation matrix*.

1.1.6 If V is a G-representation then a K-subspace, W, of V which is preserved by the action of G is called a *subrepresentation*. For example, in the permutation representation of 1.1.5(ii) the subspace given by

$$W = \left\{ \sum_{i=1}^{n} \lambda_i x_i \mid \sum \lambda_i = 0 \right\}$$

yields a subrepresentation.

1.1.7 Functorial operations on vector spaces induce corresponding operations on G-representations. The direct sum, tensor product and the exterior power operations are three fundamental examples.

If V_1 and V_2 are two vector spaces with K-linear G-actions then so is the *direct sum*, $V_1 \oplus V_2$, if we define

$$g(v_1 \oplus v_2) = gv_1 \oplus gv_2 \quad (g \in G, v_i \in V_i).$$

In terms of matrices the direct sum of $\rho_i : G \longrightarrow GL_{n_i}(K)$ $(i = 1, 2)$ is given by the homomorphism

$$\rho_1 \oplus \rho_2 : G \longrightarrow GL_{n_1+n_2}(K)$$

$$(\rho_1 \oplus \rho_2)(g) = \begin{pmatrix} \rho_1(g) & 0 \\ 0 & \rho_2(g) \end{pmatrix}.$$

The *tensor product*, $V_1 \otimes V_2$, is the vector space $Hom_K(Bil_K(V_1 \times V_2, K), K)$ in the finite-dimensional context. Here $Hom_K(V, W)$ denotes the space of K-linear maps from V to W and $Bil_K(V \times W, Z)$ denotes the space of K-bilinear maps. The dimension of $V_1 \otimes V_2$ is equal to $dim_K(V_1) \cdot dim_K(V_2)$, the product of the dimensions of V_1 and V_2.

Similarly we may define the tth *exterior power* of V, $\lambda^t(V)$, by setting

$$\lambda^t(V) = Hom_K(Alt_t(V), K),$$

where $Alt_K(V)$ denotes the subspace of multilinear maps,

$$f : V \times V \times \cdots \times V \longrightarrow K \ (t \text{ factors}),$$

such that

$$f(v_1, v_2, \ldots, v_t) = (-1)^{sign(\sigma)} f(v_{\sigma(1)}, v_{\sigma(2)}, \ldots, v_{\sigma(t)}) \qquad (v_i \in V_i)$$

for any permutation, σ, of $\{1,\ldots,t\}$. When $char(K) \neq 2$, the dimension of $\lambda^t(V)$ is given by the binomial coefficient, $\binom{n}{t}$, where $dim_K(V) = n$.

In terms of matrix homomorphisms the matrix representing $V_1 \otimes V_2$ is the Kronecker product of the representing matrices, ρ_1 and ρ_2. When $t = dim_K(V)$ the matrix representing $\lambda^n(V)$ is given by $det(\rho) : G \longrightarrow K^*$ for ρ as in 1.1.4.

Theorem 1.1.8 (*Maschke's theorem*) *Suppose that the characteristic of K is prime to the order of G. Let V be a K-representation of G and let W be a subrepresentation. Then there exists a subrepresentation, W_1, such that $W \oplus W_1 = V$.*

Proof There exists a K-linear map, $j : V \longrightarrow W$, such that $j(w) = w$ for all $w \in W$. If $j(g \cdot v) = g \cdot (j(v))$ for all $g \in G, v \in V$ then we could set $W_1 = ker(j)$, the kernel of j. However, j may not commute with the G-action so we replace it by $j_1 : V \longrightarrow W$, defined by $j_1(g) = (\#(G))^{-1}(\sum_{g \in G} g(j(g^{-1}(v))))$. Clearly $j_1(g \cdot v) = g \cdot (j_1(v))$ for all $g \in G, v \in V$ and if $w \in W$ then

$$j_1(w) = (\#(G))^{-1} \left(\sum_{g \in G} g(g^{-1}(v)) \right) = w,$$

as required. □

1.1.9 A K-representation, V, of G is called *indecomposable* if an isomorphism of the form $V = W \oplus W_1$, for subrepresentations W and W_1, implies either that $W = 0$ or $W = V$. V is called *irreducible* if V has no subrepresentations except $\{0\}$ and V. Theorem 1.1.8 states that these notions coincide when the order of G is prime to the characteristic of K, which is the situation which will mainly concern us.

1.1.10 We will often be dealing with representations afforded by vector spaces over the field, $\mathbf{C}$, of complex numbers. The group, $GL_n(\mathbf{C})$ has a number of compact subgroups which are of special interest. The *unitary group*, $U(n)$, is defined to be

1.1.11 $$U(n) = \{X \in GL_n(\mathbf{C}) | XX^* = I_n\},$$

where I_n is the $n \times n$ identity matrix and X^* is the matrix whose (i, j)th entry is $\overline{X}_{ji}$, where $\overline{z}$ denotes the complex conjugate of $z \in \mathbf{C}$. A *unitary representation* will mean the $U(n)$-conjugacy class of a homomorphism of the form

1.1.12 $$\rho : G \longrightarrow U(n).$$

Clearly, each unitary representation gives rise to a $\mathbf{C}$-representation of G and this induces a one-one correspondence between $U(n)$-representations and n-dimensional $\mathbf{C}$-representations (see 1.3.1). Note that $U(n)$ is the subgroup of $GL_n(\mathbf{C})$ consisting of matrices which preserve the semi-linear inner product on $\mathbf{C}^n$ given by $< \underline{x}, \underline{y} >= \sum_{i=1}^n x_i \cdot \overline{y}_i$, where $\underline{x} = (x_1, \ldots, x_n)$ and $\underline{y} = (y_1, \ldots, y_n)$.

Similarly, if $\mathbf{R}$ is the field of real numbers then the *orthogonal group*, $O(n)$, is the subgroup of matrices which preserve the inner product $< \underline{x}, \underline{y} >= \sum_{i=1}^n x_i \cdot y_i$ so that

1.1.13 $$O(n) = GL_n(\mathbf{R}) \bigcap U(n).$$

An *orthogonal representation* is an $O(n)$-conjugacy class of a homomorphism of the form

1.1.14 $$\rho : G \longrightarrow O(n).$$

As in the complex case there is a one–one correspondence between $O(n)$-representations and n-dimensional $\mathbf{R}$-representations of G (see 1.3.2).

Let $\mathbf{H}$ denote the quaternion skew-field. If $z = a + ib + jc + kd$ is a quaternion then $\overline{z} = a - ib - jc - kd$. On $\mathbf{H}^n$ we have an inner product given by $< \underline{x}, \underline{y} >= \sum_{i=1}^{n} x_i \cdot \overline{y}_i$ for $\underline{x} = (x_1, \ldots, x_n)$ and $\underline{y} = (y_1, \ldots, y_n)$. The group of (left) $\mathbf{H}$-semilinear automorphisms of $\mathbf{H}^n$ which preserve this form is called the *symplectic group*, $Sp(n)$. As a $\mathbf{C}$-vector space $\mathbf{H}$ is isomorphic to $\mathbf{C}^2$ so that $Sp(n)$ is a subgroup of $U(2n)$. A *symplectic representation* is a conjugacy class of homomorphisms of the form

1.1.15 $$\rho : G \longrightarrow Sp(n).$$

1.2 Complex representations

In this section we shall restrict our attention to $\mathbf{C}$-representations of G. Given a representation of the form of 1.1.4

1.2.1 $$\rho : G \longrightarrow GL_n(\mathbf{C}),$$

we define the *character* of ρ, χ_ρ, to be the $\mathbf{C}$-valued function given by

1.2.2 $$\chi_\rho(g) = Trace(\rho(g)) = \sum_{i=1}^{n} \rho(g)_{ii} \quad (g \in G).$$

Proposition 1.2.3 (i) *χ_ρ depends only on the class of ρ in* 1.2.1 *as a representation.*

(ii) *$\chi_\rho(g)$ is the sum of the eigenvalues of $\rho(g)$, counted with their multiplicities.*

(iii) $\chi_\rho(1) = dim(\rho) = n$.

(iv) $\chi_\rho(g^{-1}) = \overline{\chi_\rho(g)}$.

Proof If $X \in GL_n(\mathbf{C})$ set $\phi = X\rho X^{-1}$ then we must verify that $\chi_\phi = \chi_\rho$. However, if t is an indeterminate,

$$\begin{aligned} det(tI_n - \phi(g)) &= det(X(tI_n - \rho(g))X^{-1}) \\ &= det(tI_n - \rho(g)) \\ &= t^n - Trace(\rho(g))t^{n-1} + \ldots, \end{aligned}$$

which proves part (i).

Part (ii) follows by conjugating $\rho(g)$ into its Jordan canonical form, which is an upper triangular matrix whose diagonal entries are the eigenvalues. Each eigenvalue, λ, appears on the diagonal m_λ times, where m_λ is the multiplicity of λ.

Part (iii) is clear, since $Trace(I_n) = n$ and part (iv) follows from part (ii), since the eigenvalues of $\rho(g^{-1}) = \rho(g)^{-1}$ are the inverses of those for $\rho(g)$. However, since $\rho(g)$ has finite order, these eigenvalues are complex numbers of unit norm and therefore the inverse of each eigenvalue is equal to its complex conjugate. □

Proposition 1.2.4 *Let* $\rho_i : G \longrightarrow GL(V_i)$ $(i = 1, 2)$ *be two representations with character functions,* $\chi_i = \chi_{\rho_i}$. *Then*

(i) $\chi_1 + \chi_2$ *is the character of* $V_1 \oplus V_2$ *and*

(ii) $(\chi_1) \cdot (\chi_2)$ *is the character of* $V_1 \otimes V_2$.

Proof For $g \in G$ let $\rho_1(g) = X$ and $\rho_2(g) = Y$ then $(\rho_1 \oplus \rho_2)(g)$ is given by the matrix

$$\begin{pmatrix} X & 0 \\ 0 & Y \end{pmatrix}$$

whose trace is clearly equal to $Trace(X) + Trace(Y)$.

The tensor product, $\rho_1(g) \otimes \rho_2(g)$, is given by the Kronecker product of X and Y, by 1.3.3. This is the matrix of the form

$$\begin{pmatrix} X \cdot y_{11} & X \cdot y_{12} & \ldots & X \cdot y_{1n} \\ X \cdot y_{21} & X \cdot y_{22} & \ldots & X \cdot y_{2n} \\ \vdots & \vdots & \vdots & \vdots \\ X \cdot y_{n1} & X \cdot y_{n2} & \ldots & X \cdot y_{nn} \end{pmatrix},$$

where $dim_{\mathbf{C}}(\rho_2) = n$. The trace of this matrix is clearly equal to

$$Trace(X) \cdot [y_{11} + y_{22} + \ldots + y_{nn}].$$

□

Lemma 1.2.5 *(Schur's Lemma) Let V_1 and V_2 be irreducible representations of G. Let $Hom_G(V_1, V_2) =$*

$$\{f : V_1 \longrightarrow V_2 \text{ linear} \mid f(gv_1) = gf(v_1);\ \ g \in G, v_1 \in V_1\}.$$

Then

(i) *if $V_1 \neq V_2$, then $Hom_G(V_1, V_2) = 0$,*
(ii) *if $V_1 = V_2$,*

$$Hom_G(V_1, V_1) = \{f \mid f(x) = \lambda \cdot x \text{ for some } \lambda \in \mathbf{C}\}.$$

Proof If $f : V_1 \longrightarrow V_2$ is non-zero then $ker(f)$ and $im(f)$ are subrepresentations of V_1 and V_2 respectively. Since $f \neq 0$, $ker(f) \neq V_1$ and $im(f) \neq 0$ so that $ker(f) = 0$ and $im(f) = V_2$, which means that f is an isomorphism. This means that $V_1 = V_2$ as representations. However, when $V_1 = V_2$, let λ be an eigenvalue for f. The subspace

$$W = \{v_1 \in V_1 \mid f(v_1) = \lambda \cdot v_1\}$$

is a non-zero subrepresentation of V_1 so that $V_1 = W$, as required. □

Corollary 1.2.6 *Let V_1,V_2 be as in 1.2.5. Let $f : V_1 \longrightarrow V_2$ be a linear map. Define $F : V_1 \longrightarrow V_2$ by*

$$F(v_1) = \#(G)^{-1}\left(\sum_{g \in G} g \cdot f(g^{-1} \cdot v_1)\right).$$

Then

(i) $F = 0$ *if $V_1 \neq V_2$ and*
(ii) $F(v_1) = (dim V_1)^{-1} Trace(f) \cdot v_1$ *if $V_1 = V_2$.*

Proof Clearly $F(g \cdot v_1) = g \cdot F(v_1)$ for all $g \in G$ so that $F = 0$ unless $V_1 = V_2$. If $V_1 = V_2$ then $F(v_1) = \lambda \cdot v_1$ for all $v_1 \in V_1$ and, by taking traces,

$$(dim V_1) \cdot \lambda = \#(G)^{-1}\left(\sum_{g \in G} Trace(g \cdot f(g^{-1} \cdot -))\right) = Trace(f)$$

since $Trace(g \cdot f(g^{-1} \cdot -)) = Trace(f)$ for all $g \in G$. □

Definition 1.2.7 Suppose that W_1, W_2 are two representations with characters χ_1, χ_2 respectively. The *Schur inner product* $< W_1, W_2 > =$

$<\chi_1, \chi_2>$ is defined by

$$<\chi_1, \chi_2> = \#(G)^{-1}\left(\sum_{g \in G} \chi_1(g)\overline{\chi_2(g)}\right).$$

Theorem 1.2.8 *In* 1.2.7

(i) $<\chi_1, \chi_2> = dim_{\mathbf{C}} Hom_G(W_1, W_2)$,

(ii) $<\chi_1, \chi_2> = <\chi_2, \chi_1>$,

(iii) $<\chi_1, \chi_1> = 1$ *if and only if* W_1 *is irreducible.*

Proof Suppose that $A_1, \ldots, A_s, B_1, \ldots, B_t$ are irreducible representations such that

$$W_1 = \oplus_{i=1}^s A_i, W_2 = \oplus_{j=1}^t B_j,$$

then $Hom_G(W_1, W_2) \cong \oplus_{i,j} Hom_G(A_i, B_j)$. Now let χ_{A_i}, χ_{B_j} denote the characters of these irreducible representations. By 1.2.4, $\chi_1 = \sum_{i=1}^s \chi_{A_i}$ and $\chi_2 = \sum_{j=1}^t \chi_{B_j}$ so that

$$<\chi_1, \chi_2> = \sum_{i,j} <\chi_{A_i}, \chi_{B_j}>.$$

Therefore, in order to prove part (i), we may assume that W_1 and W_2 are irreducible.

Choose bases for W_1 and W_2 so that the representation, W_i, corresponds to $\rho_i : G \longrightarrow GL_{n_i}(\mathbf{C})$. In Corollary 1.2.6 assume that f is represented by a matrix, X. In this notation 1.2.6 becomes

1.2.9
$$\#(G)^{-1} \textstyle\sum_{a,b\, g \in G} \rho_2(g)_{ia} X_{ab} \rho_1(g)^{-1}_{bj}$$

$$= \begin{cases} 0 & \text{if } i \neq j \text{ or } W_1 \neq W_2 \\ (dim(W_1))^{-1} \cdot Trace(X) & \text{if } W_1 = W_2 \text{ and } i = j. \end{cases}$$

When $W_1 \neq W_2$ choose f with $X_{ab} = 0$ except for $X_{ij} = 1$ so that 1.2.9 implies that

$$\sum_{g \in G} \rho_2(g)_{ii} \rho_1(g^{-1})_{jj} = 0 \text{ for all } i, j.$$

Therefore

$$\begin{aligned}\#(G) < \chi_1, \chi_2 > &= \textstyle\sum_{g\in G} \chi_1(g) \cdot \overline{\chi_2(g)} \\ &= \textstyle\sum_{g\in G} \sum_i \sum_j \rho_1(g)_{ii} \cdot \rho_2(g^{-1})_{jj}, \quad \text{by 1.2.3(iv)}, \\ &= \textstyle\sum_{ij} \sum_{g\in G} \rho_1(g^{-1})_{ii} \rho_2(g)_{jj} \\ &= 0 \\ &= dim_{\mathbf{C}} Hom_G(W_1, W_2).\end{aligned}$$

When $W_1 = W_2$ we may again take $X_{ab} = 0$ except for $X_{ij} = 1$ to obtain

$$\sum_{g\in G} \rho_2(g)_{ii} \cdot \rho_1(g_{jj}^{-1}) = \begin{cases} 0 & \text{if } i \neq j, \\ \#(G) \cdot (dim W_1)^{-1} & \text{if } i = j. \end{cases}$$

Therefore

$$\begin{aligned}\#(G) \cdot < \chi_1, \chi_2 > &= \textstyle\sum_{i,j} \sum_{g\in G} \rho_1(g^{-1})_{ii} \cdot \rho_2(g)_{jj} \\ &= \#(G)\end{aligned}$$

and

$$1 = < \chi_1, \chi_2 > = < \chi_1, \chi_1 >,$$

as required.

To prove part (ii) we remark that $< \chi_2, \chi_1 >$ is the complex conjugate of $< \chi_1, \chi_2 >$, by definition, but $< \chi_1, \chi_2 >$ is a positive integer and therefore is fixed by complex conjugation.

Part (iii) follows from part (i), 1.2.5 and the observation that

$$dim_{\mathbf{C}} Hom_G(W_1, W_1) \geq \sum_{i=1}^{s} Hom_G(A_i, A_i) = s.$$

□

Corollary 1.2.10 *In the notation of* 1.2.7, $W_1 = W_2$ *as representations if and only if the characters* χ_1 *and* χ_2 *are equal.*

Proof Firstly, observe that

$$dim_{\mathbf{C}} W_1 = \chi_1(1) = \chi_2(1) = dim_{\mathbf{C}} W_2,$$

so that we may prove this by induction on dimension. When $dim_{\mathbf{C}} W_1 = 1$ both representations are irreducible and

$$dim_{\mathbf{C}} Hom_G(W_1, W_2) = dim_{\mathbf{C}} Hom_G(W_1, W_1) = 1,$$

by 1.2.8(i) so that, by 1.2.5, there is a G-isomorphism between W_1 and W_2. Now let A_1 be an irreducible summand of W_1 and set

$$W_2 = \oplus_{j=1}^{t} B_j,$$

as in the proof of 1.2.8. Hence

$$dim_{\mathbf{C}} Hom_G(A_1, W_2) = dim_{\mathbf{C}} Hom_G(A_1, W_1) \geq 1,$$

by 1.2.8(i). Therefore, for some j, $Hom_G(A_1, B_j) \neq 0$. By 1.1.8, $W_2 \cong A_1 \oplus W_2'$ and $W_1 \cong A_1 \oplus W_1'$. The characters of W_1' and W_2' are therefore equal, by 1.2.4, so that $W_1' \cong W_2'$ as representations, which completes the proof. □

Corollary 1.2.11 *If W and V are representations of G such that V is irreducible then $< W, V >$ is equal to the* multiplicity of V in W (*i.e. the number of times that V appears in a decomposition of W as a sum of irreducibles*).

Proof If $W = nV \oplus V_1 \oplus \ldots \oplus V_t$ with V_i irreducible and not equivalent to V then

$$\begin{aligned} < W, V > &= < V, W > \\ &= dim_{\mathbf{C}}(Hom_G(V, W) \\ &= n dim_{\mathbf{C}}(Hom_G(V, V) + \textstyle\sum_{i=1}^{t} dim_{\mathbf{C}}(Hom_G(V, V_i)) \\ &= n, \qquad \text{by 1.2.5} \end{aligned}$$

□

Definition 1.2.12 The *regular representation* of G is the permutation representation obtained, by the method of 1.1.5(ii), from the set G together with the G-action given by left multiplication. The regular representation will be denoted by $Ind_{\{1\}}^{G}(1)$ in recognition of its construction as an induced representation (see 1.2.31).

Proposition 1.2.13 *Let r_G denote the character of the regular representation. Then*

$$r_G(g) = \begin{cases} 0 & if\ g \neq 1, \\ \#(G) & if\ g = 1. \end{cases}$$

Proof By definition (see 1.1.4(ii)) a basis for $Ind_{\{1\}}^{G}(1)$ consists of $\{x \mid x \in G\}$. If $g \neq 1$ then $g \cdot x \neq x$ so that the trace of multiplication by g is trivial. When $g = 1$,

$$r_G(1) = dim_{\mathbf{C}}(Ind_{\{1\}}^G(1)) = \#(G).$$

□

Corollary 1.2.14 *Let V be an irreducible representation of G with character, χ_V, and dimension equal to n_V, then the multiplicity of V in r_G is equal to n_V.*

Proof By 1.2.7,

$$\begin{aligned} < \chi_V, r_G > &= \#(G)^{-1}(\textstyle\sum_{g \in G} \chi_V(g)\overline{r_G(g)}) \\ &= \#(G)^{-1}\chi_V(1)\#(G), \qquad \text{by 1.2.13,} \\ &= n_V, \end{aligned}$$

as required. □

Corollary 1.2.15 *Let $V_1, V_2, \ldots, V_t$ denote the distinct irreducible representations of G then* (i)

$$\#(G) = \sum_{i=1}^{t} dim_{\mathbf{C}}(V_i)^2,$$

and

(ii) *if $g \neq 1$ then*

$$0 = \sum_{i=1}^{t} dim_{\mathbf{C}}(V_i) \cdot \chi_{V_i}(g).$$

Proof By 1.2.14 the regular representation is equal to

$$\oplus_{i=1}^{t} dim_{\mathbf{C}}(V_i) \cdot V_i$$

so that part (i) follows by taking dimensions. Part (ii) follows by taking characters at $1 \neq g \in G$, using 1.2.13. □

Example 1.2.16 Corollary 1.2.15 may be used in finding all the irreducible representations of G. For example, let D_{2n} denote the *dihedral group of order* $2n$ given by

1.2.17 $$D_{2n} = \{X, Y \mid X^n = Y^2 = 1, YXY = X^{-1}\}.$$

We may define one-dimensional representations

$$x_1, x_2 : D_{2n} \longrightarrow \{\pm 1\} \subset \mathbf{C}^*$$

(x_2 will be non-trivial only when n is even) by the formulae

1.2.18 $$x_i(X^aY^b) = \begin{cases} (-1)^b & \text{if } i = 1, \\ (-1)^a & \text{if } i = 2. \end{cases}$$

We may define a homomorphism

1.2.19 $$v_i : D_{2n} \longrightarrow GL_2(\mathbf{C})$$

by

$$v_i(Y) = \begin{pmatrix} 0 & 1 \\ 1 & 0 \end{pmatrix},$$

$$v_i(X) = \begin{pmatrix} \xi_n^i & 0 \\ 0 & \overline{\xi_n^i} \end{pmatrix}$$

where $\xi_n = exp(2\pi i/n)$.

Clearly $\chi_{v_i} = \chi_{v_{n-1}}$ and otherwise the characters of the $\{v_i\}$ are distinct so that we obtain two-dimensional representations $v_1, v_2, \ldots, v_m$ where m is the largest integer smaller than $(n/2) - 1$. For these values of i, v_i is irreducible, by 1.2.8(iii), since

$$\begin{aligned} < v_i, v_i > &= (2n)^{-1} \textstyle\sum_{g \in D_{2n}} |\chi_{v_i}(g)|^2 \\ &= (2n)^{-1} \textstyle\sum_{j=1}^{n} (\xi_n^{ij} + \overline{\xi_n}^{ij})^2 \\ &= 1 + n^{-1} (\textstyle\sum_{j=1}^{n} \xi_n^{ij}) \\ &= 1, \end{aligned}$$

as required. These are all the irreducible representations of D_{2n}. When $n = 2t$, we have four one-dimensional representations, $(t-1)$ two-dimensional ones and

$$4 \cdot 1^2 + (t-1) \cdot 2^2 = 4t = 2n.$$

When $n = 2t + 1$ we have two one-dimensional representations, t two-dimensional ones and

$$2 \cdot 1^2 + t \cdot 2^2 = 4t + 2 = 2n.$$

Definition 1.2.20 The (complex) *representation ring of G*, denoted by $R(G)$, is defined to be the free abelian group on the irreducible, complex representations of G. The ring structure on $R(G)$ is induced from the tensor product in the following manner. In the free abelian group on the irreducible representations, $\{V_i\}$, we may identify the formal sum,

$$x = \sum_i \alpha_i V_i,$$

with the formal difference of representations

$$x = W_+ - W_-$$

$$W_+ = \oplus_{\alpha_i positive} \alpha_i V_i, \qquad W_- = \oplus_{\alpha_i negative}(-\alpha_i) V_i.$$

If $y = U_+ - U_-$ then we may define

1.2.21 $xy = [(W_+ \otimes U_+) \oplus (W_- \otimes U_-)] - [(W_+ \otimes U_-) \oplus (W_- \otimes U_+)]$.

The product defined by 1.2.21 is well-defined and makes $R(G)$ into a commutative ring.

1.2.22 *Conjugacy class functions*

Let $\mathscr{C}_\mathscr{G}$ denote the set of complex-valued functions defined on the conjugacy classes of G. Hence $f \in \mathscr{C}_\mathscr{G}$ if and only if $f : G \longrightarrow \mathbf{C}$ satisfies $f(hgh^{-1}) = f(g)$ for all $g, h \in G$. Clearly $\mathscr{C}_\mathscr{G}$ is a subspace of the complex vector space of all functions on G. If $\rho : G \longrightarrow GL(V)$ is a representation then its character, χ_ρ, is an element of $\mathscr{C}_\mathscr{G}$. Hence, by sending an irreducible representation to its character and extending $\mathbf{C}$-linearly, we obtain a map of $\mathbf{C}$-vector spaces

1.2.23 $$c_G : R(G) \otimes_{\mathbf{Z}} \mathbf{C} \longrightarrow \mathscr{C}_\mathscr{G}.$$

Theorem 1.2.24 (i) *In 1.2.23 c_G is an isomorphism.*

(ii) *The number of distinct irreducible, complex representations of G is equal to the number of conjugacy classes of elements of G.*

Proof Part (i) implies part (ii) immediately, since the rank of $R(G)$ equals the number of distinct irreducible representations of G while a basis for $\mathscr{C}_\mathscr{G}$ is given by the characteristic functions of the set of conjugacy classes of G — that is, the functions which are equal to one on one conjugacy class and zero on all the others.

To prove part (i) let $\chi_1, \chi_2, \ldots, \chi_t$ denote the characters of the distinct irreducible representations of G. Define a semi-linear inner product on $\mathscr{C}_\mathscr{G}$ by the formula,

$$< f_1, f_2 >= \#(G)^{-1} \sum_{g \in G} f_1(g)\overline{f_2(g)}.$$

By 1.2.5/1.2.8(i) the $\{\chi_i\}$ are orthonormal with respect to this inner product and hence they are linearly independent in $\mathscr{C}_{\mathscr{G}}$.

To prove that the $\{\chi_i\}$ span $\mathscr{C}_{\mathscr{G}}$ we must verify that, if $f \in \mathscr{C}_{\mathscr{G}}$ and $< f, \chi_i >= 0$ for all i then $f = 0$.

Let $\rho_i : G \longrightarrow GL(V_i)$ be the irreducible representation whose character is χ_i. The map

1.2.25 $$F_i = \sum_{g \in G} f(g)\rho_i(g) : V_i \longrightarrow V_i$$

satisfies

$$\begin{aligned} F_i(hv) &= \sum_{g \in G} f(g)\rho_i(g)\rho_i(h)v \\ &= \sum_{g \in G} f(g)\rho_i(h)\rho_i(h^{-1}gh)v \\ &= \rho_i(h)(\sum_{g \in G} f(h^{-1}gh)\rho_i(h^{-1}gh)v) \\ &= h(F_i(v)) \end{aligned}$$

for all $v \in V_i$ and $h \in H$. Hence, by 1.2.5,

1.2.26 $$F_i(v) = (dim_{\mathbf{C}}(V_i))^{-1}(\sum_{g \in G} f(g)\chi_i(g)) \cdot v.$$

If $< f, \chi_i >= 0$ for all i then $< f, \overline{\chi_i} >= 0$ also, since complex conjugation induces a permutation of the irreducible characters of G. Hence, for each i, $F_i = 0$ since

$$F_i = dim_{\mathbf{C}}(V_i)^{-1} < f, \overline{\chi_i} >,$$

by 1.2.26. Now, temporarily, write ρ for the regular representation so that

$$\rho = \sum_i dim_{\mathbf{C}}(V_i)\rho_i.$$

Let V denote the vector space whose basis is $\{g \in G\}$. Define $F : V \longrightarrow V$ by replacing ρ_i by ρ in 1.2.25. Clearly

$$F = \sum_i dim_{\mathbf{C}}(V_i) \cdot F_i = 0.$$

On the other hand, if $e_g \in V$ is the basis vector corresponding to $g \in G$ then, by definition of the regular presentation,

$$F(e_1) = \sum_{g \in G} f(g) \cdot g(e_1) = \sum_{g \in G} f(g) \cdot e_g,$$

so that $f(g) = 0$ for each $g \in G$, as required. □

Example 1.2.27 Let D_{2n} denote the dihedral group of 1.2.16.

When $n = 2t$ there are $t+3$ distinct, irreducible representations. Clearly x^i is conjugate to x^{-i} and $x^i y, x^{-i} y$ and $x^{i+2} y$ are conjugate. Therefore, by Theorem 1.2.24, the conjugacy classes of D_{4t} are represented by

$$1, x, x^2, \ldots, x^t, y \quad \text{and} \quad xy.$$

When $n = 2t + 1$ there are $t + 2$ distinct, irreducible representations and the conjugacy class representatives are

$$1, x, x^2, \ldots, x^t \quad \text{and} \quad y.$$

Theorem 1.2.28 *Let G be a finite group, then G is an abelian group if and only if all the irreducible representations of G are one-dimensional.*

Proof If G is abelian then there are $\#(G)$ distinct conjugacy classes in G. By Theorem 1.2.24 the number of distinct, irreducible representations is also $\#(G)$, of which each must be one-dimensional in order to satisfy 1.2.15(i).

Conversely, if each irreducible representation is one-dimensional then, by 1.2.15(i), there are $\#(G)$ of them and therefore, by 1.2.24, the number of conjugacy classes is equal to $\#(G)$, too. This means that no two distinct elements of G are conjugate. Thus the set

$$\{ghg^{-1} \mid g \in G\}$$

consists merely of h and hence $hg = gh$ for all $g, h \in G$. □

Corollary 1.2.29 *Let H be an abelian subgroup of G. If V is an irreducible representation of G then* $dim_{\mathbf{C}}(V) \leq \#(G)/\#(H)$.

Proof Consider V as an H-representation, by restricting the action from G to H. By 1.2.28 there exists an H-subrepresentation, $W \leq V$,with $dim_{\mathbf{C}}(W) = 1$. Hence the subspace

$$U = \sum_{g \in G} g(W) = \sum_{g \in G/H} g(W)$$

has dimension less than or equal to $\#(G)/\#(H)$. However, $g(U) \subseteq U$ for all $g \in G$ so that $U = V$, since V is irreducible. □

Theorem 1.2.30 (i) *Let V_i be an irreducible representation of G_i $(i = 1, 2)$ then $V_1 \otimes V_2$ is an irreducible representation of $G_1 \times G_2$.*

(ii) *Each irreducible representation of $G_1 \times G_2$ is of the form given in* (i).

Proof Let χ_i denote the character of V_i then

$$\begin{aligned}
&< \chi_1 \otimes \chi_2, \chi_1 \otimes \chi_2 > \\
&= \#(G_1)^{-1}\#(G_2)^{-1} \textstyle\sum_{(g_1,g_2)\in G_1\times G_2} \chi_1(g_1)\chi_2(g_2)\overline{\chi_1(g_1)\chi_2(g_2)}, \\
&\quad \text{by 1.2.4,} \\
&= < \chi_1, \chi_1 >< \chi_2, \chi_2 > \\
&= 1, \text{ by 1.2.8(iii)}
\end{aligned}$$

so that $V_1 \otimes V_2$ is irreducible, by 1.2.8(iii) again. This proves part (i).

By 1.2.24, we need only verify that if V_i' is another irreducible representation of G_i then $V_1' \otimes V_2'$ is distinct from $V_1 \otimes V_2$ unless $V_1 \cong V_1'$ and $V_2 \cong V_2'$. However, this follows at once from the relation

$$< V_1 \otimes V_2, V_1' \otimes V_2' >=< V_1, V_1' >< V_2, V_2' >$$

together with 1.2.5–1.2.8. □

Definition 1.2.31 *Induced representations* Let $\rho : H \longrightarrow GL(W)$ be a representation of H, where W is a K-vector space. The *group ring*, $K[H]$, is defined to be the K-vector space on a basis $\{h \in H\}$. An arbitrary element of $K[H]$ has the form $x = \sum_{h\in H} k_h h$, where each $k_h \in K$. In this case, $K[H]$ becomes a ring when endowed with the following sum and product:

1.2.32
$$(\textstyle\sum k_h h) + (\sum q_h h) = \sum (k_h + q_h)h,$$
$$\left(\sum k_h h\right) \cdot \left(\sum q_g g\right) = \sum_{g,h} (k_h q_g)hg.$$

In this case, W may be viewed as a (left) KH-module by means of the product ($w \in W$)

$$\left(\sum k_h h\right)(w) = \sum k_h \rho(h)(w).$$

Now suppose that H is a subgroup of G. The group-ring, $K[G]$, is a left $K[G]$-module and a right $K[H]$-module so that the tensor product, $K[G] \otimes_{K[H]} W$, is a left $K[G]$-module or, equivalently, a representation. This representation is called the *representation of G induced from ρ* and is denoted by $Ind_H^G(\rho)$. If $w_1, \ldots, w_t$ is a basis for W then $K[G] \otimes_{K[H]} W$ may be constructed by first forming the K-vector space whose basis consists of symbols $\{g \otimes w_i; g \in G, 1 \le i \le t\}$ and then dividing out by the subspace generated by elements of the form

1.2.33 $$gh \otimes w_i - \sum_{j=1}^{t} \rho(h)_{ji} g \otimes w_j$$

where $h \in H, g \in G$ and $\rho(h)(w_i) = \sum_j \rho_{ji} w_j$. Equivalently, $K[G] \otimes_{K[H]} W$ may be formed by taking one copy of W — denoted by $g \otimes W$ — for each $g \in G$ and dividing $\sum_{g \in G}(g \otimes W)$ by the subspace generated by elements of the form

1.2.34 $$gh \otimes w - g \otimes \rho(h)(w) \quad (h \in H).$$

The left G-action on $KG \otimes_{K[H]} W$ is given in these terms by the formula

1.2.35 $$(\sum_{g \in G} k_g g)(g' \otimes w) = \sum_{g \in G} gg' \otimes k_g w$$

Clearly,

$$dim_K(Ind_H^G(\rho)) = [G : H] dim_K(\rho).$$

Example 1.2.36 (i) Let $\rho = 1$ denote the one-dimensional trivial representation of the trivial group, $H = \{1\}$. The regular representation of 1.2.13 is the induced representation, $Ind_{\{1\}}^G(1)$.

(ii) Let D_{2n} denote the dihedral group of 1.2.27:

$$D_{2n} = \{X, Y \mid X^n = 1 = Y^2, YXY = X^{-1}\}.$$

Set $H =< X >$ and define $\phi : H \longrightarrow \mathbf{C}^*$ by $\phi(X) = exp(2\pi i/n)$. The representation

$$v_i : D_{2n} \longrightarrow GL_2\mathbf{C}$$

is equal to $Ind_H^{D_{2n}}(\phi^i)$. This is seen as follows. Let $e \in \mathbf{C}^*$ then a basis for $Ind_H^{D_{2n}}(\phi^i)$ is given by $1 \otimes e$ and $Y \otimes e$. The action of X and Y is given by

$$X(1 \otimes e) = X \otimes e = 1 \otimes \phi^i(X)e = 1 \otimes \xi_n^i e,$$

$$X(Y \otimes e) = XY \otimes e = Y(YXY) \otimes e = Y \otimes \phi^i(X^{-1})e = Y \otimes \xi_n^{-i} e,$$

$$Y(1 \otimes e) = Y \otimes e$$

and

$$Y(Y \otimes e) = Y^2 \otimes e = 1 \otimes e.$$

This coincides with the formulae of 1.2.27.

Definition 1.2.37 If $H \leq G$ and $\rho : G \longrightarrow GL(V)$ is a representation then we may consider V as an H-representation. This representation is the *restriction of ρ to H* and will be denoted by $Res_H^G(\rho)$.

Proposition 1.2.38 *Let H be a subgroup of G. Let W, V be K-representations of H, G respectively. There are isomorphisms of the following form:*
(i) $Hom_G(Ind_H^G(W), V) \cong Hom_H(W, Res_H^G(V))$,
(ii) $Ind_H^G(Hom(W,K)) \cong Hom(Ind_H^G(W),K)$ *as G-representations and*
(iii) $Hom_G(V, Ind_H^G(W)) \cong Hom_H(Res_H^G(V), W)$.

Proof Let $g_1, \ldots, g_d$ be a set of coset representatives for G/H. Therefore, as a vector space, $Ind_H^G(W)$ is $\sum_{i=1}^d (g_i \otimes W)$. Suppose that $g_1 = 1$, then we may assign to a G-map, $f : \sum_{i=1}^d (g_i \otimes W) \longrightarrow V$, the H-map, $f : 1 \otimes W \cong W \longrightarrow V$. This is an H-map since

$$\begin{aligned} f(h \cdot w) &= f(1 \otimes h \cdot w) \\ &= f(h(1 \otimes w)) \\ &= h \cdot (f(1 \otimes w)) \\ &= h \cdot f(w). \end{aligned}$$

This map is injective, since $f(g_i \otimes w) = g_i(f(1 \otimes w))$, so that the map is an isomorphism, by counting dimensions. This proves part (i).

To define a G-isomorphism in (ii) it suffices, by (i), to define an H-map

$$F : Hom(W,K) \longrightarrow Hom(Ind_H^G(W),K).$$

Define $F(f)(g_i \otimes w) = f(w)$ if $g_i = 1$ and zero otherwise. Clearly $F(f) = 0$ if and only if $f = 0$, so that we obtain an injective G-map

$$Ind_H^G Hom(W,K) \longrightarrow Hom(Ind_H^G(W),K),$$

which must, by dimensions, be an isomorphism.

Let $\hat{V}$ denote the *contragredient* of V, given by $Hom(V,K)$, as in 1.3.5. Part (iii) is deduced from (i)–(ii) in the following manner:

$$\begin{aligned} Hom_G(V, Ind_H^G(W)) &\cong Hom_G(Hom(Ind_H^G(W),K), \hat{V}) \\ &\cong Hom_G(Ind_H^G(\hat{W}), \hat{V}) \\ &\cong Hom_H(\hat{W}, Res_H^G(\hat{V})) \\ &\cong Hom_H(V, W). \end{aligned}$$

□

Theorem 1.2.39 (*Frobenius reciprocity*) *Let $H \leq G$ and let W, V be K-representations of H, G respectively. There is a G-isomorphism*

$$Ind_H^G(W \otimes Res_H^G(V)) \cong Ind_H^G(W) \otimes V.$$

Proof Define

$$\phi : Ind_H^G(W \otimes Res_H^G(V)) \longrightarrow Ind_H^G(W) \otimes V$$

by $\phi(g_i \otimes w \otimes v) = (g_i \otimes w) \otimes g_i v$ $(w \in W, v \in V)$. This is well-defined since

$$\begin{aligned} \phi(g_i h^{-1} \otimes hw \otimes hv) &= (g_i h^{-1} \otimes hw) \otimes g_i h^{-1} hv \\ &= g_i \otimes w \otimes g_i v \\ &= \phi(g_i \otimes w \otimes v). \end{aligned}$$

Define

$$\phi^{-1}((g_i \otimes w) \otimes v) = g_i \otimes w \otimes g_i^{-1} v.$$

These are clearly inverse isomorphisms and ϕ is a G-map since

$$\begin{aligned} g(\phi(g_i \otimes w \otimes v)) &= g((g_i \otimes w) \otimes g_i v) \\ &= (gg_i \otimes w) \otimes gg_i v \\ &= \phi(gg_i \otimes w \otimes v). \end{aligned}$$

□

Theorem 1.2.40 *Let $J, H \leq G$ and let W be a representation of H, then*

$$Res_J^G Ind_H^G(W) = \bigoplus_{z \in J\backslash G/H} Ind_{J \cap zHz^{-1}}^J((z^{-1})^*(W))$$

where $(z^{-1})^*(W)$ is W with the $(J \cap zHz^{-1})$-action given by

$$(zhz^{-1})(w) = h \cdot w.$$

Proof Let $g_1, \ldots, g_d$ be coset representatives, as in the proof of 1.2.38. Hence

$$Ind_H^G(W) = \oplus_{i=1}^d (g_i \otimes W),$$

the sum of one copy of W for each element of G/H. Since $g \in G$ sends $g_i \otimes W$ to $gg_i \otimes W$, $\oplus_{i=1}^d (g_i \otimes W)$ breaks up into J-subrepresentations, one for each J-orbit in G/H. Hence, as a J-representation, $Ind_H^G(W)$ decomposes into a sum of J-subspaces — one for each element of $J\backslash G/H$. The subspace corresponding to $g_i \otimes W$ is given by $E_i = \sum_{j \in J} jg_i \otimes W$. Now $jg_i \otimes W = g_i \otimes W$ if and only if $g_i^{-1} jg_i \in H$, in which case $j \in J \cap g_i H g_i^{-1}$. If $j_1, \ldots, j_u$ are coset representatives of $J/(J \cap g_i H g_I^{-1})$ then

$$E_i = \bigoplus_{s=1}^{u} (j_s g_i \otimes W).$$

We have to verify that the J-action on E_i corresponds to

$$Ind_{J\cap g_iHg_i^{-1}}((g_i^{-1})^*(W)).$$

However, this follows since, for $j \in J$, $j(j_sg_i \otimes w) = jj_sg_i \otimes w$ and if $j \in J \cap g_iHg_i^{-1}$ then

$$j_sg_i \otimes (g_i^{-1}jg_i)(w) = j_sg_ig_i^{-1}jg_i \otimes w = j_sjg_i \otimes w.$$

□

Corollary 1.2.41 (*Mackey's irreducibility criterion*) *Let $H \leq G$ and let W be a* **C***-representation of H. Then $Ind_H^G(W)$ is irreducible if and only if*

(i) *W is irreducible and*

(ii) *for each $g \in G - H$,*

$$< W, (g^{-1})^* Res^H_{H\cap gHg^{-1}}(W) >= 0.$$

Proof By 1.2.8(iii), $Ind_H^G(W)$ is irreducible if and only if $< Ind_H^G(W), Ind_H^G(W) >= 1$. However, by 1.2.8(i) and 1.2.40

$$\begin{aligned} < Ind_H^G(W), Ind_H^G(W) > &= dim_{\mathbf{C}} Hom_G(Ind_H^G(W), Ind_H^G(W)) \\ &= dim_{\mathbf{C}} Hom_H(W, Res_H^G Ind_H^G(W)) \\ &= \textstyle\sum_{z\in H\backslash G/H} < W, Res^H_{H\cap zHz^{-1}}(W) > . \end{aligned}$$

One of the terms in this sum is $< W, W >$, which is at least one. Hence $Ind_H^G(W)$ is irreducible if and only if $< W, W >= 1$ and all the other terms are zero. □

Proposition 1.2.42 *If $J \leq H \leq G$ then*

$$Ind_H^G Ind_J^H(U) = Ind_J^G(U).$$

Proof An isomorphism is given by sending $g \otimes h \otimes u$ to $gh \otimes u$ ($g \in G, h \in H, u \in U$). This map is evidently a G-map. It is injective, as is seen by expressing it explicitly in terms of chosen coset representatives for G/H and H/J, but both representations have the same dimension. □

Theorem 1.2.43 *Let $\rho : H \longrightarrow GL(V)$ be a complex representation with character, χ_ρ. If $H \leq G$ then the character of $\psi = Ind_H^G(\rho)$ is given by*

$$\chi_\psi(g) = \#(H)^{-1} \sum_{y\in G, ygy^{-1}\in H} \chi_\rho(ygy^{-1}).$$

Proof We may write $Ind_H^G(V)$ as $\oplus_{i=1}^s(g_i \otimes V)$ where $g_1, \ldots, g_s$ is a set of coset representatives for G/H. Since multiplication by $g \in G$ sends $g_i \otimes V$ to $gg_i \otimes V$ the only summands on which the trace of multiplication by g can contribute a non-zero amount are those for which $gg_i \otimes V = g_i \otimes V$. This can only happen when $g_i^{-1}gg_i \in H$. In this case, if $v \in V$,

$$g(g_i \otimes v) = g_i \otimes \rho(g_i^{-1}gg_i)v$$

so that we obtain

$$\chi_\psi(g) = \sum_{i, g_i^{-1}gg_i \in H} \chi_\rho(g_i^{-1}gg_i).$$

However, $\chi(h^{-1}g_i^{-1}gg_ih) = \chi(g_i^{-1}gg_i)$ if $h \in H$ so that we may allow the sum to run over all $y \in G$ such that $ygy^{-1} \in H$ provided that we divide by $\#(H)$. This completes the proof of Theorem 1.2.43. □

1.3 Exercises

1.3.1 (i) Let G be a finite group and let $\rho : G \longrightarrow GL_n(\mathbf{C})$ be a homomorphism. Show that there exists $X \in GL_n(\mathbf{C})$ such that $X\rho(G)X^{-1} \leq U(n)$.

(ii) Given homomorphisms, $\rho_i : G \longrightarrow U(n)$ $(i = 1, 2)$, which are conjugate in $GL_n(\mathbf{C})$, show that they are conjugate in $U(n)$. (*Hint:* For (i), by averaging, construct an inner product on $\mathbf{C}^n$ which is preserved by $\rho(g)$ for each $g \in G$. For (ii), let Z^* denote the conjugate transpose of Z. If $X\rho_1X^{-1} = \rho_2$ then XX^* is Hermitian. Find a suitable Z such that $Z^2 = X^*X$ and set $U = XZ^{-1}$.)

1.3.2 Prove the analogous result in which $GL_n(\mathbf{C})$ and $U(n)$ are replaced by $GL_n(\mathbf{R})$ and $O(n)$.

1.3.3 Show that the tensor product, $V_1 \otimes V_2$, may be represented in terms of matrices by the Kronecker product (see 1.2.4).

1.3.4 Let V be a $\mathbf{C}$-representation of G. Let $Sym^2(V)$ be the representation afforded by the vector subspace of $V \otimes V$ consisting of elements which are fixed by the involution that interchanges the two factors (cf. 1.1.6).

(i) Show that $\lambda^2(V)$ is given by the (-1)-eigenspace of the involution.

(ii) Show that $V \otimes V \cong \lambda^2(V) \oplus Sym^2(V)$.

(iii) If χ is the character of V, show that

$$\chi_{Sym^2(V)}(g) = 1/2[(\chi(g))^2 + \chi(g^2)]$$

and that

$$\chi_{\lambda^2(V)}(g) = 1/2[(\chi(g))^2 - \chi(g^2)].$$

1.3.5 Let $\rho : G \longrightarrow GL(V)$ be a representation. The *contragredient* of ρ is the representation, $\hat{\rho}$, afforded by $W = Hom_K(V, K)$, where the action is given by the formula

$$\hat{\rho}(g)(f)(v) = f(\rho(g^{-1})v),$$

where $g \in G, v \in V, f \in W$.

(i) Find the matrix of $\hat{\rho}$ in terms of that of ρ.

(ii) If $K = \mathbf{C}$ and ρ is unitary, show that $\chi_{\hat{\rho}} = \overline{\chi_\rho}$.

1.3.6 Let 1 denote the one-dimensional representation given by $\mathbf{C}$ with the trivial G-action. Show that the multiplicity of 1 in ρ is equal to

$$\#(G)^{-1} \sum_{g \in G} \chi_\rho(g)$$

where χ_ρ is the character of ρ.

1.3.7 Let $\{V_i; 1 \leq i \leq s\}$ be a set of distinct irreducible representations of G and set $dim_{\mathbf{C}}(V_i) = n_i$.

(i) If $\sum_{i=1}^{t} n_i^2 = \#(G)$, prove that $\{V_i\}$ consists of all the distinct irreducible representations of G.

(ii) Apply part (i) to find all the irreducible representations of the *generalised quaternion group*

$$Q_{4n} = \{X, Y \mid X^n = Y^2, Y^4 = 1, YXY^{-1} = X^{-1}\}.$$

1.3.8 Verify, in 1.2.20, that $R(G)$ is a ring.

1.3.9 For complex representations, give proofs of Frobenius reciprocity and the Double Coset Formula of 1.2.40 by means of characters.

2
Induction theorems

Introduction

The chapter concerns induction theorems; that is, theorems which express arbitrary representations as linear combinations of induced representations within the representation ring, $R(G)$, tensored with a suitable ring of coefficients.

We begin the chapter with a proof of Brauer's canonical form for Artin's Induction Theorem. Artin's result, in this form, gives a canonical rational form for the identity representation within $R(G) \otimes \mathbf{Q}$, the rationalised representation ring, in terms of representations which are induced from cyclic groups. By Frobenius reciprocity Artin's theorem yields a similar rational canonical form for any representation. Brauer's Induction Theorem, in its original (non-canonical) form states that any representation can be expressed as an *integral* linear combination of representations which are obtained by induction from one-dimensional representations of elementary subgroups. This celebrated result has been proved by a variety of methods, to which we add a new topological proof here. This proof involves replacing the given representation of G by the unique, equivalent, unitary representation and using the latter to perform some elementary algebraic topology on the resulting action of G on the compact manifold given by the unitary group modulo the normaliser of its maximal torus. This topological proof of the existence part of Brauer's theorem was the basis of the first derivation of a canonical form for Brauer's theorem, which appeared in Snaith (1988b). This topologically derived canonical form possessed two important properties: namely, of naturality and of being the identity on one-dimensional representations. These two properties, together with additivity, were taken as axioms in Robert Boltje's algebraic construction of a canonical form for Brauer's theorem.

In Section 2 we introduce the free abelian group, $R_+(G)$, on conjugacy classes of one-dimensional subhomomorphisms. This is the group within which the canonical form lies. There is a natural map, b_G, to the representation ring, and an Explicit Brauer Induction homomorphism is simply a section for b_G, which satisfies the two axioms. After developing the formal properties of $R_+(G)$ we follow Boltje's analysis of the coefficients which must appear in an Explicit Brauer Induction homomorphism. We prove that these coefficients, if they exist, are uniquely determined rational numbers, that many of them are forced to vanish and, finally, that they must be integral. All this is accomplished by means of the Boltje's (1989) algebraic arguments.

In Section 3 we show that one may inflict a non-degenerate bilinear form upon $R_+(G)$ in such a manner that the required rational-valued Explicit Brauer Induction homomorphism is simply the adjoint of the natural map from $R_+(G)$ to $R(G)$. In addition, the algebraic derivation which we have followed yields Boltje's explicit formula, a_G, in terms of Schur inner products and Möbius functions of the partially ordered set of one-dimensional subhomomorphisms of G. In the canonical form for Artin's induction theorem one finds Möbius functions from the partially ordered set of cyclic subgroups playing a similar role. The section closes with a number of examples together with an explanation of the relationship between Boltje's formula and Snaith's (1988b) topological formula. In fact, as explained in 2.5.16, Snaith's (1988b) construction can be made to yield a_G when G is a p-group. The integrality of the coefficients in a_G is the deepest property in the algebraic analysis and may be obtained, in general, by a topological method due to Peter Symonds. Symonds' construction, which uses a group action in a manner similar to the method of Snaith (1988b) (this time the action is on the projective space of the representation), is also described in Section 3.

In Section 4 another rational-valued Explicit Brauer Induction homomorphism, d_G, is constructed. In the course of the construction of a_G one sees the manner in which an idempotent may be used, together with Möbius inversion, to construct different types of such homomorphisms. This one, which was discovered by Robert Boltje, uses an idempotent on $R(G)$ which was first exploited by Andreas Dress. The homomorphism d_G is shown to be natural with respect to inclusions of subgroups and to have the remarkable property that it commutes with induction. Many properties of d_G are derived: for example, the fact that d_G commutes with taking fixed-points and the relationship between d_G and a_G. Most of these properties are established for further use in Chapter 6. However, in order that d_G should have the convenient inductive property it is necessary that its coefficients be non-integral and that it cannot be natural with respect to the inflation maps which are induced by surjective homomorphisms of groups.

Section 5 consists of a batch of exercises concerning the Explicit Brauer Induction homomorphisms.

2.1 Induction theorems of Artin and Brauer

In this chapter we will be concerned with the development of the induction theorem of Brauer in a canonical form. This was first accomplished by topological methods in Snaith (1988b). However, in the later sections of this chapter, we will present an improved approach to Explicit Brauer Induction, which is due to R. Boltje (1990). I would have liked to incorporate further topological improvements, due to P. Symonds (1991) which facilitate the verification of the integrality of Boltje's canonical form. However, I have had to content myself with a few remarks at the end of the chapter (see 2.3.28), because the introduction of the requisite topological background would take us too far afield.

In this section, however, we will prove the induction theorems of Artin and Brauer as preliminary illustrations of the various approaches. Firstly we will consider Artin's induction theorem and the manner in which it was given in a canonical form by R. Brauer (1951). Boltje's formula, involving as it does Möbius functions, is reminiscent of Brauer's formula. Secondly we will prove Brauer's Induction Theorem by the use of a minute amount of the algebraic topology which constitutes the basis of the construction in Snaith (1988b).

Let us begin by recalling the Möbius function.

Definition 2.1.1 The *Möbius function*, $\mu(n)$, is defined by

$$\mu(n) = \begin{cases} 1 & \text{if } n = 1 \\ (-1)^r & \text{if } n = p_1 \dots p_r,\ p_i \text{ distinct primes} \\ 0 & \text{if } n \text{ is not square-free.} \end{cases}$$

Recall that the Möbius function is extremely useful for inverting certain formulae (see 2.5.2). Later we will require the generalisation of this well-known fact to the Möbius functions of a *poset* (a partially ordered set).

Lemma 2.1.2 *Let $\rho : G \longrightarrow GL(V)$ be a complex representation whose character, χ_ρ, is rational-valued. Suppose that $x, y \in G$ generate the same cyclic subgroup of G, then $\chi_\rho(x) = \chi_\rho(y)$.*

Proof Let m be the order of the cyclic subgroup generated by x or y and suppose that $x^j = y$ where $HCF(j, m) = 1$. By 1.2.3(ii), if $\lambda_1, \ldots, \lambda_s$ are the eigenvalues of $\rho(x)$ then $\chi_\rho(x) = \sum_{i=1}^{s} \lambda_i$ and $\chi_\rho(y) = \sum_{i=1}^{s} \lambda_i^j$. However, there exists a Galois automorphism, σ, of $\mathbf{Q}(\xi_m)/\mathbf{Q}$ which sends $\xi_m = exp(2\pi i/m)$ to ξ_m^j. Since each λ_i is a power of ξ_m we find that $\sigma(\chi_\rho(x)) = \chi_\rho(y)$ but $\chi_\rho(x)$ is rational-valued so that $\sigma(\chi_\rho(x)) = \chi_\rho(x)$, which completes the proof. □

Theorem 2.1.3 (*Artin induction theorem*) *Let $\rho : G \longrightarrow GL(V)$ be a complex representation with rational-valued character, χ_ρ. In $R(G) \otimes \mathbf{Q}$*

$$\chi_\rho = \sum \alpha_C Ind_C^G(1),$$

where C ranges over cyclic subgroups of G and where

$$\alpha_C = [G : C]^{-1} \sum_{C \leq B} \mu([B : C])\chi_\rho(b).$$

The sum is taken over all cyclic subgroups, $C \leq B$ and $b \in B$ is any generator.

Proof By 2.1.2 the coefficients, α_C, are independent of the choice of generators, $b \in B$. Let χ_C denote the character of $Ind_C^G(1)$. By 1.2.42

$$\chi_C(g) = \#\{y \in G \mid ygy^{-1} \in C\} \cdot \#(C)^{-1}.$$

Hence, if $\chi_C(g) \neq 0$, there exists $ygy^{-1} \in C$ and $\chi_C(g) = \chi_C(ygy^{-1})$ so that we may assume that $g \in C$. In this case

$$\chi_C(g) = \#(N_G < g >)\#(C)^{-1} = \#(N_G < ygy^{-1} >)\#(C)^{-1},$$

where $N_G H$ denotes the normaliser of H.

Hence

2.1.4 $$\sum\nolimits_{C cyclic} \alpha_C \chi_C(g) =$$

$$\sum_{C cyclic, some\ ygy^{-1} \in C} \#(N_G < g >)\#(G)^{-1} \sum_{C \leq B} \mu([B : C])\chi_\rho(b).$$

However, the number of cyclic groups which are conjugate to $< g >$ is equal to $\#(G)\#(N_G < g >)^{-1}$. Furthermore, an arbitrary cyclic subgroup can contain at most one subgroup which is isomorphic to $< g >$. Therefore we may rewrite 2.1.4 as a sum over cyclic groups which contain $< g >$:

$$\begin{aligned} &\sum_{C cyclic, g \in C} \sum_{C \leq B} \mu([B : C]) \chi_\rho(b) \\ &\qquad = \sum_{g \in B, cyclic} \chi_\rho(b) \sum_{g \in C \leq B} \mu([B : C]) \\ &\qquad = \sum_{g \in B, cyclic} \chi_\rho(b) (\sum_{d \mid [B : <g>]} \mu(d)) \\ &\qquad = \chi_\rho(b), \end{aligned}$$

where $< b > = < g >$, since $\sum_{d|n} \mu(d) = 0$ unless $n = 1$ (Hunter, 1964). By 2.1.2, $\chi_\rho(b) = \chi_\rho(g)$ so that we have shown that $\sum \alpha_C Ind_C^G(1)$ has the same character-value as ρ on $g \in G$, which completes the proof of 2.1.3. □

Definition 2.1.5 A complex representation, ρ, of G is called a *monomial representation* if $\rho = Ind_H^G(\phi)$ for some $\phi : H \longrightarrow \mathbf{C}^*$. An *M-group* is a group all of whose irreducible complex representations are monomial. A group, G, is *solvable* if there exists a chain of subgroups

$$\{1\} = G_0 < G_1 < \ldots < G_n = G,$$

with $G_{i-1} \lhd G_i$ and G_i/G_{i-1} abelian. A solvable group is *supersolvable* if, in addition, the quotients G_i/G_{i-1} are cyclic and the G_{i-1} are normal in G, not merely in G_i. A supersolvable group is *nilpotent* if, in addition, each G_i/G_{i-1} is central in G/G_{i-1}.

Lemma 2.1.6 As a representation, $\rho : G \longrightarrow GL_n\mathbf{C}$, is equal to the sum of monomial representations if and only if there exists $X \in GL_n\mathbf{C}$ such that $X\rho(G)X^{-1} \leq NT^n$ where NT^n is the normaliser of the subgroup of diagonal matrices in the unitary group, $U(n)$.

Proof Firstly let us determine the matrix form of a monomial representation, $Ind_H^G(\phi)$. We note that the direct sum of $A \in NT^n$ and $B \in NT^m$ is $A \oplus B \in NT^{n+m}$ (cf. 1.1.7).

Let $g_1, \ldots, g_d$ be coset representatives for G/H. If $g \in G$ write

2.1.7 $$g g_i = g_{\sigma(g)(i)} h(i, g),$$

where $h(i, g) \in H$ and $\sigma : G \longrightarrow \Sigma_d$ is a homomorphism to the symmetric group. A basis for $Ind_H^G(\phi)$ consists of vectors of the form $\{g_i \otimes 1; 1 \leq i \leq d\}$ with G-action given by

2.1.8 $$g(g_i \otimes 1) = g_{\sigma(g)(i)} \otimes \phi(h(i, g)) = \phi(h(i, g))(g_{\sigma(g)(i)} \otimes 1).$$

Hence, if we consider $\sigma(g)$ as a permutation matrix, then, with respect to this basis,

2.1.9
$$\rho(g) = \sigma(g) \begin{pmatrix} \phi(h(1,g)) & 0 & \dots \\ 0 & \phi(h(2,g)) & \\ 0 & 0 & \\ \vdots & & \end{pmatrix}.$$

However, $NT^d \leq GL_d\mathbf{C}$ is generated by the diagonal matrices with non-zero entries of unit length and by permutation matrices. Hence 2.1.9 is a matrix lying in NT^d.

Conversely, suppose that we are given a representation in the matrix form

$$\rho : G \longrightarrow NT^n.$$

Let $\pi : NT^n \longrightarrow \sum_n$ denote the quotient map whose kernel is the diagonal, maximal torus of $U(n)$. Consider the action of G on $\{1, 2, \dots, n\}$ via $\pi\rho$. The matrix form of ρ breaks into the direct sum of homomorphisms, $\rho_i : G \longrightarrow NT^{s_i}$, where $\sum_{i=1}^{t} s_i = n$ and $\{1, 2, \dots, n\}$ consists of t G-orbits of sizes $s_1, \dots, s_t$. We may therefore assume that $t = 1$, which means that G acts transitively on $\{1, 2, \dots, n\}$. Let $H = \{g \in G \mid \pi\rho(g)(1) = 1\}$ and let $\phi : H \longrightarrow \mathbf{C}^*$ denote the homomorphism given by the $(1, 1)$-entry of $Res_H^G(\rho)$. Set $\psi = Ind_H^G(\phi)$.

We will now verify that the characters of ψ and ρ are equal. Denote these characters by χ_ψ and χ_ρ, respectively. Since $\rho(g) \in NT^n$ we have

$$\begin{aligned} Trace\rho(g) &= \textstyle\sum_{\pi\rho(g)(m)=m} \rho(g)_{m,m} & \\ &= \textstyle\sum_{gzH=zH} \rho(z^{-1}gz)_{1,1} & \text{where } H = stab_{\pi\rho}(1) \\ &= \chi_\psi(g), & \text{since } \phi(z^{-1}gz) = \rho(z^{-1}gz)_{1,1}. \end{aligned}$$

This completes the proof of 2.1.6. □

Corollary 2.1.10 *Let $\rho : G \longrightarrow U(n)$ be a unitary representation. Let G act on the left of $U(n)/NT^n$ by $g(zNT^n) = \rho(g)zNT^n$. Then ρ is equal to a sum of monomial matrices if and only if there is a G-fixed point.*

Proof If, for all $g \in G$, $\rho(g)zNT^n = zNT^n$ then $z^{-1}\rho z$ maps G to NT^n. Therefore $z^{-1}\rho z$, which is equivalent to ρ , is equal to a sum of monomial representations, by 2.1.6. Conversely, if ρ is the sum of monomial representations it may be represented by $\rho : G \longrightarrow NT^n$, which fixes the identity coset. □

2.1.11 *Topological digression*

Whenever a finite group, G, acts smoothly upon a compact manifold, X (as in 2.1.10, for example), it is possible to triangulate X so that it becomes a simplicial complex in such a manner that the G-action either fixes a simplex pointwise or translates it by a simplicial map to a different simplex. Triangulated thus, X is the union of a finite number of simplices; the boundary of each simplex being the union of simplices of lower dimension (Maunder, 1970). From this triangulation one obtains the simplicial chain groups, $C_i(X)$, given by the free abelian group on the i-dimensional simplices of X. As a module, $C_i(X)$ is a sum of permutation modules for the ring $\mathbf{Z}[G]$, since G acts simplicially on X. The simplicial chain complex of X is a sequence of $\mathbf{Z}[G]$-maps

2.1.12 $$\ldots \rightarrow C_i(X) \xrightarrow{\delta_i} C_{i-1}(X) \rightarrow \ldots \xrightarrow{\delta_1} C_0(X) \rightarrow 0.$$

We may tensor this sequence with the complex numbers to obtain a sequence of complex representations each of which is a sum of permutation representations

2.1.13 $$\ldots \rightarrow C_i(X) \otimes \mathbf{C} \xrightarrow{\delta_i} C_{i-1}(X) \otimes \mathbf{C} \rightarrow \ldots \xrightarrow{\delta_1} C_0(X) \otimes \mathbf{C} \rightarrow 0.$$

Proposition 2.1.14 *Set $X = U(n)/NT^n$ upon which G acts as in* 2.1.10. *Then X may be triangulated so that:*

(i) *In* 2.1.13, *$C_i(X) \otimes \mathbf{C}$ is the sum of permutation representations, $Ind_{H(\sigma^i)}^G(1)$, where $H(\sigma^i)$ runs through the stabilisers of orbit representatives of the set of i-dimensional simplices.*

(ii) *In* 2.1.13 *the sequence is exact ($ker\delta_i = im\delta_{i+1}$) except at $C_0(X) \otimes \mathbf{C}$ where*

$$\frac{C_0(X) \otimes \mathbf{C}}{im(\delta_1)} \cong 1 \in R(G),$$

the trivial, one-dimensional representation.

Proof Choose an i-simplex, σ^i. In $C_i(X) \otimes \mathbf{C}$ it is clear from the discussion of 2.1.11 that the $\mathbf{C}[G]$-module generated by σ^i is isomorphic to $Ind_{H(\sigma^i)}^G(1)$. This proves part (i), since $C_i(X) \otimes \mathbf{C}$ is the direct sum of such $\mathbf{C}[G]$-modules. Part (ii) follows from the fact that the homology with complex coefficients, $H_*(U(n)/NT^n; \mathbf{C})$, is isomorphic to the complex homology of a point (Snaith, 1989b, p. 208, section 2.4). □

Proposition 2.1.15 *A p-group is an M-group. In fact, any nilpotent group is an M-group.*

Proof Let p be a prime and let G be a p-group. Let $\rho : G \longrightarrow U(n)$ be an irreducible, unitary representation (recall from 1.3.1 that every complex representation is equivalent to a unique unitary representation). By 2.1.10 we must show that G acts on $U(n)/NT^n$ with a fixed point. Suppose not, then each $C_i(X) \otimes \mathbf{C}$ is the direct sum of permutation representations of the form $Ind_{H(\sigma^i)}^G(1)$ with $H(\sigma^i) \neq G$. Hence p divides each dimension, $dim_{\mathbf{C}}(C_i(X) \otimes \mathbf{C})$. However, this contradicts 2.1.14(ii).

If G is an arbitrary nilpotent group then G is the product of its Sylow subgroups,

$$G = G_1 \times \ldots \times G_s.$$

An irreducible representation of G has the form $Ind_{H_1}^{G_1}(\phi_1) \otimes \ldots \otimes Ind_{H_s}^{G_s}(\phi_s)$, by 1.2.29. Each ϕ_i has dimension one and, by 2.5.5, this representation equals

$$Ind_{H_1 \times \ldots \times H_s}^{G_1 \times \ldots \times G_s}(\phi_1 \otimes \ldots \otimes \phi_s),$$

which completes the proof. □

Definition 2.1.16 An *elementary group* is a finite group of the form $C \times P$ where P is a p-group and C is a cyclic group of order prime to p.

Proposition 2.1.17 *Let G be any finite group. There exist M-groups, $H_\alpha \leq G$, such that*

$$1 = \sum_\alpha n_\alpha Ind_{H_\alpha}^G(1) \in R(G)$$

for suitable integers, $\{n_\alpha\}$.

Proof If G is an M-group there is nothing to prove. If G is not an M-group we may choose an irreducible representation

$$\rho : G \longrightarrow U(n),$$

so that the resulting action of G on $U(n)/NT^n$, in 2.1.14, has no fixed point, by 2.1.10. Hence, in 2.1.14(i),

$$C_i(X) \otimes \mathbf{C} \cong \oplus Ind_{H(\sigma^i)}^G(1) \otimes \mathbf{C},$$

with $H(\sigma^i) \neq G$. From 2.1.14(ii) and 1.1.8 one easily shows that (see 2.5.6)

$$1 = \sum (-1)^i C_i(X) \otimes \mathbf{C} \in R(G).$$

Hence there exist integers, m_β, and subgroups, $J_\beta \neq G$, such that

2.1.18 $$1 = \sum_\beta m_\beta Ind_{J_\beta}^G(1) \in R(G).$$

By induction, for each J_β which is not an M-group,

2.1.19 $$1 = \sum_\gamma m_{\beta,\gamma} Ind_{L_{\beta,\gamma}}^{J_\beta}(1) \in R(J_\beta),$$

with $L_{\beta,\gamma}$ an M-group. The result follows by substituting 2.1.19 into 2.1.18, where necessary, and using transitivity of induction (1.2.42). □

Theorem 2.1.20 (*Brauer induction theorem*) *Let G be a finite group. Given $x \in R(G)$ there exist one-dimensional representations of elementary subgroups, H_i, $\phi_i : H_i \longrightarrow \mathbf{C}^*$ and integers, n_i, such that*

$$x = \sum_i n_i Ind_{H_i}^G(\phi_i) \in R(G).$$

Proof If we multiply x by the expression in 2.1.17 we obtain, in $R(G)$,

$$\begin{aligned} x &= x \cdot 1 \\ &= \sum_\alpha n_\alpha x Ind_{H_\alpha}^G(1) \\ &= \sum_\alpha n_\alpha Ind_{H_\alpha}^G(Res_{H_\alpha}^G(x)), \end{aligned}$$

by 1.2.39. Therefore, by 1.2.42, we may assume that G is an M-group.

Let B_G denote the $\mathbf{Z}$-linear span in $R(G)$ of representations of the form $Ind_H^G(\phi)$ with ϕ one-dimensional and H an elementary subgroup. Since a subgroup of an elementary group is again elementary, Frobenius reciprocity (1.2.39) shows that B_G is an ideal of $R(G)$.

Suppose that ρ is an irreducible representation of the M-group, G. If $\rho = Ind_H^G(\phi)$ with $H \neq G$ then, by induction on the order of G, we see that $\rho \in B_G$. Let $\hat{\rho}$ denote the contragredient of ρ (1.3.5) and suppose that λ is a one-dimensional representation of G. We have

$$< \lambda, \rho \otimes \hat{\rho} > = < \lambda \otimes \rho, \rho >$$

$$= \begin{cases} 1, & \text{if } \lambda = 1, \\ 0, & \text{if not.} \end{cases}$$

This is because ρ is irreducible and because the character of $\lambda \otimes \rho$ is $(g \longmapsto \lambda(g)\chi_\rho(g))$ which is not equal to $(g \longmapsto \chi_\rho(g))$ unless $\lambda = 1$. Hence, if $dim_{\mathbf{C}}(\rho) \geq 2$,

$$\rho \otimes \hat{\rho} = 1 + \sum_i \rho_i \in R(G),$$

where $dim_{\mathbf{C}}(\rho_i) \geq 2$ for each i. Since ρ and the $\{\rho_i\}$ belong to the ideal, B_G, then so does 1. However, if $1 \in B_G \lhd R(G)$ then $B_G = R(G)$, as required.

There remains the case in which all the irreducible representations of G are one-dimensional, so that G is abelian, by 1.2.28.

Suppose that $G = G_1 \times \ldots \times G_r$ where G_i is an abelian group of order $p_i^{u_i}$ and where $p_1, \ldots, p_r$ are distinct primes. By 1.2.30 there is a ring isomorphism

2.1.21 $$R(G) \cong R(G_1) \otimes \ldots \otimes R(G_r).$$

Also, by the Artin Induction Theorem (2.1.3)

$$p_i^{u_i} \in B_{G_i} \lhd R(G_i).$$

By 2.5.5, if $M = p_1^{u_1} \ldots p_r^{u_r} = \#(G)$, then each of $M/(p_i^{u_i}) \in B_G$ for $i = 1, 2, \ldots, r$. Hence

$$1 = HCF(M/(p_i^{u_i})) \in B_G \lhd R(G),$$

and so $B_G = R(G)$, as required. □

2.2 Brauer induction in canonical rational form

We must begin with some definitions.

Definition 2.2.1 Let G be a finite group. Let $R_+(G)$ denote the free abelian group on G-conjugacy classes of characters, $\phi : H \longrightarrow \mathbf{C}^*$, where $H \leq G$. We shall denote this character by (H, ϕ) and its G-conjugacy class by $(H, \phi)^G \in R_+(G)$.

If $J \leq G$ we define a *restriction homomorphism*

2.2.2 $$Res_J^G : R_+(G) \longrightarrow R_+(J)$$

by the double coset formula

2.2.3 $$Res_J^G((H, \phi)^G) = \sum_{z \in J \backslash G / H} (J \cap zHz^{-1}, (z^{-1})^*(\phi))^J,$$

where $(z^{-1})^*(\phi)(u) = \phi(z^{-1}uz) \in \mathbf{C}^*$.

If $\pi : J \longrightarrow G$ is a surjection then we define

2.2.4 $$\pi^* : R_+(G) \longrightarrow R_+(J)$$

by $\pi^*((H,\phi)^G) = (\pi^{-1}(H), \phi\pi)^J$.

By means of 2.2.3–2.2.4 we may define $f^* : R_+(G) \longrightarrow R_+(J)$ for any $f : J \longrightarrow G$ by factorising f as $f : J \longrightarrow im(f) \subset G$ and setting

2.2.5 $$f^* = \pi^* Res^G_{im(f)} : R_+(G) \longrightarrow R_+(im(f)) \longrightarrow R_+(J).$$

One may also define an induction map, $Ind_J^G : R_+(J) \longrightarrow R_+(G)$, and a product which makes $R_+(G)$ into a ring-valued functor satisfying Frobenius reciprocity (see 2.5.7).

Define a homomorphism, which is surjective by 2.1.20,

2.2.6 $$b_G : R_+(G) \longrightarrow R(G)$$

by

$$b_G((H,\phi)^G) = Ind_H^G(\phi).$$

This is a ring homomorphism for the ring structure of 2.5.7.

Remark 2.2.7 The objective of this section is to construct a natural homomorphism

$$a_G : R(G) \longrightarrow R_+(G)$$

such that $b_G a_G = 1$. This amounts to a canonical form for Brauer induction.

The first canonical form was given in Snaith (1988b). The method was based upon the group action of G on $U(n)/NT^n$ which was used in 2.1.6 and 2.1.10. The details, together with several applications, are given at length in Snaith (1989b) and further elaborated upon in Snaith (1989a). The topological procedure of Snaith (1988b) automatically gives a functorial association of an element of $R_+(G)$ to a representation. Furthermore, the simplicity of the group action in the one-dimensional case ensures that a one-dimensional representation, $\phi : G \longrightarrow \mathbf{C}^*$, is associated with $(G,\phi)^G \in R_+(G)$. In this section we will follow the method, due to R. Boltje (1990), which starts by taking these two properties as axioms and deduces the formula for a_G algebraically from these axioms.

2.2.8 *Axioms for a_G*

(i) For $H \leq G$ the following diagram commutes:

$$\begin{array}{ccc} R(G) & \xrightarrow{a_G} & R_+(G) \\ \Big\downarrow Res^G_H & & \Big\downarrow Res^G_H \\ R(H) & \xrightarrow{a_H} & R_+(H) \end{array}$$

(ii) Let $\rho : G \longrightarrow GL_n(\mathbf{C})$ be a representation and suppose that

$$a_G(\rho) = \sum \alpha_{(H,\phi)^G}(H,\phi)^G \in R_+(G),$$

then $\alpha_{(G,\phi)^G} =< \rho, \phi >$ for each $(H,\phi)^G$ such that $H = G$.

2.2.9 Let G be any finite group and denote by $\mathcal{M}_G$ the set of characters on subgroups, (H,ϕ), where $H \leq G$ and $\phi : H \longrightarrow \mathbf{C}^*$. $\mathcal{M}_G$ is a *poset* (a *partially ordered set*) if we define the partial ordering by

2.2.10
$$\begin{cases} (H,\phi) \leq (H',\phi') \\ \text{if and only if} \\ H \leq H' \text{ and } Res^{H'}_H(\phi') = \phi. \end{cases}$$

In addition, G acts on $\mathcal{M}_G$ by the formula

2.2.11
$$g(H,\phi) = (gHg^{-1}, (g^{-1})^*(\phi)) \quad (g \in G),$$

where $(g^{-1})^*(\phi)(u) = \phi(g^{-1}ug)$.

We may define a partial order on the orbit space, $(\mathcal{M}_G)/G$, by

2.2.12
$$\begin{cases} (H,\phi)^G \leq (H',\phi')^G \\ \text{if and only if there exists } g \in G \text{ such that} \\ (H,\phi) \leq g(H',\phi') = (gH'g^{-1}, (g^{-1})^*(\phi')). \end{cases}$$

Define a positive integer

2.2.13
$$\begin{cases} \gamma_{(H,\phi),(H',\phi')} \in \mathbf{Z}^+ \text{ by the formula} \\ \gamma_{(H,\phi),(H',\phi')} = \#\{g \in H\backslash G/H' \mid (H,\phi) \leq g(H',\phi')\}. \end{cases}$$

Lemma 2.2.14 (i) $\gamma_{(H,\phi),(H',\phi')}$ *depends only on the conjugacy classes,* $(H,\phi)^G$ *and* $(H',\phi')^G$.

(ii) $\gamma_{(H,\phi),(H',\phi')} > 0$ *if and only if* $(H,\phi)^G \leq (H',\phi')^G$.

Proof Part (ii) is clear from the definition. For part (i), suppose that $u, v \in G$ then $\gamma_{u(H,\phi),v(H',\phi')} =$

$$\#\{g \in uHu^{-1}\backslash G/vH'v^{-1} \mid \ (uHu^{-1},(u^{-1})^*(\phi)) \leq (gvHv^{-1}g^{-1},((gv)^{-1})^*(\phi'))\}.$$

However, $uHu^{-1} \leq gvH'v^{-1}g^{-1}$ and $\phi'((gv)^{-1}agv) = \phi(u^{-1}au)$ for all $a \in uHu^{-1}$ if and only if $H \leq u^{-1}gvH'v^{-1}g^{-1}u$ and $\phi'(v^{-1}g^{-1}ubu^{-1}gv) = \phi(b)$ for all $b \in B$. Hence sending $g \in uHu^{-1}\backslash G/vH'v^{-1}$ to $u^{-1}gv \in H\backslash G/H'$ shows that $\gamma_{u(H,\phi),v(H',\phi')} = \gamma_{(H,\phi),(H',\phi')}$, as required. □

Theorem 2.2.15 *There is at most one homomorphism,* a_G*, satisfying the axioms of* 2.2.8. *In fact, if*

$$a_G(\rho) = \sum_{(H,\phi)^G \in \mathcal{M}_{\mathcal{G}}/\mathcal{G}} \alpha_{(H,\phi)^G}(\rho)(H,\phi)^G \in R_+(G),$$

then the coefficients, $\alpha_{(H,\phi)^G}(\rho)$ *are inductively determined by the equations*

2.2.16
$$\alpha_{(H,\phi)^H}(Res^G_H(\rho)) = \sum_{(H,\phi)^G \leq (H',\phi')^G \in \mathcal{M}_{\mathcal{G}}/\mathcal{G}} \gamma^G_{(H,\phi),(H',\phi')}\alpha_{(H',\phi')^G}(\rho).$$

The equations of 2.2.16 *have at most one solution in the rational numbers.*

Proof First consider 2.2.16 and suppose that $\alpha_{(H',\phi')^G}(\rho) \in \mathbf{Q}$ is determined for all $H' > H$ then all the terms in the right side of 2.2.16 are determined except for the term

$$\gamma^G_{(H,\phi),(H,\phi)}\alpha_{(H,\phi)^G}(\rho).$$

However, $\gamma^G_{(H,\phi),(H,\phi)} = [N_G(H,\phi) : H]$, where

$$N_G(H,\phi) = \{g \in G \mid \phi(ghg^{-1}) = \phi(h) \ for \ all \ h \in H\}.$$

Hence $\gamma^G_{(H,\phi),(H,\phi)} \neq 0$ (this also follows from 2.2.14(ii)) and, since $\alpha_{(H,\phi)^H}(Res^G_H(\rho)) =< Res^G_H(\rho), \phi >$, we see that $\alpha_{(H,\phi)^G}(\rho)$ is inductively uniquely determined by 2.2.16. It remains to derive 2.2.16.

By definition, the $(H,\phi)^H$-term in $a_H(Res^G_H(\rho)) = Res^G_H(a_G(\rho))$ is given by

$$\sum_{(H',\phi')^G} \sum_{\substack{H=H\cap zHz^{-1} \\ z\in H\backslash G/H',(z^{-1})^*(\phi')=\phi}} \alpha_{(H',\phi')^G}(\rho)(H\cap zHz^{-1},(z^{-1})^*(\phi'))^H$$

$$= \textstyle\sum_{H\backslash G/H',(H,\phi)\leq z(H',\phi')} \alpha_{(H',\phi')^G}(\rho)(H,\phi)$$
$$= [\textstyle\sum_{(H,\phi)^G\leq (H',\phi')^G} \gamma_{(H,\phi),(H',\phi')}\alpha_{(H',\phi')^G}(\rho)](H,\phi),$$

which establishes 2.2.16. □

Remark 2.2.17 We will construct the homomorphism, a_G, by first constructing a map,

$$a_G : R(G) \longrightarrow R_+(G)\otimes \mathbf{Q},$$

which satisfies 2.2.8(i)–(ii) and then showing that the unique such map must land in $R_+(G)$. Therefore we will temporarily (for the remainder of this section) assume that we have constructed the map into $R(G)\otimes\mathbf{Q}$.

We will require some preparatory results before we can show the integrality of $a_G(\rho) \in R_+(G)\otimes\mathbf{Q}$.

Suppose that $\rho : G \longrightarrow GL(V)$ is an irreducible, complex representation. Let $H \lhd G$ be a normal subgroup and let $W \subset Res^G_H(V)$ be an irreducible H-subrepresentation. For each $g \in G/H$ the subspace, $W_g = \rho(g)(W) \subset V$, is well-defined. Also, W_g is an H-subrepresentation since, for $h \in H$ and $w \in W$,

$$\rho(h)\rho(g)(w) = \rho(hg)(w) = \rho(g)(\rho(g^{-1}hg)(w)) \in W_g.$$

W_g is called a *G-conjugate* of W.

Lemma 2.2.18 *In the situation of* 2.2.17

$$V = \oplus_{g\in\Delta} W_g,$$

for some subset, $\Delta \subset G/H$.

Proof Clearly $\sum_{g\in G/H} W_g$ is a G-subrepresentation of V, which is irreducible, so that $\sum_{g\in G/H} W_g = V$. Also, since W is an irreducible H-representation then so is W_g for each g.

Therefore, as an H-representation, V must be the direct sum of a subset of the $\{W_g ; g \in G/H\}$, which completes the proof. □

Lemma 2.2.19 *Let H be a subgroup of G and let $\phi : H \longrightarrow \mathbf{C}^*$ be a one-dimensional representation. If ρ is a representation of G then*

$$< \phi, Res_H^G(\rho) >= 0 \textit{ implies that } \alpha_{(H,\phi)^G}(\rho) = 0.$$

Proof When $H = G$ this is guaranteed by the axiom 2.2.8(ii). We will prove the result by downward induction on $\#(H)$. We have

2.2.20
$$\begin{cases} < \phi, Res_H^G(\rho) > \\ = \{\text{coefficient of } (H,\phi)^H \text{ in } a_H(Res_H^G(\rho))\} \\ = \sum_{\substack{(H,\phi)^G \leq (H',\phi')^G \\ \in (\mathscr{M})_{\mathscr{G}}/\mathscr{G}}} \gamma^G_{(H,\phi),(H',\phi')} \alpha_{(H',\phi')^G}(\rho), \end{cases}$$

by 2.2.16.

If $H < G$ then $< \phi, Res_H^G(\rho) >= 0$ implies that $< \phi', Res_{H'}^G(\rho) >= 0$ for all $(H,\phi) \leq (H',\phi')$ and so, by induction, $\alpha_{(H',\phi')^G}(\rho) = 0$ in this case. Hence 2.2.20 becomes

$$0 =< \phi, Res_H^G(\rho) >= \gamma^G_{(H,\phi),(H,\phi)} \alpha_{(H,\phi)^G}(\rho),$$

so that $\alpha_{(H,\phi)^G}(\rho) = 0$, as required. □

Definition 2.2.21 Let $\rho : G \longrightarrow GL(V)$ be a complex representation and let $(H,\phi) \in \mathscr{M}_G$. Define a subspace, $F(H,\phi) \leq V$, by

2.2.22 $$F(H,\phi) = \{v \in V \mid \rho(h)(v) = \phi(h)v \; for \; all \; h \in H\}.$$

Conversely, given a subspace, $0 \neq W \leq V$, define $P(W) \in \mathscr{M}_G$ by

2.2.23
$$P(W)$$
$$= sup\{(H,\phi) \in \mathscr{M}_G \mid \rho(h)(w) = \phi(h)w for \; all \; h \in H, w \in W\}.$$

Lemma 2.2.24 *In 2.2.23, $P(W)$ is well-defined.*

Proof We have to show that the supremum exists in 2.2.23.

Firstly we observe that $(\{1\}, 1)$ satisfies the condition of 2.2.23.

Secondly, if (H,ϕ) and (K,ψ) satisfy the condition of 2.2.23, then $Res^H_{H\cap K}(\phi) = Res^K_{H\cap K}(\psi)$, since $W \neq 0$. Therefore we can extend ϕ and ψ uniquely to the group, $U =< H, K >$, generated by H and K. If we call this extension $\mu : U \longrightarrow \mathbf{C}^*$ then (U,μ) satisfies the condition of 2.2.23.

Hence $P(W)$ exists and is equal to the character on the group generated by all the H's on the right of 2.2.23 given by the unique extension of all the ϕs on the right of 2.2.23. □

Definition 2.2.25 We call $(H,\phi) \in \mathcal{M}_G$ *admissible* for V (or ρ) if $F(H,\phi) \neq 0$ or, equivalently, if $< \phi, Res^G_H(\rho) > \neq 0$. Denote by $A(V)$ (or $A(\rho)$) the set of elements of $\mathcal{M}_G$ which are admissible for V. Denote by $S(V)$ the set of non-zero subspaces of V. Therefore we have maps

2.2.26 $$F : A(V) \longrightarrow S(V) \text{ and } P : S(V) \longrightarrow A(V).$$

Proposition 2.2.27 (i) *Let* $(H,\phi), (K,\psi) \in A(V)$ *and* $W, W' \in S(V)$.

(a) *If* $(H',\phi') \leq (H,\phi)$ *then* $(H',\phi') \in A(V)$.

(b) *If* $g \in G$ *then* $g(H,\phi) \in A(V)$ *where* $g(H,\phi)$ *is as in* 2.2.11.

(ii) $F(g(H,\phi)) = \rho(g)(F(H,\phi))$, $P(\rho(g)W) = g(P(W))$ *for all* $g \in G$.

(iii) $(K,\psi) \leq (H,\phi)$ *implies that* $F(K,\psi) \geq F(H,\phi)$ *and* $W \leq W'$ *implies that* $P(W') \leq P(W)$.

(iv) $(H,\phi) \leq PF(H,\phi)$ *and* $W \leq FP(W)$.

(v) $F(H,\phi) = FPF(H,\phi)$ *and* $P(W) = PFP(W)$.

Proof If $K \leq H$ and $Res^H_K(\phi) = \psi$ then $< \phi, Res^G_H(\rho) > \neq 0$ implies that $< \psi, Res^G_K(\rho) > \neq 0$, which proves part (i)(a). Also,

$$\begin{aligned} < (g^{-1})^*(\phi), Res^G_{gHg^{-1}}(\rho) > \; &= < (g^{-1})^*(\phi), Res^G_{gHg^{-1}}((g^{-1})^*\rho) > \\ &= < \phi, Res^G_H(\rho) >, \end{aligned}$$

since $(g^{-1})^*(\rho) = \rho$, which proves part (i)(b).

For part (ii) we note that

$$F(g(H,\phi)) = \{v \in V \mid \phi(g^{-1}zg)(v) = \rho(z)(v) \text{ for all } z \in gHg^{-1}\}$$

so that $v \in F(g(H,\phi))$ if and only if $\rho(ghg^{-1})(v) = \phi(h)v$, which is equivalent to $\rho(g)^{-1}(v) \in F(H,\phi)$. This proves the first part of (ii). Similarly, $(gHg^{-1}, (g^{-1})^*(\phi))$ satisfies $\rho(ghg^{-1})(w) = \phi(h)w$ for all $w \in W$ if and only if (H,ϕ) satisfies $\rho(g^{-1})(w') = \phi(h)w'$ for all $w' \in \rho(g^{-1})(W)$. The second part of (ii) follows by taking suprema over these sets, as in 2.2.23.

If $K \leq H$ and $\psi = Res^H_K(\phi)$ then the condition of belonging to $F(H,\phi)$ is stronger than that of belonging to $F(K,\psi)$, which proves the first part of (iii). The second part of (iii) is similar.

Since (H,ϕ) is one of the objects over which the supremum is taken when forming $PF(H,\phi)$ the first part of (iv) is clear. Also, for every

(K,ψ) in the set over which the supremum is taken in 2.2.23, every $w \in W$ satisfies $\rho(k)(w) = \psi(k)w$ for all $k \in K$ so that $w \in FP(W)$.

Finally, by (iv), $(H,\phi) \le PF(H,\phi)$ and $W \le FP(W)$ so that, by (iii), $F(H,\phi) \ge FPF(H,\phi)$ and $P(W) \ge PFP(W)$, which proves part (v), since, by (iv), $F(H,\phi) \le FP\{F(H,\phi)\}$ and $P(W) \le PF\{P(W)\}$. □

Definition 2.2.28 For $(H,\phi) \in A(V)$ and $W \in S(V)$ define their *closures*, $cl(H,\phi)$ and $cl(W)$, respectively, by

$$cl(H,\phi) = PF(H,\phi) \text{ and } cl(W) = FP(W).$$

An object will be called *closed* if it equals its own closure.

Proposition 2.2.29 *Let $(H,\phi),(K,\psi) \in A(V)$ and $W, W' \in S(V)$.*

(i) *$cl(H,\phi)$ and $cl(W)$ are closed.*

(ii) *$F(cl(H,\phi)) = F(H,\phi)$ and $P(cl(W)) = P(W)$.*

(iii) *If $(K,\psi) \le (H,\phi)$ then $cl(K,\psi) \le cl(H,\phi)$ and if $W \le W'$ then $cl(W) \le cl(W')$.*

(iv) *The following are equivalent:*

(a) *(H,ϕ) (respectively W) is closed.*

(b) *$g(H,\phi)$ (respectively $\rho(g)W$) is closed.*

(c) *(H,ϕ) is in the image of P (respectively, W is in the image of F).*

(v) *F and P are inverse bijections between the closed pairs in $A(V)$ and the closed subspaces in $S(V)$.*

(vi) *$inf\{P(W),P(W')\} = P(W+W')$. In particular the infimum of closed pairs is closed.*

Proof Parts (i)–(v) follow at once from 2.2.27. To prove part (vi) we observe that $(H,\phi) \le P(W)$ and $(H,\phi) \le P(W')$ if and only if H acts on both W and W' by ϕ and this is equivalent to H acting on $W + W'$ by ϕ. Therefore $P(W + W')$ is clearly the largest pair, (H,ϕ), such that $(H,\phi) \le P(W)$ and $(H,\phi) \le P(W')$, which is, by definition, the infimum of (vi). □

Lemma 2.2.30 *In the notation of* 2.2.13, *if $(K,\psi) \le (H,\phi)$ and (H,ϕ) is closed then*

$$\gamma^G_{(K,\psi),(H,\phi)} = \gamma^G_{cl(K,\psi),(H,\phi)}.$$

Proof Set $cl(K,\psi) = (U,\mu)$ then $(U,\mu) \leq (H,\phi)$, by 2.2.29(iii) since (H,ϕ) is closed. As in the proof of 2.2.15,

$$Res_U^G(H,\phi)^G = \gamma^G_{(U,\mu),(H,\phi)}(U,\mu)^U + \sum_z (U \cap zHz^{-1},(z^{-1})^*(\phi))^U,$$

where the sum is taken over $z \in U\backslash G/H$ such that

$$(U \cap zHz^{-1},(z^{-1})^*(\phi))^U \neq (U,\mu)^U.$$

Applying Res_K^U and using $Res_K^G = Res_K^U Res_U^G$ we obtain

2.2.31 $$Res_K^G(H,\phi)^G$$
$$= \gamma^G_{(U,\mu),(H,\phi)} Res_K^U(U,\mu)^U + \sum_z Res_K^U(U \cap zHz^{-1},(z^{-1})^*(\phi))^U.$$

Since $Res_K^U(U,\mu)^U = (K,\psi)^K$ it is sufficient to verify that, in the last sum of 2.2.31, no pair $(K,\psi)^K$ appears. For this it suffices to verify for all z in the sum that

$$(K,\psi)^U \not\leq (U \cap zHz^{-1},(z^{-1})^*(\phi))^U.$$

Therefore we will prove that

2.2.32 $$\begin{cases} (K,\psi)^U \leq (U \cap zHz^{-1},(z^{-1})^*(\phi))^U \text{ implies} \\ (U \cap zHz^{-1},(z^{-1})^*(\phi))^U = (U,\mu)^U. \end{cases}$$

If the condition of 2.2.32 holds then there exists $u \in U$ such that

2.2.33 $$(K,\psi) \leq (U \cap uzHz^{-1}u^{-1},((uz)^{-1})^*(\phi)).$$

From 2.2.33 we deduce that

2.2.34 $$(U,\mu) = cl(K,\psi) \leq cl(U \cap uzHz^{-1}u^{-1},((uz)^{-1})^*(\phi)).$$

Setting (H',ϕ') equal to the right side of 2.2.33 we obtain

$$Res^U_{U \cap uzH(uz)^{-1}}(\mu) = \phi' = ((uz)^{-1})^*(\phi),$$

which implies that

$$(K,\psi) \leq (U \cap uzHz^{-1}u^{-1},((uz)^{-1})^*(\phi)) \leq (U,\mu) = cl(K,\psi).$$

Therefore

2.2.35 $$(U,\mu) = cl(U \cap uzHz^{-1}u^{-1},((uz)^{-1})^*(\phi)).$$

However,

$$(U \cap uzHz^{-1}u^{-1}, (uz^{-1})^*(\phi)) = inf((U,\mu), (uzH(uz)^{-1}, ((uz)^{-1})^*(\phi)))$$

and both (U,μ) and $uz(H,\phi)$ are closed, by 2.2.29(i) and 2.2.27(ii), so that $(U \cap uzHz^{-1}u^{-1}, ((uz)^{-1})^*(\phi))$ is also closed, by 2.2.29(vi). Hence 2.2.32 follows from 2.2.35, which completes the proof. □

Proposition 2.2.36 *Let $\rho : G \longrightarrow GL(V)$ be a representation and let $(H,\phi) \in \mathcal{M}_G$. In 2.2.15, if $\alpha_{(H,\phi)^G}(\rho) \neq 0$ then (H,ϕ) is admissible and closed for V.*

In particular, if there exists $(H',\phi') \overset{>}{\neq} (H,\phi)$ in $\mathcal{M}_G$ with

$$< \phi', Res^G_{H'}(\rho) > = < \phi, Res^G_H(\rho) >$$

then $\alpha_{(H,\phi)^G}(\rho) = 0$.

Proof By 2.2.19, $\alpha_{(H,\phi)^G}(\rho) \neq 0$ implies that $< \phi, Res^G_H(\rho) > \neq 0$; hence $(H,\phi) \in A(V)$. We will show that (H,ϕ) is closed by descending induction on the order of H.

If $H = G$ then (G,ϕ) is closed because it is maximal in $A(V)$.

Now assume that $H \overset{<}{\neq} G$ and set $(U,\psi) = cl(H,\phi)$. Now consider the coefficient of (H,ϕ) in $a_H(Res^G_H(\rho))$ and of (U,ψ) in $a_U(Res^G_U(\rho))$. These are equal to

2.2.37
$$\begin{cases} < \phi, Res^G_H(\rho) > & = \gamma^G_{(H,\phi),(H,\phi)}\alpha_{(H,\phi)^G}(\rho) \\ & + \sum_{\substack{(H,\phi)^G < (H',\phi')^G \\ in \mathcal{M}^G/G}} \gamma^G_{(H,\phi),(H',\phi')}\alpha_{(H',\phi')^G}(\rho) \end{cases}$$

and

2.2.38
$$\begin{cases} < \psi, Res^G_U(\rho) > \\ = \sum_{\substack{(U,\psi)^G \leq (U',\psi')^G \\ in \mathcal{M}^G/G}} \gamma^G_{(U,\psi),(U',\psi')}\alpha_{(U',\psi')^G}(\rho). \end{cases}$$

By induction the sums in 2.2.37 and 2.2.38 may only be taken over admissible pairs.

If (H,ϕ) is not closed and $cl(H,\phi) = (U,\psi)$ the closed pairs which are strictly greater than (H,ϕ) are the closed pairs which are greater than or equal to (U,ψ). Therefore, by 2.2.30, the two sums are equal in 2.2.37

and 2.2.38. However, by 2.2.29(ii),

$$\begin{aligned} <\psi, Res_U^G(\rho)> &= dim_{\mathbf{C}} F(U,\psi) \\ &= dim_{\mathbf{C}} F(cl(H,\phi)) \\ &= dim_{\mathbf{C}} F(H,\phi) \\ &=<\phi, Res_H^G(\rho)>, \end{aligned}$$

so that, subtracting 2.2.38 from 2.2.37, we obtain

$$0 = \gamma^G_{(H,\phi),(H,\phi)} \alpha_{(H,\phi)^G}(\rho)$$

and therefore $\alpha_{(H,\phi)^G}(\rho) = 0$, as required.

For the final statement there is nothing to prove if $<\phi, Res_H^G(\rho)>= 0$, by 2.2.19. If not then

$$\begin{aligned} dim_{\mathbf{C}} F(H',\phi') &=<\phi', Res_{H'}^G(\rho)> \\ &=<\phi, Res_H^G(\rho)> \\ &= dim_{\mathbf{C}} F(H,\phi), \end{aligned}$$

so that $F(H,\phi) = F(H',\phi')$. Hence $cl(H',\phi') = cl(H,\phi)$ and therefore (H,ϕ) is not closed, since $(H,\phi) \underset{\neq}{<} (H',\phi') \leq cl(H',\phi')$. □

Definition 2.2.39 Let $\rho : G \longrightarrow GL(V)$ be a representation. The *centre* of ρ, $Z(\rho)$, is defined to be the largest subgroup, H, such that $Res_H^G(\rho) = dim_{\mathbf{C}}(\rho)\psi$ for some $\psi : H \longrightarrow \mathbf{C}^*$.

Corollary 2.2.40 (i) *Suppose that* $Res_{Z(\rho)}^G(\rho) = dim_{\mathbf{C}}(\rho)\psi$, *then for all* $(H,\phi) \in \mathscr{M}_G$, *in* 2.2.15,

$$\alpha_{(H,\phi)^G}(\rho) \neq 0 \textit{ implies that } (Z(\rho),\psi) \leq (H,\phi).$$

(ii) *For all* $\chi \in R(G)$

$$\alpha_{(H,\phi)^G}(\chi) \neq 0 \textit{ implies that } Z(G) \leq H.$$

Proof If $\alpha_{(H,\phi)^G}(\rho) \neq 0$ then (H,ϕ) is admissible and closed, by 2.2.36. Also $F(Z(\rho),\psi) = V \geq F(H,\phi)$ so that

$$(H,\phi) = cl(H,\phi) = PF(H,\phi) \geq PF(Z(\rho),\psi) \geq (Z(\rho),\psi).$$

This proves part (i).

For part (ii) we may assume that χ is irreducible so that $Z(G) \leq Z(\chi)$ and the result follows from part (i). □

Proposition 2.2.41 *Suppose that the homomorphism of* 2.2.15,

$$a_G : R(G) \longrightarrow R_+(G) \otimes \mathbf{Q},$$

satisfies the axioms of 2.2.8. *Then, for all* $\chi \in R(G)$,

2.2.42 $$\sum_{(H,\phi)^G \in \mathcal{M}_G/G} \alpha_{(H,\phi)^G}(\chi) Ind_H^G(1) = dim_{\mathbf{C}}(\chi) \in R(G) \otimes \mathbf{Q}.$$

Proof Let $\eta_G : R_+(G) \longrightarrow R_+(G)$ be the natural homomorphism defined by $\eta_G(H,\phi)^G = (H,1)^G$. The left side of 2.2.42 is the image of χ under

$$b_G \eta_G a_G : R(G) \longrightarrow R(G) \otimes \mathbf{Q}.$$

This map is natural for restrictions to subgroups so that, in order to evaluate the character of this expression at $g \in G$, it suffices to verify the result for the cyclic group, $< g >$. However, for G cyclic the result is immediate from axiom 2.2.8(ii). □

Theorem 2.2.43 *In* 2.2.15,

$$a_G(\rho) \in R_+(G) \subset R_+(G) \otimes \mathbf{Q} \ for \ all \ \rho \in R(G).$$

Proof Assume for all $H \stackrel{<}{\neq} G$ and for all $\chi \in R(H)$ that $a_H(\chi) \in R_+(H)$. This inductive statement is certainly true for all abelian groups, H. In order to verify that $a_G(\rho) \in R_+(G)$ it suffices to show this when ρ is irreducible.

Therefore suppose that ρ is irreducible and that $a_G(\rho) \notin R_+(G)$. Also suppose that $(H,\phi) \in \mathcal{M}_G$ is maximal such that $\alpha_{(H,\phi)^G}(\rho) \notin Z$. Under these conditions we will first show that $H \lhd G$. Let $U = N_G(H,\phi)$, the normaliser of (H,ϕ) in G.

Consider the coefficient of $(H,\phi)^U$ in $Res_U^G(a_G(\rho)) = a_U(Res_U^G(\rho))$.

This is given by

$$\begin{aligned} &\alpha_{(H,\phi)^U}(Res_U^G(\rho)) \\ &= \sum_{(H',\phi')^G \in \mathcal{M}_G} \alpha_{(H',\phi')^G}(\rho) \\ &\sum_{\substack{z \in U\backslash G/H' \\ H = U \cap zH'z^{-1}, \phi = (z^{-1})^*(\phi')}} (U \cap zH'z^{-1}, (z^{-1})^*(\phi'))^U \\ &= x\alpha_{(H,\phi)^G}(\rho) + \sum_{(H,\phi)^G \stackrel{<}{\neq} (H',\phi')^G} \delta_{(H,\phi)^G}^{(H',\phi')^G} \alpha_{(H',\phi')^G}(\rho), \end{aligned}$$

since only $(H',\phi') \geq (H,\phi)$ can contribute a multiple of $(H,\phi)^U$ to the above expression. Also, by maximality of (H,ϕ), the sum of such coefficients in the above expression is an integer. On the other hand, $x = \#\{z \in U\backslash G/H \mid zHz^{-1} = H, (z^{-1})^*(\phi) = \phi\}$ and taking $z = 1$ we see that $x \geq 1$; but since $zHz^{-1} = H$ and $(z^{-1})^*(\phi) = \phi$ if and only if $z \in U$ we find that $x = 1$. Since $\alpha_{(H,\phi)^U}(Res^G_U(\rho)) \in Z$ if $U \stackrel{<}{\neq} G$ we find that $U = G$, because

$$x \cdot \alpha_{(H,\phi)^G}(\rho) = \alpha_{(H,\phi)^G}(\rho) \notin \mathbf{Z}.$$

Therefore $H \lhd G$ and $(z^{-1})^*(\phi) = \phi$ for all $z \in G$.

Next we will show that $H = Z(G)$, the centre of G.

Since $H \lhd G$ and ρ is irreducible, we may apply 2.2.18 to assert that

$$Res^G_H(\rho) = \oplus_{g\in\Delta} W_g$$

for some $\Delta \subset G/H$. Since $\alpha_{(H,\phi)^G}(\rho) \neq 0$ we have, by 2.2.19, that $< \phi, Res^G_H(\rho) > \neq 0$ and therefore ϕ is one of the W_g's. However, if $W = \phi$ in 2.2.17–2.2.18 then each $W_g = \phi$ also, because ϕ is G-invariant. Hence

2.2.44 $$Res^G_H(\rho) = (dim_{\mathbf{C}}(\rho))\phi \in R(H).$$

This means that $Ker(\phi) \leq Ker(\rho)$ and that $Ker(\phi) \lhd G$, since ϕ is G-invariant. Therefore there are factorisations of the following form:

2.2.45 $$\begin{cases} \rho : G \longrightarrow G/(Ker(\phi)) \longrightarrow GL(V) \\ \\ \phi : H \longrightarrow H/(Ker(\phi)) \longrightarrow \mathbf{C}^*. \end{cases}$$

since $Ker(\phi) \lhd H$.

By 2.5.9,

2.2.46 $$\alpha_{(H,\phi)^G}(\rho) = \alpha_{(H/Ker(\phi),\rho)^{G/Ker(\phi)}}(\rho).$$

By minimality of G we must have $Ker(\phi) = \{1\}$ so that $\phi : H \longrightarrow \mathbf{C}^*$ is G-invariant and injective. Hence, if $h \in H$ and $g \in G$ then $\phi(ghg^{-1}) = \phi(h)$ and therefore $ghg^{-1} = h$ so that $H \leq Z(G)$, as required.

Let $Z(\rho)$ denote the centre of ρ, in the sense of 2.2.39 and suppose that $Res^G_{Z(\rho)}(\rho) = (dim_{\mathbf{C}}(\rho))\psi$ then $(Z(\rho), \psi) \leq (H', \phi')$ if $\alpha_{(H',\phi')^G} \neq 0$, by 2.2.40(i). On the other hand, $(H, \phi) = (Z(G), \phi) \leq (Z(\rho), \psi)$. Therefore, by maximality of (H, ϕ), $\alpha_{(H,\phi)^G}$ is the only non-integral coefficient in $a_G(\rho)$. By 2.2.41, in $R(G) \otimes \mathbf{Q}$ we have

2.2.47 $$dim_{\mathbf{C}}(\rho)$$

$$= \alpha_{(H,\phi)^G}(\rho) Ind^G_H(1) + \sum_{(H,\phi)^G \neq (H',\phi')^G \text{in} \mathcal{M}_G/G} \alpha_{(H',\phi')^G} Ind^G_{H'}(1).$$

Since Frobenius reciprocity implies that $< Ind_H^G(1), 1 >= 1$, we obtain

$$dim_{\mathbf{C}}(\rho) = \alpha_{(H,\phi)^G}(\rho) + \sum_{(H,\phi)^G \neq (H',\phi')^G \text{in} \mathscr{M}_G/G} \alpha_{(H',\phi')^G} \in \mathbf{Z},$$

so that $\alpha_{(H,\phi)^G} \in \mathbf{Z}$, which is a contradiction. □

2.3 Brauer induction in canonical integral form

In this section we will complete the construction of the homomorphism

2.3.1
$$a_G : R(G) \longrightarrow R_+(G),$$

which satisfies the axioms of 2.2.8(i)–(ii). We will begin by recapitulating the results of the previous section, which culminates in 2.2.43.

Proposition 2.3.2 *Let $a_G : R(G) \longrightarrow R_+(G) \otimes \mathbf{Q}$ be a homomorphism which satisfies the following two conditions:*

(i) *For all $H \leq G$ the following diagram commutes:*

$$\begin{array}{ccc} R(G) & \xrightarrow{a_G} & R_+(G) \otimes \mathbf{Q} \\ {\scriptstyle Res_H^G}\downarrow & & \downarrow{\scriptstyle Res_H^G \otimes 1} \\ R(H) & \xrightarrow{a_H} & R_+(H) \otimes \mathbf{Q} \end{array}$$

(ii) *Let $\rho : G \longrightarrow GL_n(\mathbf{C})$ be a representation and suppose that*

$$a_G(\rho) = \sum \alpha_{(H,\phi)^G}(H,\phi)^G \in R_+(G) \otimes \mathbf{Q}$$

then $\alpha_{(G,\phi)^G} =< \rho, \phi >$ for each $(H,\phi)^G$ such that $H = G$.

Then $a_G(R(G)) \subset R_+(G)$ and a_G satisfies the axioms of 2.2.8(i)–(ii). *In addition, if b_G is the homomorphism of* 2.2.6 *then* (see 2.2.7)

$$b_G a_G = 1 : R(G) \longrightarrow R(G).$$

Proof Only the last statement requires to be proved. However, this is easy since a natural endomorphism of $R(G)$, such as $b_G a_G$, which is the identity when G is cyclic, must be the identity for all G. □

Definition 2.3.3 Let us define a bilinear form

2.3.4 $$[-,-] : (R_+(G) \otimes \mathbf{Q}) \times (R_+(G) \otimes \mathbf{Q}) \longrightarrow \mathbf{Q}$$

by means of the formula

2.3.5 $$[(H,\phi)^G, (H',\phi')^G] = \gamma^G_{(H,\phi),(H',\phi')}.$$

By 2.2.14(i) this is a well-defined bilinear form, which is not symmetric. In fact, if we take the poset $\{(H,\phi)^G \in (\mathcal{M}_G)/G\}$ as a basis then the matrix representing 2.3.4 with respect to this basis is 'upper triangular' with respect to the partial ordering. Therefore 2.3.4 is a non-singular bilinear form.

Proposition 2.3.6 *Let J be a subgroup of G.*
If $x \in R_+(J)$ and $y \in R_+(G)$ then

$$[Ind_J^G(x), y] = [x, Res_J^G(y)],$$

where $Ind_J^G((U,\psi)^J) = (U,\psi)^G \in R_+(G)$.

Proof It will suffice to take $x = (U,\psi)^J$ and $y = (H,\phi)^G$. In this case

$$\begin{aligned}[Ind_J^G(x), y] &= [(U,\psi)^G, (H,\phi)^G] \\ &= \#\{g \in U\backslash G/H \mid (U,\psi) \leq g(H,\phi)\}.\end{aligned}$$

On the other hand,

$$\begin{aligned}&[x, Res_J^G(y)] \\ &= \textstyle\sum_{z \in J\backslash G/H} [(U,\psi)^J, (J \cap zHz^{-1}, (z^{-1})^*(\phi))^J] \\ &= \textstyle\sum_{z \in J\backslash G/H} \#\{w \in U\backslash J/J \cap (zHz^{-1}) \mid \\ &\qquad (U,\psi) \leq w(J \cap zHz^{-1}, (z^{-1})^*(\phi))\}.\end{aligned}$$

To see that these expressions are equal we first observe that $w(J \cap zHz^{-1}, (z^{-1})^*(\phi)) = (J \cap wzH(wz)^{-1}, ((wz)^{-1})^*(\phi))$ and that, since $U \leq J$, $(U,\psi) \leq (gHg^{-1}, (g^{-1})^*(\phi))$ if and only if $(U,\psi) \leq (J \cap gHg^{-1}, (g^{-1})^*(\phi))$. Therefore the two expressions which we wish to prove equal are both sums over the same type of object. The result now follows from the fact that there is a bijection

$$\{(z,w) \in (J\backslash G/H) \times (U\backslash J/(J \cap zHz^{-1})\} \longleftrightarrow \{g \in U\backslash G/H\}$$

given by sending (z,w) to wz. This bijection is derived from enumerating the J-orbits and then counting the decomposition of each J-orbit into U-orbits, using the fact that $J \cap zHz^{-1}$ is the J-stabiliser of zH in G/H. □

Definition 2.3.7 Since $[-,-]$ is a non-singular bilinear form we may define a homomorphism

$$a_G : R(G) \otimes \mathbf{Q} \longrightarrow R_+(G) \otimes \mathbf{Q}$$

as the adjoint of $b_G : R_+(G) \longrightarrow R(G)$ of 2.2.6. That is, for all $x \in R_+(G) \otimes \mathbf{Q}$, $\rho \in R(G) \otimes \mathbf{Q}$,

2.3.8
$$[x, a_G(\rho)] = < b_G(x), \rho > .$$

Theorem 2.3.9 *There exists a homomorphism*

$$a_G : R(G) \longrightarrow R_+(G)$$

which satisfies the axioms of 2.2.8(i)–(ii).

Also, if b_G *is the map of* 2.2.6 *then* $b_G a_G = 1$.

Proof It suffices to verify, for a_G of 2.3.7, the two conditions of 2.3.2. For condition (i), let $\rho \in R(G)$ and suppose that $H \leq G$ is a subgroup. For all $x \in R_+(H)$ we have

$$\begin{aligned}
[x, Res^G_H(a_G(\rho))] &= [Ind^G_H(x), a_G(\rho)], && \text{by 2.3.6} \\
&=< b_G(Ind^G_H(x)), \rho >, && \text{by 2.3.8} \\
&=< Ind^G_H(b_H(x)), \rho > && \\
&=< b_H(x), Res^G_H(\rho) >, && \text{by 1.2.8, 1.2.38} \\
&= [x, a_H(Res^G_H(\rho))], && \text{by 2.3.8 again}
\end{aligned}$$

so that $Res^G_H(a_G(\rho)) = a_H(Res^G_H(\rho))$ by the non-singularity of 2.3.4.

Now let $\phi : G \longrightarrow \mathbf{C}^*$ be a one-dimensional representation and let $(H, \psi)^G \in \mathcal{M}_G$. Therefore, by 1.2.38,

$$< b_G(H, \psi)^G, \phi >= \begin{cases} 1 & \text{if } Res^G_H(\phi) = \psi, \\ 0 & \text{otherwise.} \end{cases}$$

However,

$$[(H, \psi)^G, (G, \phi)^G] = \#\{g \in H\backslash G/G \mid (H, \psi) \leq (G, \phi)\},$$

so that

$$[(H, \psi)^G, a_G(\phi)] =< b_G((H, \psi)^G), \phi >= [(H, \psi)^G, (G, \phi)^G]$$

for all $(H, \psi)^G$ and therefore $a_G(\phi) = (G, \phi)^G$. If ρ is irreducible and not one-dimensional then the multiplicity of $(G, \phi)^G$ in $a_G(\rho)$ is equal to

$$[(G, \phi)^G, a_G(\rho)] =< \phi, \rho >= 0,$$

which completes the proof of Theorem 2.3.9. □

Definition 2.3.10 (*See also* 2.5.1) Let $\mathcal{M}$ be a finite partially ordered set (a poset). The *Möbius function* of $\mathcal{M}$ is an integer-valued function, μ, on $\mathcal{M} \times \mathcal{M}$ which is defined in the following manner. A *chain of length i* in $\mathcal{M}$ is a totally ordered subset of elements of $\mathcal{M}$,

2.3.11
$$M_0 \overset{<}{\neq} M_1 \overset{<}{\neq} \ldots \overset{<}{\neq} M_i.$$

We define $\mu_{A,B}$, for $A, B \in \mathcal{M}$, by

2.3.12
$$\mu_{A,B}$$
$$= \sum_i (-1)^i \#\{\text{chains of length } i \text{ with } M_0 = A, M_i = B \text{ in 2.3.11}\}.$$

2.3.13 *The formula for a_G in terms of Möbius functions*

Let $\mu^{\mathcal{M}_G}$ denote the Möbius function for the poset, $\mathcal{M}_G$, of pairs, (H, ϕ), of 2.2.9.

We will now give, without proof, the formula for a_G in 2.3.7. In order to do this one considers the formula of 2.2.16:

2.3.14
$$< \phi, Res_H^G(\rho) >= \sum_{\substack{(H,\phi)^G \leq (H',\phi')^G \\ in \mathcal{M}_G/G}} \gamma^G_{(H,\phi),(H',\phi')} \alpha_{(H',\phi')^G}(\rho).$$

In order to solve 2.3.14 for the $\alpha_{(H',\phi')^G}(\rho)$s we have to invert the *incidence matrix*, $(\gamma^G_{(H,\phi),(H',\phi')})$, of the poset $\mathcal{M}_G/G$. The inverse of this incidence matrix is the Möbius matrix for the poset $\mathcal{M}_G/G$. The relationship between the Möbius matrix for $\mathcal{M}_G$, $(\mu^{\mathcal{M}_G}_{(H,\phi),(H',\phi')})$, and that for $\mathcal{M}_G/G$, consists of the insertion of correction factors, $\#(H)/\#(G)$. Further details of the solution of 2.3.14 are to be found in Boltje (1990) and Boltje (1989 (2.35) *et seq.*). We will not need the following explicit formula, except in some illustrative examples.

Theorem 2.3.15 *The homomorphism*
$$a_G : R(G) \longrightarrow R_+(G)$$
of 2.3.9 *is given by the formula* $a_G(\rho)$
$$= \#(G)^{-1} \sum_{\substack{(H,\phi) \leq (H',\phi') \\ in \mathcal{M}_G}} \#(H) \mu^{\mathcal{M}_G}_{(H,\phi),(H',\phi')} < \phi', Res^G_{H'}(\rho) > (H, \phi)^G.$$

Table 2.1. *Character table for Q_8*

	1	2	4^1	4^2	4^3
1	1	1	1	1	1
μ_1	1	1	1	-1	-1
μ_2	1	1	-1	1	-1
μ_3	1	1	-1	-1	1
ν	2	-2	0	0	0

Table 2.2. *Table of $(Res_H^G(\chi))_{ab}$ for Q_8*

H	$\hat{H}$	1	μ_1	μ_2	μ_3	ν
Q_8	$1, \mu_1, \mu_2, \mu_3$	1	μ_1	μ_2	μ_3	0
C_4^1	$1, \phi \sim \phi^3, \phi^2$	1	1	ϕ^2	ϕ^2	$\phi + \phi^3$
C_4^2	$1, \phi \sim \phi^3, \phi^2$	1	ϕ^2	1	ϕ^2	$\phi + \phi^3$
C_4^3	$1, \phi \sim \phi^3, \phi^2$	1	ϕ^2	ϕ^2	1	$\phi + \phi^3$
C_2	$1, \phi$	1	1	1	1	$2 \cdot \phi$
$\{1\}$	1	1	1	1	1	$2 \cdot 1$

2.3.16 *Example*: $G = Q_8$ (*cf*. 1.3.7)

We denote the three conjugacy classes in Q_8 which contain elements of order four by $4^1, 4^2$ and 4^3. Similarly let 2 denote the conjugacy class of the non-trivial central elements, $X^2 = Y^2$, in the notation of 1.3.7. The character table of Q_8 is presented in Table 2.1 (the (i, j)th entry is the value of the ith irreducible character on the jth conjugacy class).

Now let us tabulate the conjugacy classes of subgroups, H, of Q_8 together with the conjugacy classes of one-dimensional characters, $(\phi : H \longrightarrow \mathbf{C}^*) \in Hom(H, \mathbf{C}^*) = \hat{H}$. We denote the three cyclic subgroups of order four by C_4^1, C_4^2, C_4^3 and denote the centre by C_2. For each group ϕ is an injective homomorphism in $\hat{H}$. For each irreducible representation, χ, of G we also tabulate the sum of all the one-dimensional representations which appear in $Res_H^G(\chi)$. We call this sum of one-dimensional characters the *abelian part* of $Res_H^G(\chi)$ and denote it by $(Res_H^G(\chi))_{ab}$. (See Table 2.2.)

Now consider the formula for $a_{Q_8}(\nu)$. By 2.2.36–2.2.40 and Table 2.2 this must take the form

2.3.17 $$a_{Q_8}(\nu) = a(C_4^1, \phi)^{Q_8} + b(C_4^2, \phi)^{Q_8} + c(C_4^3, \phi)^{Q_8} + d(C_2, \phi)^{Q_8}.$$

Table 2.3. *Character table for* A_5 ($\alpha_1 = (1+\sqrt{5})/2, \alpha_2 = (1-\sqrt{5})/2$)

	1	2	3	5^1	5^2
1	1	1	1	1	1
$v_{3,1}$	3	-1	0	α_1	α_2
$v_{3,2}$	3	-1	0	α_2	α_1
v_4	4	0	1	-1	-1
v_5	5	1	-1	0	0

However, by maximality of (C_4^i, ϕ) in $\mathcal{M}_{Q_8}$ we have, by 2.5.10,

$$a = b = c = < \phi, Res^{Q_8}_{C_4^i}(v) > /(|N_{Q_8}(C_4^i, \phi) : C_4^i|) = 1.$$

By applying $b_{Q_8} : R_+(Q_8) \longrightarrow R(Q_8)$ to 2.3.17 and taking dimensions we find that $d = -1$ and we obtain

2.3.18 $$a_{Q_8}(v) = \sum_{i-1}^{3}(C_4^i, \phi)^{Q_8} - (C_2, \phi)^{Q_8} \in R_+(Q_8).$$

2.3.19 *Example:* $G = A_5$

The character table of the alternating group, A_5, has the form shown in Table 2.3, in which the elements $2, 3, 5^1$ and 5^2 have orders $2, 3, 5$ and 5, respectively.

In Table 2.4, D_6 and D_{10} are the dihedral subgroups of orders six and ten, respectively, $V_4 \cong \mathbf{Z}/2 \times \mathbf{Z}/2$ and C_n denotes a cyclic group of order n. The characters, $\phi : H \longrightarrow \mathbf{C}^*$, can be taken to be any of the obvious choices.

By 2.2.36–2.2.40, Table 2.4 and 2.5.10

$$a_{A_5}(v_{3,1}) = (D_{10}, \phi)^{A_5} + (D_6, \phi)^{A_5} + (C_5, \phi)^{A_5} + (V_4, \mu_1)^{A_5} + (C_3, \phi)^{A_5} + x(C_2, \phi)^{A_5} + y(\{1\}, 1)^{A_5}.$$

From 2.2.42

$$3 = Ind^{A_5}_{D_{10}}(1) + Ind^{A_5}_{D_6}(1) + Ind^{A_5}_{C_5}(1) + Ind^{A_5}_{V_4}(1) + Ind^{A_5}_{C_3}(1) + xInd^{A_5}_{C_2}(1) + yInd^{A_5}_{\{1\}}(1)$$

in $R(A_5)$, from which one finds that $x = -2$ and $y = 0$ to give

2.3.20 $$a_{A_5}(v_{3,1}) = (D_{10}, \phi)^{A_5} + (D_6, \phi)^{A_5} + (C_5, \phi)^{A_5} + (V_4, \mu_1)^{A_5} + (C_3, \phi)^{A_5} - 2(C_2, \phi)^{A_5}$$

in $R_+(A_5)$.

Table 2.4. *Table of* $(Res_H^G(\chi))_{ab}$ *for* A_5

H	$\hat{H}$	1	$v_{3,1}$	$v_{3,2}$	v_4	v_5
A_5	1	1	0	0	0	0
A_4	$1, \phi, \phi^2$	1	0	0	1	$\phi + \phi^2$
D_{10}	$1, \phi$	1	ϕ	ϕ	0	1
D_6	$1, \phi$	1	ϕ	ϕ	$1 + \phi$	1
C_5	$1, \phi \sim \phi^4, \phi^2 \sim \phi^3$	1	$1 + \phi + \phi^4$	$1 + \phi^2 + \phi^3$	$\phi + \phi^2 + \phi^3 + \phi^4$	$1 + \phi + \phi^2 + \phi^3 + \phi^4$
V_4	$1, \mu_1 \sim \mu_2 \sim \mu_3$	1	$\mu_1 + \mu_2 + \mu_3$	$\mu_1 + \mu_2 + \mu_3$	$1 + \mu_1 + \mu_2 + \mu_3$	$2 \cdot 1 + \mu_1 + \mu_2 + \mu_3$
C_3	$1, \phi \sim \phi^2$	1	$1 + \phi + \phi^2$	$1 + \phi + \phi^2$	$2 \cdot 1 + \phi + \phi^2$	$1 + 2\phi + 2\phi^2$
C_2	$1, \phi$	1	$1 + 2\phi$	$1 + 2\phi$	$2 \cdot 1 + 2\phi$	$3 \cdot 1 + 2\phi$
$\{1\}$	1	1	$3 \cdot 1$	$3 \cdot 1$	$4 \cdot 1$	$5 \cdot 1$

2.3.21 *Topological versions of explicit Brauer induction*

I will close this section with some remarks relating Robert Boltje's Explicit Brauer Induction formula, a_G, to my original topological Explicit Brauer Induction formula (Snaith, 1988b) and to Peter Symonds' topological construction of a_G (Symonds, 1991).

Suppose that $v : G \longrightarrow U(n)$ is an n-dimensional representation. Define $R_+(G, NT^n)$ to be the free abelian group on $(G - NT^n)$-conjugacy classes of homomorphisms of the form $(\phi : H \longrightarrow NT^n)$, where $H \leq G$ and NT^n is the normaliser of the diagonal maximal torus, as in 2.1.10. Hence $R_+(G, NT^1) = R_+(G)$.

Let G act, via v, on the left of $U(n)/NT^n = X$, as in 2.1.10. For each conjugacy class of a subgroup, $H \leq G$, let $Y_{(H)}$ denote the subspace of the orbit space, $Y = G\backslash X$, consisting of orbits which are isomorphic to G/H as G-sets. Let $\{Y_\alpha \mid \alpha \in \mathscr{A}\}$ denote the set of connected components of the $Y_{(H)}$s as H varies. If $H^*_{cpt}(-; \mathbf{C})$ denotes *compactly supported cohomology* with complex coefficients, we may define an integer, $\chi^\#_\alpha$, to be the *Euler characteristic* of Y_α

2.3.22 $$\chi^\#_\alpha = \textstyle\sum_i (-1)^i \ dim_{\mathbf{C}} H^i_{cpt}(Y_\alpha; \mathbf{C}) \ \in \mathbf{Z}.$$

For each $\alpha \in \mathscr{A}$ choose $g_\alpha \in U(n)$ which lies above Y_α. Therefore we have a homomorphism

$$g_\alpha^{-1} v g_\alpha : (g_\alpha^{-1} v g_\alpha)^{-1}(NT^n) \longrightarrow NT^n$$

and, if $Y_\alpha \subset Y_{(H)}$, then H is conjugate to $H_\alpha = (g_\alpha^{-1} v g_\alpha)^{-1}(NT^n)$. Hence we may define

2.3.23 $$\tau_G(v) = \textstyle\sum_{\alpha \in \mathscr{A}} \chi^\#_\alpha (H_\alpha \stackrel{g_\alpha^{-1} v g_\alpha}{\longrightarrow} NT^n) \in R_+(G, NT^n).$$

Another formulation of $\tau_G(v)$ may be given by means of a triangulation. If we triangulate X in such a manner that G acts simplicially, as in 2.1.11, 2.1.14, we may choose an element, $g_\sigma \in U(n)$, above each i-dimensional simplex in $Y = G\backslash X$. Defining H_σ as $(g_\sigma^{-1} v g_\sigma)^{-1}(NT^n)$, which is the stabiliser of the simplex above σ and containing g_σ, we may express $\tau_G(v)$ as

2.3.24 $$\tau_G(v) = \textstyle\sum_{\sigma in Y} (-1)^{dim(\sigma)} (H_\sigma \longrightarrow NT^n) \in R_+(G, NT^n).$$

We may define a homomorphism

2.3.25 $$\rho_G : R_+(G, NT^n) \longrightarrow R_+(G)$$

by applying the procedure used in the proof of 2.1.6 which *canonically* decomposes a conjugacy class of maps, $\phi : H \longrightarrow NT^n$, into a sum of monomial representations, $\sum_j Ind_{H_j}^H(\psi_j)$. Define

$$\rho_G(H \xrightarrow{\phi} NT^n) = \sum_j (H_j, \psi_j)^G \in R_+(G).$$

This homomorphism is natural if we define

$$Res_J^G : R_+(G, NT^n) \longrightarrow R_+(J, NT^n)$$

by the double coset formula, as for $R_+(G)$ in 2.2.1.

Define

2.3.26
$$T_G(v) = \rho_G(\tau_G(v)) \in R_+(G).$$

Also define, in the notation of 2.3.23,

2.3.27
$$\epsilon_G(v) = \sum_{\alpha \in \mathscr{A}} \chi_\alpha^{\#}(H_\alpha, 1)^G \in R_+(G).$$

We collect the properties of these constructions from Snaith (1988b, 1989a, 1989b).

Theorem 2.3.28 (i) *T_G, τ_G and ϵ_G depend only on the class of the representation, v, in $R(G)$.*

(ii) *If $b_n : R_+(G, NT^n) \longrightarrow R(G)$ is given by*

$$b_n(H \xrightarrow{\phi} NT^n) = Ind_H^G(\phi)$$

then

$$b_n(\tau_G(v)) = v \in R(G),$$

where $n = dim_{\mathbf{C}}(v)$.

(iii) *If $b_G = b_1 : R_+(G) \longrightarrow R(G)$ is as in 2.2.6 then*

$$b_1(\epsilon_G(v)) = 1 \in R(G).$$

(iv) *T_G, τ_G and ϵ_G are natural with respect to homomorphisms of G.*

(v) *If $R_+(G)$ is given the ring structure of 2.5.7 then*

$$T_G(v \oplus \mu) = T_G(v)\epsilon_G(\mu) + \epsilon_G(v)T_G(\mu)$$

and

$$\epsilon_G(v \oplus \mu) = \epsilon_G(v) \cdot \epsilon_G(\mu).$$

(vi) (*See* 2.5.11.)

If a_G is as in 2.2.43, then for all representations, v,

$$T_G(v) = a_G(v) \cdot \epsilon_G(v) \in R_+(G).$$

2.3.29 *Symond's description of a_G*

In Symonds (1991) one finds the following topological construction, which is similar in flavour to 2.3.24. Given $v : G \longrightarrow GL(V)$ we may let G act, via v, on $P(V)$, the projective space of V. Triangulate $P(V)$ so that G acts simplicially on $P(V)$. For each simplex, σ, of $G\backslash P(V)$ let $H(\sigma)$ denote the stabiliser of a simplex, $\tilde{\sigma}$, chosen above σ. The points of $\tilde{\sigma}$ correspond to lines in V which are preserved by $H(\sigma)$. Let $\phi_\sigma : H(\sigma) \longrightarrow \mathbf{C}^*$ be the resulting one-dimensional representation, given by $H(\sigma)$ acting on one of these lines.

Symonds (1991) defines

2.3.30 $$L_G(v) = \sum_{\sigma \in G\backslash P(V)} (-1)^{dim(\sigma)} (H(\sigma), \phi_\sigma)^G \in R_+(G).$$

Theorem 2.3.31 $L_G(v) = a_G(v)$ *for all representations, v.*

In Symonds (1991) all the properties of a_G, which we derived algebraically, are derived topologically for L_G.

2.4 Inductive explicit Brauer induction

In 2.3.9 we established the existence of an Explicit Brauer Induction homomorphism

$$a_G : R(G) \longrightarrow R_+(G)$$

for any finite group, G. The most essential property of a_G is that it commutes with restriction to subgroups, $H \leq G$. Slightly less important is the property of commuting with inflation maps. However, a_G does *not* commute with induction. In many applications (see 6.3.8, for example) it would be desirable to have an Explicit Brauer Induction homomorphism which commutes with induction.

In this section we shall describe an Explicit Brauer Induction homomorphism with rational coefficients,

$$d_G : R(G) \longrightarrow R_+(G) \otimes \mathbf{Q},$$

which is natural both with respect to induction from and restriction to subgroups. In order to have this inductive property one must forego something, namely, the integrality of the coefficients and *naturality with respect to inflation.*

The discovery of this remarkable homomorphism is due to R. Boltje, who gives the following characterisation of d_G.

Theorem 2.4.1 *In the notation of 2.2.1–2.2.8, there is a unique family of homomorphisms*

$$d_G : R(G) \longrightarrow R_+(G) \otimes \mathbf{Q},$$

as G ranges over all finite groups, which satisfies the following properties:

(i) *For all $H \leq G$ the following diagrams commute:*

$$\begin{array}{ccc} R(G) & \xrightarrow{d_G} & R_+(G) \otimes \mathbf{Q} \\ \downarrow Res_H^G & & \downarrow Res_H^G \\ R(H) & \xrightarrow{d_H} & R_+(H) \otimes \mathbf{Q} \end{array}$$

$$\begin{array}{ccc} R(H) & \xrightarrow{d_H} & R_+(H) \otimes \mathbf{Q} \\ \downarrow Ind_H^G & & \downarrow Ind_H^G \\ R(G) & \xrightarrow{d_G} & R_+(G) \otimes \mathbf{Q} \end{array}$$

(ii) *If $\phi : G \longrightarrow \mathbf{C}^*$ is a one-dimensional representation and $\chi \in R(G)$ then*

$$d_G(\phi \cdot \chi) = (G, \phi)^G \cdot d_G(\chi) \in R_+(G) \otimes \mathbf{Q}.$$

(iii) $b_G d_G = 1 \otimes \mathbf{Q} : R(G) \longrightarrow R(G) \otimes \mathbf{Q}.$

2.4.2 We shall prove Theorem 2.4.1 in a series of steps, culminating in 2.4.13. We begin with some preliminary discussion.

Let G be a *cyclic* group. Denote by e_G the class function on G which is given at $x \in G$ by

2.4.3
$$e_G(x) = \begin{cases} 1 & \text{if } <x>= G, \\ 0 & \text{otherwise.} \end{cases}$$

When G is non-cyclic we define e_G to be identically zero.

Lemma 2.4.4 *Let G be a finite cyclic group. In* $R(G) \otimes \mathbf{Q}$

(i)

$$e_G = \#(G)^{-1} \sum_{K \leq G} \#(K) \mu_{K,G} Ind_K^G(1)$$

and

(ii)

$$1 - e_G = -(\#(G)^{-1}) \sum_{K \underset{\neq}{<} G} \#(K) \mu_{K,G} Ind_K^G(1),$$

where (see 2.5.1) $\mu_{K,G}$ *is given in terms of the Möbius function,* μ, *by* $\mu_{K,G} = \mu([G:K])$.

Proof Part (ii) follows at once from part (i).

If $x \in G$ then the character of $Ind_K^G(1)$ at x is equal to $[G:K]$ if $x \in K$ and is equal to zero otherwise. Hence the character value at x of the right-hand side of (i) is equal to

$$\sum_{<x> \leq K \leq G} \mu([G:K]),$$

which, by 2.5.2, is equal to 1 if $<x>= G$ and is equal to zero otherwise, as required. □

Corollary 2.4.5 *For any finite group, G:*

(i)

$$e_G \in \left(\bigcap_{K \underset{\neq}{<} G} Ker(Res_K^G) \right) \lhd R(G) \otimes \mathbf{Q},$$

(ii)

$$1 - e_G \in \left(\sum_{K \underset{\neq}{<} G} Im(Ind_K^G) \right) \lhd R(G) \otimes \mathbf{Q},$$

(iii)

$$R(G) \otimes \mathbf{Q} = \left(\bigcap_{K \overset{<}{\neq} G} Ker(Res_K^G) \right) \oplus \left(\sum_{K \overset{<}{\neq} G} Im(Ind_K^G) \right),$$

(iv)

$$\left(\bigcap_{K \overset{<}{\neq} G} Ker(Res_K^G) \right) = e_G(R(G) \otimes \mathbf{Q}),$$

(v)

$$\left(\sum_{K \overset{<}{\neq} G} Im(Ind_K^G) \right) = (1 - e_G)(R(G) \otimes \mathbf{Q}).$$

Proof The result is trivial when G is not a cyclic group, since $e_G = 0$ in this case. Hence we may assume that G is cyclic. In this case e_G is clearly an idempotent of $R(G) \otimes \mathbf{Q}$ that vanishes on proper subgroups, which proves (i). Part (ii) follows from 2.4.5(ii). Parts (iii)–(v) follow easily from the fact that both the summands in (iii) are ideals of $R(G) \otimes \mathbf{Q}$. □

2.4.6 Let $\hat{G}$ denote the multiplicative character group of G, $\hat{G} = Hom(G, \mathbf{C}^*)$. Define a ring homomorphism

2.4.7
$$\pi_G : R_+(G) \otimes \mathbf{Q} \longrightarrow \mathbf{Q}[\hat{G}]$$

by the formula

$$\pi_G((H, \phi)^G) = \begin{cases} \phi & \text{if } H = G \\ 0 & \text{otherwise.} \end{cases}$$

Define a ring homomorphism

2.4.8
$$\rho_G : R_+(G) \otimes \mathbf{Q} \longrightarrow (\textstyle\prod_{H \leq G} \mathbf{Q}[\hat{H}])^G$$

by setting the $\hat{H}$-component of ρ_G equal to

$$\pi_H Res_H^G : R_+(G) \otimes \mathbf{Q} \longrightarrow R_+(H) \otimes \mathbf{Q} \longrightarrow \mathbf{Q}[\hat{H}],$$

so that $\rho_G = \prod_H \pi_H Res_H^G$. In 2.4.8, G acts on $\prod_{H \leq G} \mathbf{Q}[\hat{H}]$ by means of its (left) conjugation action on the poset of subgroups, $H \leq G$.

Proposition 2.4.9 *In 2.4.8 ρ_G is a ring isomorphism.*

Proof In the proof of 2.2.15 we showed that $\rho \in R_+(G)$ is determined uniquely by its 'leading terms' (i.e. those terms of the form $(H, \phi)^G$ with $H = G$) and by its restriction to $R_+(H)$ for all proper subgroups, $H < G$. This shows that ρ_G is injective. The fact the ρ_G is onto also follows from the formulae in the proof of 2.2.15 which were used to show that a rational solution existed to 2.2.16. □

Definition 2.4.10 Define a homomorphism

$$p_G : R(G) \otimes \mathbf{Q} \longrightarrow \mathbf{Q}[\hat{G}]$$

by the formula, for $\chi \in R(G)$,

2.4.11
$$p_G(\chi) = e_G\chi$$

where e_G is the idempotent of 2.4.2. In 2.4.11 we note that $p_G(\chi)$ is zero unless G is cyclic and in this case there is a tautological isomorphism

$$R(G) \otimes \mathbf{Q} \xrightarrow{\cong} \mathbf{Q}[\hat{G}],$$

so that $e_G\chi$ may be interpreted as an element of $\mathbf{Q}[\hat{G}]$. Define a homomorphism

$$r_G = \prod_{H \leq G} p_H Res^G_H : R(G) \otimes \mathbf{Q} \longrightarrow \left(\prod_{H \leq G} \mathbf{Q}[\hat{H}] \right)^G .$$

Define d_G by means of the following commutative diagram:

2.4.12

$$\begin{array}{ccc} R(G) \otimes \mathbf{Q} & \xrightarrow{d_G} & R_+(G) \otimes \mathbf{Q} \\ & {\scriptstyle r_G} \searrow \quad \swarrow {\scriptstyle \rho_G} & \\ & (\prod_{H \leq G} \mathbf{Q}[\hat{H}])^G & \end{array}$$

Hence, by definition, for all $H \leq G$ and $\chi \in R(G)$,

$$p_H(Res_H^G(\chi)) = \pi_H(Res_H^G(d_G(\chi))) \in \mathbf{Q}[\hat{H}].$$

2.4.13 *Proof of Theorem* 2.4.1
For each $J \leq H$ and $\chi \in R(G)$ we have

$$\begin{aligned}
&\pi_J(Res_J^H(Res_H^G(d_G(\chi)))) \\
&= \pi_J(Res_J^G(d_G(\chi))) \\
&= p_J(Res_J^G(\chi)) \qquad \text{by } 2.4.12 \\
&= p_J(Res_J^H(Res_H^G(\chi))) \\
&= \pi_J(Res_J^H(d_H(Res_H^G(\chi)))) \qquad \text{by } 2.4.12,
\end{aligned}$$

so that $\rho_H(Res_H^G(d_G(\chi))) = \rho_H(d_H(Res_H^G(\chi)))$. By 2.4.9, ρ_H is injective so that $Res_H^G(d_G(\chi)) = d_H(Res_H^G(\chi))$, as required.

Suppose now that $\chi \in R(H)$ for $H \leq G$ and let $J \leq G$ be any other subgroup. In this case we have

$$\begin{aligned}
&\pi_J(Res_J^G(Ind_H^G(d_H(\chi)))) \\
&= \textstyle\sum_{z \in J\backslash G/H} \pi_J(Ind_{J\cap zHz^{-1}}^J(Res_{J\cap zHz^{-1}}^{zHz^{-1}}((z^{-1})^*(d_H(\chi))))) \\
&\qquad \text{by } 2.5.17 \\
&= \textstyle\sum_{z \in J\backslash G/H} \pi_J(Ind_{J\cap zHz^{-1}}^J(Res_{J\cap zHz^{-1}}^{zHz^{-1}}(d_{zHz^{-1}}((z^{-1})^*(\chi))))) \\
&= \sum_{\substack{z \in J\backslash G/H \\ J \leq zHz^{-1}}} \pi_J(Res_{J\cap zHz^{-1}}^{zHz^{-1}}(d_{zHz^{-1}}((z^{-1})^*(\chi)))) \\
&= \sum_{\substack{z \in J\backslash G/H \\ J \leq zHz^{-1}}} \pi_J(d_J(Res_{J\cap zHz^{-1}}^{zHz^{-1}}((z^{-1})^*(\chi)))).
\end{aligned}$$

In this calculation we have used the fact that

$$d_{zHz^{-1}}((z^{-1})^*(\chi)) = (z^{-1})^*(d_H(\chi))$$

and that

$$\pi_J Ind_K^J : R_+(K) \longrightarrow R_+(J) \longrightarrow \mathbf{Q}[\hat{J}]$$

is trivial when K is a proper subgroup of J.

When $J \cap zHz^{-1}$ is not equal to J then

$$\pi_J(d_J(Ind^J_{J\cap zHz^{-1}}(Res^{zHz^{-1}}_{J\cap zHz^{-1}}((z^{-1})^*(\chi)))))$$

$$= p_J(d_J(Ind^J_{J\cap zHz^{-1}}(Res^{zHz^{-1}}_{J\cap zHz^{-1}}((z^{-1})^*(\chi)))))$$

$$= 0,$$

so that we may rewrite $\pi_J(Res^G_J(Ind^G_H(d_H(\chi))))$ in the following manner:

$$\pi_J(Res^G_J(Ind^G_H(d_H(\chi))))$$

$$= \sum_{z\in J\backslash G/H} \pi_J(d_J(Ind^J_{J\cap zHz^{-1}}(Res^{zHz^{-1}}_{J\cap zHz^{-1}}((z^{-1})^*(\chi)))))$$

$$= \pi_J(d_J(Res^G_J(Ind^G_H(\chi))))$$

$$= \pi_J(Res^G_J(d_G(Ind^G_H(\chi)))),$$

which implies that $\rho_G(Ind^G_H(d_H(\chi))) = \rho_G(d_G(Ind^G_H(\chi)))$ and hence

$$Ind^G_H(d_H(\chi)) = d_G(Ind^G_H(\chi)),$$

as required.

For $\phi : G \longrightarrow \mathbf{C}^*$, $\chi \in R(G)$ and $H \leq G$ observe that

$$\pi_H(Res^G_H(d_G(\phi \cdot \chi)))$$

$$= p_H(Res^G_H(\phi) \cdot Res^G_H(\chi))$$

$$= Res^G_H(\phi) \cdot p_H(Res^G_H(\chi)),$$

since $e_H Res^G_H(\phi) Res^G_H(\chi) = Res^G_H(\phi) \cdot (e_H Res^G_H(\chi))$ when H is cyclic.

On the other hand,

$$\pi_H(Res^G_H((G,\phi)^G \cdot d_G(\chi)))$$

$$= \pi_H((H, Res^G_H(\phi))^H \cdot Res^G_H(d_G(\chi)))$$

$$= Res^G_H(\phi) \cdot \pi_H(d_H(Res^G_H(\chi))) \qquad \text{by part (i)}$$

$$= Res^G_H(\phi) \cdot p_H(Res^G_H(\chi)),$$

which proves part (ii).

For part (iii), by 2.1.3 and the inductivity part of (i), it suffices to assume that G is cyclic. In this case, by 2.4.5(v), it suffices to verify for $\chi \in e_G R(G) \otimes \mathbf{Q}$ that

$$b_G(d_G(\chi)) = \chi \in R(G) \otimes \mathbf{Q}.$$

For such a χ suppose that $\chi = \sum_i a_i \phi_i$ where the $\{\phi_i\}$ are one-dimensional representations. In this case,

$$\begin{aligned}
&\rho_G(\textstyle\sum_i a_i (G, \phi_i)^G) \\
&= \{\textstyle\sum_i a_i (H, Res^G_H(\phi_i))^H\}_{H \leq G} \\
&= \{Res^G_H(\chi)\}_{H \leq G} \\
&= \begin{cases} \chi & \text{if } H = G, \\ 0 & \text{otherwise,} \end{cases}
\end{aligned}$$

while

$$\begin{aligned}
&r_G(\chi) \\
&= \{p_H(Res^G_H(\chi))\}_{H \leq G} \\
&= \begin{cases} \chi & \text{if } H = G, \\ 0 & \text{otherwise.} \end{cases}
\end{aligned}$$

Therefore, for such a χ, $d_G(\chi) = \sum_i a_i (G, \phi_i)^G$ and hence $b_G(d_G(\chi)) = \sum_i a_i \phi_i = \chi$, as required.

It remains to prove the uniqueness part of Theorem 2.4.1. However, as in the proof of part (iii), one is easily reduced to studying the case in which G is cyclic and $\chi \in e_G R(G) \otimes \mathbf{Q}$. In this case the argument which was used to verify part (iii) for our construction of d_G also shows that, if $\chi = \sum_i a_i \phi_i$, then we must have $d_G(\chi) = \sum_i a_i (G, \phi_i)^G$ so that there is at most one choice for d_G, which completes the proof of 2.4.1. □

The following result establishes some more useful properties of the homomorphism, d_G, of 2.4.1.

Theorem 2.4.14 *Let d_G be the homomorphism of* 2.4.1, *which was constructed in* 2.4.12–2.4.13. *Then*

(i) *d_G is a ring homomorphism.*

(ii) *For* $\chi \in R(G)$

$$d_G(\chi) = \#(G)^{-1} \sum_{\substack{K \leq H \leq G \\ K,H \ \text{cyclic}}} \#(K)\mu_{K,H}(H, Ind_K^H(Res_K^G(\chi)))^G,$$

where $(H, Ind_K^H(Res_K^G(\chi)))^G = \sum_i \epsilon_i(H, \phi_i)^G$ *if* $\sum_i \epsilon_i \phi_i = Res_K^G(\chi) \in R(K)$ *for* $\epsilon_i = \pm 1$ *and* $\phi_i : H \longrightarrow \mathbf{C}^*$.

(iii) *The coefficient of* $(\{1\}, 1)^G$ *in* $d_G(\chi)$ *is equal to* $dim(\chi)\#(G)^{-1}$.

(iv) *Define* $\eta_G : R_+(G) \longrightarrow R_+(G)$ *by* $\eta_G((H, \phi)^G) = (H, 1)^G$ *(see also* 6.3.1), *then, for* $\chi \in R(G)$,

$$b_G((\eta_G \otimes 1)(d_G(\chi))) = dim(\chi)\#(G)^{-1} Ind_{\{1\}}^G(1) \in R(G).$$

(v) *If* $d_G(\chi) = \sum_i \alpha_i(H_i, \phi_i)^G \in R_+(G) \otimes \mathbf{Q}$ *then*

$$\sum_i \alpha_i = dim(\chi)\#(G)^{-1}.$$

(vi) *If* $a_G : R(G) \longrightarrow R_+(G)$ *is the homomorphism of* 2.3.2 *and* 2.3.9 *then*

$$d_G(\chi) = a_G(\chi) d_G(1).$$

Proof Observe that p_G is a ring homomorphism, being multiplication by an idempotent, so that r_G is a ring homomorphism. Since $d_G = \rho_G^{-1} r_G$, part (i) follows from 2.4.9.

Part (ii) follows by Möbius inversion of the equation $\rho_G(d_G(\chi)) = r_G(\chi)$. This is similar to the manner in which the explicit formula for $a_G(\chi)$ of 2.3.15 is derived. We shall omit the details (see Boltje, 1990; Boltje, 1989, (2.35) *et seq.*; 2.5.18).

Part (iii) follows at once from part (ii).

Next we shall prove part (vi). It suffices to show that

$$\rho_G(d_G(\chi)) = \rho_G(a_G(\chi) d_G(1)) = \rho_G(a_G(\chi)) \rho_G(d_G(1)).$$

By construction, $\rho_G(d_G(\chi)) = r_G(\chi) = \{p_H(Res_H^G(\chi))\}_{H \leq G}$, where $p_H(Res_H^G(\chi)) = 0$ if H is non-cyclic and is equal to $e_H Res_H^G(\chi)$ if H is cyclic. On the other hand,

$$\begin{aligned} &\rho_G(a_G(\chi)) \rho_G(d_G(1)) \\ &= \{\pi_H(a_H(Res_H^G(\chi)))\}_{H \leq G} \cdot \{e_H\}_{H \leq G} \\ &= \{(\textstyle\sum_i a_i(H, \phi_i)^H) e_H\}_{H \leq G}, \end{aligned}$$

where $a_i = < \phi_i, Res^G_H(\chi) >$ for $\phi_i : H \longrightarrow \mathbf{C}^*$, by 2.3.2(ii). Hence, by 2.4.1(ii),

$$\rho_G(a_G(\chi)d_G(1)) = \{e_H Res^G_H(\chi)\}_{H \leq G},$$

which is $\{p_H(Res^G_H(\chi))\}_{H \leq G}$, as required.

In order to prove part (iv) we shall use part (ii). However, since we omitted the details of the general case of (ii) we will use only an easy special case. We shall use only the formula for $d_G(1)$ when G is cyclic. This formula is straightforward to prove directly by induction on $\#(G)$.

The homomorphisms, b_G, η_G and d_G, commute with Ind^G_H for $H \leq G$. Suppose that H is a cyclic subgroup of G and $\phi : H \longrightarrow \mathbf{C}^*$ is a one-dimensional representation which satisfies $b_H((\eta_H \otimes 1)(d_H(\phi))) = \#(H)^{-1} Ind^H_{\{1\}}(1)$. Therefore

$$\begin{aligned} & b_G((\eta_G \otimes 1)(d_G(Ind^G_H(\phi)))) \\ &= Ind^G_H(\#(H)^{-1} Ind^H_{\{1\}}(1)) \\ &= dim(Ind^G_H(\phi))\#(G)^{-1} Ind^G_{\{1\}}(1), \end{aligned}$$

as required. Therefore, by 2.1.3, we may suppose that G is cyclic. Hence

$$\begin{aligned} & b_G((\eta_G \otimes 1)(d_G(\chi))) \\ &= b_G(\eta_G(a_G(\chi))) \cdot b_G((\eta_G \otimes 1)(d_G(1))) \qquad \text{by (vi)} \\ &= dim(\chi) \cdot b_G((\eta_G \otimes 1)(d_G(1))) \qquad \text{by 2.2.41} \\ &= dim(\chi)\#(G)^{-1} \textstyle\sum_{K \leq H \leq G} \#(K)[H : K]\mu_{K,H} Ind^G_H(1) \qquad \text{by (ii)} \\ &= dim(\chi)\#(G)^{-1} \textstyle\sum_{K \leq H \leq G} \#(H)\mu_{K,H} Ind^G_H(1) \\ &= dim(\chi)\#(G)^{-1} Ind^G_{\{1\}}(1), \end{aligned}$$

since, as remarked in the proof of 2.4.4, for H fixed,

$$\sum_{K \leq H} \mu_{K,H}$$

vanishes when H is not equal to the trivial subgroup.

Finally, for part (v), we observe that

$$\begin{aligned}\sum_i \alpha_i &= < b_G(\eta_G(a_G(\chi))), 1 >_G \qquad \text{by } 1.2.39\\ &= dim(\chi)\#(G)^{-1} < Ind_{\{1\}}^G(1), 1 >_G \qquad \text{by (iv)}\\ &= dim(\chi)\#(G)^{-1},\end{aligned}$$

as required. This completes the proof of Theorem 2.4.14. □

We will close this section with some results concerning the behaviour of d_G with respect to fixing under normal subgroups. These results will be used in our study of Swan conductors of Galois representations of complete, discrete valuation fields in the non-separable residue field case (see 6.3.20).

Definition 2.4.15 Let $N \lhd G$ be a normal subgroup of the finite group, G. If V is a complex representation of G, define $Fix_N(V)$ to be equal to the N-fixed points of V,

$$Fix_N(V) = V^N = \{v \in V \mid n(v) = v \ for \ all \ n \in N\}.$$

Since G/N acts on V^N, this defines homomorphisms

$$Fix_N : R(G) \longrightarrow R(G/N)$$

and

$$Fix_N : R(G) \otimes \mathbf{Q} \longrightarrow R(G/N) \otimes \mathbf{Q}.$$

We may also define homomorphisms

$$Fix_{+,N} : R_+(G) \longrightarrow R_+(G/N)$$

and

$$Fix_{+,N} : R_+(G) \otimes \mathbf{Q} \longrightarrow R_+(G/N) \otimes \mathbf{Q}$$

by the formula

$$Fix_{+,N}((H,\phi)^G) = \begin{cases} (HN/N, \hat{\phi})^{G/N}, & \text{if } Res^G_{H\cap N}(\phi) = 1,\\ 0 & \text{otherwise,}\end{cases}$$

where $\hat{\phi}(nh) = \phi(h)$ for $n \in N$, $h \in H$.

Proposition 2.4.16 *The following diagrams commute:*

$$\begin{array}{ccc} R_+(G) & \xrightarrow{b_G} & R(G) \\ \downarrow Fix_{+,N} & & \downarrow Fix_N \\ R_+(G/N) & \xrightarrow{b_{G/N}} & R(G/N) \end{array}$$

$$\begin{array}{ccc} R(G) & \xrightarrow{d_G} & R_+(G)\otimes \mathbf{Q} \\ \downarrow Fix_N & & \downarrow Fix_{+,N} \\ R(G/N) & \xrightarrow{d_{G/N}} & R_+(G/N)\otimes \mathbf{Q} \end{array}$$

Proof If $Res^G_{H\cap N}(\phi) = 1$ then, by 2.5.14

$$\begin{aligned} Fix_N(b_G((H,\phi)^G)) &= Ind^G_{HN}(\hat{\phi}) \\ &= b_G(Fix_{+,N}((H,\phi)^G)), \end{aligned}$$

as required.

When $Res^G_{H\cap N}(\phi) \neq 1$ we must show that $Fix_N(b_G((H,\phi)^G)) = 0$. In this case, by 2.5.13,

$$Fix_N(b_G((H,\phi)^G)) = \mathbf{C}[G/N] \otimes_{\mathbf{C}[H]} V,$$

where V is the field of complex numbers upon which $h \in H$ acts by multiplication by $\phi(h)$. If $n \in H \cap N$ and $\phi(n) \neq 1$ then, in $\mathbf{C}[G/N] \otimes_{\mathbf{C}[H]} V$,

$$gN \otimes_{\mathbf{C}[H]} v = gnN \otimes_{\mathbf{C}[H]} v = gN \otimes_{\mathbf{C}[H]} \phi(n)v,$$

so that $gN \otimes_{\mathbf{C}[H]} v = 0$ for all $g \in G$ and $v \in V$, as required.

Now we shall prove by induction on $\#(G)$ that the second diagram commutes.

First we shall consider

$$\chi \in \left(\sum_{K \overset{<}{\neq} G} Im(Ind^G_K) \right) = (1 - e_G)(R(G) \otimes \mathbf{Q}),$$

in which case, by linearity, it suffices to set $\chi = Ind_H^G(\theta)$ for some proper subgroup, $H < G$, and for $\theta \in R(H)$. Let $Res^{HN/N}_{H/(H\cap N)}(-)$ denote the homomorphism induced by the isomorphism, $HN/N \xrightarrow{\cong} H/(H \cap N)$.

We will also need to know that there is an isomorphism of complex G/N-representations of the form, where M is a complex representation of H,

$$\mathbf{C}[G/N] \otimes_{\mathbf{C}[HN/N]} Res^{HN/N}_{H/(H\cap N)}(M^{H\cap N}) \cong (Ind_H^G(M))^N.$$

This can easily be seen from 2.5.13, which yields an isomorphism of the form

$$\mathbf{C}[G/N] \otimes_{\mathbf{C}[H]} M \cong (Ind_H^G(M))^N.$$

Now we may decompose the H-representation, M, as $M \cong M^{H\cap N} \oplus M'$ and observe that, by the argument used in the first part of this proof,

$$\mathbf{C}[G/N] \otimes_{\mathbf{C}[H]} M' = 0.$$

Hence we have

$$\begin{aligned}
Fix_{+,N}(d_G(\chi)) &= Fix_{+,N}(d_G(Ind_H^G(\theta))) \\
&= Fix_{+,N}(Ind_H^G(d_H(\theta))) \\
&= Ind^{G/N}_{HN/N}(Res^{HN/N}_{H/(H\cap N)}(Fix_{+,H\cap N}(d_H(\theta)))) \\
&= d_{G/N}(Fix_N(Ind_H^G(\theta))) \\
&= d_{G/N}(Fix_N(\chi)),
\end{aligned}$$

as required.

Finally we may assume, by 2.4.5, that

$$\chi \in \left(\bigcap_{K \stackrel{<}{\neq} G} Ker(Res_K^G \right)) = e_G(R(G) \otimes \mathbf{Q})$$

and that G is cyclic, since such a χ would be trivial otherwise. By the argument used in the proof of 2.4.1 in 2.4.13 we have $d_G(\chi) = (G, \chi)^G$, so that

$$Fix_{+,N}(d_G(\chi)) = (G/N, Fix_N(\chi))^{G/N}.$$

However, $d_{G/N}(Fix_N(\chi)) = (G/N, Fix_N(\chi))^{G/N}$ since

$$Fix_N(\chi) \in e_{G/N}R(G/N) \otimes \mathbf{Q},$$

by 2.5.20. □

2.5 Exercises

2.5.1 Let $A \le B$ be cyclic groups such that $[B : A]$ is finite. Define $\mu_{A,B}$ by

$$\mu_{A,B} = \sum_{i \ge 0} (-1)^i \#\{\text{chains of subgroups of length } i \text{ from } A \text{ to } B\}.$$

Show that $\mu_{A,B} = \mu([B : A])$, where $\mu(n)$ is the Möbius function of 2.1.1. (See 2.3.10 for the general definition of the Möbius function of a finite poset.)

2.5.2 *Möbius inversion* Let f be any arithmetical function defined on positive integers. Show that, if $F(n) = \sum_{d|n} f(d)$ then $f(n) = \sum_{d|n} \mu(n/d)F(d)$.

2.5.3 Show that every finite p-group is nilpotent. Conversely, show that every finite nilpotent group is the product of its Sylow subgroups.

2.5.4 Let G be an M-group and let $1 = f_1 < f_2 \ldots < f_k$ be the distinct dimensions of the irreducible complex representations of G. Let $\rho_i = Ind_{H_i}^G(\phi_i)$ be irreducible with $f_i = [G : H_i]$.

(i) Show that $ker(\rho_i)$ contains the ith group, $G^{(i)}$, in the *derived series* of G (the series obtained by taking successive commutator subgroups).

(ii) Show that an M-group is solvable.

(*Hint:* Use induction on i. Compare the kernel of ρ_i with the kernel of $Ind_{H_i}^G(1)$.)

2.5.5 Let $H_i \le G_i$ be finite groups and let $\rho_i : H_i \longrightarrow GL(V_i)$ be K-representations for $i = 1, 2$. Show that

$$Ind_{H_1}^{G_1}(\rho_1) \otimes Ind_{H_2}^{G_2}(\rho_2) \cong Ind_{H_1 \times H_2}^{G_1 \times G_2}(\rho_1 \otimes \rho_2).$$

2.5.6 Let

$$0 \to V_n \xrightarrow{d} V_{n-1} \to \ldots \xrightarrow{d} V_0 \to 0$$

be a sequence of maps of $\mathbf{C}[G]$-modules. Suppose that $dd = 0 : V_i \longrightarrow V_{i-2}$ for each i. Set

$$H_i = \frac{(ker(d) : V_i \longrightarrow V_{i-1})}{(im(d) : V_{i+1} \longrightarrow V_i)}$$

and show that $\sum_i (-1)^i H_i = \sum_i (-1)^i V_i \in R(G)$.

2.5.7 Let $J \leq G$ and define $Ind_J^G : R_+(J) \longrightarrow R_+(G)$ by $Ind_J^G((H,\phi)^J) = (H,\phi)^G$. Also define a product on $R_+(G)$ by the formula

$$(K,\phi)^G \cdot (H,\psi)^G = \sum_{w \in K\backslash G/H} ((w^{-1}Kw) \cap H, w^*(\phi)\psi)^G,$$

where $w^*(\phi)(z) = \phi(wzw^{-1})$.

With these definitions, verify that $R_+(G)$ is a ring-valued functor. Also verify that, if $A \in R_+(G)$ and $B \in R_+(J)$ then

$$Ind_J^G(Res_J^G(A) \cdot B) = A \cdot Ind_J^G(B) \in R_+(G).$$

2.5.8 Verify that b_G, in 2.2.6, is a ring homomorphism which commutes with the induction and restriction maps of 2.2.1.

2.5.9 In the notation of Theorem 2.2.15 show that, if $\rho : G \longrightarrow GL(V)$ factorises as $\rho : G \rightarrow G/N \xrightarrow{\rho'} GL(V)$ then

(i) $\alpha_{(H,\phi)^G}(\rho) \neq 0$ implies that $N \leq H$ and $Res_N^H(\phi) = 1$.
(ii) $\alpha_{(H,\phi)^G}(\rho) = \alpha_{(H/N,\phi)^{G/N}}(\rho')$.

2.5.10 Let $\chi \in R(G)$ and suppose that $(H,\phi) \in \mathcal{M}_G$ is an element which is maximal among those satisfying $< \phi, Res_H^G(\chi) > \neq 0$. Prove that

$$\alpha_{(H,\phi)^G}(\chi) =< \phi, Res_H^G(\chi) > \cdot([N_G(H,\phi) : H])^{-1}.$$

2.5.11 In the notation of 2.3.28, prove that

$$a_G(v) \cdot \epsilon_G(v) = T_G(v)$$

for all $v : G \longrightarrow U(n)$.

2.5.12 Let $v \in R(G)$ and let $\lambda : G \longrightarrow \mathbf{C}^*$ be a one-dimensional representation. Show that

$$a_G(\lambda \otimes v) = (G,\lambda)^G \cdot a_G(v) \in R_+(G).$$

That is, if

$$a_G(v) = \sum_i n_i(H_i, \phi_i)^G$$

then

$$a_G(\lambda \otimes v) = \sum_i n_i(H_i, \lambda \cdot \phi_i)^G.$$

2.5.13 Let G be a finite group. Let $N \lhd G$ and let V be a finite-dimensional K-representation of G. Suppose that $char(K)$ is coprime to $\#(G)$. Define a G/N-representation, $\natural(V)$, by

$$\natural(V) = K[G/N] \otimes_{K[G]} V.$$

Show that $\natural(V)$ and $V^N = \{v \in V \mid n(v) = v \ for\ all\ n \in N\}$ are isomorphic G/N-representations.

2.5.14 Let $\phi : J \longrightarrow \mathbf{C}^*$ be a one-dimensional complex representation of a subgroup, $J \leq G$. Suppose that $N \lhd G$ and that ϕ is trivial on $N \cap J$. Define $\hat{\phi} : NJ \longrightarrow \mathbf{C}^*$ by $\hat{\phi}(nj) = \phi(j)$ for $n \in N, j \in J$.

In the notation of 2.5.13 show that

$$\natural(Ind_J^G(\phi)) = Ind_{NJ}^G(\hat{\phi}).$$

(*Hint:* Use 2.5.13 and the discussion of $Ind_J^G(-)^N$ in 5.3.18 (proof) to obtain a convenient basis.)

2.5.15 Let $N \lhd G$ be finite groups and let V be a finite-dimensional, complex representation of G. Prove that the character, $\chi_\natural$, of $\natural(V)$ is given by

$$\chi_\natural(gN) = \#(N)^{-1} \sum_{n \in N} \chi_V(gn)$$

where χ_V is the character of V.

(*Hint:* Reduce to the case in which $V = Ind_J^G(\phi)$, as in 2.5.14. Use 2.5.14 when $Res^J_{N \cap J}(\phi) = 1$ and in the remaining case show directly that both expressions are zero.)

2.5.16 (*A topological construction of a_G for p-groups*) Let G be a finite p-group. Let $\epsilon : R_+(G) \longrightarrow \mathbf{Z}$ be the augmentation defined by $\epsilon((H, \phi)^G) = [G : H]$ and set $IR_+(G) = ker(\epsilon)$.

(i) Show that the $IR_+(G)$-adic completion of $R_+(G)$ is isomorphic to $\mathbf{Z} \oplus (IR_+(G) \otimes \mathbf{Z}_p)$, where $\mathbf{Z}_p$ denotes the p-adic integers. (For the notion of the completion of a ring with respect to an ideal see Matsumura, 1990.)

(ii) In the notation of 2.5.11, show that $T_G \cdot \epsilon_G^{-1}$ gives a well-defined homomorphism from $R(G)$ to the $IR_+(G)$-adic completion of $R_+(G)$ satisfying the conditions of 2.2.8 *save for the fact that the homomorphism has p-adic coefficients.*

(iii) Use the formula of 2.2.16 together with the fact that $\mathbf{Z}_p \cap \mathbf{Z}[1/p] = \mathbf{Z}$, the integers, to show that $T_G \cdot \epsilon_G^{-1}$ has integral coefficients.

This verifies that a_G exists and satisfies the conditions of 2.2.8 and 2.5.11 when G is a p-group.

2.5.17 Let $H, J \leq G$ be subgroups of a finite group, G. Prove that the composition

$$Res_J^G Ind_H^G : R_+(H) \longrightarrow R_+(G) \longrightarrow R_+(J)$$

is given by the double coset formula

$$Res_J^G(Ind_H^G(w)) = \sum_{z \in J\backslash G/H} Ind_{J\cap zHz^{-1}}^J(Res_{J\cap zHz^{-1}}^{zHz^{-1}}((z^{-1})^*(w))).$$

2.5.18 Prove 2.4.14(ii) directly. (*Hint:* Prove that the right-hand side of the formula is inductive and then verify the formula when $\chi \in e_G R(G) \otimes \mathbf{Q}$.)

2.5.19 Let d_G be the homomorphism of Theorem 2.4.1 and 2.4.12.

(i) If $\phi : \mathbf{Z}/p \longrightarrow \mathbf{C}^*$ is a non-trivial character of the cyclic group of prime order, p, show that

$$d_{\mathbf{Z}/p}(\phi) = (p-1)p^{-1}(\mathbf{Z}/p, \phi)^{\mathbf{Z}/p} - p^{-1}\sum_{\psi \neq \phi}(\mathbf{Z}/p, \psi)^{\mathbf{Z}/p} + p^{-1}(\mathbf{Z}/p, 1)^{\mathbf{Z}/p}.$$

(ii) Show that

$$\begin{aligned} d_{\mathbf{Z}/4}(1) &= 2^{-1}(\mathbf{Z}/4, 1)^{\mathbf{Z}/4} - 2^{-1}(\mathbf{Z}/4, \chi)^{\mathbf{Z}/4} + 4^{-1}(\mathbf{Z}/2, 1)^{\mathbf{Z}/4} \\ &\quad -4^{-1}(\mathbf{Z}/2, \tau)^{\mathbf{Z}/4} + 4^{-1}(\{1\}, 1)^{\mathbf{Z}/4}, \end{aligned}$$

where χ and τ are one-dimensional characters of order two.

(iii) If χ is the two-dimensional irreducible representation of the dihedral group of order six show that

$$d_{D_6}(\chi) = 3^{-1}(\mathbf{Z}/3, \phi)^{D_6} - 3^{-1}(\mathbf{Z}/3, 1)^{D_6} + 3^{-1}(\{1\}, 1)^{D_6}.$$

2.5.20 Let $N \lhd G$ be a normal subgroup of the finite group, G. Prove that, if $\chi \in e_G R(G) \otimes \mathbf{Q}$, then

$$Fix_N(\chi) \in e_{G/N} R(G/N) \otimes \mathbf{Q}.$$

2.5.21 Derive the following alternative formula for the homomorphism of 2.3.15:

$$a_G : R(G) \longrightarrow R_+(G).$$

For $\rho \in R(G)$,

$$a_G(\rho) = \sum_{\underline{\alpha}} (-1)^n < \phi_n, Res^G_{H_n}(\rho) > (H_0, \phi_0)^G,$$

where $\underline{\alpha}$ runs through the set of G-orbits of chains in $\mathcal{M}_G$ of the form $(H_0, \phi_0) < (H_1, \phi_1) < \ldots < (H_n, \phi_n)$.

3

GL_2F_q

Introduction

In this chapter we evaluate some of the terms in the Explicit Brauer Induction formula, $a_{GL_2F_q}$, applied to a finite-dimensional, irreducible complex representation of GL_2F_q. I have included these examples because the family of irreducible representations of GL_nF_q (as n and q vary) exhibit a number of interesting features which one would like to understand from the viewpoint of Explicit Brauer Induction. These features mirror the conjectural properties of admissible representations of GL_n of a local field, which form part of the scheme of conjectures posed by R.P. Langlands and others (Gérardin & Labesse, 1979). This similarity is reason enough to wish to examine GL_nF_q in a new light.

One of the most elegant properties of GL_nF_q is the Shintani correspondence, which assigns to each $G(F_{q^s}/F_q)$-invariant irreducible representation of $GL_nF_{q^s}$ an irreducible representation of GL_nF_q. When $n = 1$ this correspondence follows at once from Hilbert's Theorem 90. For higher values of n the correspondence is much more difficult to establish. However, in Section 2, we describe the Shintani correspondence for GL_2 entirely in terms of Explicit Brauer Induction and Hilbert's Theorem 90.

Section 1 is devoted to the construction of the cuspidal representations. We follow a method, using the Fourier transform, which was used by André Weil to construct cuspidal (admissible) representations of matrix groups of local fields.

Section 2 gives all the other irreducible representations of GL_nF_q and evaluates sufficiently many of the terms in the Explicit Brauer Induction formulae to enable us to describe the Shintani correspondence.

Section 3 consists of a set of exercises and a research problem connected with the Shintani correspondence for GL_nF_q for larger values of n.

3.1 Weil representations

In this chapter we shall calculate some of the *maximal terms* in the Explicit Brauer Induction formula, $a_{GL_2F_q}(\rho)$, when ρ is an irreducible representation of GL_2F_q, the group of invertible 2×2 matrices with entries in the finite field, F_q. All such irreducible representations are well-known. In fact, all the irreducible representations of GL_nF_q were described in Green (1955) see also Macdonald (1979, chapter IV) for all values of n and q. However, for completeness and convenience, we shall recall here the explicit construction of the irreducible representations of GL_2F_q. We shall begin with the *cuspidal* or *Weil representations* which are the most difficult ones to construct. These representations are originally due to A. Weil. The construction works in greater generality than we will need. For example, in Gérardin & Labesse (1979, p. 122) the Weil representation is described for the case in which F_q is replaced by a local field.

3.1.1 Let F be any field. Define the *Borel subgroup*, $B \leq GL_2F$, to be

$$B = \left\{ X \in GL_2F \mid X = \begin{pmatrix} \alpha & \beta \\ 0 & \delta \end{pmatrix} \right\}.$$

Define the *unitriangular subgroup*, $U \leq B$, to be

$$U = \left\{ Y \in B \mid Y = \begin{pmatrix} 1 & \beta \\ 0 & 1 \end{pmatrix} \right\}.$$

Define $w \in SL_2F$ to be given by

$$w = \begin{pmatrix} 0 & 1 \\ -1 & 0 \end{pmatrix}.$$

The *Bruhat decomposition* of GL_2F takes the form

$$GL_2F = B \bigsqcup BwU.$$

This is elementary in the case of 2×2 matrices and an explicit formula which gives this decomposition is to be found in 3.3.1.

Proposition 3.1.2 *Let F be any field, then GL_2F is generated by matrices of the form $(\alpha, \delta \in F^*, u \in F)$*

$$\begin{pmatrix} \alpha & 0 \\ 0 & \delta \end{pmatrix}, \quad \begin{pmatrix} 1 & u \\ 0 & 1 \end{pmatrix} \quad \textit{and} \quad w = \begin{pmatrix} 0 & 1 \\ -1 & 0 \end{pmatrix},$$

subject to the following relations:

(i)

$$\begin{pmatrix} \alpha & 0 \\ 0 & \delta \end{pmatrix}\begin{pmatrix} 1 & u \\ 0 & 1 \end{pmatrix}\begin{pmatrix} \alpha^{-1} & o \\ 0 & \delta^{-1} \end{pmatrix} = \begin{pmatrix} 1 & \alpha u \delta^{-1} \\ 0 & 1 \end{pmatrix},$$

(ii)

$$\begin{pmatrix} 1 & u_1 \\ 0 & 1 \end{pmatrix}\begin{pmatrix} 1 & u_2 \\ 0 & 1 \end{pmatrix} = \begin{pmatrix} 1 & u_1 + u_2 \\ 0 & 1 \end{pmatrix},$$

(iii)

$$w\begin{pmatrix} \alpha & 0 \\ 0 & 1 \end{pmatrix}w^{-1} = \begin{pmatrix} 1 & 0 \\ 0 & \alpha \end{pmatrix},$$

(iv)

$$w\begin{pmatrix} 1 & u \\ 0 & 1 \end{pmatrix}w = \begin{pmatrix} -u^{-1} & 0 \\ 0 & -u \end{pmatrix}\begin{pmatrix} 1 & -u \\ 0 & 1 \end{pmatrix}w\begin{pmatrix} 1 & -u^{-1} \\ 0 & 1 \end{pmatrix},$$

and

(v) $$w^4 = 1.$$

Proof See 3.3.2. □

3.1.3 We will now discuss the $(q-1)$-dimensional complex vector space upon which we are going to inflict the Weil representation of GL_2F_q. We will require some preliminary notation. Let F_{q^2} denote the field of order q^2 so that the Galois group, $G(F_{q^2}/F_q)$, is cyclic of order two generated by the Frobenius automorphism (see also 4.3.7):

$$F : F_{q^2} \longrightarrow F_{q^2}$$

given by $F(z) = z^q$ for all $z \in F_{q^2}$.

In order to construct a Weil representation we shall need a character of the form

$$\Theta : F_{q^2}^* \longrightarrow \mathbf{C}^*,$$

which we shall generally assume to be distinct from its conjugate by the Frobenius, $\Theta \neq F^*(\Theta)$ where $F^*(\Theta)(z) = \Theta(F(z))$.

Let $\mathscr{H}$ denote the following complex vector space:

$$\mathscr{H} = \{f : F_{q^2}^* \longrightarrow \mathbf{C} \mid f(t^{-1}x) = \Theta(t)f(x) \;\; if \;\; N(t) = 1\},$$

where $N = N_{F_{q^2}/F_q} : F_{q^2}^* \longrightarrow F_q^*$ is the norm.

Let $H \leq B$ denote the abelian subgroup which consists of matrices whose diagonal entries are equal. Hence there is an isomorphism of the form

$$\gamma : H \xrightarrow{\cong} F_q^* \times F_q$$

given by

$$\gamma \begin{pmatrix} z & y \\ 0 & z \end{pmatrix} = (z, yz^{-1}).$$

Define the *additive character* of F_q to be the homomorphism

3.1.4 $$\Psi = \Psi_{F_q} : F_q \longrightarrow \mathbf{C}^*$$

by the formula ($char(F_q) = p$)

$$\Psi_{F_q}(y) = exp(2\pi i(Trace_{F_q/F_p}(y))/p).$$

Hence we may obtain a one-dimensional representation of H, denoted by $\Theta \otimes \Psi$, by the composition

$$\Theta \otimes \Psi = (\Theta \otimes \Psi_{F_q})\gamma : H \longrightarrow \mathbf{C}^*.$$

By induction we obtain a $(q-1)$-dimensional induced representation of B, $Ind_H^B(\Theta \otimes \Psi)$. We may identify the underlying vector space of this as a mapping space in the following manner. There is an isomorphism

$$\lambda : W \xrightarrow{\cong} Ind_H^B(\Theta \otimes \Psi),$$

where

$$W = \left\{ g : B \longrightarrow \mathbf{C} \mid g\left(X \begin{pmatrix} \alpha & \beta \\ 0 & \alpha \end{pmatrix}\right) = \overline{\Theta}(\alpha)\overline{\Psi}(\beta/\alpha)g(X) \right\}$$

where $\overline{\Theta}$ denotes the complex conjugate of Θ. Explicitly λ is given by

$$\lambda(g) = \sum_{X \in B/H} X \otimes g(X) \in \mathbf{C}[B] \otimes_{\mathbf{C}[H]} \mathbf{C}_{\Theta\otimes\Psi},$$

where $\mathbf{C}_{\Theta\otimes\Psi}$ denotes the complex numbers with the H-action via $\Theta \otimes \Psi$. The isomorphism, λ, is well-defined since

$$X\begin{pmatrix} \alpha & \beta \\ 0 & \alpha \end{pmatrix} \otimes g\left(X\begin{pmatrix} \alpha & \beta \\ 0 & \alpha \end{pmatrix}\right)$$

$$= X \otimes \Theta(\alpha)\Psi(\beta/\alpha)g\left(X\begin{pmatrix} \alpha & \beta \\ 0 & \alpha \end{pmatrix}\right)$$

$$= X \otimes g(X).$$

Define an action of B on W by the formula ($g \in W, X \in B$)

$$\left(\begin{pmatrix} \alpha & \beta \\ 0 & \delta \end{pmatrix} g\right)(X) = g\left(\begin{pmatrix} \alpha & \beta \\ 0 & \delta \end{pmatrix}^{-1} X\right) = g\left(\begin{pmatrix} \alpha^{-1} & -\beta(\alpha\delta)^{-1} \\ 0 & \delta^{-1} \end{pmatrix} X\right).$$

Proposition 3.1.5 *With this B-action on W*

$$\lambda : W \xrightarrow{\cong} Ind_H^B(\Theta \otimes \Psi)$$

is an isomorphism of B-representations.

Proof This result will be left as an easy exercise for the reader (3.3.4). □

3.1.6 Since matrices of the type

$$\begin{pmatrix} \alpha & 0 \\ 0 & 1 \end{pmatrix}$$

form a set of coset representatives for B/H we may define an isomorphism of vector spaces

$$A : \mathscr{H} \longrightarrow W$$

by the formula, where $b \in F_{q^2}^*$ and $N_{F_{q^2}/F_q}(b) = \alpha^{-1}$,

3.1.7
$$A(h)\left(\begin{pmatrix} \alpha & 0 \\ 0 & 1 \end{pmatrix}\right) = \Theta(b)h(b).$$

Notice that 3.1.7 is well-defined because, if $N_{F_{q^2}/F_q}(t) = 1$,

$$\Theta(tb)h(tb) = \Theta(t)\Theta(b)\Theta(t)^{-1}h(b).$$

Proposition 3.1.8 *Define a B-action on* $\mathscr{H}$ *by the formula*

$$\begin{pmatrix} \alpha & \beta \\ 0 & \delta \end{pmatrix}(h) = \{x \mapsto \Theta(\delta)\Psi(\beta N(x)\delta^{-1})h(\lambda x)\Theta(\lambda)\},$$

where $N = N_{F_{q^2}/F_q}$ *is the norm and* $N(\lambda) = \alpha\delta^{-1}$. *Then, with this B-action, A yields an isomorphism of B-representations*

$$A : \mathscr{H} \xrightarrow{\cong} Ind_H^B(\Theta \otimes \Psi).$$

Proof First we note that the proposed formula does indeed define an action which sends functions in $\mathscr{H}$ to other functions in $\mathscr{H}$.

We shall work in terms of the map $A : \mathscr{H} \longrightarrow W$. By definition,

$$\left(\begin{pmatrix} \alpha & \beta \\ 0 & \delta \end{pmatrix} A(h)\right)\left(\begin{pmatrix} v & 0 \\ 0 & 1 \end{pmatrix}\right)$$

$$= A(h)\left(\begin{pmatrix} \alpha^{-1} & -\beta(\alpha\delta)^{-1} \\ 0 & \delta^{-1} \end{pmatrix}\begin{pmatrix} v & 0 \\ 0 & 1 \end{pmatrix}\right)$$

$$= A(h)\left(\begin{pmatrix} v\alpha^{-1} & -\beta(\alpha\delta)^{-1} \\ 0 & \delta^{-1} \end{pmatrix}\right)$$

$$= A(h)\left(\begin{pmatrix} v\delta\alpha^{-1} & 0 \\ 0 & 1 \end{pmatrix}\begin{pmatrix} \delta^{-1} & -\beta(v\delta^2)^{-1} \\ 0 & \delta^{-1} \end{pmatrix}\right)$$

$$= \left[A(h)\left(\begin{pmatrix} v\delta\alpha^{-1} & 0 \\ 0 & 1 \end{pmatrix}\right)\right]\Theta(\delta)\Psi(\beta(v\delta)^{-1})$$

$$= \Theta(b)h(b)\Theta(\delta)\Psi(\beta(v\delta)^{-1}),$$

where $N(b) = \alpha(v\delta)^{-1}$. Now define $g \in \mathscr{H}$ by the formula

$$g(x) = \Theta(\delta)\Psi(\beta N(x)\delta^{-1})h(\lambda x)\Theta(\lambda),$$

where $N(\lambda) = \alpha\delta^{-1}$. We have, if $N(c) = v^{-1}$,

$$A(g)\left(\begin{pmatrix} v & 0 \\ 0 & 1 \end{pmatrix}\right)$$

$$= \Theta(c)g(c)$$

$$= \Theta(c)\Theta(\delta)\Psi(\beta N(c)\delta^{-1})h(\lambda c)\Theta(\lambda)$$

$$= \Theta(\lambda c)\Theta(\delta)\Psi(\beta(v\delta)^{-1})h(\lambda c)$$

$$= \left(\begin{pmatrix} \alpha & \beta \\ 0 & \delta \end{pmatrix} A(h)\right)\left(\begin{pmatrix} v & 0 \\ 0 & 1 \end{pmatrix}\right)$$

since $N(\lambda c) = \alpha(v\delta)^{-1}$. This shows that the vector space isomorphism, A, commutes with the B-actions, as required. □

Remark 3.1.9 From 3.1.8 we find that the B-action on $\mathscr{H}$ is given by

$$\left(\begin{pmatrix} 1 & u \\ 0 & 1 \end{pmatrix} h\right)(x) = \Psi(uN(x))h(x),$$

$$\left(\begin{pmatrix} \alpha & 0 \\ 0 & 1 \end{pmatrix} h\right)(x) = h(\lambda x)\Theta(\lambda),$$

where $N(\lambda) = \alpha$ and

$$\left(\begin{pmatrix} 1 & 0 \\ 0 & \delta \end{pmatrix} h\right)(x) = h(xF(\mu)^{-1})\Theta(\mu),$$

where $N(\mu) = \delta$ since, if $N(\lambda) = \delta^{-1}$,

$$\begin{aligned} \Theta(\delta)h(\lambda x)\Theta(\lambda) &= \Theta(\delta^{-1})\Theta(F(\delta)^{-1})h(\lambda x)\Theta(\lambda) \\ &= h(xF(\mu)^{-1})\Theta(\mu). \end{aligned}$$

These formulae coincide with those of Gérardin & Labesse (1979, p. 122).

In order to manipulate the Fourier transform on $\mathscr{H}$ we will repeatedly make use of the following result.

Lemma 3.1.10 *As in 3.1.3, let $f \in \mathscr{H}$ and suppose that $\Theta : F_{q^2}^* \longrightarrow \mathbf{C}^*$ is non-trivial. Then*

$$\sum_{v \in F_{q^2}^*} f(v) = 0.$$

Proof Let $x_1, \ldots, x_{q-1} \in F_{q^2}^*$ comprise a set of coset representatives for $F_{q^2}^*/Ker(N)$ where $N = N_{F_{q^2}/F_q}$ is the norm. We may rewrite the sum as

$$\begin{aligned} \textstyle\sum_{v \in F_{q^2}^*} f(v) &= \textstyle\sum_{i=1}^{q-1} \sum_{t \in Ker(N)} f(tx_i) \\ &= \textstyle\sum_{i=1}^{q-1} [\sum_{t \in Ker(N)} \Theta(t)^{-1}] f(x_i) \\ &= 0 \end{aligned}$$

by 3.3.3. □

3.1.11 *The Fourier transform on $\mathscr{H}$*

In the notation of 3.1.1 and 3.1.3 suppose that $f \in \mathscr{H}$. We define the *Fourier transform*, $\hat{f} \in \mathscr{H}$, of f by means of the formula

3.1.12 $$\hat{f}(z) = -q^{-1} \sum_{y \in F^{*}_{q^2}} f(y)\Psi_{F_q}(yF(z) + zF(y)),$$

where Ψ_{F_q} is the additive character of 3.1.4.

Lemma 3.1.13 *The map which sends $f \in \mathscr{H}$ to its Fourier transform is a* **C***-linear endomorphism of order four.*

Proof If $z \in F^{*}_{q^2}$ and $t \in Ker(N)$ then, by 3.1.12,

$$\hat{f}(tz) = -q^{-1} \sum_{y \in F^{*}_{q^2}} f(y)\Psi_{F_q}(yF(tz) + tzF(y))$$

$$= -q^{-1} \sum_{y \in F^{*}_{q^2}} f(y)\Psi_{F_q}(yt^{-1}F(z) + zF(t^{-1}y))$$

since $F(t) = t^{-1}$,

$$= -q^{-1} \sum_{v \in F^{*}_{q^2}} f(tv)\Psi_{F_q}(vF(z) + zF(v))$$

setting $v = yt^{-1}$,

$$= -q^{-1} \sum_{v \in F^{*}_{q^2}} \Theta(t)^{-1} f(v)\Psi_{F_q}(vF(z) + zF(v))$$

$$= \Theta(t)^{-1}\hat{f}(v).$$

Hence the Fourier transform, which is evidently **C**-linear, yields an endomorphism of $\mathscr{H}$. To show that the Fourier transform gives an endomorphism of order four we will establish the identity

3.1.14 $$\hat{\hat{f}}(z) = f(-z) \qquad \text{for all } z \in F^{*}_{q^2}, f \in \mathscr{H}.$$

This is seen as follows:

$$\hat{\hat{f}}(z) = -q^{-1} \sum_{y \in F^{*}_{q^2}} \hat{f}(y)\Psi_{F_q}(yF(z) + zF(y))$$

$$= q^{-2} \sum_{y,v \in F^{*}_{q^2}} f(v)\Psi_{F_q}(yF(z) + zF(y))\Psi_{F_q}(yF(v) + vF(y))$$

$$= q^{-2} \sum_{y,v \in F^{*}_{q^2}} f(v)\Psi_{F_q}(yF(v + z) + (v + z)F(y)).$$

When $v+z\neq 0$ we have, by 3.1.10,

$$\sum_{y\in F^*_{q^2}} \Psi_{F_q}(yF(v+z)+(v+z)F(y)) = -1$$

and when $v+z=0$ we have

$$\sum_{y\in F^*_{q^2}} \Psi_{F_q}(0) = q^2-1.$$

Therefore we obtain

$$\hat{\hat{f}}(z) = q^{-2}\{(q^2-1)f(-z) - \sum_{-z\neq v\in F^*_{q^2}} f(v)\} = f(-z),$$

by 3.1.10. □

Definition 3.1.15 Let $\Theta : F^*_{q^2} \longrightarrow \mathbf{C}^*$ be a non-trivial character, as in 3.1.3. The following three formulae characterise the Weil representation associated to Θ:

$$r(\Theta) : GL_2F_q \longrightarrow Aut_C(\mathscr{H}) \cong GL_{q-1}\mathbf{C}$$

(i)

$$(r(\Theta)(w)f)(x) = \hat{f}(x) \qquad (f \in \mathscr{H}, x \in F^*_{q^2}),$$

where w is as in 3.1.1,
(ii)

$$(r(\Theta)\begin{pmatrix} 1 & u \\ 0 & 1 \end{pmatrix} f)(x) = \Psi_{F_q}(uN_{F_{q^2}/F_q}(x))f(x)$$

and
(iii)

$$\left(r(\Theta)\begin{pmatrix} \alpha & 0 \\ 0 & 1 \end{pmatrix} f\right)(x) = \Theta(\beta)f(\beta x),$$

where $\alpha \in F_q, \beta \in F_{q^2}$ and $N_{F_{q^2}/F_q}(\beta) = \alpha$.

Theorem 3.1.16 *The formulae of* 3.1.15 *characterise a unique, well-defined $(q-1)$-dimensional, irreducible representation, $r(\Theta)$, of GL_2F_q.*

Proof Notice that, by 3.1.2, 3.1.9 and 3.1.13, these formulae do define a unique automorphism of $\mathscr{H}$ for each element of GL_2F_q. Therefore we must verify that the formulae respect the relations (i)–(v) of 3.1.2. Relations (i) and (ii) follow from 3.1.8–3.1.9 since they are relations between elements of the Borel subgroup, B. Also relation (v) follows from the fact that the Fourier transform is of order four, by 3.1.13. Relation (iii) is straightforward to verify, using the well-known identity

$$\hat{h}(x) = \hat{f}(xF(\beta)^{-1})$$

if $f \in \mathscr{H}; x, \beta \in F_{q^2}$ and $h(x) = f(\beta x)$.

In order to verify relation (iv) we will prove the following identity, in which $r(\Theta)$ has been abbreviated to r:

3.1.17

$$\begin{cases} \left(r(w)r\begin{pmatrix} 1 & u \\ 0 & 1 \end{pmatrix} r(w)f\right)(x) \\ = q^{-1}\sum_{v\in F^*_{q^2}} f(v)\Psi_{F_q}(-N(x+v)u^{-1}) \\ = \left(r\begin{pmatrix} -u^{-1} & 0 \\ 0 & -u \end{pmatrix} r\begin{pmatrix} 1 & -u \\ 0 & 1 \end{pmatrix} r(w)r\begin{pmatrix} 1 & -u^{-1} \\ 0 & 1 \end{pmatrix} f\right)(x). \end{cases}$$

We have

$$\left(r(w)r\begin{pmatrix} 1 & u \\ 0 & 1 \end{pmatrix} r(w)f\right)(x)$$

$$= \left(r(w)r\begin{pmatrix} 1 & u \\ 0 & 1 \end{pmatrix} \hat{f}\right)(x)$$

$$= (r(w)\{z \mapsto \Psi_{F_q}(uN(z))\hat{f}(z)\})(x)$$

$$= -q^{-1}\sum_{y\in F^*_{q^2}} \Psi_{F_q}(uN(y))\hat{f}(y)\Psi_{F_q}(yF(x)+xF(y))$$

$$= q^{-2}\sum_{y,v\in F^*_{q^2}} \Psi_{F_q}(uyF(y))\Psi_{F_q}(yF(x)+xF(y)) \times\Psi_{F_q}(yF(v)+vF(y))f(v)$$

$$= q^{-2}\sum_{y,v\in F^*_{q^2}} \Psi_{F_q}(uyF(y)+yF(x+v)+(x+v)F(y))f(v).$$

If $a = x + v$ and $N(z) = u \neq 0$, where N denotes the norm, we observe that

$$N(yz + aF(z)^{-1})$$
$$= (yz + aF(z)^{-1})(F(y)F(z) + F(a)z^{-1})$$
$$= uyF(y) + aF(y) + yF(a) + N(a)u^{-1}.$$

Therefore we have

$$\sum_{y\in F^*_{q^2}} \Psi_{F_q}(uyF(y) + yF(x+v) + (x+v)F(y))$$

$$= \sum_{y\in F^*_{q^2}} \Psi_{F_q}(N(yz + aF(z)^{-1}))\Psi_{F_q}(-N(a)u^{-1})$$

$$= -1 + \sum_{y\in F_{q^2}} \Psi_{F_q}(N(yz + aF(z)^{-1}))\Psi_{F_q}(-N(a)u^{-1})$$

$$= -1 + \{\sum_{s\in F_{q^2}} \Psi_{F_q}(N(s))\}\Psi_{F_q}(-N(a)u^{-1})$$

$$= -1 + \Psi_{F_q}(-N(a)u^{-1})\{\Psi_{F_q}(0) + (q+1)\sum_{z\in F^*_q} \Psi_{F_q}(z)\}$$

$$= -1 + \Psi_{F_q}(-N(a)u^{-1})\{1 + (q+1)(-1)\},$$

by 3.3.3,

$$= -1 - q\Psi_{F_q}(-N(a)u^{-1}).$$

Therefore, substituting into the previous formula,

$$\left(r(w)r\begin{pmatrix} 1 & u \\ 0 & 1 \end{pmatrix} r(w)f\right)(x)$$

$$= -q^{-2}\{\sum_{v\in F^*_{q^2}} f(v)\} - q^{-1}\{\sum_{v\in F^*_{q^2}} f(v)\Psi_{F_q}(-N(x+v)u^{-1})\}$$

$$= -q^{-1}\{\sum_{v\in F^*_{q^2}} f(v)\Psi_{F_q}(-N(x+v)u^{-1})\},$$

by 3.3.3, which establishes the first half of 3.1.17.

On the other hand, by 3.1.9, one easily deduces that

$$\left(r\begin{pmatrix} -u^{-1} & 0 \\ 0 & -u \end{pmatrix} r\begin{pmatrix} 1 & -u \\ 0 & 1 \end{pmatrix} r(w) r\begin{pmatrix} 1 & -u^{-1} \\ 0 & 1 \end{pmatrix} f\right)(x)$$

$$= \left(r\begin{pmatrix} 1 & -u \\ 0 & 1 \end{pmatrix} r(w) r\begin{pmatrix} 1 & -u^{-1} \\ 0 & 1 \end{pmatrix} f\right)(-xu^{-1}).$$

Finally,

$$\left(r\begin{pmatrix} 1 & -u \\ 0 & 1 \end{pmatrix} r(w) r\begin{pmatrix} 1 & -u^{-1} \\ 0 & 1 \end{pmatrix} f\right)(-xu^{-1})$$

$$= \Psi_{F_q}(-uN(-xu^{-1}))\{r(w)(z \mapsto \Psi_{F_q}(-N(z)u^{-1})f(z))\}(-xu^{-1})$$

$$= q^{-1}\Psi_{F_q}(-uN(-xu^{-1}))\{-\textstyle\sum_{v\in F^{\bullet}_{q^2}} f(v)\Psi_{F_q}(-N(v)u^{-1})$$
$$\times\Psi_{F_q}(-xu^{-1}F(v) - F(x)F(u)^{-1}v)\}$$

$$= -q^{-1}\textstyle\sum_{v\in F^{\bullet}_{q^2}} f(v)\Psi_{F_q}(-u^{-1}(N(v) + N(x) + vF(x) + F(v)x))$$

$$= -q^{-1}\textstyle\sum_{v\in F^{\bullet}_{q^2}} f(v)\Psi_{F_q}(-u^{-1}N(x+v)),$$

as required to establish the second half of 3.1.17.

It remains to show that $r(\Theta)$ is irreducible. By 3.1.8 it suffices to show that $Ind_H^B(\Theta \otimes \Psi)$ is an irreducible representation of B. However,

$$< Ind_H^B(\Theta \otimes \Psi), Ind_H^B(\Theta \otimes \Psi) >_B$$

$$=< \Theta \otimes \Psi, Res_H^B(Ind_H^B(\Theta \otimes \Psi)) >_H$$

$$= \textstyle\sum_{z\in F_q^{\bullet}} < \Theta \otimes \Psi, \Theta \otimes \Psi(z \cdot -) >_H,$$

by 1.2.40, where $\Psi(z \cdot -)(a) = \Psi(za)$. However, when $z \neq 1$,

$$< \Psi, \Psi(z \cdot -) >= 0,$$

by 3.3.3, so that

$$< Ind_H^B(\Theta \otimes \Psi), Ind_H^B(\Theta \otimes \Psi) >_B = 1,$$

which completes the proof of Theorem 3.1.16. □

3.1.18 For future use let us record the conjugacy class information concerning GL_2F_q.

Table 3.1. *Conjugacy classes in GL_2F_q*

Type	Minimal polynomial	Conjugacy class representative	Number in class
I	$(t-\alpha)(t-\beta)$ $\alpha \neq \beta \in F_q^*$	$\begin{pmatrix} \alpha & 0 \\ 0 & \beta \end{pmatrix}$	$q(q+1)$
II	$(t-\alpha)^2$ $\alpha \in F_q^*$	$\begin{pmatrix} \alpha & 1 \\ 0 & \alpha \end{pmatrix}$	q^2-1
III	$(t-\alpha)$ $\alpha \in F_q^*$	$\begin{pmatrix} \alpha & 0 \\ 0 & \alpha \end{pmatrix}$	1
IV	$t^2-(x+F(x))t+xF(x)$ $F(x) \neq x \in F_{q^2}^*$	$\begin{pmatrix} 0 & -xF(x) \\ 1 & x+F(x) \end{pmatrix}$	q^2-q

The conjugacy class of a matrix, $X \in GL_2F_q$, is determined by its minimal polynomial. The minimal polynomial of X must have degree one or two. The representatives of each conjugacy class, together with their minimal polynomial and the number of elements within each class, are listed in Table 3.1.

The remainder of this section will be devoted to the calculation of the character values of the Weil representation.

Lemma 3.1.19 *Let*

$$Y = \begin{pmatrix} 0 & -xF(x) \\ 1 & x+F(x) \end{pmatrix} \in GL_2F_q$$

denote the conjugacy class representative of type IV *in Table* 3.1. *Then, for $f \in \mathscr{H}$, $(r(\Theta)(Y)f)(z) =$*

$$-q^{-1}\Theta(-x)\sum_{y\in F_{q^2}^*} f(y)\Psi_{F_q}((x+F(x))N(y)+yF(xz)+xzF(y)),$$

where $N = N_{F_{q^2}/F_q}$ is the norm.

Proof We begin with the observation that, by 3.3.1,

3.1.20

$$Y = \begin{pmatrix} 0 & -xF(x) \\ 1 & x+F(x) \end{pmatrix} = \begin{pmatrix} -xF(x) & 0 \\ 0 & -1 \end{pmatrix} w \begin{pmatrix} 1 & x+F(x) \\ 0 & 1 \end{pmatrix},$$

where w is as in 3.1.1.

Write $t = x+F(x)$, $d = xF(x) = N(x)$ and let $a \in F_{q^2}^*$ satisfy $N(a) = -1$. Therefore we have

$(r(\Theta)(Y)f)(z)$

$$= \left(r\begin{pmatrix} -d & 0 \\ 0 & -1 \end{pmatrix} r(w)r\begin{pmatrix} 1 & t \\ 0 & 1 \end{pmatrix} f\right)(z)$$

$$= \left(r\begin{pmatrix} -d & 0 \\ 0 & -1 \end{pmatrix} r(w)\{s \mapsto \Psi(tN(s))f(s)\}\right)(z)$$

$$= -q^{-1}\left(r\begin{pmatrix} -d & 0 \\ 0 & -1 \end{pmatrix} \{s \mapsto \textstyle\sum_{y\in F_{q^2}^*} f(y)\Psi(tN(y))\Psi(yF(s) + sF(y))\}\right)(z)$$

$$= -q^{-1}\Theta(ax)\Theta(a)\{\textstyle\sum_{y\in F_{q^2}^*} f(y)\Psi(tN(y)+yF(axzF(a)^{-1})+F(y)axzF(a)^{-1})\}$$

$$= -q^{-1}\Theta(x)\{\textstyle\sum_{y\in F_{q^2}^*} f(ya^{-2})\Psi(tN(ya^{-2}) - ya^{-2}F(xz) - F(ya^{-2})xz)\},$$

since $\qquad aF(a) = -1$ and $N(a^{-2}) = 1,$

$$= -q^{-1}\Theta(x)\Theta(-1)\{\textstyle\sum_{v\in F_{q^2}^*} f(v)\Psi(tN(v) + vF(xz) + F(v)xz)\},$$

by setting $v = -ya^{-2}$, which completes the proof. □

Proposition 3.1.21 *With the notation of Table* 3.1, *the character-values of* $r(\Theta)$ *on conjugacy classes of type* IV *are given by*

$$Trace\left(r(\Theta)\left(\begin{pmatrix} 0 & -xF(x) \\ 1 & x+F(x) \end{pmatrix}\right)\right) = -\{\Theta(x) + \Theta(F(x))\}.$$

Proof In this and subsequent calculations it will be convenient to have chosen a basis for the representation space, $\mathscr{H}$. To this end let us choose, once and for all, $v_1, \ldots, v_{q-1} \in F_{q^2}^*$ so that the set $\{N(v_i); 1 \leq i \leq q-1\}$ is equal to F_q^*. Hence the $\{v_i\}$ are a complete set of coset representatives for $F_{q^2}^*/Ker(N)$. Let $\{f_i \in \mathscr{H}; 1 \leq i \leq q-1\}$ be the unique functions which satisfy

3.1.22
$$f_i(v_j) = \begin{cases} 1 & \text{if } i = j, \\ 0 & \text{if } i \neq j. \end{cases}$$

Hence, if $Y \in GL_2F_q$ and

3.1.23 $$r(\Theta)(Y)(f_i) = \textstyle\sum_{i=1}^{q-1} a_{ij}f_j,$$

then we see at once that

3.1.24 $$\begin{cases} a_{ij} = (r(\Theta)(Y)f_i)(v_j) \quad \text{and} \\ Trace(r(\Theta)(Y)) = \sum_{i=1}^{q-1} r(\Theta)(Y)(f_i)(v_i). \end{cases}$$

Therefore, from 3.1.19 and 3.1.24, if we abbreviate by setting $x+F(x) = t$ then we obtain

$$Trace\left(r(\Theta)\left(\begin{pmatrix} 0 & -xF(x) \\ 1 & x+F(x) \end{pmatrix}\right)\right)$$

$$= \sum_{i=1}^{q-1}\left(r(\Theta)\left(\begin{pmatrix} 0 & -xF(x) \\ 1 & x+F(x) \end{pmatrix}\right) f_i\right)(v_i)$$

$$= -q^{-1}\Theta(-x)\sum_{i=1}^{q-1}\sum_{y\in F^*_{q^2}} f_i(y)\Psi(tN(y) + yF(xv_i) + xv_iF(y))$$

$$= -q^{-1}\Theta(-x)\sum_{i=1}^{q-1}\sum_{u\in Ker(N)} f_i(uv_i)\Psi(tN(v_i)$$
$$+uF(x)N(v_i) + xF(u)N(v_i)),$$

since $\qquad f_i(uv_j) = 0$ if $i \neq j$ and $N(u) = 1$,

$$= -q^{-1}\Theta(-x)\sum_{i=1}^{q-1}\sum_{u\in Ker(N)} \Theta(u)^{-1}\Psi(N(v_i)(t + uF(x) + xF(u))).$$

Next we observe that if $t + uF(x) + xF(u) \neq 0$ then, by 3.3.3,

$$\sum_{i=1}^{q-1}\Psi(N(v_i)(t + uF(x) + xF(u))) = -1.$$

This happens for all the values of $u \in Ker(N)$ except for $u = -1$ and $u = -xF(x)^{-1}$ for which $t + uF(x) + xF(u) = 0$. Therefore the terms with $u = -1$ contribute

$$-q^{-1}\Theta(-x)(q-1)\Theta(-1)^{-1} = -(q-1)q^{-1}\Theta(x)$$

to the sum, while those for which $u = -xF(x)^{-1}$ contribute

$$-q^{-1}\Theta(-x)(q-1)\Theta(-xF(x)^{-1})^{-1} = -(q-1)q^{-1}\Theta(F(x)).$$

The remaining terms contribute, by 3.3.3,

$$-q^{-1}\Theta(-x)\{-\Theta(-1)^{-1} - \Theta(xF(x)^{-1})^{-1}\} = -q^{-1}\{\Theta(x) + \Theta(F(x))\},$$

so that adding these contributions together yields the required formula. □

Proposition 3.1.25 *With the notation of Table* 3.1, *the character-values of* $r(\Theta)$ *on conjugacy classes of types* I *and* III *are given by*

$$Trace\left(r(\Theta)\left(\begin{pmatrix} \alpha & 0 \\ 0 & \beta \end{pmatrix}\right)\right) = \begin{cases} (q-1)\Theta(\alpha) & \text{if} \quad \alpha = \beta, \\ 0 & \text{otherwise.} \end{cases}$$

Proof Let $a, b \in F_{q^2}^*$ satisfy $N(a) = \alpha$ and $N(b) = \beta$ where it is understood that if $\alpha = \beta$ then we shall choose $a = b$. Choosing the basis for $\mathscr{H}$ which was used in the proof of 3.1.21 we find, by 3.1.9, that

$$Trace\left(r(\Theta)\left(\begin{pmatrix} \alpha & 0 \\ 0 & \beta \end{pmatrix}\right)\right) = \sum_{i=1}^{q-1} \Theta(a)\Theta(b)f_i(av_iF(b)^{-1}).$$

When $\alpha \neq \beta$ then $N(aF(b)^{-1}) \neq 1$ and $f_i(av_iF(b)^{-1}) = 0$ for all i, which yields the required formula in the case when α and β are distinct. On the other hand, when $\alpha = \beta$

$$Trace\left(r(\Theta)\left(\begin{pmatrix} \alpha & 0 \\ 0 & \alpha \end{pmatrix}\right)\right)$$

$$= \textstyle\sum_{i=1}^{q-1} \Theta(a)\Theta(a)f_i(av_iF(a)^{-1})$$

$$= \textstyle\sum_{i=1}^{q-1} \Theta(a)\Theta(a)\Theta(F(a)^{-1})^{-1}\Theta(a)^{-1}$$

$$= (q-1)\Theta(\alpha),$$

as required. □

Proposition 3.1.26 *With the notation of Table* 3.1, *the character-values of* $r(\Theta)$ *on conjugacy classes of type* II *are given by*

$$Trace\left(r(\Theta)\left(\begin{pmatrix} \alpha & 1 \\ 0 & \alpha \end{pmatrix}\right)\right) = -\Theta(\alpha).$$

Table 3.2. *Character values of the Weil representation, $r(\Theta)$*

Type	Conjugacy class representative	Character value
I	$\begin{pmatrix} \alpha & 0 \\ 0 & \beta \end{pmatrix}$	0
II	$\begin{pmatrix} \alpha & 1 \\ 0 & \alpha \end{pmatrix}$	$-\Theta(\alpha)$
III	$\begin{pmatrix} \alpha & 0 \\ 0 & \alpha \end{pmatrix}$	$(q-1)\Theta(\alpha)$
IV	$\begin{pmatrix} 0 & -xF(x) \\ 1 & x+F(x) \end{pmatrix}$	$-\{\Theta(x)+\Theta(F(x))\}$

Proof As in the proof of 3.1.25, let $N(a) = \alpha$. Then we have

$$Trace\left(r(\Theta)\left(\begin{pmatrix} \alpha & 1 \\ 0 & \alpha \end{pmatrix}\right)\right)$$

$$= Trace\left(r(\Theta)\left(\begin{pmatrix} \alpha & 0 \\ 0 & \alpha \end{pmatrix}\right) r(\Theta)\left(\begin{pmatrix} 1 & \alpha^{-1} \\ 0 & 1 \end{pmatrix}\right)\right)$$

$$= \sum_{i=1}^{q-1} \Theta(a)^2 \left(r(\Theta)\left(\begin{pmatrix} 1 & \alpha^{-1} \\ 0 & 1 \end{pmatrix}\right) f_i\right)(av_iF(a)^{-1})$$

$$= \sum_{i=1}^{q-1} \Theta(a)^2 \Psi(a^{-1}N(v_iaF(a)^{-1}))f_i(v_iaF(a)^{-1})$$

$$= \sum_{i=1}^{q-1} \Theta(\alpha)\Psi(a^{-1}N(v_i)),$$

since $N(aF(a)^{-1}) = 1$. This yields the required formula since, by 3.3.3,

$$\sum_{i=1}^{q-1} \Psi(a^{-1}N(v_iaF(a)^{-1})) = -1.$$

□

For the reader's convenience we tabulate the results of 3.1.21, 3.1.25 and 3.1.26 (see Table 3.2).

3.2 Explicit Brauer induction and Shintani descent

In this section we will evaluate some of the maximal terms, $(H,\phi)^{GL_2F_q}$, in the expressions for $a_{GL_2F_q}(\rho)$, as ρ runs through the irreducible representations of GL_2F_q. In the previous section we constructed the Weil representations and, before proceeding further, we shall now construct the remaining irreducible representations of GL_2F_q.

Suppose that we are given characters of the form

$$\chi, \chi_1, \chi_2 : F_q^* \longrightarrow \mathbf{C}^*,$$

then we clearly have a one-dimensional representation, $L(\chi)$, given by

3.2.1 $$L(\chi) = \chi \cdot det : GL_2F_q \xrightarrow{det} F_q^* \xrightarrow{\chi} \mathbf{C}^*.$$

If χ_1 and χ_2 are *distinct* define

$$Inf_T^B(\chi_1 \otimes \chi_2) : B \longrightarrow \mathbf{C}^*$$

by inflating $\chi_1 \otimes \chi_2$ from the diagonal torus, T, to the Borel subgroup, B. That is,

$$Inf_T^B(\chi_1 \otimes \chi_2)\left(\begin{pmatrix} \alpha & \beta \\ 0 & \delta \end{pmatrix}\right) = \chi_1(\alpha)\chi_2(\delta).$$

Define a $(q+1)$-dimensional representation, $R(\chi_1, \chi_2)$, by

3.2.2 $$R(\chi_1, \chi_2) = Ind_B^{GL_2F_q}(Inf_T^B(\chi_1 \otimes \chi_2)).$$

When $\chi = \chi_1 = \chi_2$ we have

$$Inf_T^B(\chi \otimes \chi) = Res_B^{GL_2F_q}(L(\chi)) : B \longrightarrow \mathbf{C}^*,$$

so that there is a canonical surjection of the form

$$Ind_B^{GL_2F_q}(Inf_T^B(\chi \otimes \chi)) \longrightarrow Ind_{GL_2F_q}^{GL_2F_q}(L(\chi)) = L(\chi).$$

Therefore we may define a q-dimensional representation, $S(\chi)$, by means of the following short exact sequence of representations (which is *split*, by 1.1.8):

3.2.3 $$0 \longrightarrow S(\chi) \longrightarrow Ind_B^{GL_2F_q}(Inf_T^B(\chi \otimes \chi)) \longrightarrow L(\chi) \longrightarrow 0.$$

Theorem 3.2.4 *A complete list of all the irreducible representations of GL_2F_q is given by*

(i) $L(\chi)$ *of* 3.2.1 *for* $\chi : F_q^* \longrightarrow \mathbf{C}^*$,

(ii) $S(\chi)$ *of* 3.2.3 *for* $\chi : F_q^* \longrightarrow \mathbf{C}^*$,

(iii) $R(\chi_1,\chi_2) = R(\chi_2,\chi_1)$ *of* 3.2.2 *for any pair of distinct characters* $\chi_1,\chi_2 : F_q^* \longrightarrow \mathbf{C}^*$
and

(iv) $r(\Theta) = r(F^*(\Theta))$ *of* 3.1.15 *for any character* $\Theta : F_{q^2}^* \longrightarrow \mathbf{C}^*$ *which is distinct from its Frobenius conjugate,* $F^*(\Theta)$.

Proof Assuming that the representations in the list are distinct and irreducible, the numbers of each type are $q-1, q-1, (q-1)(q-2)/2$ and $(q^2-q)/2$, respectively. Therefore, by 1.2.15, there can be no more irreducibles since

$$\begin{aligned}&(q-1)\cdot 1^2 + (q-1)\cdot q^2 + ((q-1)(q-2)/2)(q+1)^2\\ &\quad +((q^2-q)/2)(q-1)^2\\ &= (q-1)(q^3-q)\\ &= \#(GL_2F_q).\end{aligned}$$

To see that the representations in the list are distinct and irreducible we calculate the Schur inner product, using 1.2.40, Frobenius reciprocity and the Bruhat decomposition (1.2.39 and 3.1.1)

$$\begin{aligned}&< Ind_B^{GL_2F_q}(Inf_T^B(\chi_1\otimes\chi_2)), Ind_B^{GL_2F_q}(Inf_T^B(\chi_1\otimes\chi_2)) >_{GL_2F_q}\\ &=< Inf_T^B(\chi_1\otimes\chi_2), Res_B^{GL_2F_q}(Ind_B^{GL_2F_q}(Inf_T^B(\chi_1\otimes\chi_2))) >_B\\ &=< Inf_T^B(\chi_1\otimes\chi_2), Inf_T^B(\chi_1\otimes\chi_2)\\ &\qquad + Ind_T^B(Res_T^{wBw^{-1}}((w^{-1})^*(Inf_T^B(\chi_1\otimes\chi_2)))) >_B\\ &= 1+ < Inf_T^B(\chi_1\otimes\chi_2), Ind_T^B(\chi_2\otimes\chi_1) >_B\\ &= 1+ < Res_T^B(Inf_T^B(\chi_1\otimes\chi_2)), \chi_2\otimes\chi_1 >_T\\ &= 1+ < \chi_1\otimes\chi_2, \chi_2\otimes\chi_1 >_T\\ &= \begin{cases} 2 & \text{if} \quad \chi_1=\chi_2,\\ 1 & \text{otherwise.}\end{cases}\end{aligned}$$

This calculation, together with 3.1.16, shows that each representation

Table 3.3. (*Theorem* 3.2.5)

Type	$L(\chi)$	$R(\chi_1,\chi_2)$	$S(\chi)$	$r(\Theta)$
I	$\chi(\alpha\beta)$	$\chi_1(\alpha)\chi_2(\beta)+\chi_2(\alpha)\chi_1(\beta)$	$\chi(\alpha\beta)$	0
II	$\chi(\alpha)^2$	$\chi_1(\alpha)\chi_2(\alpha)$	0	$-\Theta(\alpha)$
III	$\chi(\alpha)^2$	$(q+1)\chi_1(\alpha)\chi_2(\alpha)$	$q\chi(\alpha)^2$	$(q-1)\Theta(\alpha)$
IV	$\chi(N(x))$	0	$-\chi(N(x))$	$-\{\Theta(x)+\Theta(F(x))\}$

in the list is irreducible. A similar calculation shows that

$$< Ind_B^{GL_2F_q}(Inf_T^B(\chi_1\otimes\chi_2)), Ind_B^{GL_2F_q}(Inf_T^B(\lambda_1\otimes\lambda_2)) >_{GL_2F_q}$$

$$=< Inf_T^B(\chi_1\otimes\chi_2), Inf_T^B(\lambda_1\otimes\lambda_2) >_B + < \chi_1\otimes\chi_2, \lambda_2\otimes\lambda_1 >_T$$

and this formula, together with dimension, enables one to distinguish between the listed representation, which completes the proof. □

Theorem 3.2.5 *With the notation of Table* 3.1 *and* 3.2.4 *the character values of the irreducible representations of* GL_2F_q *are given by Table* 3.3, *where* $N = N_{F_{q^2}/F_q}$ *denotes the norm.*

Proof The character values of the Weil representation, $r(\Theta)$, were calculated in 3.1.21, 3.1.25 and 3.1.26. The character values of $L(\chi)$ are trivial to verify. From 1.2.43 we have the formula

$$Trace(Ind_H^G(\rho)(g)) = \#(H)^{-1} \sum_{y\in G, ygy^{-1}\in H} Trace(\rho(ygy^{-1}))$$

by means of which to calculate the character-values of

$$Ind_B^{GL_2F_q}(Inf_T^B(\chi_1\otimes\chi_2)).$$

When $\chi=\chi_1=\chi_2$ we may calculate the character-values of $S(\chi)$ by means of the relation

$$Ind_B^{GL_2F_q}(Inf_T^B(\chi\otimes\chi)) = S(\chi)\oplus L(\chi).$$

The representative for type I can only be conjugated into B by $y\in B$ or $y\in Bw$, the former type of conjugation preserving the diagonal entries and the latter reversing them. The representative for type II can only be conjugated into B by $y\in B$, whereas that for type III is central in GL_2F_q and that for type IV cannot be conjugated into B at all, since it has a characteristic polynomial whose roots do not lie in F_q. The entries in the table follow easily from these remarks and will be left to the reader. $\square$

3.2.6 *Maximal pairs in* $\mathcal{M}_{GL_2F_q}$

As in 2.2.9–2.2.10 let $\mathcal{M}_{GL_2F_q}$ denote the poset of pairs (J, ϕ) with $J \leq GL_2F_q$ and $\phi : J \longrightarrow \mathbf{C}^*$. From 2.3.15, if $v \in R(GL_2F_q)$ and (J, ϕ) is *maximal* in $\mathcal{M}_{GL_2F_q}$ then the coefficient of $(J, \phi)^{GL_2F_q}$ in $a_{GL_2F_q}(v)$ is given by

3.2.7
$$\begin{cases} \{\text{multiplicity of } (J, \phi)^{GL_2F_q} \text{ in } a_{GL_2F_q}(v)\} \\ = [N_{GL_2F_q}(J, \phi) : J]^{-1} < \phi, Res_J^{GL_2F_q}(v) > . \end{cases}$$

Here

$$N_{GL_2F_q}(J, \phi) = \{X \in GL_2F_q \mid (XJX^{-1}, (X^{-1})^*(\phi)) = (J, \phi)\}.$$

In other words, it is easy to calculate the multiplicity of maximal pairs, $(J, \phi)^{GL_2F_q}$, in $a_{GL_2F_q}(v)$. For this reason we will now introduce four types of maximal pairs in $\mathcal{M}_{GL_2F_q}$:

Type A: $(GL_2F_q, \chi \cdot det)$ for $\chi : F_q^* \longrightarrow \mathbf{C}^*$.
Type B: $(H, \lambda \otimes \mu)$ where $\mu : F_q \longrightarrow \mathbf{C}^*$ is *non-trivial* and $\lambda : F_q^* \longrightarrow \mathbf{C}^*$ is any homomorphism. Here

$$\gamma : H = \left\{ \begin{pmatrix} z & y \\ 0 & z \end{pmatrix} \in GL_2F_q \right\} \xrightarrow{\cong} F_q^* \times F_q$$

is as in 3.1.3.
Type C: $(B, Inf_T^B(\lambda_1 \otimes \lambda_2))$, in the notation of 3.2.2, where $\lambda_1, \lambda_2 : F_q^* \longrightarrow \mathbf{C}^*$ are *distinct*.
Type D: $(F_{q^2}^*, \rho)$ where ρ and $F^*(\rho)$ are *distinct*.

Here we consider $F_{q^2}^*$ to be the cyclic subgroup generated by the matrix

3.2.8
$$\begin{pmatrix} 0 & -xF(x) \\ 1 & x + F(x) \end{pmatrix}$$

of Table 3.1 where $x \in F_{q^2}^*$ is a generator. Up to conjugation this subgroup is independent of the choice of x. On this subgroup the Frobenius map, $F \in G(F_{q^2}/F_q)$, corresponds to conjugation by the matrix

3.2.9
$$f = \begin{pmatrix} 1 & x + F(x) \\ 0 & -1 \end{pmatrix}.$$

This is seen as follows. With respect to the F_q-basis, $\{1, x\}$, of F_{q^2} multiplication by x is represented by the matrix of 3.2.8. However

$$\begin{pmatrix} 1 & x+F(x) \\ 0 & -1 \end{pmatrix} \begin{pmatrix} 0 & -xF(x) \\ 1 & x+F(x) \end{pmatrix} \begin{pmatrix} 1 & x+F(x) \\ 0 & -1 \end{pmatrix}$$

$$= \begin{pmatrix} x+F(x) & -xF(x)+(x+F(x))^2 \\ -1 & -x-F(x) \end{pmatrix} \begin{pmatrix} 1 & x+F(x) \\ 0 & -1 \end{pmatrix}$$

$$= \begin{pmatrix} x+F(x) & xF(x) \\ -1 & 0 \end{pmatrix},$$

which represents multiplication by $F(x)$ with respect to this basis.

Proposition 3.2.10 *Each of the pairs, $(J, \phi) \in \mathcal{M}_{GL_2F_q}$, listed in types* A–D *of* 3.2.6, *is maximal. In addition, for each of types* A–D,

$$N_{GL_2F_q}(J, \phi) = J.$$

Proof The result is obvious for type A, $(GL_2F_q, \chi \cdot det)$.

From the classification of maximal subgroups of SL_2F_q, which is given in Dickson (1958, p. 286, section 262), it is straightforward to see that any proper subgroup of GL_2F_q which contains H must lie in $B = N_{GL_2F_q}H$. However,

$$\begin{pmatrix} \alpha & \beta \\ 0 & \delta \end{pmatrix} \begin{pmatrix} z & y \\ 0 & z \end{pmatrix} \begin{pmatrix} \alpha & \beta \\ 0 & \delta \end{pmatrix}^{-1}$$

$$= \begin{pmatrix} \alpha z & \alpha y + \beta z \\ 0 & \delta z \end{pmatrix} \begin{pmatrix} \alpha^{-1} & -\alpha^{-1}\beta\delta^{-1} \\ 0 & \delta^{-1} \end{pmatrix}$$

$$= \begin{pmatrix} z & \alpha z \delta^{-1} \\ 0 & z \end{pmatrix}.$$

Hence the action of this matrix on $H \cong F_q^* \times F_q$ is to multiply the second coordinate by $\alpha\delta^{-1}$. However μ and $\mu(\alpha\delta^{-1}-)$ are distinct unless $\alpha = \delta$, which shows that $N_{GL_2F_q}(H, \lambda \otimes \mu) \leq H$ and proves the result for type B.

The result for type C is immediate from the facts that $N_{GL_2F_q}B = B$ and that the conjugate, xvx^{-1} $(x, v \in B)$, has the same diagonal as v.

From the classification of maximal subgroups of SL_2F_q (Dickson, 1958, p. 286, section 262), one readily finds that $N_{GL_2F_q}F_{q^2}^* =< F_{q^2}^*, f >$ where f is as in 3.2.9. However, conjugation by f induces the Frobenius map of $F^*_{q^2}$ so that $f \notin N_{GL_2F_q}(F^*_{q^2}, \rho)$, which easily yields the result for type D. □

Corollary 3.2.11 *Suppose that* $g \in GL_2F_q$ *and that* $(J, \phi) \in \mathcal{M}_{GL_2F_q}$ *is one of the maximal pairs of type* A–D, *as in* 3.2.10. *If* $(gJg^{-1}, (g^{-1})^*(\phi)) = (J, \phi_1)$ *then* $\phi = \phi_1$ *in the case of types* A *and* C. *For type* B, $\phi = \lambda \otimes \mu$ *and* $\phi_1 = \lambda \otimes \mu(u \cdot -)$ *for some* $u \in F^*_q$ *while, for type* D, $\phi_1 = \phi$ *or* $\phi_1 = F^*(\phi)$ *where* F *is the Frobenius of* $G(F_{q^2}/F_q)$.

Proof We must have $g \in N_{GL_2F_q}J = GL_2F_q, B, B$ or $< F^*_{q^2}, f >$ for types A–D, respectively. From this observation the result follows easily from the computations which were used in the proof of 3.2.10. □

Definition 3.2.12 Suppose that $v : GL_2F_q \longrightarrow GL(V)$ is an *irreducible* representation. We will write

3.2.13

$$\begin{aligned} a_{GL_2F_q}(v) &= \textstyle\sum_r a_r(H, \lambda_r \otimes \mu_r)^{GL_2F_q} \\ &+ \textstyle\sum_s b_s(B, Inf^B_T(\lambda_{1,s} \otimes \lambda_{2,s}))^{GL_2F_q} \\ &+ \textstyle\sum_t c_t(F^*_{q^2}, \rho_t)^{GL_2F_q} \\ &+ \textstyle\sum_u d_u(GL_2F_q, \chi_u \cdot det)^{GL_2F_q} \\ &+ \cdots \end{aligned}$$

to signify that the multiplicities in $a_{GL_2F_q}(v)$ of the maximal pairs of types A–D in 3.2.6 are as shown in 3.2.13 (the ellipsis denoting the sum of all the terms of other types). In 3.2.13 the sums over r, s, t, u are taken over all the terms of types A,B,C,D respectively.

Theorem 3.2.14 *With the notation of* 3.2.4 *and* 3.2.13

(i)

$$a_{GL_2F_q}(L(\chi)) = (GL_2F_q, \chi \cdot det)^{GL_2F_q}.$$

(ii)

$$\begin{aligned} a_{GL_2F_q}(R(\chi_1,\chi_2)) &= \sum\nolimits_{1\neq\mu}(H,\chi_1\chi_2\otimes\mu)^{GL_2F_q} \\ &+(B,Inf_T^B(\chi_1\otimes\chi_2))^{GL_2F_q} \\ &+(B,Inf_T^B(\chi_2\otimes\chi_1))^{GL_2F_q} \\ &+\sum\nolimits_{Res_{F_q^*}^{F_{q^2}^*}(\rho)=\chi_1\chi_2}(F_{q^2}^*,\rho)^{GL_2F_q} \\ &+\cdots. \end{aligned}$$

(iii)

$$\begin{aligned} a_{GL_2F_q}(S(\chi)) &= \sum\nolimits_{1\neq\mu}(H,\chi^2\otimes\mu)^{GL_2F_q} \\ &+(B,Inf_T^B(\chi\otimes\chi))^{GL_2F_q} \\ &+\sum\nolimits_{Res_{F_q^*}^{F_{q^2}^*}(\rho)=\chi^2}(F_{q^2}^*,\rho)^{GL_2F_q} \\ &+\cdots. \end{aligned}$$

(iv)

$$\begin{aligned} a_{GL_2F_q}(r(\Theta)) &= \sum\nolimits_{v\in F_q^*}(H,Res_{F_q^*}^{F_{q^2}^*}(\Theta)\otimes\Psi_{F_q}(v\cdot -))^{GL_2F_q} \\ &+\sum\nolimits_{\rho\notin\{\Theta,F^*(\Theta)\},Res_{F_q^*}^{F_{q^2}^*}(\rho)=Res_{F_q^*}^{F_{q^2}^*}(\Theta)}(F_{q^2}^*,\rho)^{GL_2F_q} \\ &+\cdots. \end{aligned}$$

Proof Part (i) follows from 2.3.2(ii).

Parts (ii) and (iii) are similar and therefore we will only prove part (ii). By 3.2.7 and 3.2.10, the multiplicity of a term of type B from 3.2.6 in $a_{GL_2F_q}(R(\chi_1,\chi_2))$ is equal to

$$\begin{aligned} &< \lambda\otimes\mu, Res_H^{GL_2F_q}(Ind_B^{GL_2F_q}(Inf_T^B(\chi_1\otimes\chi_2))) > \\ &= \sum_{z\in H\backslash GL_2F_q/B} < \lambda\otimes\mu, \\ &\quad Ind_{H\cap zBz^{-1}}^H(Res_{H\cap zBz^{-1}}^{zBz^{-1}}((z^{-1})^*(Inf_T^B(\chi_1\otimes\chi_2)))) > \end{aligned}$$

$$= \sum_{z=1,w} < \lambda \otimes \mu,$$

$$Ind^H_{H\cap zBz^{-1}}(Res^{zBz^{-1}}_{H\cap zBz^{-1}}((z^{-1})^*(Inf^B_T(\chi_1 \otimes \chi_2)))) >$$

by the Bruhat decomposition,

$$=< \lambda \otimes \mu, (\chi_1\chi_2 \otimes 1) + Ind^H_{F^*_q}(\chi_2 \otimes \chi_1) >$$

$$=< \lambda \otimes \mu, \chi_1\chi_2 \otimes Ind^H_{F^*_q}(1) >$$

$$= \begin{cases} 1 & \text{if } \lambda = \chi_1\chi_2, \\ 0 & \text{otherwise.} \end{cases}$$

This accounts for the first part of the formula in part (ii).

By 3.2.7 and 3.2.10, the multiplicity of a term of type C from 3.2.6 in $a_{GL_2F_q}(R(\chi_1, \chi_2))$ is equal to

$$< Inf^B_T(\lambda_1 \otimes \lambda_2), Res^{GL_2F_q}_B(Ind^{GL_2F_q}_B(Inf^B_T(\chi_1 \otimes \chi_2))) >$$

$$=< Inf^B_T(\lambda_1 \otimes \lambda_2), Inf^B_T(\chi_1 \otimes \chi_2) + Ind^B_T(\chi_2 \otimes \chi_1) >$$

$$=< \lambda_1, \chi_1 >< \lambda_2, \chi_2 > + < \lambda_2, \chi_1 >< \lambda_1, \chi_2 >$$

$$= \begin{cases} 1 & \text{if } \{\lambda_1, \lambda_2\} = \{\chi_1, \chi_2\} \\ & \\ 0 & \text{otherwise.} \end{cases}$$

This accounts for the remaining part of the formula in part (ii).

Clearly there are no terms of type A in $a_{GL_2F_q}(R(\chi_1, \chi_2))$ and the multiplicity of a term of type D is equal to

$$< \rho, Res^{GL_2F_q}_{F^*_{q^2}}(R(\chi_1, \chi_2)) >$$

$$= (q^2-1)^{-1} \textstyle\sum_{x \in F^*_{q^2} - F^*_q} \overline{\rho}(x) \cdot 0$$
$$+ (q^2-1)^{-1} \textstyle\sum_{x \in F^*_q} (q+1)\overline{\rho}(x)\chi_1(x)\chi_2(x)$$

$$=< Res^{F^*_{q^2}}_{F^*_q}(\rho), \chi_1\chi_2 >$$

by the character values of 3.2.5. This completes the proof of part (ii).

For part (iv) we observe that, since $r(\Theta)$ is irreducible, there can be no terms of type A. Also there can be no terms of type C, since

$$< Inf_T^B(\lambda_1 \otimes \lambda_2), Ind_H^B(\Theta \otimes \Psi_{F_q}) >$$

$$=< \lambda_1\lambda_2 \otimes 1, \Theta \otimes \Psi_{F_q} >$$

$$= 0.$$

By 3.2.7 and 3.2.10, the multiplicity of a term of type B from 3.2.6 in $a_{GL_2F_q}(r(\Theta))$ is equal to

$$< \lambda \otimes \mu, Res_H^{GL_2F_q}(r(\Theta)) >$$

$$=< \lambda \otimes \mu, Res_H^B(Ind_H^B(\Theta \otimes \Psi)) >$$

$$= \sum_{z \in B/H} < \lambda \otimes \mu, (z^{-1})^*(\Theta \otimes \Psi) >$$

$$= \sum_{v \in F_q^*} < \lambda \otimes \mu, Res_{F_q^*}^{F_{q^2}^*}(\Theta) \otimes \Psi_{F_q}(v \cdot -) >,$$

which accounts for the first part of the formula in part (iv).

Finally, by 3.2.7 and 3.2.10, the multiplicity of a term of type D from 3.2.6 in $a_{GL_2F_q}(r(\Theta))$ is equal to

$$< \rho, Res_{F_{q^2}^*}^{GL_2F_q}(r(\Theta)) >$$

$$= (q^2-1)^{-1} \sum_{x \in F_q^*} \overline{\rho}(x)(q-1)\Theta(x)$$

$$-(q^2-1)^{-1} \sum_{x \in F_{q^2}^* - F_q^*} \overline{\rho}(x)(\Theta(x) + \Theta(F(x)))$$

by 1.2.7 and 3.2.5

$$= (q+1)(q^2-1)^{-1} \sum_{x \in F_q^*} \overline{\rho}(x)\Theta(x) - < \rho, \Theta > - < \rho, F^*(\Theta) >$$

$$=< Res_{F_q^*}^{F_{q^2}^*}(\rho), Res_{F_q^*}^{F_{q^2}^*}(\Theta) > - < \rho, \Theta > - < \rho, F^*(\Theta) >,$$

which accounts for the remaining terms in the formula for part (iv) and completes the proof. □

Corollary 3.2.15 *The irreducible representations, v, of GL_2F_q are uniquely characterised by the terms of types* A–D, *in the terminology of* 3.2.6, *which occur in the Explicit Brauer Induction formula*

$$a_{GL_2F_q}(v) \in R_+(GL_2F_q).$$

These (maximal) terms are given by the formulae of 3.2.14.

Proof This follows easily by inspection of the formulae of 3.2.14. For example, the Weil representations are the only ones for which a term of type D appears. The type B terms in $a_{GL_2F_q}(r(\Theta))$ determine the sum over which the type D terms are taken and the characters Θ and $F^*(\Theta)$ are characterised by being the only two characters on $F^*_{q^2}$ with the prescribed restriction to F^*_q which *do not* appear in the sum. Of course, $r(\Theta) = r(F^*(\Theta))$. □

3.2.16 *The Shintani correspondence for GL_2F_q*

Let $\Sigma \in G(F_{q^n}/F_q)$ denote the Frobenius transformation. Shintani (1976) discovered a remarkable one-one correspondence of the following form:

3.2.17
$$\left\{\begin{array}{c} \{\text{irreducible representations, } v, \text{ of} \\ GL_mF_{q^n} \text{ fixed under } \Sigma\} \\ \\ \updownarrow Sh \\ \\ \{\text{irreducible representations, } Sh(v), \text{ of } GL_mF_q\}. \end{array}\right.$$

In 3.2.17 the Frobenius, Σ, acts via its action upon the entries of a matrix. This correspondence, which was also treated by Shintani for GL_2 of a local field, is also sometimes called *Shintani descent* or *lifting* (see Gérardin & Labesse, 1979, for example).

The correspondence of 3.2.17 may be characterised by means of the *Shintani norm.* For $X \in GL_mF_{q^n}$ define

3.2.18
$$N(X) = \Sigma^{n-1}(X)\Sigma^{n-2}(X)\ldots\Sigma(X)X.$$

Although $N(X)$ lies in $GL_mF_{q^n}$, its conjugacy class contains a unique GL_mF_q-conjugacy class, which depends only on the conjugacy class of X. This gives a meaning to the equation

3.2.19 $$Trace(Sh(v)(N(X))) = Trace(v(X)).$$

The correspondence of 3.2.17 is characterised by the fact 3.2.19 holds for all $X \in GL_m F_{q^n}$.

When $m = 1$ this correspondence is a consequence of Hilbert's Theorem 90, which states that $H^1(G(L/K); L^*) = 0$. When L/K is an extension of finite fields we obtain an exact sequence of the form

3.2.20 $$F^*_{q^n} \xrightarrow{\Sigma/1} F^*_{q^n} \xrightarrow{N} F^*_q \longrightarrow \{1\}.$$

If $v : F^*_{q^n} \longrightarrow \mathbf{C}^*$ satisfies $v = \Sigma^*(v)$ then, by 3.2.20, there exists a unique $Sh(v) : F^*_q \longrightarrow \mathbf{C}^*$ such that

3.2.21 $$Sh(v)(N(x)) = v(x) \quad (x \in F^*_{q^n}).$$

3.2.22 We shall now use Explicit Brauer Induction and 3.2.14–3.2.15 to describe a correspondence of the type of 3.2.17 in the modest circumstances of $GL_2 F_{q^n}$. As it happens, our correspondence will coincide with that of 3.2.16, although no mention of the Shintani norm appears in our construction. My correspondence will be effected by applying Hilbert's Theorem 90, in the sense of 3.2.20–3.2.21, to the maximal one-dimensional characters which appear in the formula for $a_{GL_2 F_{q^n}}(v)$.

We begin by observing that, if v is irreducible and $\Sigma^*(v) = v$, then

$$a_{GL_2 F_{q^n}}(v) = a_{GL_2 F_{q^n}}(\Sigma^*(v)) = \Sigma^*(a_{GL_2 F_{q^n}}(v)),$$

where $\Sigma^* : R_+(GL_2 F_{q^n}) \longrightarrow R_+(GL_2 F_{q^n})$ is given by the formula

$$\Sigma^*(J, \phi)^{GL_2 F_{q^n}} = (\Sigma(J), \phi(\Sigma^{-1} \cdot -))^{GL_2 F_{q^n}}.$$

Since $\Sigma^*(v)$ is also irreducible the maximal terms of types A–D in $a_{GL_2 F_{q^n}}(\Sigma^*(v))$ will be obtained by applying Σ^* to the maximal terms of type A–D in $a_{GL_2 F_{q^n}}(v)$.

Now let us describe our construction of the correspondence, which will be denoted by Υ.

If $\Sigma^*(L(\chi)) = L(\chi)$ then $\Sigma^*(\chi) = \chi$ and there exists a unique $\overline{\chi} : F^*_q \longrightarrow \mathbf{C}^*$ such that $\chi(z) = \overline{\chi}(N(z))$ where N is the norm. In this case we set

3.2.23 $$\Upsilon(L(\chi)) = L(\overline{\chi}).$$

Next suppose that $\Sigma^*(R(\chi_1,\chi_2)) = R(\chi_1,\chi_2)$ then, in 3.2.14(ii),

$$\sum_{1\neq\mu}(H,\chi_1\chi_2\otimes\mu)^{GL_2F_{q^n}} = \sum_{1\neq\mu}(H,\Sigma^*(\chi_1\chi_2)\otimes\mu)^{GL_2F_{q^n}}$$

and

$$\begin{aligned}&(B,Inf_T^B(\chi_1\otimes\chi_2))^{GL_2F_{q^n}} + (B,Inf_T^B(\chi_2\otimes\chi_1))^{GL_2F_{q^n}}\\ &= (B,Inf_T^B(\Sigma^*(\chi_1)\otimes\Sigma^*(\chi_2)))^{GL_2F_{q^n}} +\\ &\qquad (B,Inf_T^B(\Sigma^*(\chi_2)\otimes\Sigma^*(\chi_1)))^{GL_2F_{q^n}}.\end{aligned}$$

By 3.2.11, these equations imply that either

(a) $\chi_1 = \Sigma^*(\chi_1)$ and $\chi_2 = \Sigma^*(\chi_2)$

or

(b) $\chi_1 = \Sigma^*(\chi_2)$, $\chi_2 = \Sigma^*(\chi_1)$ and $\chi_1\chi_2 = \Sigma^*(\chi_1\chi_2)$.

In case (a) there exist unique homomorphisms, $\overline{\chi}_i : F_q^* \longrightarrow \mathbf{C}^*$ $(i = 1,2)$ such that $\chi_i(z) = \overline{\chi}_i(N(z))$ for each $i = 1,2$. In this case we set

3.2.24 $$\Upsilon(R(\chi_1,\chi_2)) = R(\overline{\chi}_1,\overline{\chi}_2).$$

In case (b) we have a surjective homomorphism

$$\lambda : G(F_{q^n}/F_q) \cong Z/n \longrightarrow \{\pm 1\}$$

given by $\lambda(g) = (-1)^{i-1}$ if $g(\chi_1) = \chi_i$. Hence $n = 2d$, $Ker(\lambda) = G(F_{q^n}/F_{q^2})$ and each χ_i is fixed by $Ker(\lambda)$. Hence there exists a unique $\overline{\chi}_1 : F_{q^2}^* \longrightarrow \mathbf{C}^*$ such that $\chi_1(z) = \overline{\chi}_1(N(z))$, where $N : F_{q^n}^* \longrightarrow F_{q^2}^*$ is the norm. Also, if F generates $G(F_{q^2}/F_q)$ and $\overline{\chi}_2 = F^*(\overline{\chi}_1)$ then $\chi_2(z) = \overline{\chi}_2(N(z))$.

Notice also that, in case (b), $\Sigma^*(\chi_1\chi_2) = \chi_1\chi_2$ so that there exists a unique $\overline{\chi}_{1,2} : F_q^* \longrightarrow \mathbf{C}^*$ such that $\chi_1(z)\chi_2(z) = \overline{\chi}_{1,2}(N(z))$. In fact, we have

$$Res^{F_q^*}_{F_{q^2}^*}(\overline{\chi}_1) = \overline{\chi}_{1,2} = Res^{F_q^*}_{F_{q^2}^*}(\overline{\chi}_2).$$

For, if $w \in F_q^*$, $v \in F_{q^2}^*$ and $r \in F_{q^n}^*$ satisfy $N(v) = w$, $N(r) = v$ then

$$\begin{aligned}
\overline{\chi}_1(w) &= \overline{\chi}_1(vF(v)) \\
&= \overline{\chi}_1(v)\overline{\chi}_1(F(v)) \\
&= \overline{\chi}_1(v)\overline{\chi}_2(v) \\
&= \overline{\chi}_1(N(r))\overline{\chi}_2(N(r)) \\
&= \chi_1(r)\chi_2(r) \\
&= \overline{\chi}_{1,2}(N(r)) \\
&= \overline{\chi}_{1,2}(w).
\end{aligned}$$

From these characters it is natural to form

$$\begin{aligned}
&\textstyle\sum_{v \in F_q^*}(H, \overline{\chi}_{1,2} \otimes \Psi_{F_q}(v \cdot -))^{GL_2F_q} \\
&+\sum_{\rho \notin \{\overline{\chi}_1, \overline{\chi}_2\}, Res_{F_q^*}^{F_{q^2}^*}(\rho) = \overline{\chi}_{1,2}} (F_{q^2}^*, \rho)^{GL_2F_q} \\
&+ \cdots .
\end{aligned}$$

in $R_+(GL_2F_q)$.

These are the maximal terms of types A–D in $a_{GL_2F_q}(r(\overline{\chi}_1))$ and therefore, in case (b), we set

3.2.25 $$\Upsilon(R(\chi_1, \chi_2)) = r(\overline{\chi}_1) = r(\overline{\chi}_2).$$

If $\Sigma^*(S(\chi)) = S(\chi)$ then, as in case (a) above, we see that $\Sigma^*(\chi) = \chi$ and that there exists a unique $\overline{\chi} : F_q^* \longrightarrow \mathbf{C}^*$ such that $\chi(z) = \overline{\chi}(N(z))$. In this case we set

3.2.26 $$\Upsilon(S(\chi)) = S(\overline{\chi}).$$

Finally, suppose that $\Sigma^*(r(\Theta)) = r(\Theta)$. Hence, by 3.2.14(iv),

$$\begin{aligned}
&\textstyle\sum_{v \in F_{q^n}^*}(H, Res_{F_{q^n}^*}^{F_{q^{2n}}^*}(\Sigma^*(\Theta)) \otimes \Psi_{F_{q^n}}(v \cdot -))^{GL_2F_{q^n}} \\
&= \textstyle\sum_{v \in F_{q^n}^*}(H, Res_{F_{q^n}^*}^{F_{q^{2n}}^*}(\Theta) \otimes \Psi_{F_{q^n}}(v \cdot -))^{GL_2F_{q^n}}
\end{aligned}$$

and

$$\sum_{\rho\notin\{\Theta,F^*(\Theta)\},Res^{F^*_{q^{2n}}}_{F^*_{q^n}}(\rho)=Res^{F^*_{q^{2n}}}_{F^*_{q^n}}(\Theta)} (F^*_{q^{2n}},\Sigma^*(\rho))^{GL_2F_{q^n}}$$

$$=\sum_{\rho\notin\{\Theta,F^*(\Theta)\},Res^{F^*_{q^{2n}}}_{F^*_{q^n}}(\rho)=Res^{F^*_{q^{2n}}}_{F^*_{q^n}}(\Theta)} (F^*_{q^{2n}},\rho)^{GL_2F_{q^n}}.$$

The first equation implies that $Res^{F^*_{q^{2n}}}_{F^*_{q^n}}(\Sigma^*(\Theta)) = Res^{F^*_{q^{2n}}}_{F^*_{q^n}}(\Theta)$ and so there exists a unique $\tilde{\Theta} : F^*_q \longrightarrow \mathbf{C}^*$ such that, for all $z \in F^*_{q^n}$, $\Theta(z) = \tilde{\Theta}(N(z))$, where $N : F^*_{q^n} \longrightarrow F^*_q$ is the norm. However, $\Sigma^*(\Theta)$ and Θ must be distinct on $F^*_{q^{2n}}$, since $F \in G(F_{q^{2n}}/F_{q^n})$ acts non-trivially on Θ, by assumption. The second equation shows that Σ^* permutes the set

$$\{\rho \notin \{\Theta, F^*(\Theta)\}, Res^{F^*_{q^{2n}}}_{F^*_{q^n}}(\rho) = Res^{F^*_{q^{2n}}}_{F^*_{q^n}}(\Theta)\}$$

so that we must have $\Sigma^*(\Theta) = F^*(\Theta)$. Since $G(F_{q^{2n}}/F_q) \cong Z/2n$ and $F = \Sigma^n$ we have

3.2.27 $$(\Sigma^{n-1})^*(\Theta) = \Theta.$$

Since Θ is not Galois invariant $< \Sigma^{n-1} >$ must be a proper subgroup of $< \Sigma >$. However, $HCF(n-1, 2n) \in \{1, 2\}$ so that we must have $HCF(n-1, 2n) = 2$ and therefore *n must be odd* . This means that

3.2.28 $$Z/n \cong < \Sigma^{n-1} > = G(F_{q^{2n}}/F_{q^2})$$

and, by 3.2.27, there exists a unique $\overline{\Theta} : F^*_{q^2} \longrightarrow \mathbf{C}^*$ such that $\Theta(w) = \overline{\Theta}(N(w))$ for all $w \in F^*_{q^{2n}}$. If $z \in F^*_q$ and $s \in F^*_{q^n}$ satisfy $N(s) = z$ then

$$\begin{aligned} Res^{F^*_{q^2}}_{F^*_q}(\overline{\Theta})(z) &= Res^{F^*_{q^2}}_{F^*_q}(\overline{\Theta})(N_{F_{q^n}/F_q}(s)) \\ &= \overline{\Theta}(N_{F_{q^{2n}}/F_{q^2}}(s)) \\ &= Res^{F^*_{q^{2n}}}_{F^*_{q^n}}(\Theta)(s) \\ &= \tilde{\Theta}(N_{F_{q^n}/F_q}(s)) \\ &= \tilde{\Theta}(z), \end{aligned}$$

so that $Res_{F_q^*}^{F_{q^2}^*}(\overline{\Theta}) = \tilde{\Theta}$. From these characters it is natural to form

$$\begin{aligned} &\textstyle\sum_{v \in F_q^*} (H, \tilde{\Theta} \otimes \Psi_{F_q}(v \cdot -))^{GL_2F_q} \\ &+ \sum_{\rho \notin \{\overline{\Theta}, F^*(\overline{\Theta})\}, Res_{F_q^*}^{F_{q^2}^*}(\rho) = \tilde{\Theta}} (F_{q^2}^*, \rho)^{GL_2F_q} \\ &+ \cdots . \end{aligned}$$

These are the maximal terms of types A–D in $a_{GL_2F_q}(r(\overline{\Theta}))$ and therefore we set

3.2.29 $$\Upsilon(r(\Theta)) = r(\overline{\Theta}).$$

Each of these recipes is reversible and one easily sees that the process yields a one-one correspondence similar to that of 3.2.17. The discussion of 3.2.22 may be summarised as follows:

Theorem 3.2.30 *The yoga of* 3.2.22 *yields a one-one correspondence of the form*

3.2.31 $$\begin{cases} \{\text{irreducible representations, } v, \text{ of } GL_2F_{q^n} \text{ fixed under } \Sigma\} \\ \updownarrow \Upsilon \\ \{\text{irreducible representations,} \Upsilon(v), \text{ of } GL_2F_q\}. \end{cases}$$

In fact, Υ *coincides with the Shintani correspondence as described in Shintani* (1976, p. 410, section 4).

Proof The fact that Υ satisfies the characterisation of Sh which is given by 3.2.21 is easily verified by means of the table of character-values in 3.2.5. □

Remark 3.2.32 It would be very interesting to develop for GL_mF_q a yoga similar to that which is given in 3.2.22 for GL_2F_q. In such an enterprise one would have to determine suitable generalisations of types A–D of 3.2.6. In this example the types were arrived at by considering first the maximal *abelian* pairs and then, should they prove not to be self-normalising, their normalisers. In the case of GL_2F_q what we have given is merely a calculation and in general one would wish for a more intrinsic proof; preferably one which, in the presence of a suitable Explicit Brauer Induction technique, would extend to the case of GL_mF where F is a local field.

3.3 Exercises

3.3.1 Let F be any field. Suppose that

$$A = \begin{pmatrix} a & b \\ c & d \end{pmatrix} \in GL_2F.$$

Verify that

$$A = \begin{pmatrix} \alpha & \beta \\ 0 & \delta \end{pmatrix} w \begin{pmatrix} 1 & u \\ 0 & 1 \end{pmatrix}$$

where

$$\delta = -c, u = d/c, \beta = -a \ and \ \alpha = -det(A)/c.$$

3.3.2 Use 3.3.1 to prove 3.1.2.

3.3.3 Suppose that A is a finite, abelian group and that $\chi : A \longrightarrow \mathbf{C}^*$ is a non-trivial homomorphism. Show that

$$\sum_{a\in A} \chi(a) = 0.$$

3.3.4 Verify 3.1.5.

3.3.5 Verify directly from the formulae of 3.1.15 that they are compatible with those of 3.1.9 in the sense that

$$\left(r(\Theta)\begin{pmatrix} 1 & 0 \\ 0 & \delta \end{pmatrix} f\right)(x) = \Theta(\beta)f(xF(\beta)^{-1}),$$

where $N_{F_{q^2}/F_q}(\beta) = \delta$. (Use the identity

$$\hat{h}(x) = \hat{f}(xF(\beta)^{-1})$$

if $f \in \mathscr{H}; x, \beta \in F_{q^2}$ and $h(x) = f(\beta x)$.)

3.3.6 Establish the irreducibility of the Weil representation, $r(\Theta)$, by means of its character function and the Schur inner product on GL_2F_q.

3.3.7 For each partition $\mathscr{M} = (m_1, \ldots, m_t)$ of $m = \sum_i m_i$ let $U_{\mathscr{M}}$ denote the subgroup of $GL_m F_q$ consisting of upper triangular matrices which have the identity matrices, $I_{m_1}, \ldots, I_{m_1}$ block-wise down the diagonal. A representation

$$\rho : GL_m F_q \longrightarrow GL(V)$$

is called *cuspidal* if, for each partition,

$$< Res_{U_{\mathscr{M}}}^{GL_m F_q}(\rho), 1 >_{U_{\mathscr{M}}} = 0.$$

Prove that the Weil representation $r(\Theta)$, of 3.1.15 is cuspidal.

3.3.8 Complete the proof of Theorem 3.2.5.

3.3.9 Let $r(\Theta) : GL_2 F_q \longrightarrow U(q-1)$ denote a Weil representation, as in 3.1.15.

Show that the fixed-point set of $Res_B^{GL_2 F_q}(r(\Theta))$,

$$(U(q-1)/NT^{q-1})^B$$

consists of a single point. (See 2.1.10.)

3.3.10 (*Research problem*) Give a description of a correspondence of the Shintani type for $GL_m F_{q^n}$ using the maximal terms in the Explicit Brauer Induction formula together with Hilbert's Theorem 90, generalising the method of 3.2.22.

4

The class-group of a group-ring

Introduction

Section 1 shows how Adams operations, ψ^k, behave with respect to Explicit Brauer Induction. In particular it is proved that one may express $\psi^k(V)$ as an integral linear combination of monomial representations (i.e. induced from one-dimensional characters of subgroups) by applying ψ^k to each one-dimensional subhomomorphism in the Explicit Brauer Induction formula for V and then mapping the result to the representation ring, $R(G)$. This result holds for all the Explicit Brauer Induction formulae, since it depends mainly on the naturality property. The effect of this result is to give one a form of Brauer's theorem which 'commutes with Adams operations'. This result, which was first proved in Snaith (1989a) using the results of Snaith (1988b), is very convenient and rather unexpected and the remainder of this chapter consists of implications of this result.

In Section 2 we describe the adèlic Hom-description of Fröhlich, which gives the class-group of an integral group-ring of a finite group in terms of groups of Galois-equivariant functions from $R(G)$ to the idèles of a suitably large number field. Those who are familiar with algebraic K-theory will recognise the Hom-description as being equivalent to the exact K-theory sequence (at dimension zero) which was first obtained by C.T.C. Wall by applying algebraic K-theory to the canonical adèlic fibre square of group-rings. As an example, the Swan modules are introduced and their classes in the class-group are expressed in terms of the Hom-description and from this we prove the well-known result that for cyclic and dihedral groups the class of a Swan module is trivial.

Section 3 is devoted to a proof of a conjecture of Martin Taylor, which was posed in (Taylor, 1978). The Hom-description gives the class-group as a complicated quotient of the weak product of the groups of Galois-equivariant maps from $R(G)$ to all the local completions of a large number field. The quotient involves factoring out by two subgroups of which one is the group of *determinantal functions.* For the group functions into the units of an l-adic local field one may construct endomorphisms, denoted by $(F\psi^l/l)$, by means of the Adams operations. Taylor conjectured that for l-groups the result of this endomorphism on determinants would be congruent to one modulo l. Using the result of Section 1 it is easy to deduce these congruences for any group, G. From the basic congruences we proceed to deduce higher order congruences, for all G.

In Section 4 the higher order determinantal congruences are used to construct new maps out of the class-group of an integral group-ring. In the case of l-groups, M.J. Taylor (1978) had previously manufactured such maps by means of some rather weaker congruences and we give his determination of the Swan subgroup of an l-group. This calculation uses the determinantal congruence maps together with the fact that the Artin exponent of G annihilates the Swan subgroup, $T(G)$.

One of the three main steps in M.J. Taylor's proof of the tame Fröhlich conjecture (Fröhlich, 1983; Taylor, 1981) is the tame Galois descent for the determinantal functions. The original proof involved the difficult construction of the group-ring logarithm (independently discovered by R. Oliver). In Section 5 we give the simplified proof which is available once one has the determinantal congruences, for then the mysterious, non-commutative group-ring logarithm becomes merely $log(F\psi^l/l)$ (see 4.5.30), where log is the classical l-adic logarithm.

In Section 6 we prove that Adams operations preserve the subgroup of determinantal homomorphisms. Once again the determinantal congruences enable one to simplify the original proof, which is due to Ph. Cassou-Nogués and M.J. Taylor. Our proof is similar to the original in that we reduce to special types of groups, but it is significantly different in that nowhere do we mention the decomposition homomorphism, which is used in an essential manner in the proof which is to be found in Taylor (1984).

Section 7 consists of a collection of exercises which are centred about the class-group of a group-ring.

4.1 Adams operations and rationality

Definition 4.1.1 Let $\rho : G \longrightarrow GL(V)$ be a complex representation of a finite group with character, χ_ρ. Let $k \geq 0$ be an integer. The *Adams operation*, ψ^k, is defined by the character formula

4.1.2 $$\chi_{\psi^k(\rho)}(g) = \chi_\rho(g^k) \qquad \text{for all } g \in G.$$

Lemma 4.1.3 (i) *The Adams operation defines a natural ring homomorphism*

$$\psi^k : R(G) \longrightarrow R(G).$$

(ii) *For* ρ *as in* 4.1.1

$$\psi^k(\rho) = N_k(\lambda^1(\rho), \ldots, \lambda^{dim(\rho)}(\rho))$$

where N_k *is the kth* Newton polynomial *and* $\lambda^i(\rho)$ *is the ith exterior power* (*cf.* 1.1.7).

(iii) *If* $dim_{\mathbf{C}}(\rho) = 1$ *then* $\psi^k(\rho) = \rho^k$.

Proof Recall that the Newton polynomial is defined in the following manner. Let $t_1, \ldots, t_m$ denote indeterminates and let $\sigma_i(t_1, \ldots, t_m) = \sigma_i(\underline{t})$ be the ith elementary symmetric function, which is equal to the sum of all products of i-tuples of distinct indeterminates. In $\mathbf{Z}[t_1, \ldots, t_m]$ we have the identity

4.1.4 $$\sum_{i=1}^m t_i^k = N_k(\sigma_1(\underline{t}), \ldots, \sigma_k(\underline{t})).$$

In order to prove part (ii) it suffices to calculate the character of $N_k(\lambda^1(\rho), \ldots, \lambda^{dim(\rho)}(\rho))$ at $g \in G$. However, since $< g >$ is cyclic

$$Res^G_{<g>}(\rho) = L_1 \oplus \ldots \oplus L_n,$$

where the $\{L_i\}$ are one-dimensional and $n = dim_{\mathbf{C}}(\rho)$. Also,

$$\lambda^i(L_1 \oplus \ldots \oplus L_n) = \sigma_i(L_1, \ldots, L_n) \in R(< g >),$$

so that 4.1.4 implies, in $R(< g >)$, that

$$Res^G_{<g>}(N_k(\lambda^1(\rho), \ldots, \lambda^{dim(\rho)}(\rho))) = L_1^k \oplus \ldots \oplus L_n^k,$$

where $L_i^k = L_i \otimes \ldots \otimes L_i$ (k copies). Therefore the value of the character of $N_k(\lambda^1(\rho), \ldots, \lambda^{dim(\rho)}(\rho))$ at $g \in G$ is equal to

$$\begin{aligned} \sum_{i=1}^n \chi_{L_i^k}(g) &= \sum_{i=1}^n \chi_{L_i}(g)^k \\ &= \sum_{i=1}^n \chi_{L_i}(g^k) \\ &= \psi^k(g), \end{aligned}$$

which proves part (ii).

Once we know, from part (ii), that $\psi^k(\rho) \in R(G)$ then parts (i) and (iii) follow immediately by easy character calculations. □

Remark 4.1.5 (*cf.* 4.7.1) In general, $\psi^k(Ind_H^G(\rho))$ and $Ind_H^G(\psi^k(\rho))$ are not equal. For example, take $G \cong \mathbf{Z}/p$, $H = \{1\}$, $\rho = 1$ and $k = p$. In this case $Ind_{\{1\}}^{\mathbf{Z}/p}(\psi^p(1)) = Ind_{\{1\}}^{\mathbf{Z}/p}(1)$ is the regular representation. However, since $g^p = 1$ for all $g \in \mathbf{Z}/p$, $\psi^p(Ind_{\{1\}}^{\mathbf{Z}/p}(1)) = p \in R(\mathbf{Z}/p)$.

However, the proof of the following result, which was originally proved in Theorem 2.33 of Snaith (1989a), shows that ψ^k behaves well with respect to Explicit Brauer Induction.

Theorem 4.1.6 *Let $\rho : G \longrightarrow GL(V)$ be a representation of a finite group over any algebraically closed field, K, of characteristic zero. Then, for all $k \geq 0$, there exist integers, n_i, and one-dimensional representations of subgroups, $\phi_i : H_i \longrightarrow K^*$, such that*

$$\psi^k(\rho) = \sum_i n_i Ind_{H_i}^G(\phi_i^k) \in R_K(G),$$

where $R_K(G)$ is the ring of K-representations of G.

Proof In this situation $R_K(G) \cong R(G)$ and therefore we may assume that $K = \mathbf{C}$, the complex numbers. Define a homomorphism

$$\Psi^k : R_+(G) \longrightarrow R_+(G)$$

by $\Psi^k((H, \phi)^G) = (H, \phi^k)^G$. The homomorphism, Ψ^k, is a natural map with respect to the restriction to subgroups. Therefore the composition

$$R(G) \xrightarrow{a_G} R_+(G) \xrightarrow{\Psi^k} R_+(G) \xrightarrow{b_G} R(G)$$

is a natural homomorphism. Thus, in order to evaluate this composite, it suffices to evaluate the character of $b_G(\Psi^k(a_G(\rho)))$ at $g \in G$. By naturality this is the same as the character-value of

$$b_{<g>}(\Psi^k(a_{<g>}(Res_{<g>}^G(\rho))))$$

at g. Suppose that $Res_{<g>}^G(\rho) = \sum_{j=1}^t L_j$ where $dim_{\mathbf{C}}(L_j) = 1$, then the character-value is given by

$$\begin{aligned}\sum_{j=1}^t b_{<g>}(\Psi^k(a_{<g>}(L_j)))(g) &= \sum_{j=1}^t b_{<g>}(\Psi^k(< g >, L_j)^{<g>})(g) \\ &= \sum_{j=1}^t L_j^k(g) \\ &= \psi^k(Res_{<g>}^G(\rho))(g).\end{aligned}$$

Therefore, by naturality of ψ^k, $b_G(\Psi^k(a_G(\rho))) = \psi^k(\rho) \in R(G)$, which proves the result if $a_G(\rho) = \sum_i n_i(H_i, \phi_i)^G \in R_+(G)$. □

4.1.7 *Adams operations and rationality*

Let K be a subfield of the complex numbers. Denote by $R_K(G)$ the representation ring of G which is generated, in $R(G)$, by representations of the form

4.1.8
$$\rho : G \longrightarrow GL_n(K).$$

$R_K(G)$ is a subring of $R(G)$.

We say that a representation of G is K*-rational* if its class in $R(G)$ lies in $R_K(G)$. Let N be the least common multiple of the orders of the elements of G. Hence, if $\xi_N = exp(2\pi i/N)$, any one-dimensional representation, $\phi : H \longrightarrow \mathbf{C}^*$, is $\mathbf{Q}(\xi_N)$-rational and therefore so is $Ind_H^G(\phi)$.

Theorem 4.1.9 (i) *With the notation introduced above*

$$R_{\mathbf{Q}(\xi_N)}(G) = R(G).$$

(ii) *If* $n = HCF(k, N)$ *and* $m = N/n$ *then*

$$\psi^k(R(G)) \subset R_{\mathbf{Q}(\xi_m)}(G).$$

Proof Part (i) follows from part (ii) by setting $k = 1$. Part (ii) is a corollary of Theorem 4.1.6. For, if $\phi_i : H_i \longrightarrow \mathbf{C}^*$ is a one-dimensional character, then the image of ϕ^k lies in the roots of unity of $\mathbf{Q}(\xi_N^k)$. Therefore $\phi_i(g)^k$ is a power of ξ_m and therefore each term, $Ind_{H_i}^G(\phi_i^k)$ in Theorem 4.1.6, is $\mathbf{Q}(\xi_m)$-rational, which completes the proof. □

4.2 Describing the class-group by representations

4.2.1 *Adèles and idèles*

Let K be an *algebraic number field*; that is, a field extension of finite degree, $K/\mathbf{Q}$. Denote by $\mathcal{O}_K$ the ring of *algebraic integers* of K (Lang, 1970). By a *place of* K we will mean an embedding of K as a dense subfield into a *complete discrete valuation field*, $\hat{K}$. For example, $\hat{K}$ might be isomorphic to $\mathbf{R}$ or $\mathbf{C}$, the real or complex numbers, respectively. Embeddings into such fields are called *infinite or Archimedean places* of K. The *finite places* of K are comprised of all the non-Archimedean ones and they all come about by choosing a prime ideal, $P \lhd \mathcal{O}_K$, and completing K in the *P-adic topology*. This is the topology in which $\{x + P^n; n \geq 0\}$ is a base of neighbourhoods of the element, x, in $\mathcal{O}_K$. We denote this field by K_P.

Example 4.2.2 If p is a rational prime then the p-adic completion of $\mathbf{Z}$, the integers, is given by

$$\mathbf{Z}_p = \lim_{\overleftarrow{n}} \mathbf{Z}/(p^n)$$

and is called the ring of *p-adic integers*. Its field of fractions is given by $\mathbf{Q}_p = \mathbf{Z}_p[1/p]$.

4.2.3 The completed fields introduced in 4.2.1 are called *local fields*. If $P \lhd \mathcal{O}_K$ is a prime such that the integral ideal, $P \cap \mathbf{Z}$, equals $p^m\mathbf{Z}$ for a rational prime, p, then $K_P/\mathbf{Q}_p$ is a finite extension and we say that P *divides p* or that P *lies over p*. This is equivalent to the existence of a commutative square of the following form:

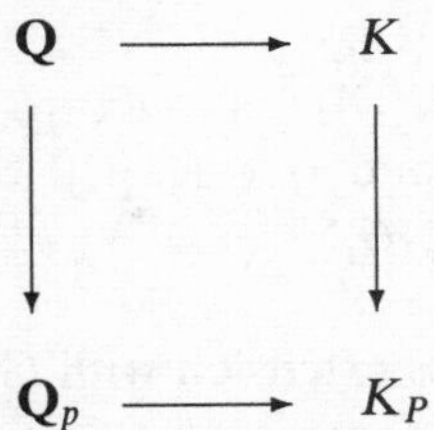

It is customary to refer to 'places' as 'primes' and vice versa. In which case, to say that one Archimedean prime divides another means that there is a corresponding commutative square

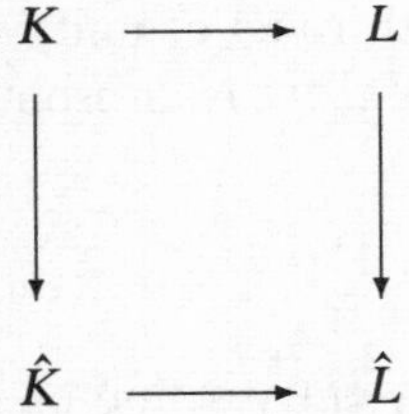

in which the vertical maps are infinite places and L/K is an extension of number fields. If P is a finite prime then the algebraic integers of K_P, $\mathcal{O}_{K_P}$, are given by the P-adic completion of $\mathcal{O}_K$. By convention, if $\hat{K}$ is Archimedean we set $\mathcal{O}_{\hat{K}}$ equal to $\hat{K}$.

The *adèle ring* of K is defined to be the ring given by the restricted product

4.2.4 $$J(K) = \prod'_{P\,prime} K_P.$$

In 4.2.4 $\prod'$ signifies that we take those elements of the topological ring, $\prod_P K_P$, for whom almost all entries lie in $\mathcal{O}_{K_P}$. The group of *idèles* is the group of units in $J(K)$,

4.2.5 $J^*(K) = \{(x_P) \in J(K) \mid x_P \neq 0$ and almost everywhere $x_P \in \mathcal{O}^*_{K_P}\}$,

where $\mathcal{O}^*_{K_P}$ denotes the multiplicative group of units in $\mathcal{O}_{K_P}$. The *unit idèles* is the subgroup

4.2.6 $$U(\mathcal{O}_K) = \prod_{P\,prime} \mathcal{O}^*_{K_P}.$$

Now let G be a finite group. We may extend the adèles and idèles to the group-rings, $\mathcal{O}_K[G]$ and $K[G]$. Define

4.2.7 $$\begin{cases} J(K[G]) & = \prod'_{P\,prime} K_P[G], \\ J^*(K[G]) & = \{(\alpha_P) \in J(K[G]) \mid \alpha_P \in \mathcal{O}_{K_P}[G]^* \text{ for} \\ & \quad \text{almost all } P \text{ and } \alpha_P \in K_P[G]^* \text{ otherwise}\}, \\ U(\mathcal{O}_K[G]) & = \prod_{P\,prime} \mathcal{O}_{K_P}[G]^*. \end{cases}$$

Now suppose that E/K is a finite Galois extension with Galois group $G(E/K)$. In this case $G(E/K)$ acts on the set of primes of E and hence acts upon the groups $J^*(E), U(\mathcal{O}_E), J^*(E[G])$ and $U(\mathcal{O}_E[G])$. If Q is a prime of E which divides the prime, P, of K then $G(E_Q/K_P)$ is a subgroup of $G(E/K)$ which is called a *decomposition group for* P and depends only on P, up to conjugation in $G(E/K)$.

If E is large enough to contain all #(G)th roots of unity we call E a *splitting field* for G. In this case, by 4.1.9(i), there is an isomorphism of the form

4.2.8 $$R_E(G) \cong R(G)$$

and $G(E/K)$ acts upon $R(G)$ by the entry-by-entry action on a representation

4.2.9 $$T : G \longrightarrow GL_n(E).$$

Therefore we may consider the group of $G(E/K)$-equivariant maps

4.2.10 $Hom_{G(E/K)}(R(G), J^*(E)) = \{f : R(G) \longrightarrow J^*(E) \mid f(g(z))$
$$= g(f(z)) \text{ for all } g \in G(E/K), z \in R(G)\}.$$

More generally, if L/K is a Galois extension which contains E/K then $G(L/K)$ acts on $R(G)$ and on $J^*(E)$, since E/K is Galois and $G(E/K)$ is a quotient of $G(L/K)$. Therefore we may pass to the *absolute Galois group*, Ω_K, which is the topological group defined by

4.2.11
$$\Omega_K = \operatorname{inv\,lim}_{\substack{K \subset L \subset K^c \\ L/K\,Galois}} G(L/K),$$

where K^c/K is a chosen (algebraic) algebraic closure of K. In this case we have

4.2.12
$$Hom_{\Omega_K}(R(G), J^*(E)) \cong Hom_{G(E/K)}(R(G), J^*(E)).$$

4.2.13 *$\mathcal{O}_K[G]$-modules and determinants*

A more extensive reference for the material of this section is Curtis & Reiner (1987, p. 334 *et seq.*).

Let M be an $\mathcal{O}_K[G]$-module of rank one which is *locally free*. This means that $M \otimes_{\mathcal{O}_K} \mathcal{O}_{K_P}$ is a free $\mathcal{O}_{K_P}[G]$-module on one generator, x_P, for each prime, P, of K and that $M \otimes_{\mathcal{O}_K} K$ is a free $K[G]$-module on one generator, x_0. Since $K[G]$ and $\mathcal{O}_{K_P}[G]$ are subrings of $K_P[G]$ this means that there is a unit, $\lambda_P \in K_P[G]^*$ which is defined by

4.2.14
$$x_0 \cdot \lambda_P = x_P \in M \otimes_{\mathcal{O}_K} K_P.$$

In fact, λ_P will almost always lie in $\mathcal{O}_{K_P}[G]^*$ so that we obtain an idèle

4.2.15
$$(\lambda_P) \in J^*(K[G]).$$

Now suppose that T is a representation, as in 4.2.9. We may apply T to each λ_P to obtain

4.2.16
$$T(\lambda_P) \in M_n(K_P \otimes_K E),$$

where $M_n(A)$ denotes the $n \times n$ matrices with entries in A. There is a ring isomorphism of the form

4.2.17
$$K_P \otimes_K E \cong \prod_{\substack{Q|P \\ Q\,prime\,of\,E}} E_Q.$$

Therefore we obtain an element

4.2.18 $$det(T(\lambda_P)) \in \prod_{\substack{Q|P \\ Q prime of E}} E_Q^*.$$

Since $(\lambda_P) \in J^*(K[G])$ we obtain an Ω_K-equivariant map, given by 4.2.18 at the primes of E which divide P,

4.2.19 $$Det((\lambda_P)) \in Hom_{\Omega_K}(R(G), J^*(E)).$$

Now let us consider the dependence of 4.2.19 upon the choices of x_0 and x_P in 4.2.14. If we replace x_P by another generator, x'_P, these choices will be related by an equation

4.2.20 $$x'_P = u_P x_P \text{ for some } u_P \in \mathcal{O}_{K_P}[G]^*$$

so that we obtain a unit idèle

4.2.21 $$u = (u_P) \in U(\mathcal{O}_K[G])$$

and 4.2.19 will be altered by multiplication by

4.2.22 $$Det(u) \in Det(U(\mathcal{O}_K[G])) \subset Hom_{\Omega_K}(R(G), J^*(E)).$$

Also there is a diagonal embedding of E^* into $J^*(E)$ which induces an inclusion

4.2.23 $$Hom_{\Omega_K}(R(G), E^*) \subset Hom_{\Omega_K}(R(G), J^*(E)).$$

By a similar argument, changing x_0 to x'_0 will change $Det((\lambda_P))$ by a function which lies in the subgroup of 4.2.23. Therefore we have associated to each locally free $\mathcal{O}_K[G]$-module of rank one, M, a well-defined element

4.2.24 $$Det[M] \in \frac{Hom_{\Omega_K}(R(G), J^*(E))}{Hom_{\Omega_K}(R(G), E^*) \cdot Det(U(\mathcal{O}_K[G]))}.$$

Now let us recall the definition of the class-group, $\mathcal{CL}(\mathcal{O}_K[G])$. This is defined to be the *Grothendieck group* of finitely generated, locally free $\mathcal{O}_K[G]$-modules. To be precise, consider the set of isomorphism classes of finitely generated, locally free $\mathcal{O}_K[G]$-modules, $Mod.l.f.(\mathcal{O}_K[G])$. This set is a monoid if we endow it with an addition operation defined by

4.2.25 $$[M] + [N] = [M \oplus N],$$

where $[-]$ denotes an isomorphism class. Define an equivalence relation on $Mod.l.f.(\mathcal{O}_K[G])$ which is generated by two types of relations

4.2.26

(i) $[M] \sim [N]$ if for some m, n

$$M \oplus (\mathcal{O}_K[G])^m \cong N \oplus (\mathcal{O}_K[G])^n$$

and

(ii) $[A] + [C] \sim [B]$ if there exists an exact sequence of $\mathcal{O}_K[G]$-modules

$$0 \longrightarrow A \longrightarrow B \longrightarrow C \longrightarrow 0.$$

With this notation

4.2.27 $$\mathcal{CL}(\mathcal{O}_K[G]) = \{Mod.l.f.(\mathcal{O}_K[G])\}/\sim.$$

It is known that $\mathcal{CL}(\mathcal{O}_K[G])$ is generated by locally free $\mathcal{O}_K[G]$-modules of rank one; the connection with 4.2.24 is the following Hom-description, due to Fröhlich.

Theorem 4.2.28 (*Curtis & Reiner*, 1987, p. 334; *Fröhlich*, 1983) *With the notation introduced above there is an isomorphism*

$$Det : \mathcal{CL}(\mathcal{O}_K[G]) \xrightarrow{\cong} \frac{Hom_{\Omega_K}(R(G), J^*(E))}{Hom_{\Omega_K}(R(G), E^*) \cdot Det(U(\mathcal{O}_K[G]))},$$

which sends a locally free module of rank one, M, to $Det[M]$ *of* 4.2.24.

We will not prove this theorem and for our purposes the reader may take the group of 4.2.24 as the definition of the class-group.

4.2.29 *The kernel group,* $D(\mathcal{O}_K[G])$

An $\mathcal{O}_K$-*order*, Λ in $K[G]$, is a subring containing $\mathcal{O}_K$, which is a finitely generated, projective $\mathcal{O}_K$-module such that $K \otimes_{\mathcal{O}_K} \Lambda = K[G]$. Suppose that Λ is a *maximal* $\mathcal{O}_K$-*order* of $K[G]$, then we may define the *kernel group*

4.2.30 $$D(\mathcal{O}_K[G]) = ker(\mathcal{CL}(\mathcal{O}_K[G]) \longrightarrow \mathcal{CL}(\Lambda)),$$

where $\mathcal{CL}(\Lambda)$ is the class-group of Λ, defined by a Grothendieck group construction analogous to that of 4.2.27. The group, $D(\mathcal{O}_K[G])$, defined in this manner, is independent of the choice of Λ. There is also a Hom-description of this group.

To describe this we need to introduce a subgroup

4.2.31 $$Hom^+_{\Omega_K}(R(G), \mathcal{O}_E^*) \subset Hom_{\Omega_K}(R(G), \mathcal{O}_E^*),$$

An irreducible representation, $T : G \longrightarrow GL_n(E)$, as in 4.2.19 gives rise to a complex representation, $T : G \longrightarrow GL_n(\mathbf{C})$, by choosing an embedding of E into $\mathbf{C}$. Let $\mathbf{H}$ denote the skewfield given by the quaternions. A (left) $\mathbf{H}$-vector space, V, of dimension m over $\mathbf{H}$, may be considered as a $2m$-dimensional complex vector space. If $GL_m(\mathbf{H})$ is the group of invertible $\mathbf{H}$-linear maps from V to itself then we obtain a map

4.2.32
$$c : GL_m(\mathbf{H}) \longrightarrow GL_{2m}(\mathbf{C}).$$

The representation, T, is called *quaternionic* or *symplectic* if $T : G \longrightarrow GL_{2m}(\mathbf{C})$ factors through the map, c, of 4.2.32. On the other hand, if T is symplectic then, by 4.7.4, complex conjugation fixes T in $R(G)$ so that if $f \in Hom_{\Omega_K}(R(G), E^*)$ and if K is a subfield of the real numbers, $\mathbf{R}$, then $f(T)$ lies in $\mathbf{R}$ for every Archimedean prime of E which divides $K \subset \mathbf{R}$. Therefore it makes sense to define $Hom^+_{\Omega_K}(R(G), E^*)$ to be

4.2.33
$$\{f \in Hom_{\Omega_K}(R(G), E^*) \mid f(T) \text{ is positive, } T \text{ symplectic}\},$$

where, in 4.2.33, *positive* means that $f(T)$ is positive under every Archimedean place of E which lies over a real place of K.

Similarly we may define

4.2.34
$$Hom^+_{\Omega_K}(R(G), \mathcal{O}^*_E) = \{f \in Hom^+_{\Omega_K}(R(G), E^*) \mid im(f) \subset \mathcal{O}^*_E\}$$

Theorem 4.2.35 (*Curtis & Reiner,* 1987, p. 334 *et seq.; Fröhlich,* 1983) *The isomorphism of* 4.2.28 *induces an isomorphism*

$$Det : D(\mathcal{O}_K[G]) \xrightarrow{\cong} \frac{Hom_{\Omega_K}(R(G), U(\mathcal{O}_E))}{Hom^+_{\Omega_K}(R(G), \mathcal{O}^*_E) \cdot Det(U(\mathcal{O}_K[G]))}.$$

Remark 4.2.36 By 4.7.3, $\mathcal{O}_{E_Q}[G]$ is a maximal $\mathcal{O}_{E_Q}$-order in $E_Q[G]$ for all primes, Q, which do not divide the order of G. Hence one may rewrite Theorem 4.2.35 as an isomorphism of the following form:

4.2.37
$$Det : D(\mathcal{O}_K[G]) \xrightarrow{\cong} \frac{Hom_{\Omega_K}(R(G), \prod_{Q|\#(G)} \mathcal{O}^*_{E_Q})}{Hom^+_{\Omega_K}(R(G), \mathcal{O}^*_E) \cdot Det(\prod_{P|\#(G)} \mathcal{O}_{K_P}[G]^*)}.$$

4.2.38 Next we will work towards the construction of new maps out of $D(\mathbf{Z}[G])$ by means of 4.2.37. In preparation, let us consider the action of

Ω_K and $G(E/K)$ on $U(\mathcal{O}_E) = \prod_{Q prime} \mathcal{O}^*_{E_Q}$. The action is induced by the action of $g \in G(E/K)$ on $K_P \otimes_K E$ by the formula

4.2.39 $$g(a \otimes b) = a \otimes g(b) \qquad (a \in K_P, b \in E).$$

Via the isomorphism of 4.2.17 the action of $G(E/K)$ permutes the primes, Q, of E which divide P and $G(E_Q/K_P)$ is the subgroup of $G(E/K)$ which preserves the factor, E_Q, in 4.2.17. From this discussion we see that, if $f \in Hom_{G(E/K)}(R(G), \prod_{Q|P} E^*_Q)$, then taking the $G(E_Q/K_P)$-map given by the E_Q-component gives an isomorphism

4.2.40 $$Hom_{G(E/K)}(R(G), \prod_{Q|P} E^*_Q) \xrightarrow{\cong} Hom_{\Omega_{K_P}}(R(G), E^*_Q).$$

On the level of idèles we have an isomorphism of rings (Curtis & Reiner, 1987, p. 331)

4.2.41 $$\mathcal{O}_{K_P} \otimes_{\mathcal{O}_K} \mathcal{O}_E \cong \prod_{Q|P} \mathcal{O}_{E_Q}$$

and a corresponding isomorphism

4.2.42 $$Hom_{\Omega_K}(R(G), \prod_{Q|P} \mathcal{O}^*_{E_Q}) \xrightarrow{\cong} Hom_{\Omega_{K_P}}(R(G), \mathcal{O}^*_{E_Q}).$$

We will make use of 4.2.42 in the next section.

Example 4.2.43 *Swan modules*

Let G be a finite group of order n. Let k be any integer which is coprime to n and denote by $S(k)$ the $\mathbf{Z}[G]$-module which is given by the ideal

4.2.44 $$S(k) = ideal\{k, \sigma\} \lhd \mathbf{Z}[G],$$

where $\sigma = \sum_{g \in G} g$. The $\mathbf{Z}[G]$-module of 4.2.44 will be called a *Swan module.* Clearly $S(k)$ is free when $k = \pm 1$. In general, $S(k)$ is a locally free (and hence projective) $\mathbf{Z}[G]$-module which is often non-trivial in $\mathcal{CL}(\mathbf{Z}[G])$.

To see that $S(k)$ is locally free we give another description of $S(k)$. Let

4.2.45 $$x = (1 - (\sigma/n)) + k \cdot (\sigma/n) \in \mathbf{Q}[G]^*.$$

In 4.2.45 x is a central unit of $\mathbf{Q}[G]^*$, since σ is central and

$$\begin{aligned}&(1 + \beta\sigma)(1 - (\sigma/n) + k(\sigma/n))\\ &= 1 + \sigma(-1/n + k/n + \beta k)\\ &= 1, \text{if } \beta k = (1 - k)/n.\end{aligned}$$

Therefore, within $\mathbf{Q}[G]$, we have the following isomorphisms of $\mathbf{Z}[G]$-modules:

$$\begin{aligned}\{k,\sigma\} &\cong (x/k)\cdot\{k,\sigma\}\\ &\cong \{x,(x\sigma)/k\}\\ &\cong \{x,\sigma\}\end{aligned}$$

since $(x\sigma)/k = \sigma^2/n = \sigma$.

Within $\mathbf{Q}_p[G]$ the $\mathbf{Z}[G]$-module, $\{x,\sigma\}$, is equal to $\mathbf{Z}_p[G] < 1 >$ if $p \not| \#(G)$, since in that case $x \in \mathbf{Z}_p[G]^*$. However, if $p | \#(G)$ then $\{x,\sigma\}$ is equal to $\mathbf{Z}_p[G] < x >$, since $(x\sigma)/k = \sigma^2/n = \sigma$. Similarly, at Archimedean primes $\{x,\sigma\}$ is free with generator given by 1. Therefore $S(k)$ is locally free of rank one and the idèle associated to such a module in 4.2.15 is equal, in $J^*(E)$, to

4.2.46 $$\lambda_Q = \begin{cases} 1 & \text{at Archimedean primes or at } Q \not| \#(G), \\ x & \text{at finite primes } Q \mid \#(G). \end{cases}$$

Now let $T : G \longrightarrow GL_n(E)$ be an irreducible representation. If $T \neq 1$ and $Q \mid \#(G)$ then

$$det(T(\lambda_Q)) = det(I_n + ((k-1)/n)\sum_{g\in G} T(g)).$$

However, by Schur's lemma (1.2.5), $\sum_{g\in G} T(g) = \mu \cdot I_n$ and

$$\begin{aligned} n\mu &= Trace(\textstyle\sum_{g\in G} T(g))\\ &= \textstyle\sum_{g\in G} Trace(T(g))\\ &= \#(G)\cdot < T, 1 >\\ &= 0,\end{aligned}$$

so that $det(T(\lambda_Q)) = 1$ if $T \neq 1$. If $T = 1$ then $det(T(\lambda_Q)) = k$ so that we have proved the following result:

Proposition 4.2.47 *In terms of the Hom-description of* 4.2.28, *$S(k)$ is represented by the element of $Hom_{\Omega_{\mathbf{Q}}}(R(G), J^*(E))$ given by*

$$S(k)(T) = \begin{cases} 1 & \text{if } T \text{ is irreducible, } T \neq 1, \text{ at any prime,} \\ k & \text{if } T = 1, \text{ at finite primes, } Q \mid \#(G), \\ 1 & \text{if } T = 1 \text{ otherwise.} \end{cases}$$

In particular, by 4.2.35, *$S(k) \in D(\mathbf{Z}[G])$.*

Proposition 4.2.48 *Let $G = C_n, D_{2n}$ be the cyclic group of order n or the* *dihedral group of order* $2n$*, respectively. If* $HCF(k, \#(G)) = 1$ *then* $S(k)$ *is trivial in* $\mathscr{CL}(\mathbf{Z}[G])$.

Proof We will prove this by means of the Hom-description of $D(\mathbf{Z}[G])$ of 4.2.35 in the form 4.2.37.

Consider first the case when $G = C_n$. By 4.7.9, $1+x+\ldots+x^{k-1} \in \mathbf{Z}_p[C_n]^*$ for each prime, p, dividing n. Let $\xi_n = exp(2\pi i/n)$ and let $y : C_n \longrightarrow \mathbf{C}^*$ be given by $y(x) = \xi_n$. Hence

$$Det(1 + x + \ldots + x^{k-1})(y^i) = \begin{cases} (1 - \xi_n^{ki})/(1 - \xi_n^i) & \text{if } y^i \neq 1 \in R(C_n), \\ k & \text{if } y^i = 1. \end{cases}$$

Therefore the function, f, given by

$$f(y^i) = \begin{cases} (1 - \xi_n^{ki})/(1 - \xi_n^i) & \text{if } y^i \neq 1, \\ 1 & \text{if } y^i = 1. \end{cases}$$

lies in $Hom_{\Omega_{\mathbf{Q}}}(R(C_n), \mathbf{Z}[\xi_n]^*)$. However, this group equals

$$Hom^+_{\Omega_{\mathbf{Q}}}(R(C_n), \mathbf{Z}[\xi_n]^*)$$

since a cyclic group has no irreducible symplectic characters. Hence, by 4.2.47, $S(k) = Det(1 + x + \ldots + x^{k-1}) \cdot f^{-1}$ at all primes dividing n and is therefore in the indeterminacy of 4.2.37.

Next we consider the case when $G = D_{2n}$. In this case, also, $Hom^+_{\Omega_{\mathbf{Q}}}$ is equal to $Hom_{\Omega_{\mathbf{Q}}}$.

Let $u_k \in \mathbf{Z}_p[D_{2n}]^*$ denote the unit of 4.7.10, when p is a prime which divides $2n$. If $\chi : D_{2n} \longrightarrow \mathbf{C}^*$ is a one-dimensional representation then $\chi(D_{2n}) \subset \{\pm 1\}$ and an easy calculation yields

$$Det(u_k)(\chi) = \begin{cases} 1 & \text{if } \chi \neq 1, \\ 2k + 1 & \text{if } \chi = 1. \end{cases}$$

Now let $\rho_i = Ind_{C_n}^{D_{2n}}(y^i)$ with $y^i \neq 1$. If $\eta = \xi_n^i$, one finds that

$$\begin{aligned} &Det(u_k)(\rho_i) \\ &= det \begin{pmatrix} 1 + \eta + \ldots + \eta^k & 1 + \eta + \ldots + \eta^{k-1} \\ 1 + \eta^{-1} + \ldots + \eta^{-(k-1)} & 1 + \eta^{-1} + \ldots + \eta^{-k} \end{pmatrix} \\ &= [(\eta^{k+1} - 1)(\eta^{-(k+1)} - 1) - (\eta^k - 1)(\eta^{-k} - 1)](\eta - 1)^{-1}(\eta^{-1} - 1)^{-1} \\ &= (\eta^{2k} - 1)(\eta^{-(k+1)} - \eta^{-(k-1)})(\eta - 1)^{-1}(\eta^{-1} - 1)^{-1} \in \mathbf{Z}[\xi_n]^*. \end{aligned}$$

Define $h \in Hom^{+}_{\Omega_{\mathbf{Q}}}(R(D_{2n}), \mathbf{Z}[\xi_n]^*)$ by

$$h(T) = \begin{cases} 1 & \text{if } dim_{\mathbf{C}}(T) = 1, \\ Det(u_k)(\rho_i) & \text{if } T = \rho_i,\ y^i \neq 1, \end{cases}$$

then $S(k) = Det(u_k) \cdot h^{-1}$ at all primes dividing $2n$ so that it is in the indeterminacy of 4.2.37. □

4.3 Determinantal congruences

Throughout this section let l be a prime and let M be an *unramified extension* of the l-adic field, $\mathbf{Q}_l$. Let $N/\mathbf{Q}_l$ be a large, finite Galois extension such that N contains M and all the nth roots of unity. Let G be a finite group of order n. We will consider the group of Ω_M-equivariant functions

4.3.1 $$Hom_{\Omega_M}(R(G), \mathcal{O}_N^*)$$

where $R(G)$ is identified with $R_N(G)$ and is therefore generated by representations of the form

4.3.2 $$T : G \longrightarrow GL_u(N).$$

We have a determinantal homomorphism

4.3.3 $$Det : \mathcal{O}_M[G]^* \longrightarrow Hom_{\Omega_M}(R(G), \mathcal{O}_N^*)$$

given by the formula of 4.2.18

4.3.4 $$Det(\textstyle\sum_\gamma \lambda_\gamma \gamma)(T) = det(\sum_\gamma \lambda_\gamma T(\gamma)) \in \mathcal{O}_N^*.$$

Since $M/\mathbf{Q}_l$ is unramified, l is a prime in $\mathcal{O}_M$ and the *residue field* of M is given by

4.3.5 $$m = \mathcal{O}_M/l.$$

The residue field of $\mathbf{Q}_l$ is F_l, the field of order l. In addition, there is an isomorphism of Galois groups

4.3.6 $$G(M/\mathbf{Q}_l) \cong G(m/F_l).$$

The right-hand group of 4.3.6 is generated by the *Frobenius map*, F, which is defined by

4.3.7 $$F(z) = z^l \qquad (z \in m),$$

so that we may lift F uniquely to $F \in G(M/\mathbf{Q}_l)$, which satisfies

4.3.8 $$F(w) \equiv w^l \qquad (mod\ l \cdot \mathcal{O}_M)$$

for all $w \in \mathcal{O}_M$.

If $\sum_\gamma \lambda_\gamma \gamma \in \mathcal{O}_M[G]$ we may therefore define

4.3.9 $$F(\textstyle\sum_\gamma \lambda_\gamma \gamma) = \sum_\gamma F(\lambda_\gamma)\gamma \in \mathcal{O}_M[G].$$

We are now in a position to state our first result concerning determinantal congruences. The following result proves a conjecture of M.J. Taylor (1978) which is also mentioned in Fröhlich (1983, p. 79).

Theorem 4.3.10 *Let $z \in \mathcal{O}_M[G]^*$. Then, for all $T \in R(G)$,*

$$Det(F(z))(\psi^l(T))/(Det(z)(T))^l \in 1 + l\mathcal{O}_N.$$

Here ψ^l denotes the Adams operation of 4.1.1.

Proof By Theorem 4.1.5 there exist integers, $\{n_i\}$, and one-dimensional representations, $\{\phi_i : H_i \longrightarrow N^*\}$, such that

4.3.11 $$\begin{cases} T = \sum_i n_i Ind_{H_i}^G(\phi_i) \in R(G) \quad \text{and} \\ \psi^l(T) = \sum_i n_i Ind_{H_i}^G(\phi_i^l) \in R(G). \end{cases}$$

By multiplicativity

4.3.12 $$\frac{Det(F(z))(\psi^l(T))}{(Det(z)(T))^l} = \prod_i \frac{Det(F(z))(Ind_{H_i}^G(\phi_i^l))^{n_i}}{(Det(z)(Ind_{H_i}^G(\phi_i)))^{ln_i}} \in \mathcal{O}_N^*$$

so that we are reduced to the modulo l comparison of the expressions

$$Det\left(\sum_\gamma \lambda_\gamma \gamma\right)(Ind_{H_i}^G(\phi_i))^l \text{ and } Det\left(\sum_\gamma F(\lambda_\gamma)\ \gamma\right)(Ind_{H_i}^G(\phi_i^l))$$

in $\mathcal{O}_N^*$, where $z = \sum_\gamma \lambda_\gamma \gamma) \in \mathcal{O}_M[G]^*$.

Let us abbreviate (H_i, ϕ_i) to (H, ϕ).

Choose coset representatives, $x_1, \ldots, x_d \in G$, for G/H. There is a homomorphism, $\sigma : G \longrightarrow \Sigma_d$, such that for $g \in G$

$$gx_i = x_{\sigma(g)(i)} h(i, g), \qquad (h(i,g) \in H).$$

With this notation we find that

$$\begin{aligned}
&Det(\textstyle\sum_\gamma \lambda_\gamma \gamma)(Ind_H^G(\phi)) \\
&= det(\textstyle\sum_\gamma \lambda_\gamma \sigma(\gamma) \cdot diag[\phi(h(1,\gamma)), \ldots, \phi(h(d,\gamma))]) \\
&= det(X), say,
\end{aligned}$$

where $diag[u_1, \ldots, u_d]$ is the diagonal matrix whose (i,i)th entry is equal to u_i. Also we have

$$\begin{aligned}
&Det(\textstyle\sum_\gamma F(\lambda_\gamma)\gamma)(Ind_H^G(\phi^l)) \\
&= det(\textstyle\sum_\gamma F(\lambda_\gamma)\sigma(\gamma) \cdot diag[\phi^l(h(1,\gamma)), \ldots, \phi^l(h(d,\gamma))]) \\
&= det(Y), say.
\end{aligned}$$

If $X_{i,j}$ is the (i,j)th entry of X then

$$det(X) = \sum_{\beta \in \Sigma_d} sign(\beta) X_{1,\beta(1)} \ldots X_{d,\beta(d)} \in \mathcal{O}_N^*$$

and, by 4.3.8 and the binomial theorem,

$$\begin{aligned}
det(X)^l &\equiv \textstyle\sum_{\beta \in \Sigma_d} sign(\beta) X_{1,\beta(1)}^l \ldots X_{d,\beta(d)}^l && (mod\ l\mathcal{O}_N) \\
&\equiv \textstyle\sum_{\beta \in \Sigma_d} sign(\beta) Y_{1,\beta(1)} \ldots Y_{d,\beta(d)} && (mod\ l\mathcal{O}_N) \\
&\equiv det(Y) && (mod\ l\mathcal{O}_N).
\end{aligned}$$

Since $det(X)^l$ and $det(Y)$ both lie in $\mathcal{O}_N^*$ we see that $det(Y)det(X)^{-l} \in 1 + \mathcal{O}_N$, as required. □

When $l = 2$ it is very convenient to work modulo four. For this reason I will record the following determinantal congruence, which is obtained by squaring the result of Theorem 4.3.10.

Corollary 4.3.13 *When $l = 2$ in 4.3.10, if $z \in \mathcal{O}_M[G]^*$, then*

$$\frac{Det(F(z^2))(\psi^2(T))}{Det(z^2)(T)^2} \in 1 + 4 \cdot \mathcal{O}_N.$$

4.3.14 *log and exp*

The congruences of 4.3.10 and 4.3.13 will permit us to take the l-adic logarithms of

$$\frac{Det(F(z))(\psi^l(T))}{(Det(z)(T))^l}$$

in 4.3.10. By analysing this construction we shall naturally be led to the group-ring logarithm of Curtis & Reiner (1987, p. 359) which was originally discovered, independently, by R. Oliver and M.J. Taylor. However, our route to the logarithm will remove some of the agony and all of the mystery.

We will begin by recalling some well-known facts from Curtis & Reiner (1987, p. 356).

Let $N/\mathbf{Q}_l$ be a finite extension of local fields. Let $\pi \in \mathcal{O}_N$ be a *uniformising element*, so that $< \pi > \lhd \mathcal{O}_N$ is the maximal ideal. Let $v : N \longrightarrow \mathbf{Q}$ be the *normalised exponential valuation* which counts the π-divisibility of elements of N. This function is normalised so that $v(l) = 1$. The *ramification index* of $N/\mathbf{Q}_l$ is the integer, e, such that

$$< \pi^e >= l \cdot \mathcal{O}_N \lhd \mathcal{O}_N.$$

Hence $v(\pi) = e^{-1}$ is the least positive value taken by v on N. Define two formal series by

4.3.15
$$\begin{cases} log(1+x) = \sum_{n=1}^{\infty}(-1)^{n-1}x^n/n, \\ \\ exp(x) = \sum_{n=0}^{\infty} x^n/(n!). \end{cases}$$

Proposition 4.3.16 (*Curtis & Reiner*, 1987, *section* 54.2) (i) *The series, $log(1+x)$, l-adically converges to an element of N for all $x \in < \pi > \lhd \mathcal{O}_N$. If $v(x) \geq 1$ then $log(1+x) \in l\mathcal{O}_N$.*

(ii) *The series, $exp(x)$, converges to an element of* $1+ < \pi > \subseteq \mathcal{O}_N^*$ *whenever $v(x) > (l-1)^{-1}$. If $l \neq 2$ and $v(x) \geq 1$ then $exp(x) \in 1 + l\mathcal{O}_N$.*

(iii) *If $l \neq 2$ and $v(x) \geq 1$ then*

$$exp(log(1+x)) = 1 + x, \qquad log(exp(x)) = x.$$

(iv) *If $l = 2$ and $v(x) \geq 2$ then*

$$exp(log(1+x)) = 1 + x, \qquad log(exp(x)) = x.$$

Corollary 4.3.17 (i) *If $\omega \in \mathcal{O}_N$ is an l-primary root of unity then $\omega - 1 \in < \pi >$ and $log(\omega) = log(1+(\omega-1))$ converges. In fact, $log(\omega) = 0$.*

(ii) *Let $x \in \mathcal{O}_N$. Suppose that $\omega \in \mathcal{O}_N$ is an l-primary root of unity such that, for some $k \geq 0$*

$$x \equiv \omega \qquad (mod\ l^{k+1}\mathcal{O}_N) \ and$$

$$x \equiv 1\ (mod\ l\mathcal{O}_N),$$

then we may choose

$$\omega = \begin{cases} 1 & if\ l \neq 2, \\ \pm 1 & if\ l = 2. \end{cases}$$

Proof Part (i) is well-known (for example, Curtis & Reiner, 1987, p. 357). For part (ii) assume first that $l \neq 2$. In this case there is nothing to prove if $k = 0$. Therefore we shall suppose that $k \geq 1$. Let us recall some facts from Curtis & Reiner (1987, pp. 356–7) concerning the valuation, v.

Let $x \in \mathcal{O}_N$ and let $n \geq 1$ be an integer then, if $v(x) \geq 1$,

$$v(x^n/n) \geq 1 \text{ and } v(x^n/(n!)) > n(v(x) - 1/(l-1)).$$

Suppose that $x/\omega = 1 + l^{k+1}u$ then, by part (i), $log(x)$ is convergent and

$$\begin{aligned} log(x) &= log(x) - log(\omega) \\ &= log(x/\omega) \\ &= \textstyle\sum_{n=1}^{\infty}(-1)^{n-1}(l^{k+1}u)^n/n, \end{aligned}$$

which lies in $l^{k+1}\mathcal{O}_N$, by 4.3.16. When $l \neq 2$ then $exp(log(x)) \in 1+l^{k+1}\mathcal{O}_N$, using 4.3.16 again, and therefore

$$x = exp(log(x)) \equiv 1 \pmod{l^{k+1}\mathcal{O}_N},$$

as required.

When $l = 2$ we may write

$$(x/\omega)^2 = 1 + 2^{k+2}u$$

from which we find that $log(x^2) \in 2^{k+2}\mathcal{O}_N$ and $exp(log(x^2)) \in 1+2^{k+2}\mathcal{O}_N$, by 4.3.16. Hence

$$x^2 \equiv 1 \pmod{2^{k+2}\mathcal{O}_N}$$

and therefore

$$x \equiv \pm 1 \pmod{2^{k+1}\mathcal{O}_N},$$

since such elements, x^2, have precisely two square roots, which differ by a sign and lie in $1 + 2^{k+1}\mathcal{O}_N$.

4.3.18 In Theorem 4.3.10 we produced a mod l determinantal congruence which holds for all units, $z \in \mathcal{O}_M[G]^*$, and for all representations, T. Theorem 4.3.10 proved the conjecture of M.J. Taylor which was posed in Fröhlich (1983, p. 76) and Taylor (1978). However, when T is congruent to zero mutually higher powers of l, in the sense of Definition 4.3.31, we can improve the determinantal congruence of 4.3.10 considerably. We shall work towards such an improvement and in the course of this process we will recover the group-ring logarithm, which was originally discovered by M.J. Taylor (1980) and R. Oliver (1980), independently.

By Theorem 4.3.10 and 4.3.16(i) we may form the following composite homomorphism:

4.3.19
$$\begin{cases} log(F\psi^l/l(Det)) : \mathcal{O}_M[G]^* \longrightarrow Hom_{\Omega_M}(R(G), 1 + l\mathcal{O}_N) \\ \qquad\qquad\qquad\qquad \longrightarrow Hom_{\Omega_M}(R(G), l\mathcal{O}_N) \end{cases}$$

given by ($z \in \mathcal{O}_M[G]^*, T \in R(G)$)

$$log(F\psi^l/l(Det(z)))(T) = log\left(\frac{DetF(z)(\psi^l(T))}{Det(z)(T)^l}\right) \in l\mathcal{O}_N.$$

We would like to rewrite the logarithm in 4.3.19 as a difference of logarithms which both lie in $l\mathcal{O}_N$. The obvious expansion of the logarithm of a quotient as a difference of the logarithms of the numerator and denominator cannot be applied to 4.3.19 for *all* $z \in \mathcal{O}_M[G]^*$. Therefore we will study, pro tem, the case in which z is congruent to one mutually the *Jacobson radical*, J.

The Jacobson radical, $J \lhd \mathcal{O}_M[G]$ (Lang, 1984, p. 636), is the left ideal which is equal to the intersection of all the maximal left ideals of $\mathcal{O}_M[G]$. In fact, J, is a two-sided ideal and $\mathcal{O}_M[G]/J$ is semi-simple. Hence, by Wedderburn's theorem (Lang, 1984, p. 629) $\mathcal{O}_M[G]/J$ is a product of matrix rings over *division rings*. However, in this finitely generated, local situation some power of J lies in $l\mathcal{O}_M[G]$ (Curtis & Reiner, 1981, section 5.22, p. 112). Hence the division algebras have characteristic l. The division algebras whose centre is the field, K, are measured by the *Brauer group*, $Br(K)$, and for a finite field the Brauer group vanishes (Serre, 1979, p. 161) so that there is an isomorphism of the form

4.3.20
$$\mathcal{O}_M[G]/J \cong \textstyle\prod_{i=1}^{s} M_{n_i}(F_{l^{d_i}}).$$

Lemma 4.3.21 *Let $T \in R(G)$ and let $r \in J$, the Jacobson radical of $\mathcal{O}_M[G]$ in* 4.3.10. *Then $1 - r \in \mathcal{O}_M[G]^*$ and both $log(Det(1 - r)(T))$ and $log(Det(1 - F(r))(\psi^l(T)))$ are l-adically convergent in $\mathcal{O}_N$. In addition,*

$$log(Det(1 - F(r))(\psi^l(T))/[Det(1 - r)(T)]^l)$$

$$= log(Det(1 - F(r))(\psi^l(T))) - l log(Det(1 - r)(T)).$$

Proof Since $r^n \in l\mathcal{O}_M[G]$ for some n (Curtis & Reiner, 1981, section 5.22, p. 112) the series for $(1 - r)^{-1}$ converges l-adically and $1 - r \in \mathcal{O}_M[G]^*$. Since 4.3.19 is l-adically convergent it suffices to show that

$log(Det(1-r)(T))$ is l-adically convergent and, by additivity, we may assume that T is a representation of the form of 4.3.2.

Since $r^n \in l\mathcal{O}_M[G]$ we find, for large m, that

$$\begin{aligned} T(1-r)^{l^m} &\equiv 1 - T(r^{l^m}) \quad (mod\ l) \\ &\equiv 1 \quad (mod\ l). \end{aligned}$$

On the other hand, $T(1-r) \in \mathcal{O}_N$ whose residue field, $\mathcal{O}_N/ < \pi >$, has no l-primary roots of unity so that $T(1-r) = 1 - T(r) \equiv 1\ (mod\ < \pi >)$. Therefore $det(T(1-r)) \equiv 1 (mod\ < \pi >)$ and $log(Det(1-r)(T))$ is l-adically convergent in $\mathcal{O}_N$. □

4.3.22 If $r \in J$ as in 4.3.21, define the logarithm, $L_0(1-r)$, by

4.3.23 $L_0(1-r) = l\sum_{n=1}^{\infty} r^n/n - \sum_{n=1}^{\infty} \hat{F}(r^n)/n \in \mathcal{O}_M[G]$,

where $\hat{F}(\sum_{\gamma \in G} \lambda_\gamma \gamma) = \sum_{\gamma \in G} F(\lambda_\gamma)\gamma^l$, F being the lifted Frobenius of 4.3.8.

If $T \in R(G)$ has characteristic function, χ_T, so that

$$\chi_T(g) = Trace(T(g))$$

for all $g \in G$, then we may define a homomorphism

4.3.24
$$\xi_T : \mathcal{O}_M[G] \longrightarrow \mathcal{O}_N$$

by the formula

$$\xi_T\left(\sum_{\gamma \in G} \lambda_\gamma \gamma\right) = \sum_{\gamma \in G} \lambda_\gamma \chi_T(\gamma).$$

Proposition 4.3.25 *Let $r \in J$ and $T \in R(G)$ be as in* 4.3.10–4.3.21. *Then, if ξ_T is as in* 4.3.24,

$$log\left(\frac{Det(1-F(r))(\psi^l(T))}{(Det(1-r)(T))^l}\right) = \xi_T(L_0(1-r)) \in \mathcal{O}_N.$$

Proof By additivity we may suppose that T is a representation, as in 4.3.2. Let $\lambda_1, \ldots, \lambda_u$ denote the eigenvalues of $T(r)$. In the course of proving Lemma 4.3.21 we saw that each λ_i lies in the maximal ideal, $< \pi > \lhd \mathcal{O}_N$, if N is large enough. Therefore the following series converges:

$$\begin{aligned} log(Det(1-T(r))) &= log(\textstyle\prod_{i=1}^{u}(1-\lambda_i)) \\ &= -\textstyle\sum_{i=1}^{u}\sum_{m=1}^{\infty} \lambda_i^m/m \\ &= -\textstyle\sum_{m=1}^{\infty}\sum_{i=1}^{u} \lambda_i^m/m \\ &= -\textstyle\sum_{m=1}^{\infty} Trace(T(r^m))/m. \end{aligned}$$

Similarly, since the eigenvalues of $\psi^l(T)(r)$ are $\{\lambda_i^l\}$, we find that

$$log(Det(1 - F(r))(\psi^l(T))) = -\sum_{m=1}^{\infty} Trace(T(\hat{F}(r^m)))/m,$$

which completes the proof. □

Definition 4.3.26 Let

$$\Lambda_G = \mathcal{O}_M[G]/\left(\sum_{x,y \in G} \mathcal{O}_M(xy - yx)\right).$$

Thus Λ_G is an $\mathcal{O}_M$-module and we may define, for $r \in J \triangleleft \mathcal{O}_M[G]$,

4.3.27 $$L(1 - r) = \pi(L_0(1 - r)) \in \Lambda_G,$$

where $\pi : \mathcal{O}_M[G] \longrightarrow \Lambda_G$ is the canonical quotient map.

Proposition 4.3.28 *Let G be any finite group. If $r \in J$ then*

$$L(1 - r) = \pi(L_0(1 - r)) \in l\Lambda_G.$$

Proof Consider the series

$$L(1 - r) = l\sum_{m=1}^{\infty} \pi(r^m/m) - \sum_{m=1}^{\infty} \pi(\hat{F}(r^m)/m) \in \Lambda_G.$$

If l does not divide m then $lr^m/m \in l\mathcal{O}_M[G]$ and the π-image of such a term is l-divisible in Λ_G. Now consider the remaining terms

4.3.29

$$\textstyle\sum_{m=1}^{\infty} l\pi(r^{lm}/lm) - \pi(\hat{F}(r^m)/m)$$

$$= \textstyle\sum_{m=1}^{\infty} \pi(r^{lm} - \hat{F}(r^m))/m.$$

Suppose that $m = l^{s-1}q$ with $HCF(q, l) = 1$ then we may set $t = r^q$ so that

$$(r^{lm} - \hat{F}(r^m))/m = (t^{l^s} - \hat{F}(t^{l^{s-1}}))/m.$$

Therefore we must show that, if $r \in J$,

4.3.30 $$\pi(r^{l^s} - \hat{F}(r^{l^{s-1}})) \in l^s\Lambda_G$$

for all $s \geq 1$.

Suppose that $r = \sum_i a_i g_i$ (summed over all $g_i \in G$). Therefore

$$r^{l^s} = \sum a_{j_1} \ldots a_{j_{l^s}} g_{j_1} \ldots g_{j_{l^s}},$$

where $(j_1, \ldots, j_{l^s})$ ranges over all possible l^s-tuples, which we shall think of as 'permutations'. Fix a permutation, $\underline{i} = (i_1, \ldots, i_{l^s})$, and consider all the terms, $\underline{j} = (j_1, \ldots, j_{l^s})$, which are obtained from $\underline{i}$ by a cyclic permutation. The products $g_1 \ldots g_v$ and $g_2 \ldots g_v g_1$ are conjugate in G so that each term in the subsum of $\underline{j}$s will have the same π-image in Λ_G. Let H denote the cyclic subgroup of order l^s which cyclically permutes $\underline{i} = (i_1, \ldots, i_{l^s})$ and suppose that the stabiliser of $\underline{i}$ in H has order l^u. In this case there are l^{s-u} terms in the subsum of $\underline{j}$'s and, in Λ_G, we obtain l^{s-u} times their common π-image.

When $u = 0$ the contribution of these terms to 4.3.30 will lie in $l^s \Lambda_G$. Now suppose that $u \geq 1$ so that we may write

$$\underline{i} = (i_1, \ldots, i_{l^{s-1}}, i_1, \ldots, i_{l^{s-1}}, i_1, \ldots)$$

and we may consider the terms in $r^{l^{s-1}}$ which are of the form

$$a_{j_1} \ldots a_{j_{l^{s-1}}} g_{j_1} \ldots g_{j_{l^{s-1}}},$$

where $\hat{\underline{j}} = (j_1, \ldots, j_{l^{s-1}})$ is a cyclic permutation of $\hat{\underline{i}} = (i_1, \ldots, i_{l^{s-1}})$. The stabiliser of $\hat{\underline{i}}$ in the cyclic group of order l^{s-1} is of order l^{u-1} so that there are l^{s-u} terms in this subsum of $\hat{F}(r^{l^{s-1}})$ which map to the same image in Λ_G.

Now, in Λ_G, let us collect together the terms from these two troublesome subsums when $u \geq 1$. From the subsum of $\underline{j}$s we obtain a contribution of the form $l^{s-u} \alpha^{l^u} \pi(\overline{g}^l)$ while from the subsum of $\hat{\underline{j}}$s we obtain $-l^{s-u} F(\alpha^{l^{u-1}}) \pi(\overline{g}^l)$ where $\overline{g}$ is the monomial, $(g_{i_1} g_{i_2} \ldots)$. However, by induction starting with 4.3.8, we find that

$$\alpha^{l^u} - F(\alpha^{l^{u-1}}) \in l^u \mathcal{O}_M$$

so that the difference of the two subsums contributes an element of $l^{s-u}(l^u \Lambda_G)$, which establishes 4.3.30 and completes the proof. □

Definition 4.3.31 Let $T \in R(G)$. We shall say that $T \equiv 0$ $(mod\ l^k)$ if, for all $g \in G$, $\chi_T(g) \equiv 0$ $(mod\ l^k \mathcal{O}_N)$ where χ_T is the character of T.

Theorem 4.3.32 *Let G be a finite group and let J denote the Jacobson radical of $\mathcal{O}_M[G]$, in the notation of* 4.3.10–4.3.21. *Let $r \in J$ and suppose* *that* $T \in R(G)$ *satisfies* $T \equiv 0 \ (mod\ l^k)$ *then*

$$\frac{Det(1-F(r))(\psi^l(T))}{(Det(1-r)(T))^l} \in \epsilon + l^{k+1}\mathcal{O}_N$$

where $\epsilon = 1$ *if* l *is odd and* $\epsilon = \pm 1$ *if* $l = 2$.

Proof There is nothing to prove if $k = 0$ so we may assume that $k \geq 1$. The map, ξ_T of 4.3.24, factorises as

$$\xi_T : \mathcal{O}_M[G] \xrightarrow{\pi} \Lambda_G \xrightarrow{\hat{\xi}_T} \mathcal{O}_N,$$

where $\hat{\xi}_T(\pi(\sum \lambda_\gamma \gamma)) = \sum \lambda_\gamma \chi_T(\gamma)$.

By 4.3.28, $L(1-r) = \pi(L_0(1-r)) \in l\Lambda_G$ so that $\xi_T(L_0(1-r)) \in l^{k+1}\mathcal{O}_N$. Therefore, by 4.3.25,

4.3.33

$$log\left(\frac{Det(1-F(r))(\psi^l(T))}{(Det(1-r)(T))^l}\right) \in l^{k+1}\mathcal{O}_N.$$

From 4.3.33 we may now finish by means of the argument which was used to prove Corollary 4.3.17(ii).

Suppose first that $l \neq 2$, then we know from 4.3.10 and 4.3.16 that

4.3.34

$$exp\left(log\left[\frac{Det(1-F(r))(\psi^l(T))}{(Det(1-r)(T))^l}\right]\right)$$

$$= \frac{Det(1-F(r))(\psi^l(T))}{(Det(1-r)(T))^l}.$$

On the other hand we know, from 4.3.17(ii)(proof) that

$$exp(l^{k+1}\mathcal{O}_N) \subset 1 + l^{k+1}\mathcal{O}_N,$$

which completes the proof for odd primes, l.

When $l = 2$ we may, as in 4.3.17, apply the previous argument to the square of 4.3.34 and then extract the square root in $1 + 2^{k+2}\mathcal{O}_N$. □

4.3.35 We will close this section with a strengthening of Theorem 4.3.32. Firstly let us define some subgroups, $W_G(k) \subseteq 1 + l\mathcal{O}_N$, in the notation of 4.3.10–4.3.21.

Set

4.3.36

$$W_G(k) = \begin{cases} 1 + l^{k+1}\mathcal{O}_N & \text{if } l \neq 2, \text{for all } G \text{ or} \\ & \text{if } l = 2 \text{ and } k = 0, \\ \{\pm 1\} + 2^{k+1}\mathcal{O}_N & \text{if } l = 2, G \text{ has no quotient} \\ & \text{isomorphic to } D_6, \\ \{\pm 1, \pm\sqrt{-1}\} + 2^k\mathcal{O}_N & \text{if } l = 2 \text{ and } G \text{ has a} \\ & \text{quotient isomorphic to } D_6. \end{cases}$$

Notice that N is a 'large' field so that $\sqrt{-1} \in N$.

With this preparatory notation we are now ready to state and prove an extension of Theorem 4.3.32 to apply to all units of $\mathcal{O}_M[G]$.

Theorem 4.3.37 *Let G be a finite group and, in the notation of* 4.3.10–4.3.21, *let* $z \in \mathcal{O}_M[G]^*$. *Suppose that* $T \in R(G)$ *satisfies* $T \equiv 0 \ (mod\ l^k)$ *then*

4.3.38
$$\frac{Det(F(z))(\psi^l(T))}{(Det(z)(T))^l} \in W_G(k) \subset 1 + l\mathcal{O}_N.$$

Proof By 4.3.10 we may assume that $k \geq 1$. Fixing $T \in R(G)$, the expression of 4.3.38 defines a homomorphism on $z \in \mathcal{O}_M[G]^*$ which, by 4.3.10–4.3.32, factorises to give a homomorphism

$$\Psi_l : \frac{\mathcal{O}_M[G]^*}{1+J} \longrightarrow \frac{1 + l\mathcal{O}_N}{1 + l^{k+1}\mathcal{O}_N} \quad \text{when } l \neq 2$$

and

$$\Psi_2 : \frac{\mathcal{O}_M[G]^*}{1+J} \longrightarrow \frac{1 + 2\mathcal{O}_N}{\{\pm 1\} + 2^{k+1}\mathcal{O}_N} \quad \text{when } l = 2.$$

The map, Ψ_l, sends the coset, $z(1+J)$ to 4.3.38. In each case the range of Ψ_l is an abelian l-group, as is seen by means of the logarithm of 4.3.15 and 4.3.16.

However, by 4.3.20, we have an isomorphism of the form

4.3.39
$$\mathcal{O}_M[G]^*/(1+J) \cong \prod_{i=1}^s GL_{n_i}(F_{l^{d_i}}).$$

We will use 4.3.39 to show that Ψ_l is trivial unless $l = 2$ and G has a quotient which is isomorphic to the dihedral group, D_6. In the remaining case we will show that Ψ_2^2 is the trivial map. When G has no D_6 quotient this will obviously show that 4.3.38 lies in $W_G(k)$ of 4.3.36. When $l = 2$ and $G/H \cong D_6$ for some $H \lhd G$ then the square of 4.3.38 will lie in $(\{\pm 1\} + 2^{k+1}\mathcal{O}_N) \subset N^*$ and the square root of any such element exists and belongs to $\{\pm 1, \pm\sqrt{-1}\} + 2^k \mathcal{O}_N$, which equals $W_G(k)$ in this case, also.

By 4.3.39 it will suffice to show that $Hom(GL_n(F_{l^d}), \mathbf{Z}/l)$ is zero when $l \neq 2$ or when $l = 2$ and $d > 1$. For the remaining case $GL_n(F_{l^d}) = GL_2(F_2) \cong D_6$ and the only homomorphisms from D_6 to a finite abelian 2-group are of order two.

When $n = 1$, $GL_1(F_{l^d}) = F_{l^d}^*$ has no l-torsion and so $Hom(F_{l^d}^*, \mathbf{Z}/l) = 0$. More generally,

$$Hom(GL_n(F_{l^d}), \mathbf{Z}/l) \cong H^1(GL_n(F_{l^d}); \mathbf{Z}/l),$$

which vanishes if $d(l-1) > 1$, by Quillen (1972, theorem 6, p. 578). Alternatively, when $l^d \geq 4$ or $n \geq 3$ $SL_n(F_{l^d})$ is simple (Lang, 1984, pp. 472–480) so that $Hom(SL_n(F_{l^d}), \mathbf{Z}/l) = 0$ and therefore

$$Hom(GL_n(F_{l^d}), \mathbf{Z}/l) \cong Hom(F_{l^d}^*, \mathbf{Z}/l) = 0,$$

too.

These remarks rule out all cases except $GL_2(F_2) \cong D_6$, which completes the proof of Theorem 4.3.37. □

4.4 Detecting elements in the class-group

4.4.1 Suppose now that $K/\mathbf{Q}$ is a finite Galois extension ($\mathbf{Q}$ is the field of rationals) and that l is a prime which is unramified in $K/\mathbf{Q}$. Let $P \lhd \mathcal{O}_K$ be a prime which divides l. Therefore

$$G(K_P/\mathbf{Q}_l) \subseteq G(K/\mathbf{Q}).$$

Let F_P denote the Frobenius of $G(K_P/\mathbf{Q}_l)$, which is characterised by the congruence of 4.3.8

$$F_P(x) \equiv x^l \pmod{P} \quad (x \in \mathcal{O}_K).$$

Now let $E/\mathbf{Q}$ be a 'large' finite Galois extension, containing K, as in 4.2.28. Let us suppose that there exists

4.4.2 $$\begin{cases} \tilde{F}_P \in G(E/\mathbf{Q}) \text{ such that } \tilde{F}_P = F_P \\ \text{on } K \text{and } \tilde{F}_P \text{ acts like the identity} \\ \text{on } T(g) \text{ for all } g \in G, T \in R(G). \end{cases}$$

In 4.4.2 it is sufficient that $\tilde{F}_P$ acts trivially on the matrix entries of $T(g)$ where T is a representation of the form 4.2.9 and T runs through a set of generators for $R(G) \cong R_E(G)$. The condition of 4.4.2 is vacuously fulfilled when $K = \mathbf{Q}$ for then $\tilde{F}_P$ is the identity. It may also be fulfilled, for example, when $\#(G) = l^e$ for then we may take $E = K \cdot \mathbf{Q}(exp(2\pi i/l^e))$. In that case $K/\mathbf{Q}$, being unramified at l, is linearly disjoint from $\mathbf{Q}(exp(2\pi i/l^e))/\mathbf{Q}$.

4.4.3 *The detection homomorphisms*

Let $T \in R(G)$ and suppose that $K/\mathbf{Q}$ satisfies the conditions of section 4.4.1, so that $\tilde{F}_P \in G(E/\mathbf{Q})$ exists in 4.4.2. Consider the map

4.4.4

$$Hom_{\Omega_K}(R(G), (K_P \otimes_K E)^*) \xrightarrow{(\tilde{F}_P\psi^l)/l} Hom_{\Omega_K}(R(G), (K_P \otimes_K E)^*)$$

given by

$$(\tilde{F}_P\psi^l)(f)(T) = \tilde{F}_P(f(\psi^l(T)))/(f(T))^l.$$

Notice that if $\omega \in \Omega_K$ then $z_P = \tilde{F}_P^{-1}\omega\tilde{F}_P \in \Omega_K$ so that

$$\begin{aligned} \omega(\tilde{F}_P(f(\psi^l(T)))) &= \tilde{F}_P(z_P(f(\psi^l(T)))) \\ &= \tilde{F}_P f(z_P(\psi^l(T))) \\ &= \tilde{F}_P f(\tilde{F}_P^{-1}\omega(\psi^l(T))), \quad \text{by 4.4.2,} \\ &= \tilde{F}_P f(\tilde{F}_P^{-1}(\psi^l(\omega(T)))) \\ &= \tilde{F}_P f(\psi^l(\omega(T))), \quad \text{by 4.4.2,} \end{aligned}$$

so that 4.4.4 takes values in Ω_K-equivariant homomorphisms.

For each $T \in R(G)$ we also have an evaluation homomorphism

4.4.5 $$eval_T : Hom_{\Omega_K}(R(G), (K_P \otimes_K E)^*) \longrightarrow (K_P \otimes_K E)^*.$$

By Galois invariance 4.4.5 must take values in $(K_P \otimes_K K(\chi_T))^*$, where $K(\chi_T)$ is the smallest Galois extension of K which contains all the character values of T, $\{\chi_T(g); g \in G\}$.

Finally, by means of the isomorphism of 4.2.17,

4.4.6 $$(K_P \otimes_K K(\chi_T))^* \cong \prod_{Q|P} K(\chi_T)^*_Q,$$

where Q runs through the primes of $K(\chi_T)$ which divide P. We therefore have a projection map

4.4.7 $$\pi_Q : (K_P \otimes_K K(\chi_T))^* \longrightarrow K(\chi_T)^*_Q.$$

Combining 4.4.4–4.4.7 we obtain a homomorphism

4.4.8
$$S_{T,Q} = \pi_Q(eval_T(\tilde{F}_P\psi^l/l)) :$$
$$Hom_{\Omega_K}(R(G), \textstyle\prod_{P'|P} E_{P'}^*) \longrightarrow K(\chi_T)_Q^*.$$

Now consider the effect of $S_{T,Q}$ on a function of the form

$$Det(u) : R(G) \longrightarrow \prod_{P'|P} E_{P'}^*,$$

where $u \in \mathcal{O}_{K_P}[G]^*$. If $u = \sum_{\gamma \in G} \lambda_\gamma \gamma$ then, by the assumptions of 4.4.2,

$$(\tilde{F}_P\psi^l/l)Det(u)(T) = \frac{Det(\sum F_P(\lambda_\gamma)\gamma)(\psi^l(T))}{(Det(\sum \lambda_\gamma\gamma)(T))^l}.$$

Therefore, if $T \equiv 0 \pmod{l^k}$ then Theorem 4.3.37 implies that

4.4.9 $$S_{T,Q}(Det(u)) \in W_G(k) \cap K(\chi_T)_Q^* = V(T)_P, \text{ say.}$$

Also note that $Hom_{\Omega_K}(R(G), E^*)$ and $Hom_{\Omega_K}^+(R(G), \mathcal{O}_E^*)$ are mapped by $S_{T,Q}$ into $K(\chi_T)^*$ and $\mathcal{O}_{K(\chi_T)}^*$ respectively.

We may now apply the maps $\{S_{T,Q}\}$ to obtain homomorphisms out of $\mathcal{CL}(\mathbf{Z}[G])$ and $D(\mathbf{Z}[G])$.

For $T \in R(G)$, define

4.4.10 $$V(T) = \textstyle\prod_P V(T)_P \subseteq U(\mathcal{O}_{\mathbf{Q}(\chi_T)}),$$

where $V(T)_P$ is as in 4.4.9. With this notation 4.4.9 together with the Hom-descriptions of 4.2.28, 4.2.35 and 4.2.37 yields the following result:

Theorem 4.4.11 *Let G be a finite group.*

(i) *With the preceding notation there is a well-defined homomorphism* $(T \in R(G))$

$$S_T = \{S_{T,Q}\} : \mathcal{CL}(\mathbf{Z}[G]) \longrightarrow \frac{J_{fin}^*(\mathbf{Q}(\chi_T))}{V(T) \cdot S_T(Hom)}$$

where Hom denotes $Hom_{\Omega_{\mathbf{Q}}}(R(G), E^*)$ *and* $J_{fin}^*(E)$ *denotes the adèlic group obtained by deleting the Archimedean places from* $J^*(E)$.

(ii) *There is a well-defined homomorphism*

$$S_T = \{S_{T,Q}\} : D(\mathbf{Z}[G]) \longrightarrow \frac{\prod_{P \in \mathcal{A}} \mathcal{O}_{\mathbf{Q}(\chi_T)_P}^*}{\prod_{P \in \mathcal{A}} V(T)_P \cdot S_T(Hom^+)},$$

where Hom^+ *denotes* $Hom_{\Omega_{\mathbf{Q}}}^+(R(G), \mathcal{O}_E^*)$ *and* $\mathcal{A}$ *is the set of primes of* $\mathbf{Q}(\chi_T)$ *which divide* $\#(G)$.

Example 4.4.12 Let l be a prime and suppose that $\#(G) = l^k$. Take T to be the regular representation, $T = Ind_{\{1\}}^G(1) \in R(G)$. Hence

$$\chi_T(g) = \begin{cases} 0 & \text{if } g \neq 1, \\ \#(G) & \text{if } g = 1, \end{cases}$$

so that $T \equiv 0$ (modulo l^k). Since the character values of T are rational, we obtain in this case

$$S_T : D(\mathbf{Z}[G]) \longrightarrow \frac{\mathbf{Z}_l^*}{\{\pm 1\} \cdot (1 + l^{k+1}\mathbf{Z}_l)}$$

from Theorem 4.4.11(ii), since D_6 is not a quotient of G. The $\{\pm 1\}$ in the denominator of the target group arises because $S_T(Hom^+) \subseteq \mathbf{Z}^* = \{\pm 1\}$. In terms of the Hom-description S_T is given here by

$$S_T(f) = f(\psi^l(Ind_{\{1\}}^G(1)) - lInd_{\{1\}}^G(1)),$$

where $f \in Hom^+_{\Omega_{\mathbf{Q}_l}}(R(G), \mathcal{O}^*_{E_P})$ and P is any prime of E which divides l.

4.4.13 *The Swan subgroup*

The *Swan subgroup* of G is defined to be the subgroup of $D(\mathbf{Z}[G])$ which is generated by the classes of the Swan modules, $S(k)$ (with $HCF(k, \#(G)) = 1$), which were introduced in 4.2.43. The Swan subgroup will be denoted by $T(G)$. By 4.7.6–4.7.8 we may define a surjective homomorphism

4.4.14 $$S : (\mathbf{Z}/\#(G))^*/\{\pm 1\} \longrightarrow T(G)$$

by

$$S(k) = [ideal\{k, \sigma\}] \quad \text{of} \quad 4.2.44.$$

We will use the map, S_T, of Example 4.4.12 to determine $T(G)$ when $\#(G) = l^k$. The proof is originally due to M.J. Taylor (1978), who did not have available all the maps of Theorem 4.4.11 but did have the particular case of 4.4.12.

4.4.15 *The Artin exponent*

The *Artin exponent* of G, which we will denote by $A(G)$, is defined (Curtis & Reiner, 1987, p. 782) to be the smallest integer such that there is an equation in the rational representation ring, $R_{\mathbf{Q}}(G)$, of the form

4.4.16 $$A(G)\cdot 1 = \sum_i Ind_{C_i}^G(\phi_i) \in R_{\mathbf{Q}}(G),$$

where C_i is a cyclic subgroup of G and $\phi_i \in R_{\mathbf{Q}}(C_i)$.

By 2.1.3, $A(G)$ divides the order of G.

Lemma 4.4.17 *The Artin exponent annihilates $T(G)$. In particular, $\#(G)\cdot T(G) = 0$.*

Proof Let C be a subgroup of G. We have maps

$$Res_C^G : R(G) \longrightarrow R(C)$$

and

$$Ind_C^G : R(C) \longrightarrow R(G).$$

These maps induce homomorphisms, via the Hom-description of 4.2.35,

4.4.18

$$\begin{array}{ll} Ind_C^G : D(\mathbf{Z}[C]) \longrightarrow D(\mathbf{Z}[G]) & \text{and} \\ Res_C^G : D(\mathbf{Z}[G]) \longrightarrow D(\mathbf{Z}[C]) & \text{respectively.} \end{array}$$

In terms of modules, $Ind_C^G[M] = [\mathbf{Z}[G] \otimes_{\mathbf{Z}[C]} M]$ and $Res_C^G[M] = [M]$ (Curtis & Reiner, 1987, pp. 339–341; Fröhlich, 1983).

Now suppose that $f : R(G) \longrightarrow U(\mathcal{O}_E)$ is an Ω_Q-equivariant homomorphism whose class in 4.2.35 represents $S(k) \in T(G)$. In this case $A(G)S(k) = A(G)[f]$.

From Curtis & Reiner (1987, section 76.8, p. 784) there exists an equation, which apparently improves upon 4.4.16, of the form

$$A(G)\cdot 1 = \sum_C n_C Ind_C^G(1) \in R(G),$$

where the sum is taken over cyclic subgroups, C, and the $\{n_C\}$ are integers.

Hence, if $T \in R(G)$, we have

4.4.19

$$\begin{aligned} f(A(G)T) &= f(\textstyle\sum_C n_C(T \otimes Ind_C^G(1))) \\ &= \textstyle\prod_C (f(Ind_C^G(Res_C^G(T))))^{n_C} \quad \text{by 1.2.39.} \end{aligned}$$

However, Res_C^G of 4.4.18 maps $T(G)$ to $T(C)$ (Curtis & Reiner, 1987, p. 345) so that $[f \cdot Ind_C^G] \in T(C)$. However, $T(C)$ is the trivial group, by 4.2.48. Hence $[f(A(G)\cdot -)]$ is trivial in $D(\mathbf{Z}[G])$, by 4.4.19. □

Table 4.1. (*Theorem* 4.4.20)

$G = \{l\text{-group of order } l^k\}$	$\#(T(G))$
cyclic	1
non-cyclic , $l \neq 2$	l^{k-1}
generalised quaternion of order $2^m, m \geq 3$	2
dihedral of order $2^m, m \geq 2$	1
semi-dihedral of order $2^m, m \geq 4$	2
all other non-cyclic groups of orders 2^k	2^{k-2}

Theorem 4.4.20 *Let l be a prime and let $\#(G) = l^k$. The Swan subgroup, $T(G)$, is a cyclic l-group whose order is given by Table* 4.1.

Proof The l-primary torsion in $(\mathbf{Z}/l^k)^*/\{\pm 1\}$ is cyclic so that $T(G)$ is a cyclic l-group, by 4.4.17. It remains to determine $\#(T(G))$.

The cyclic and dihedral Swan subgroups are both trivial, by 4.2.48. The Swan subgroup of the generalised quaternion group, Q_8, is of order two, by 4.7.11. Let Q_{2^n} denote the generalised quaternion group of order 2^n. By 4.7.12, the restriction map, $T(Q_{2^n}) \longrightarrow T(Q_8)$ is onto, so that it suffices to show that $2 \cdot T(Q_{2^n}) = 0$. This will follow from Theorem 5.3.3, since Q_{2^n} has a cyclic subgroup of index two, $C_{2^{n-1}}$, and there is no element of order two in $Q_{2^n} - C_{2^{n-1}}$. The case of the semi-dihedral group is dealt with similarly.

The semi-dihedral group, SD_n, has order 2^{n+2} and is given by

$$SD_n = \{a, b \mid a^{2^{n+1}} = 1 = b^2, bab = a^{2^n - 1}\}.$$

SD_n has a subgroup which is isomorphic to Q_8, namely the subgroup generated by $\{a^{2^{n-1}}, ba\}$. The restriction map, $T(SD_n) \longrightarrow T(Q_8)$ is onto. Therefore it suffices to show that $2 \cdot T(SD_n) = 0$. If we apply Theorem 5.3.3 to SD_n and its subgroup of index two, $J = \{a\}$, we obtain a relation of the form, where k is any odd integer,

$$0 = 2 \cdot S(k) - \sum_g Ind_{C(g)}^{SD_n}(S(k)) \in D(\mathbf{Z}[SD_n]).$$

This sum ranges over some elements of order two in $SD_n - J$ and $C(g)$ denotes the centraliser of g. Such elements of order two have the form $a^{2j}b$ and have a copy of the dihedral group, D_8, for a centraliser. Since $0 = S(k) \in T(D_8)$ the relation shows that $S(k)$ is annihilated by two, as required.

Now let us suppose that G belongs to one of the remaining cases; therefore G is not a cyclic, dihedral, semi-dihedral or generalised quaternion group. For these remaining cases

$$(\mathbf{Z}/\#(G))^*/\{\pm 1\} = (\mathbf{Z}/l^k)^*/\{\pm 1\}$$

has an l-primary part which is cyclic of order l^{k-1} when $l \neq 2$ and 2^{k-2} when $l = 2$ so that it will suffice to detect classes, $S(k)$, which are of at least this order. To do this we will evaluate

$$S_T(S(1+l)) \in \frac{\mathbf{Z}_l^*}{\{\pm 1\}(1 + l^{k+1}\mathbf{Z}_l)}$$

where S_T is as in 4.4.12. By definition this is equal to

4.4.21 $$S(1+l)(\psi^l(Ind_{\{1\}}^G(1)) - l \cdot Ind_{\{1\}}^G(1)),$$

where $S(1+l)$ is the idèlic-valued function given in 4.2.47. Therefore 4.4.21 is equal to

$$(1+l)^e \in \frac{\mathbf{Z}_l^*}{\{\pm 1\}(1 + l^{k+1}\mathbf{Z}_l)},$$

where

$$\begin{aligned} e &= < 1, \psi^l(Ind_{\{1\}}^G(1)) - l \cdot Ind_{\{1\}}^G(1) > \\ &= < 1, \psi^l(Ind_{\{1\}}^G(1)) > -l \\ &= [\#(G)^{-1} \textstyle\sum_{g \in G} \chi_{Ind_{\{1\}}^G(1)}(g^l)] - l, \qquad \text{by 4.1.2 and 1.2.7,} \\ &= \#\{g \in G \mid g^l = 1\} - l. \end{aligned}$$

However, by a theorem of Kulakoff (when $l \neq 2$) and of Alperin, Feit and Thompson (when $l = 2$) (see Huppert, 1967, p. 314; Isaacs, 1976, p. 52)

$$\#\{g \in G \mid g^l = 1\} \equiv 0 (mod\ l^2).$$

Therefore $e = lq$ with $HCF(l, q) = 1$ so that

$$S_T(S(1+l)) = (1+l)^{lq} \in \frac{\mathbf{Z}_l^*}{\{\pm 1\}(1 + l^{k+1}\mathbf{Z}_l)},$$

which has order l^{k-1} when $l \neq 2$ and 2^{k-2} when $l = 2$. This completes the proof of Theorem 4.4.20. □

Corollary 4.4.22 *Let l be an odd prime.*
If G is a finite l-group then the Artin exponent, $A(G)$, is given by

$$A(G) = \begin{cases} 1 & \text{if } G \text{ is cyclic,} \\ \#(G) \cdot l^{-1} & \text{if } G \text{ is non-cyclic.} \end{cases}$$

Proof The result is trivial when G is cyclic and otherwise it is clear from 2.1.3 that $A(G)$ divides $\#(G) \cdot l^{-1}$. Therefore, by 4.4.20, $A(G) = \#(G) \cdot l^{-1}$ since $A(G) \cdot T(G) = 0$, by 4.4.17. □

Remark 4.4.23 (i) When $\#(G) = 2^k$, $A(G) = 2^{k-2}$ if G is not cyclic, dihedral, semi-dihedral or generalised quaternionic and is $1, 2, 4, 2$, respectively, in the exceptional cases (Curtis & Reiner, 1987, p. 365).

(ii) In the course of proving Theorem 4.4.20 we used Kulakoff's theorem that, if $l \neq 2$ and G is an l-group, then

$$\#\{g \in G \mid g^l = 1\} \equiv 0(mod\ l^2).$$

In fact the proof shows that for all k satisfying $HCF(k, l) = 1$

$$S_T(S(k)) = k^{A-l} \in \frac{\mathbf{Z}_l^*}{\{\pm 1\}(1 + l^{k+1}\mathbf{Z}_l)},$$

where $A = \alpha l^2 = \#\{g \in G \mid g^l = 1\}$.

However, if k has order $(l-1)/2$ in $(\mathbf{Z}/l^k)^*/\{\pm 1\}$ then $S_T(S(k)) = 1$ so that we find that $A - l$ is a multiple of $(l-1)/2$.

This is equivalent to the following:

Proposition 4.4.24 *Let $\#(G) = l^k$ with l an odd prime. Then*

$$\#\{g \in G \mid g^l = 1\} = (1 + t(l-1)/2)l^2$$

for some positive integer, t.

4.5 Galois properties of local determinants

4.5.1 Let $M/\mathbf{Q}_l$ be a finite Galois extension and let G be a finite group. If $N/\mathbf{Q}_l$ is a 'large' Galois extension, in the sense of 4.2.10, which contains $M/\mathbf{Q}_l$ then we have the determinantal homomorphism

4.5.2 $$Det : \mathcal{O}_M[G]^* \longrightarrow Hom_{\Omega_M}(R(G), \mathcal{O}_N^*),$$

whose image is $Det(\mathcal{O}_M[G]^*)$.

In Fröhlich (1983, p. 84) we find the following result of M.J. Taylor:

Theorem 4.5.3 (*Fröhlich*, 1983, *Theorem* 10A, p. 84; *Taylor*, 1980, p. 93) *Let $M/\mathbf{Q}_l$ be a tame, finite Galois extension with Galois group, $H = G(M/\mathbf{Q}_l)$. Then*

$$(Det(\mathcal{O}_M[G]^*))^H \cong Det(\mathbf{Z}_l[G]^*).$$

Theorem 4.5.3 is one of the three main steps in Taylor's proof of the Fröhlich conjecture, which is the subject of the book (Fröhlich, 1983). We will not go into the details of the Fröhlich conjecture here (see Chapter 7). Suffice it to say that the ring of algebraic integers in a *tame* Galois extension of number fields is a projective module over the integral group-ring of the Galois group and Fröhlich's conjecture gives an analytic description of the class of this module in the class-group.

Incidentally, in Theorem 4.5.3, $M/\mathbf{Q}_l$ is tame if and only if $\mathcal{O}_M$ is a projective $\mathbf{Z}_l[H]$-module.

As an application of the material of Section 3 we will prove Theorem 4.5.3 when G is an l-group and $M/\mathbf{Q}_l$ is *unramified* (see Theorem 4.5.39) — a case to which 4.5.3 is readily reducible (see Fröhlich, 1983, pp. 85–89; see also 4.6.4). Our proof will follow the basic lines of that given in Fröhlich (1983) but is much more natural, being simplified by the fact that we start with the new determinantal congruences of 4.3.10. En route we shall encounter several maps and isomorphisms which are of interest in their own right.

Definition 4.5.4 Denote by $M\{G\}$ the M-vector space whose basis consists of the conjugacy classes of elements of G. Let

4.5.5
$$c : M[G] \longrightarrow M\{G\}$$

denote the M-linear map which sends $g \in G$ to its conjugacy class. $M\{G\}$ may be identified with the centre of $M[G]$ (see Lang, 1984, section 4.1, p. 647), but we will not use that description here.

Let $N/\mathbf{Q}_l$ be a finite Galois extension which contains M and all the n-th roots of unity ($\#(G) = n$). Hence N is 'large' in the sense of Section 3. Suppose that $\gamma_1, \ldots, \gamma_e$ denote the distinct conjugacy classes of G and let $\hat{\gamma}_1, \ldots, \hat{\gamma}_e$ be the corresponding *characteristic functions* given by

$$\hat{\gamma}_i(\gamma_j) = \begin{cases} 1 & \text{if } i = j, \\ 0 & \text{if } i \neq j. \end{cases}$$

Finally, let $T_1, \ldots, T_e$ denote the distinct,irreducible N-representations of G so that $T_j \in R_N(G) \cong R(G)$. Let χ_j denote the character function of T_j.

Lemma 4.5.6 *Let $n = \#(G)$ and let $\xi_n = exp(2\pi i/n)$. Let $\mathscr{C}_G$ denote the space of N-valued class functions on G (cf. 1.2.22). There exists a matrix*

$$A = (\alpha_{ij}) \in GL_e(\mathbf{Q}(\xi_n)) \subseteq GL_e(N),$$

such that

$$\hat{\gamma}_i = \sum_{j=1}^{e} \alpha_{ij}\chi_j \in \mathscr{C}_G.$$

Proof By 1.2.23 we know that the matrix, A, exists with entries in some large field. We must show that the entries lie in $\mathbf{Q}(\xi_n)$.

By 2.1.3 there exist cyclic subgroups, $\{C_s\}$, and one-dimensional representations, $\{\phi_s : C_s \longrightarrow N^*\}$, such that (for suitable $a_s \in \mathbf{Z}$)

$$n = \#(G) = \sum_s a_s(Ind_{C_s}^G(\phi_s)) \in R(G).$$

Therefore, by Frobenius reciprocity for class functions (1.2.39)

$$\begin{aligned} n\hat{\gamma}_i &= \textstyle\sum_s a_s\hat{\gamma}_i \otimes Ind_{C_s}^G(\phi_s) \\ &= \textstyle\sum_s a_s Ind_{C_s}^G(\phi_s \otimes Res_{C_s}^G(\hat{\gamma}_i)). \end{aligned}$$

Therefore we are reduced to proving that any class function on a cyclic group is a $\mathbf{Q}(\xi_n)$-linear combination of characters of representations.

Let C be a cyclic group of order t with generator, x. Let $y : C \longrightarrow N^*$ be given by $y(x) = \xi_t$. Define an invertible matrix in $M_t(\mathbf{Q}(\xi_t))$ $(\xi = \xi_t)$

4.5.7
$$X = (X_{ij}) = \begin{pmatrix} 1 & 1 & 1 & \dots & 1 \\ 1 & \xi & \xi^2 & & \xi^t \\ 1 & \xi^2 & \xi^4 & & \xi^{2t} \\ 1 & \xi^3 & \xi^6 & & \xi^{3t} \\ 1 & & & & \\ 1 & & & & \end{pmatrix}$$

so that, for $1 \leq i, j \leq t$,

4.5.8
$$X_{ij} = y^{i-1}(x^{j-1}) = \xi_t^{(i-1)(j-1)}.$$

From 4.5.8

$$\begin{aligned} &\textstyle\sum_i (X^{-1})_{s,i} y^{i-1}(x^u) \\ &= \textstyle\sum_i (X^{-1})_{s,i} X_{i,u+1} \\ &= \begin{cases} 1 & \text{if } s = u, \\ 0 & \text{if } s \neq u. \end{cases} \end{aligned}$$

The result follows, since $X^{-1} \in GL_t(\mathbf{Q}(\xi_t))$. □

4.5.9 Define a homomorphism

$$\phi : Hom_{\Omega_M}(R(G), N) \longrightarrow N\{G\}$$

by the formula

4.5.10 $$\phi(f) = \sum_s f(\hat{\gamma}_s)\gamma_s = \sum_{s=1}^{e} \sum_{j=1}^{e} \alpha_{sj} f(\chi_j)\gamma_s,$$

where (α_{ij}) is the matrix of 4.5.6.

The map of 4.5.10 is the composition of two maps of the following form:

4.5.11 $$Hom_{\Omega_M}(R(G), N) \longrightarrow Hom_{\Omega_M}(R(G) \otimes N, N)$$

$$f \longmapsto mult(f \otimes 1)$$

(where 'mult' denotes multiplication in N) and

4.5.12 $$Hom_{\Omega_M}(R(G) \otimes N, N) \longrightarrow N\{G\}$$

$$h \longmapsto \sum_{s=1}^{e} h(\hat{\gamma}_s)\gamma_s.$$

If $\omega \in \Omega_M$ then

$$\begin{aligned} \omega(\phi(f)) &= \sum_s \omega(f(\hat{\gamma}_s))\gamma_s \\ &= \sum_s f(\omega(\hat{\gamma}_s))\gamma_s \\ &= \sum_s f(\hat{\gamma}_s)\gamma_s \\ &= \phi(f) \end{aligned}$$

so that the image of 4.5.10 is contained within $N\{G\}^{\Omega_M} = M\{G\}$.

On the other hand, we may define another map

4.5.13 $$\psi : M\{G\} \longrightarrow Hom_{\Omega_M}(R(G), N)$$

by

$$\psi\left(\sum_\gamma m_\gamma \gamma\right)(T) = \sum_\gamma m_\gamma Trace T(\gamma).$$

Proposition 4.5.14 *The homomorphisms of* 4.5.10–4.5.13 *define (inverse) isomorphisms*

$$\phi : Hom_{\Omega_M}(R(G), N) \xrightarrow{\cong} M\{G\}.$$

and

$$\psi : M\{G\} \xrightarrow{\cong} Hom_{\Omega_M}(R(G), N).$$

Proof Consider the composition

$$\begin{aligned}\phi(\psi(\textstyle\sum_\gamma m_\gamma \gamma)) &= \phi(T \longmapsto \textstyle\sum_\gamma m_\gamma Trace(T(\gamma))) \\ &= \textstyle\sum_s \sum_\gamma m_\gamma Trace(\hat{\gamma}_s)(\gamma)\gamma_s \\ &= \textstyle\sum_{\gamma_s} m_{\gamma_s}\gamma_s \\ &= \textstyle\sum_\gamma m_\gamma \gamma,\end{aligned}$$

so that $\phi\psi = 1$.

On the other hand,

$$\begin{aligned}\psi(\phi(f)) &= \psi(\sum_{s=1}^{e}\sum_{j=1}^{e} \alpha_{sj} f(\chi_j)\gamma_s) \\ &= (\chi_t \longmapsto \sum_{s=1}^{e}\sum_{j=1}^{e} \alpha_{sj} f(\chi_j)\chi_t(\gamma_s)) \\ &= (\chi_t \longmapsto \sum_{s=1}^{e}\sum_{j=1}^{e} \alpha_{sj} f(\chi_j)(A^{-1})_{ts})\end{aligned}$$

where

$$A = (\alpha_{ij}) \text{ is as in 4.5.6,}$$

$$\begin{aligned}&= (\chi_t \longmapsto f(\chi_t)) \\ &= f,\end{aligned}$$

so that $\psi\phi = 1$ also. □

Remark 4.5.15 If one identifies $M\{G\}$ with the centre of $M[G]$ (see 4.5.4) then Proposition 4.5.14 may be proved by means of the reduced norm map, $nr : M[G]^* \longrightarrow M\{G\} \cong Centre(M[G])$, (Curtis & Reiner, 1987, section 52.9, p. 332).

4.5.16 *Explicit Brauer Induction and* $M\{G\}$

Recall from 2.2.15 the explicit Brauer induction homomorphism

$$a_G : R(G) \longrightarrow R_+(G)$$

given by

$$a_G(\rho) = \sum_{(H,\phi)^G \in \mathcal{M}_{\mathcal{G}}/\mathcal{G}} \alpha_{(H,\phi)^G}(\rho)(H,\phi)^G \in R_+(G),$$

where $(H,\phi)^G$ denotes the G-conjugacy class of a one-dimensional representation, $\phi : H \longrightarrow N^*$.

The absolute Galois group, Ω_M, acts on $R_+(G)$ by the formula

4.5.17 $$\omega((H,\phi)^G) = (H, \omega(\phi))^G \qquad (\omega \in \Omega_M).$$

Since a_G and $\omega(a_G(\omega^{-1}(-)))$ satisfy the axioms of 2.2.8 they must be equal, so that a_G is Ω_M-equivariant

4.5.18 $$a_G \in Hom_{\Omega_M}(R(G), R_+(G)).$$

If H is a subgroup of G let N_GH denote the normaliser of H in G and set $W_GH = N_GH/H$, the *Weyl group* of H in G. Let $\phi : H \longrightarrow N^*$ be a homomorphism so that we may uniquely factorise ϕ as

$$\phi : H \xrightarrow{\pi_H} H^{ab} \longrightarrow N^*,$$

where H^{ab} denotes the abelianisation of H. When there is no risk of confusion we will also denote the resulting map, $H^{ab} \longrightarrow N^*$, by ϕ. The Weyl group, W_GH, acts upon the set of characters, $\phi : H \longrightarrow N^*$, by conjugation. Since $(H,\phi)^G$ is the G-conjugacy class of $\phi : H \longrightarrow N^*$ then, once we have chosen a conjugacy class representative for H, ϕ is determined up to the action of W_GH.

Therefore a_G induces a homomorphism

4.5.19 $$A_G : R(G) \longrightarrow \oplus_{(H)} R(H^{ab})_{W_GH},$$

where (H) runs through the conjugacy classes of subgroups of G and $X_{W_GH} = X/\{z(x) - x \mid x \in X, z \in W_GH\}$ denotes the coinvariants of the W_GH-action on X. The homomorphism, A_G, is given by the same formula as a_G. Moreover, the canonical map gives an isomorphism

4.5.20 $$Hom_{\Omega_M}(R(H^{ab})_{W_GH}, N^*) \xrightarrow{\cong} Hom_{\Omega_M}(R(H^{ab}), N^*)^{W_GH}$$

since W_GH permutes the basis of the free abelian group, $R(H^{ab})$, given by the one-dimensional representations.

In addition, 4.5.19 is a split surjection whose inverse, by 4.2.7, is given by the map

$$\sum_{(H)} R(H^{ab})_{W_G H} \xrightarrow{\sum \pi_H^*} \sum_{(H)} R(H)_{N_G H} \xrightarrow{\sum Ind_H^G} R(G).$$

Thus, by 4.5.14 and the foregoing discussion, the dual of 4.5.19 yields a split surjection of the form

4.5.21 $$A_G^* : \oplus_{\substack{(H)\\ H \leq G}} (M[H^{ab}]^{W_G H}) \longrightarrow M\{G\},$$

where the sum in 4.5.21 is taken over the conjugacy classes, (H), of subgroups of G.

Now let us derive the naturality property of 4.5.21, which results from the naturality of A_G in 4.5.19, in the sense of 2.2.8.

Suppose that $J \leq G$, then we may define an M-linear map

4.5.22 $$IND_J^G : \sum_{\substack{(I)\\ I \leq J}} (M[I^{ab}]^{W_J I}) \longrightarrow \sum_{\substack{(H)\\ H \leq G}} (M[H^{ab}]^{W_G H})$$

by the formula (for $y \in M[I^{ab}]^{W_J I}$)

$$IND_J^G(y)_{(H)} = \sum_{s=1}^{e} \sum_{\substack{z \in J\backslash G/H \\ I \subset zHz^{-1}}} \hat{\gamma}_s(z^{-1}yz)\gamma_s \in M[H^{ab}]^{W_G H}.$$

In 4.5.22 $IND_J^G(y)_{(H)}$ denotes the (H)-component of $IND_J^G(y)$. Notice also that in 4.5.22 the $\{\gamma_s\}$ run through the conjugacy class representatives for H, *not* for H^{ab}. The facts that 4.5.22 is well-defined and that it does indeed lie within the $W_G H$-invariants may either be proved directly or follows from (the proof of) the following result.

Theorem 4.5.23 *Let M be any l-adic local field or any number field. Let G be a finite group. There is a split-surjective, M-linear map, defined by the construction of* 4.5.19–4.5.21,

$$A_G^* : \oplus_{\substack{(H)\\ H \leq G}} M[H^{ab}]^{W_G H} \longrightarrow M\{G\}.$$

Let $i : J \longrightarrow G$ be the inclusion of a subgroup, then the following diagram commutes:

$$\begin{array}{ccc}
\oplus_I M[I^{ab}]^{W_J I} & \xrightarrow{A_J^*} & M\{J\} \\
\Big\downarrow IND_J^G & & \Big\downarrow M\{i\} \\
\oplus_H M[H^{ab}]^{W_G H} & \xrightarrow{A_G^*} & M\{G\}
\end{array}$$

Proof The preceding discussion, which applies equally well to number fields as to local fields, establishes the existence of A_G^*. It remains to prove the commutativity of the diagram.

From 2.2.8(i) we know that there is a commutative diagram

$$\begin{array}{ccc}
R(G) & \xrightarrow{A_G} & \oplus_{(H)} R(H^{ab})_{W_G H} \\
\Big\downarrow Res_J^G & & \Big\downarrow Res_J^G \\
R(J) & \xrightarrow{A_J} & \oplus_{(I)} R(I^{ab})_{W_J I}
\end{array}$$

where

4.5.24 $$Res_J^G(\lambda : I^{ab} \longrightarrow N^*) = \sum_{z \in J \backslash G / I} ((J \cap zIz^{-1})^{ab} \stackrel{(z^{-1})^*(\lambda)}{\longrightarrow} N^*).$$

Therefore we have only to show the commutativity of the following diagram,in which the $\{\phi_{H^{ab}}\}$ and $\{\psi_{I^{ab}}\}$ are induced by the maps of 4.5.13 and 4.5.10 when G is set equal to H and I, respectively:

$$\begin{array}{ccc}
\oplus_I M[I^{ab}]^{W_J I} & \xrightarrow[\cong]{\oplus \psi_{I^{ab}}} & \oplus_{(I)} Hom_{\Omega_M}(R(I^{ab})_{W_J I}, N) \\
\Big\downarrow IND_J^G & & \Big\downarrow (Res_J^G)^* \\
\oplus_H M[H^{ab}]^{W_G H} & \xleftarrow[\cong]{\oplus \phi_{H^{ab}}} & \oplus_{(H)} Hom_{\Omega_M}(R(H^{ab})_{W_G H}, N)
\end{array}$$

Table 4.2. *Character table of Q_8*

	1	L_1	L_2	L_1L_2	v
1	1	1	1	1	2
x^2	1	1	1	1	−2
x	1	−1	1	−1	0
y	1	1	−1	−1	0
xy	1	−1	−1	1	0

If $y \in M[I^{ab}]^{W_J I}$ we have

$$(\oplus_{(H)}\phi_{H^{ab}})(Res_J^G)^*(\psi_{I^{ab}}(y))$$

$$= (\oplus_{(H)}\phi_{H^{ab}})(Res_J^G)^*((\lambda : I^{ab} \longrightarrow N^*) \mapsto \lambda(y))$$

$$= (\oplus_{(H)}\phi_{H^{ab}})\left(\sum_{\substack{z\in J\backslash G/H \\ J\cap zHz^{-1}=I}} ((\mu : H^{ab} \longrightarrow N^*) \mapsto \mu(z^{-1}yz))\right)$$

$$= (\oplus_{(H)}\phi_{H^{ab}})\left(\sum_{\substack{z\in J\backslash G/H \\ zHz^{-1}\geq I}} ((\mu : H^{ab} \longrightarrow N^*) \mapsto \mu(z^{-1}yz))\right)$$

so that the (H)-component of this map is given by

$$\phi_{H^{ab}}\left(\sum_{\substack{z\in J\backslash G/H \\ zHz^{-1}\geq I}} ((\mu : H^{ab} \longrightarrow N^*) \mapsto \mu(z^{-1}yz))\right)$$

$$= \sum_{\substack{z\in J\backslash G/H \\ zHz^{-1}\geq I}} \sum_s \hat{\gamma}_s(z^{-1}yz)\gamma_s$$

$$= IND_J^G(y)_{(H)},$$

as required. □

Example 4.5.25 Let $G = Q_8 = \{x, y \mid x^2 = y^2, y^4 = 1, yxy^{-1} = x^{-1}\}$ be the quaternion group of order eight. We may choose one-dimensional representations, L_1 and L_2 of Q_8, together with an irreducible two-dimensional representation, v, so that the character values of Q_8 are given by Table 4.2.

From Table 4.2 one finds that

$$\hat{1} = (1 + L_1 + L_2 + L_1L_2 + 2v)/8,$$
$$\widehat{x^2} = (1 + L_1 + L_2 + L_1L_2 - 2v)/8,$$
$$\hat{x} = (1 - L_1 + L_2 - L_1L_2)/4,$$
$$\hat{y} = (1 + L_1 - L_2 - L_1L_2)/4,$$

and

$$\widehat{xy} = (1 - L_1 - L_2 + L_1L_2)/4.$$

The homomorphism, $a_{Q_8} : R(Q_8) \longrightarrow R_+(Q_8)$, is given by

$$A_{Q_8}(L) = (Q_8, L)^{Q_8}$$

for $L = 1, L_1, L_2$ or L_1L_2 and

$$a_{Q_8}(v) = (\{x\}, \lambda_1)^{Q_8} + (\{y\}, \lambda_2)^{Q_8} + (\{xy\}, \lambda_3)^{Q_8} - (\{x^2\}, \lambda_4)^{Q_8}$$

where $\lambda_1, \lambda_2, \lambda_3, \lambda_4$ are all injective homomorphisms into the fourth roots of unity in N^*. From this, one may calculate that $A^*_{Q_8}$, which is defined on the Weyl group invariant elements of $M\{Q_8^{ab}\} \oplus M\{x\} \oplus M\{y\} \oplus M\{xy\} \oplus M\{1\}$ and maps to $M\{Q_8\}$, is given by the following formula. On $M\{Q_8^{ab}\}$, if $\overline{x}, \overline{y}$ are the images of x and y then

$$A^*_{Q_8}(1) = (1 + x^2)/4, A^*_{Q_8}(\overline{x}) = x, A^*_{Q_8}(\overline{y}) = y \text{ and } A^*_{Q_8}(\overline{x} \cdot \overline{y}) = xy.$$

On $M\{x\}$, if $z = (1 - x^2)/4$, then

$$A^*_{Q_8}(1) = z, A^*_{Q_8}(x) = iz, A^*_{Q_8}(x^2) = -z \text{ and } A^*_{Q_8}(x^3) = -iz.$$

On $M\{y\}$, $A^*_{Q_8}(y^i)$ is equal to the image of x^i under the map, $A^*_{Q_8}$, on $M\{x\}$ and similarly for $A^*_{Q_8}$ on $M\{xy\}$. On $M\{x^2\}$, $A^*_{Q_8}(1) = 2z$ and $A^*_{Q_8}(x^2) = -2z$. Finally, $A^*_{Q_8}$ is trivial on $M\{1\}$.

Remark 4.5.26 Whenever one is given a functor of G which has a Hom-description in terms of $R(G)$ then one can make a map, induced by A_G, which is analogous to 4.5.21. For $\mathcal{CL}(\mathcal{O}_K[G])$ and $D(\mathcal{O}_K[G])$ this analogue is called a *restricted determinant* and will be described in 5.1.4.

Definition 4.5.27 For the remainder of this section we shall suppose that $M/\mathbf{Q}_l$ is an unramified extension of local fields and that G is a finite l-group.

Define $\mathcal{A}_M(G)$ to be equal to the kernel of the natural map from $\mathcal{O}_M[G]$ to $\mathcal{O}_M[G^{ab}]$,

4.5.28 $$\mathscr{A}_M(G) = ker(\mathcal{O}_M[G] \longrightarrow \mathcal{O}_M[G^{ab}]).$$

Therefore $\mathscr{A}_M(G) \subset J$, the Jacobson radical of $\mathcal{O}_M[G]$, which was introduced in 4.3.19. We are going to study the subgroup $Det(1+\mathscr{A}_M(G))$ of $Hom_{\Omega_M}(R(G), \mathcal{O}_N^*)$. We recall a result of C.T.C. Wall (1979).

Proposition 4.5.29 *$Det(1 + \mathscr{A}_M(G))$ is torsion free.*

4.5.30 We may define a natural map, α_G, by means of the following commutative diagram of $G(M/Q_l)$-maps:

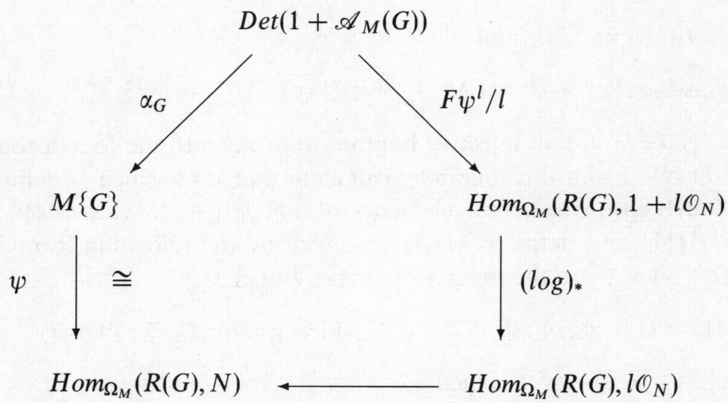

The lower map in this diagram is the injection induced by the inclusion of $l\mathcal{O}_N$ into N. Here F is the Frobenius of 4.3.7 and $F\psi^l/l$ is the map given by 4.3.10. The map ψ is 4.5.13, whose inverse is ϕ of 4.5.10.

Lemma 4.5.31 *If $\omega \in 1 + l\mathcal{O}_N$ then $log(\omega) = 0$ if and only if ω is an l-primary root of unity.*

Proof Firstly we note that if ω is an l-primary root of unity then $log(\omega)$ is defined and is zero by 4.3.17. Conversely, there is an integer, m, such that $\omega^{l^m} \in 1 + l^2\mathcal{O}_N$ so that

$$\begin{aligned} \omega^{l^m} &= exp(log(\omega^{l^m})) \\ &= exp(l^m log(\omega)) \\ &= exp(0) \\ &= 1. \end{aligned}$$

□

Proposition 4.5.32 *The $G(M/Q_l)$-map of* 4.5.30

$$\alpha_G : Det(1+\mathscr{A}_M(G)) \longrightarrow M\{G\}$$

is injective.

Proof When G is abelian there is nothing to prove; therefore we will assume that G is non-abelian.

Let $u \in 1+\mathscr{A}_M(G)$ and suppose that $\alpha_G(Det(u)) = 0$. This occurs if and only if the homomorphism

$$T \mapsto log\left(\frac{Det(F(u))\psi^l(T)}{(Det(u)(T))^l}\right)$$

is zero for all $T \in R(G)$. Since $R(G)$ is finitely generated, 4.5.31 implies that there exists an integer, m, such that $Det(u)^{l^m} = Det(u^{l^m})$ is in the kernel of $(F\psi^l/l)$. However, since $Det(1+\mathscr{A}_M(G))$ is torsion free, it will suffice to show that the kernel of $(F\psi^l/l)$ is trivial. Assume, therefore, that $(F\psi^l/l)(Det(u)) = 1$ and, furthermore, since α_G is a $G(M/\mathbf{Q}_l)$-map, we also have

$$(F\psi^l/l)(Det(F^j(u))) = 1$$

for all $j \geq 0$. This means that we have equations of the form

$$Det(F(u))(\psi^l(T)) \quad = Det(u)(T)^l,$$

$$\begin{aligned} Det(F^2(u))(\psi^{l^2}(T)) &= Det(F^2(u))(\psi^l(\psi^l(T))) \\ &= (Det(F(u))(\psi^l(T)))^l \\ &= (Det(u)(T))^{l^2}, \end{aligned}$$

and, in general, that for all $T \in R(G)$

$$Det(F^{l^m}(u))(\psi^{l^m}(T)) \quad = (Det(u)(T))^{l^m}.$$

Now suppose that $\#(G) = l^m$ so that

$$\psi^{l^m}(T) = dim(T) \in R(G).$$

This means that, if $\epsilon : \mathcal{O}_M[G] \longrightarrow \mathcal{O}_M[\{1\}] = \mathcal{O}_M$ is the augmentation map, then

$$\begin{aligned} Det(u^{l^m})(T) &= Det(\epsilon(F^{l^m}(u)))^{dim(T)} \\ &= 1 \end{aligned}$$

since $\epsilon(1+\mathscr{A}_M(G)) = \{1\}$. Therefore $Det(u)^{l^m} = 1$ and so $Det(u) = 1$, by 4.5.29, which completes the proof. □

4.5.33 Recall from 4.5.5 that there is a $G(M/\mathbf{Q}_l)$-map, $c : M[G] \longrightarrow M\{G\}$, which sends a group element to its conjugacy class. Therefore $c(\mathscr{A}_M(G))$ is an $\mathcal{O}_M$-submodule of $M\{G\}$. In particular, $l \cdot c(\mathscr{A}_M(G))$ is also an $\mathcal{O}_M$-submodule of $M\{G\}$ which is preserved by the $G(M/\mathbf{Q}_l)$-action.

Theorem 4.5.34 *The map, α_G, of* 4.5.30 *induces an isomorphism*

$$\alpha_G : Det(1 + \mathscr{A}_M(G)). \xrightarrow{\cong} l \cdot c(\mathscr{A}_M(G))$$

Proof We begin by proving that $\mathrm{Im}(\alpha_G)$ lies in $l \cdot c(\mathscr{A}_M(G))$. Let $u \in 1 + \mathscr{A}_M(G)$ then we may define, as in 4.3.23,

$$L_0(u) \in \mathcal{O}_M[G],$$

which, by naturality, lies in $\mathscr{A}_M(G)$ so that we have

$$L_0(u) \in \mathscr{A}_M(G).$$

By 4.3.28 the image of $L_0(u)$ in $M\{G\}$ lies in $l \cdot \mathcal{O}_M\{G\}$, where $\mathcal{O}_M\{G\}$ denotes the free $\mathcal{O}_M$-module on the conjugacy classes of G. The map, ψ, of 4.5.13 is given by

$$\psi\left(\sum m_\gamma \gamma\right)(T) = \sum m_\gamma Trace(T(\gamma)) = \xi_T\left(\sum m_\gamma \gamma\right)$$

where ξ_T is the map of 4.3.24. Therefore Proposition 4.3.25 translates into the statement that

$$\alpha_G(Det(u)) = c(L_0(u)) \in l \cdot \mathcal{O}_M\{G\}.$$

Hence

$$\alpha_G(Det(u)) \in l \cdot c(\mathcal{O}_M[G]) \cap c(\mathscr{A}_M(G)) = l \cdot c(\mathscr{A}_M(G)),$$

which completes the first part of the proof.

Now we must prove that α_G maps onto $l \cdot c(\mathscr{A}_M(G))$. We shall do this by induction on the order of G.

Choose a central element $z \in G$ which is of order l and which is a commutator. We shall prove next that

4.5.35 $$c(L_0(1 - (1 - z)\mathcal{O}_M[G])) = l \cdot c((1 - z)\mathcal{O}_M[G]).$$

If $x \in \mathcal{O}_M[G]$ then, since z is central, $\hat{F}((1 - z)x) = 0$ because $z^l = 1$. Therefore

$$\begin{aligned} L_0(1 - (1 - z)x) &= l(\textstyle\sum_{n=1}^{\infty}(1 - z)^n x^n/n) \\ &\equiv l(1 - z)x - (1 - z)^l x^l \quad (mod\, l(1 - z)^2 x^2) \end{aligned}$$

but $(1-z)^l \in l(1-z)\mathcal{O}_M[G]$ so that

$$L_0(1-(1-z)x) \in l(1-z)\mathcal{O}_M[G]$$

and therefore $c(L_0(1-(1-z))\mathcal{O}_M[G]) \subset l \cdot c((1-z))\mathcal{O}_M[G])$.

If $x \in J$ then one sees that

$$lc((1-z)x) \in c(L_0(1-(1-z)\mathcal{O}_M[G]))$$

by means of a standard approximation argument (see 4.7.13).

Hence we claim that

4.5.36 $$c((1-z)\mathcal{O}_M[G]) \subseteq c((1-z)J),$$

which will complete the proof of 4.5.35. Write $z = a^{-1}b^{-1}ab$ for $a, b \in G$ and let $v = \sum m_\gamma \gamma \in \mathcal{O}_M[G]$. We have to show that

4.5.37 $$c((1-z)v) \in c((1-z)J).$$

Rewrite v as $v = \sum m_\gamma(\gamma - a) + \sum m_\gamma a$ so that

$$\begin{aligned} c((1-z)v) &= c((1-z)(\textstyle\sum m_\gamma(\gamma-a) + \sum m_\gamma a)) \\ &= c((1-z)\textstyle\sum m_\gamma(\gamma-a)), \end{aligned}$$

since $c(a-za) = c(a-b^{-1}ab) = 0$. However, $\gamma - a \in Ker(\mathcal{O}_M[G] \longrightarrow \mathcal{O}_M)$, the kernel of the augmentation, and since G is an l-group the kernel of the augmentation lies inside J.

Finally, we prove the surjectivity of α_G by induction on the order of G. Set $H = G/\{z\}$ and assume that α_H is surjective.

The group, $\{z\}$, acts (by multiplication in G) on $\mathcal{O}_M[G]$, $\mathcal{O}_M\{G\}$, $M[G]$ and $M\{G\}$, since z is central. It is easily verified that

$$Ker(\mathcal{O}_M\{G\} \stackrel{\pi}{\longrightarrow} \mathcal{O}_M\{H\}) = (1-z)\mathcal{O}_M\{G\}$$

and that

$$Ker(M\{G\} \stackrel{\pi}{\longrightarrow} M\{H\}) = (1-z)M\{G\}.$$

Therefore, if we are given $y \in l \cdot c(\mathcal{A}_M(G))$ then we know that there exists $v \in 1 + \mathcal{A}_M(H)$ such that

4.5.38 $$\pi(y) = \alpha_H(Det(v)) = c(L_0(v)) \in l \cdot c(\mathcal{A}_M(H)).$$

Since $\mathcal{O}_M[G]$ is local we may lift v to $v' \in 1 + \mathcal{A}_M(G)$, using the fact that $G \longrightarrow G^{ab}$ factors through H. Thus

$$\begin{aligned} \alpha_G(Det(v')) - y &\in Ker(\pi : l \cdot c(\mathcal{O}_M[G]) \longrightarrow l \cdot c(\mathcal{O}_M[H])) \\ &= l \cdot Ker(\pi : c(\mathcal{O}_M[G]) \longrightarrow c(\mathcal{O}_M[H])) \end{aligned}$$

so that

$$\alpha_G(Det(v')) - y \in l \cdot c((1-z)\mathcal{O}_M[G])$$

and, by 4.5.35, there exists $v'' \in \mathcal{O}_M[G]$ with

$$\begin{aligned}\alpha_G(Det(v')) - y &= c(L_0(1-(1-z)v'')) \\ &= \alpha_G(Det(1-(1-z)v'')).\end{aligned}$$

Since $1-(1-z)v'' \in 1 + \mathscr{A}_M(G)$ we have shown that $y \in Im(\alpha_G)$, which completes the proof of Theorem 4.5.34. □

We shall now close this section by proving the following weak form of Theorem 4.5.3, from which one may easily deduce Theorem 4.5.3 by the argument of Fröhlich (1983, pp. 85–89).

Theorem 4.5.39 *Let $M/\mathbf{Q}_l$ be an unramified Galois extension of local fields. Let G be a finite l-group. Then*

$$Det(\mathcal{O}_M[G]^*)^{G(M/\mathbf{Q}_l)} \cong Det(\mathbf{Z}_l[G]^*).$$

Proof Consider the following diagram, whose rows are easily seen to be short exact (that is, the left map is injective and the right one is surjective, while the kernel equals the image in the middle):

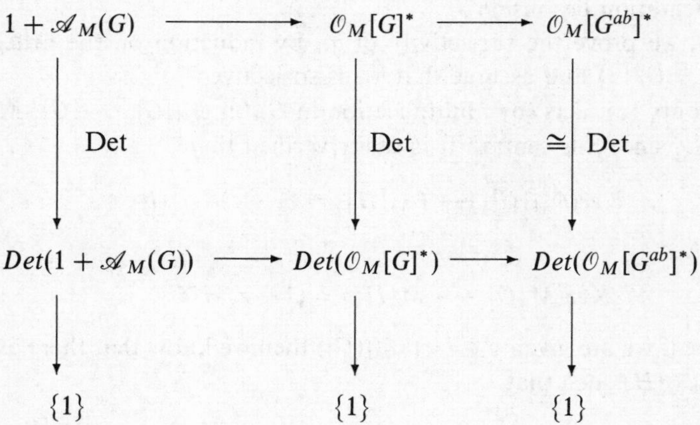

$$\begin{array}{ccccc} 1+\mathscr{A}_M(G) & \longrightarrow & \mathcal{O}_M[G]^* & \longrightarrow & \mathcal{O}_M[G^{ab}]^* \\ \downarrow \text{Det} & & \downarrow \text{Det} & & \cong \downarrow \text{Det} \\ Det(1+\mathscr{A}_M(G)) & \longrightarrow & Det(\mathcal{O}_M[G]^*) & \longrightarrow & Det(\mathcal{O}_M[G^{ab}]^*) \\ \downarrow & & \downarrow & & \downarrow \\ \{1\} & & \{1\} & & \{1\} \end{array}$$

in which the vertical maps are induced by the determinant, which is an isomorphism for abelian groups.

Let $H = G(M/\mathbf{Q}_l)$ then we may compare the bottom row for $\mathbf{Q}_l$ with the H-invariants of the bottom row for M.

$$\begin{array}{ccccc} Det(1+\mathscr{A}_{Q_l}(G)) & \longrightarrow & Det(\mathbf{Z}_l[G]^*) & \longrightarrow & Det(\mathbf{Z}_l[G^{ab}]^*) \\ \cong \downarrow \beta_1 & & \downarrow \beta_2 & & \cong \downarrow \beta_3 \\ Det(1+\mathscr{A}_M(G))^H & \longrightarrow & Det(\mathcal{O}_M[G]^*)^H & \longrightarrow & Det(\mathcal{O}_M[G^{ab}]^*)^H \end{array}$$

The map, β_3, is an isomorphism because

$$(\mathcal{O}_M[G^{ab}]^*)^H \cong \mathbf{Z}_l[G^{ab}]^*$$

and therefore the lower sequence is short exact. However, by 4.5.34, β_1 may be identified with the isomorphism

$$l \cdot c(\mathscr{A}_{\mathbf{Q}_l}(G)) \cong l \cdot (c(\mathscr{A}_M(G)))^H,$$

so that β_2 is an isomorphism, by the five-lemma, which completes the proof. □

4.6 Adams operations and determinants

4.6.1 As in the previous section let $M/\mathbf{Q}_l$ be a finite Galois extension and suppose, as in 4.5.39, that this extension is unramified. Let G be a finite group and let $N/\mathbf{Q}_l$ be a 'large' Galois extension, as in 4.3.1, which contains $M/\mathbf{Q}_l$. For any integer, h, we have the Adams operation, $\psi^h : R(G) \longrightarrow R(G)$, of 4.1.1 and an induced homomorphism

4.6.2 $$\psi^h : Hom_{\Omega_M}(R(G), N^*) \longrightarrow Hom_{\Omega_M}(R(G), N^*)$$

given by

$$\psi^h(f)(\chi) = f(\psi^h(\chi))$$

for all $\chi \in R(G)$.

In this section we shall be concerned with the proof of the following result, which is originally due to Ph. Cassou-Noguès and M.J. Taylor and is described in chapter 9 of Taylor (1980).

Theorem 4.6.3 *Let $Det(\mathcal{O}_M[G]^*) \subset Hom_{\Omega_M}(R(G), N^*)$ denote the image of the determinant homomorphism of* 4.5.2 *then, for all* $0 \leq h \in \mathbf{Z}$ *in* 4.6.2

$$\psi^h(Det(\mathcal{O}_M[G]^*)) \subset Det(\mathcal{O}_M[G]^*).$$

Remark 4.6.4 (a) Theorem 4.6.3 will be proved in a series of steps, of which (as was the case with Theorem 4.5.3) the most difficult is the case in which G is an l-group. In the previous section we proved only the l-group case of Theorem 4.5.3, claiming that reduction to that case is not difficult. In this section we shall carry out the analogous reduction to the l-group case. The assiduous and independently minded reader is invited to use this pattern of reduction to derive Theorem 4.5.3 from Theorem 4.5.39.

Here and there our proof of Theorem 4.6.3 will differ considerably from that of chapter 9 of Taylor (1980) by virtue of the fact that we have available better determinantal congruences; in particular, we do not require the use of the decomposition homomorphism (cf. Taylor, 1980, p. 106).

(b) The main application of Theorem 4.6.3 is to show that the Adams operation, ψ^h, induces an endomorphism of the class group

$$\psi^h : \mathcal{CL}(\mathcal{O}_K[G]) \longrightarrow \mathcal{CL}(\mathcal{O}_K[G]).$$

These endomorphisms exist when K is a number field which is unramified at the prime divisors of $\#(G)$ and when, in addition, h is odd or G has no irreducible symplectic characters. This last condition is required to take care of the infinite places in the Hom-description of 4.2.28. It is necessary in order that ψ^h should preserve the subgroup of $R(G)$ which is generated by symplectic characters, which in turn ensures that ψ^h preserves the Hom-group, $Hom^+_{\Omega_K}(R(G), E^*)$ of 4.2.33. For further details the reader is referred to Taylor (1980, p. 102).

Let us first deal with a very trivial case of 4.6.3.

Lemma 4.6.5 *Theorem* 4.6.3 *is true when* $HCF(l, \#(G)) = 1$.

Proof In this case $\mathcal{O}_M[G]$ is a maximal order in $M[G]$ (see 4.7.3) and therefore, by Fröhlich (p. 23),

$$Det(\mathcal{O}_M[G]^*) = Hom_{\Omega_M}(R(G), \mathcal{O}_N^*),$$

which is clearly mapped to itself by ψ^h. □

Lemma 4.6.6 *Let p be any prime (not necessarily equal to the residue characteristic, l) and let G be a finite p-group. If*

$$\alpha_G : Det(\mathcal{O}_M[G]^*) \longrightarrow M\{G\}$$

is the homomorphism defined in 4.5.30 *then*

$$\alpha_G(Det(\mathcal{O}_M[G]^*)) \subset l\mathcal{O}_M\{G\}.$$

Proof Let $J \triangleleft \mathcal{O}_M[G]$ denote the Jacobson radical. If $r \in J$ then, by the argument which was employed in the proof of Theorem 4.5.34, $\alpha_G(1+r) \in l\mathcal{O}_M\{G\}$. Hence α_G induces a map

4.6.7 $$\hat{\alpha} : \frac{\mathcal{O}_M[G]^*}{(1+J)} \longrightarrow \frac{M\{G\}}{l\mathcal{O}_M\{G\}}.$$

The image of $\hat{\alpha}$ must be a finite, abelian l-group and, since when $l = 2$ the group G has no homomorphic image equal to D_6, $\hat{\alpha}$ must be trivial, by the argument which was used in the proof of Theorem 4.3.37. □

Lemma 4.6.8 *Let G be a finite l-group. Suppose that*

$$f \in Ker(Hom_{\Omega_M}(R(G), N^*) \longrightarrow Hom_{\Omega_M}(R(G^{ab}), N^*))$$

satisfies

$$\tilde{F}(\psi^l(f)) = f^l,$$

where, as in 4.4.2, $\tilde{F}$ *extends the Frobenius in* $G(M/\mathbf{Q}_l)$ *and is trivial on the l-primary roots of unity. Then, for some* $m \geq 1$,

$$f^{l^m} = 1.$$

Proof We use an argument from the proof of 4.5.32. By that argument there exists an integer, t, such that

$$f(dim(T)) = \tilde{F}^t(f(dim(T))) = f(T)^{l^t}$$

for all $T \in R(G)$. However, since f lies in the kernel of the abelianisation map, $f(dim(T)) = 1$, which completes the proof. □

Lemma 4.6.9 *Let* $\mu_{l^\infty}(N)$ *denote the subgroup of* N^* *consisting of l-primary roots of unity. If* $M/\mathbf{Q}_l$ *is unramified then for any group, G,*

$$Hom_{\Omega_M}(R(G), \mu_{l^\infty}(N)) = Hom_{\Omega_{\mathbf{Q}_l}}(R(G), \mu_{l^\infty}(N)).$$

Proof Since Ω_M is a subgroup of $\Omega_{\mathbf{Q}_l}$ we have an inclusion of $Hom_{\Omega_{\mathbf{Q}_l}}$ into Hom_{Ω_M}. However, since $M/\mathbf{Q}_l$ is unramified, Ω_M and $\tilde{F}$ generate $\Omega_{\mathbf{Q}_l}$. Since $\tilde{F}$ acts trivially on $\mu_{l^\infty}(N)$ it also acts trivially on $R(G)$ and therefore any Ω_M-equivariant map, $h : R(G) \longrightarrow \mu_{l^\infty}(N)$, is automatically an $\Omega_{\mathbf{Q}_l}$-map. □

Definition 4.6.10 Let G be any finite l-group. Suppose that $M_1 \geq M \geq \mathbf{Q}_l$ is a chain of finite, unramified Galois extensions. Define the *norm* homomorphism

4.6.11
$$N_{M_1/M} : Hom_{\Omega_{M_1}}(R(G), N^*) \longrightarrow Hom_{\Omega_M}(R(G), N^*)$$

by

$$N_{M_1/M}(f) = \prod_{g \in G(M_1/M)} g(f),$$

where we identify $G(M_1/M)$ with Ω_M/Ω_{M_1}.

Lemma 4.6.12 *Under the circumstances of* 4.6.10

$$N_{M_1/M}(Det(\mathcal{O}_{M_1}[G]^*)) = Det(\mathcal{O}_M[G]^*)$$

for any l-group, G.

Proof As in the proof of Theorem 4.5.39 we have two horizontal short exact sequences in the following commutative diagram:

4.6.13

$$\begin{array}{ccccc}
l \cdot \mathcal{O}_M\{G\} & \longrightarrow & Det(\mathcal{O}_{M_1}[G]^*) & \longrightarrow & M_1[G^{ab}]^* \\
\downarrow Tr_{M_1/M} & & \downarrow N_{M_1/M} & & \downarrow N_{M_1/M} \\
l \cdot \mathcal{O}_{M_1}\{G\} & \longrightarrow & Det(\mathcal{O}_M[G]^*) & \longrightarrow & M[G^{ab}]^*
\end{array}$$

In 4.6.13 the trace map on the left is surjective, because M_1/M is unramified (Serre, 1979), and so is the right-hand norm map (from the case when G is abelian; see 4.7.16). A diagram chase completes the proof. □

Proposition 4.6.14 *Let G be a finite l-group. For any integer, h,*

$$\psi^h(Det(\mathcal{O}_M[G]^*)) \subset Det(\mathcal{O}_M[G]^*).$$

Proof Let $X = \sum \alpha_\gamma \gamma \in \mathcal{O}_M[G]^*$ and let $Y = \sum \alpha_\gamma \gamma^h$. Since the Jacobson radical is preserved by the function induced by sending γ ($\gamma \in G$) to γ^h and since $(\mathcal{O}_M[G])/J$ is finite we see that $Y \in \mathcal{O}_M[G]^*$. Therefore we may consider

4.6.15 $$f = \frac{\psi^h(Det(X))}{Det(Y)} \in Hom_{\Omega_M}(R(G), N^*).$$

Notice that f lies in the kernel of the abelianisation map of 4.6.8, since ψ^h on an abelian group is induced by the hth power map.

Now let us introduce the temporary notation

4.6.16 $$B = (\tilde{F}\psi^l/l)^{-1}(Hom_{\Omega_M}(R(G), 1 + l \cdot \mathcal{O}_N)),$$

where $\tilde{F}$ is the extended Frobenius of 4.4.2 and 4.6.8 and $(\tilde{F}\psi^l/l)$ is the endomorphism of $Hom_{\Omega_M}(R(G), N^*)$ which was introduced in 4.4.3. Therefore, as in 4.5.30, we may define a homomorphism

4.6.17 $$\alpha_G = \psi^{-1}(log(\tilde{F}\psi^l/l)) : B \longrightarrow M\{G\},$$

which extends the map of 4.5.30

$$\alpha_G : Det(1 + \mathscr{A}_M(G)) \longrightarrow M\{G\}.$$

Since ψ^l and ψ^h commute we find, from the determinantal congruences of 4.3.10, that

4.6.18 $$f \in B \cap ker(Hom_{\Omega_M}(R(G), N^*) \longrightarrow Hom_{\Omega_M}(R(G^{ab}), N^*)).$$

By 4.6.6, $\alpha_G(Det(X))$ and $\alpha_G(Det(Y))$ both lie in $l \cdot \mathcal{O}_M\{G\}$. Hence we may write

4.6.19 $$\alpha_G(Det(X)) = \sum_{[\gamma]} \beta_{[\gamma]}[\gamma] \in M\{G\}$$

where $[\gamma]$ runs through the conjugacy class of G and $\beta_{[\gamma]} \in l \cdot \mathcal{O}_M$. Therefore, for $T \in R(G)$,

$$\psi(\alpha_G(Det(X)))(\psi^h(T)) = \sum_{[\gamma]} \beta_{[\gamma]} Trace(T(\gamma^h))$$

and

$$\begin{aligned}\psi(\alpha_G(Det(X)))(\psi^h(T)) &= log(\tilde{F}\psi^l/l(Det(X)))(\psi^h(T)) \\ &= log(\tilde{F}\psi^l/l(\psi^h(Det(X))))(T), \\ &= \psi(\alpha_G(\psi^h(Det(X))))(T),\end{aligned}$$

since ψ^h commutes with $\tilde{F}\psi^l$.

Therefore

$$\alpha_G(\psi^h(Det(X))) = \sum_{[\gamma]} \beta_{[\gamma]}[\gamma^h] \in l \cdot \mathcal{O}_M\{G\}$$

and we find that

$$\alpha_G(f) \in l \cdot \mathcal{O}_M\{G\} \cap ker(M\{G\} \longrightarrow M\{G^{ab}\})$$

and, by Theorem 4.5.34, there exists

4.6.20 $$u \in Det(1 + \mathcal{A}_M(G))$$

which satisfies

$$\alpha_G(f/u) = 1.$$

By the properties of log (see 4.3.16) we may apply 4.6.10 and 4.6.11 to $f/u \in Hom_{\Omega_M}(R(G), N^*)$ (or to $(f/u)^2$ when $l = 2$) to conclude that

4.6.21 $$f/u \in Hom_{\Omega_M}(R(G), \mu_{l^\infty}(N)) = Hom_{\Omega_{\mathbf{Q}_l}}(R(G), \mu_{l^\infty}(N)).$$

Consider the operation which assigns to $Det(X)$ the l-primary torsion element, f/u, of 4.6.21. If both

$$\frac{\psi^h(Det(X))}{Det(\alpha)}, \frac{\psi^h(Det(X))}{Det(\beta)} \in Hom_{\Omega_M}(R(G), \mu_{l^\infty}(N))$$

for $\alpha, \beta \in \mathcal{O}_M[G]^*$ then

$$Det(\alpha/\beta) \in Det(\mathcal{O}_M[G]^*) \cap Hom_{\Omega_M}(R(G), \mu_{l^\infty}(N))$$

which is trivial, by a result of Wall (1979) see also Taylor (1980, p. 45). Hence we obtain a homomorphism

4.6.22 $\Gamma_M : Det(\mathcal{O}_M[G]^*) \longrightarrow Hom_{\Omega_M}(R(G), \mu_{l^\infty}(N))$ *given by* $\Gamma_M(Det(X)) = f/u$.

We must show that $f/u = 1$ to complete the proof. Suppose that $Hom_{\Omega_M}(R(G), \mu_{l^\infty}(N))$ has exponent l^e. Let M_1/M denote the unique unramified extension of degree l^e. Since the norm maps commute with ψ^h we obtain a commutative diagram in which the vertical maps are norms.

4.6.23

$$\begin{array}{ccc} Det(\mathcal{O}_{M_1}[G]^*) & \xrightarrow{\Gamma_{M_1}} & Hom_{\Omega_{M_1}}(R(G), \mu_{l^\infty}(N)) \\ \downarrow N_{M_1/M} & & \downarrow N_{M_1/M} \\ Det(\mathcal{O}_M[G]^*) & \xrightarrow{\Gamma_M} & Hom_{\Omega_M}(R(G), \mu_{l^\infty}(N)) \end{array}$$

In 4.6.23 the left-hand norm is surjective while the right-hand norm is the l^eth power map, by 4.6.9. Hence there exists $V \in Det(\mathcal{O}_{M_1}[G]^*)$ such that

$$\begin{aligned} f/u &= \Gamma_M(Det(X)) \\ &= \Gamma_M(N_{M_1/M}(V)) \\ &= N_{M_1/M}(\Gamma_{M_1}(V)) \\ &= (\Gamma_{M_1}(V))^{l^e} \\ &= 1, \end{aligned}$$

which completes the proof of Proposition 4.6.14. □

Definition 4.6.24 Let p be a prime. A finite group, G, is called **Q**-*p-elementary* (or *p-hyperelementary*) if it is a semi-direct product of the form, $H \propto C$, where H is a p-group and C is a cyclic group whose order is prime to p. Thus G contains C as a normal subgroup and is generated by C and H with $G/C \cong H$.

Let m be a divisor of $\#(C)$ and suppose that $\chi : C \longrightarrow N^*$ is a character of order m. Set

4.6.25 $$H_m = ker(H \longrightarrow Aut(\chi(C)))$$

so that H_m is the kernel of the conjugation action by H on C and thence on $\chi(C)$. Also set

4.6.26 $$A_m = H/H_m, \qquad G_m = H_m C \lhd G.$$

Lemma 4.6.27 *Let h be any positive integer.*

Let G be a **Q**-*p-elementary group, as in 4.6.24. Let $a \in \mathcal{O}_M[G]^*$ then there exists $b \in \mathcal{O}_M[G]^*$ such that, for all $\alpha \in R(C)$ and $\theta_m \in R(H_m)$ (where m divides $\#(C)$),*

$$Det(b)(Ind_{G_M}^G(\alpha\theta_m)) = Det(a)(Ind_{G_M}^G(\psi^h(\alpha)\theta_m)).$$

Here $\alpha\theta$ is the representation which is additive in α and when α is one-dimensional is given by the extension of $\theta : H_m \longrightarrow GL(V)$ to $G = H_mC$ by demanding that $c \in C$ acts via scalar multiplication by $\alpha(c)$.

Proof For $x \in C$ and $y \in H$ we may define a homomorphism, $\rho : G \longrightarrow G$, by $\rho(xy) = x^h y$. Hence ρ induces a ring endomorphism of $\mathcal{O}_M[G]$ and we may set $b = \rho(a)$. It is sufficient to verify the relation when $dim(\alpha) = 1$. Since $G/G_m \cong H/H_m$ we may choose coset representatives of G/G_m of the form $y_1, \ldots, y_t$ ($y_i \in H$). If $g = xy$ $(x \in C, y \in H)$ then there is a homomorphism

$$\sigma : G \longrightarrow G/C \longrightarrow \Sigma_t$$

such that

$$gy_i = y_{\sigma(g)(i)} h(i, g) = y_{\sigma(y)} h(i, g),$$

where $h(i, g) \in H_m$. In addition,

$$h(i, g) = h(i, y)(y_i^{-1} yxyy_i),$$

so that, by the formula of 2.1.9,

$$Ind_{G_m}^G(\psi^h(\alpha)\theta)(g)$$

$$= \sigma(g) \begin{pmatrix} \alpha(y_1^{-1}y^{-1}xyy_1)^h\theta(h(1, y)) & 0 & 0 & \ldots \\ 0 & \alpha(y_2^{-1}y^{-1}xyy_2)^h\theta(h(2, y)) & 0 & \\ 0 & 0 & & \\ \vdots & & & \end{pmatrix}$$

$$= Ind_{G_m}^G(\alpha\theta)(x^h y),$$

which establishes the required formula. □

Definition 4.6.28 Let $G = H \propto C$ be a **Q**-l-elementary group. For each divisor, m, of $\#(C)$ let $\xi_m = exp(2\pi i/m)$ and let $\lambda_1, \ldots, \lambda_{k(m)} : C \longrightarrow N^*$ be a complete set of Ω_M-orbits of abelian characters of order m. The group, A_m, of 4.6.26 acts by conjugation to permute the members of the set $\{\lambda_1, \ldots, \lambda_{k(m)}\}$.

Define a map

4.6.29

$$Hom_{\Omega_M}(R(G), \mathcal{O}_N^*) \longrightarrow \prod_{m|\#(C)} \left(\prod_{i=1}^{k(m)} Hom_{\Omega_{M(\xi_m)}}(R(H_m), \mathcal{O}_N^*) \right)$$

by $\alpha_{m,i}(f)(\theta_m) = f(Ind_{G_m}^G(\lambda_i \theta_m))$ for all $\theta_m \in R(H_m)$.

From Taylor (1980, p. 108 and p. 70, section 3.9) we have the following result.

Proposition 4.6.30 *In the notation of* 4.6.28 *the map,* α, *of* 4.6.29 *induces an isomorphism of the form*

$$\alpha : Det(\mathcal{O}_M[G]^*) \longrightarrow \prod_{m|\#(C)} \left(\prod_{i=1}^{k(m)} Det(\mathcal{O}_{M(\xi_m)}[H_m]^*) \right)^{A_m} .$$

Proposition 4.6.31 *Let G be a finite* **Q**-*l-elementary group. For any integer, h,*

$$\psi^h(Det(\mathcal{O}_M[G]^*)) \subset Det(\mathcal{O}_M[G]^*).$$

Proof Let $a, b \in \mathcal{O}_M[G]^*$ be as in 4.6.27. We are going to show that, for each $m \mid \#(C)$ and for each $1 \leq i \leq k(m)$, in 4.6.29

$$\alpha_{m,i}(\psi^h(Det(a))) = \psi^h(\alpha_{m,i}(Det(b))).$$

By 4.6.14 and 4.6.30, this will establish Proposition 4.6.31.

Suppose that $\theta_m \in R(H_m)$ and that, as in 2.3.9,

$$a_{H_m}(\theta_m) = \sum_j n_j(V_j, \phi_j)^{H_m} \in R_+(H_m).$$

Therefore, by 4.1.6, for all t,

4.6.32 $$\psi^t(\theta_m) = \textstyle\sum_j n_j Ind_{V_j}^{H_m}(\phi_j^t) \in R(H_m).$$

Also, if $\lambda_{m,i} : C \longrightarrow N^*$ is a character of order m then $\lambda_{m,i}\theta_m \in R(G_m)$ is a one-dimensional character such that, by a slight extension of 2.5.12,

$$a_{G_m}(\lambda_{m,i}\theta_m) = \sum_j n_j(CV_j, \lambda_{m,i}\phi_j)^{G_m} \in R_+(G_m).$$

Therefore, for all t,

4.6.33 $$\psi^t(\lambda_{m,i}\theta_m) = \textstyle\sum_j n_j Ind_{CV_j}^{G_m}(\lambda_{m,i}^t \phi_j^t) \in R(G_m).$$

Hence we have

$$\alpha_{m,i}(\psi^h(Det(a))(\theta_m))$$

$$= \prod_j [Det(a)(Ind_{CV_j}^G(\lambda_{m,i}^h \phi_j^h))]^{n_j}, \qquad \text{by 4.6.33 and 4.6.29,}$$

$$= \prod_j [Det(b)(Ind_{G_m}^G(\lambda_{m,i}^h Ind_{V_j}^{H_m}(\phi_j^h)))]^{n_j}$$

$$= Det(b)(Ind_{G_m}^G(\lambda_{m,i}\psi^h(\theta_m))), \qquad \text{by 4.6.32,}$$

$$= \psi^h(\alpha_{m,i}(Det(b)))(\theta_m),$$

as required. □

Proposition 4.6.34 *Let h be any positive integer. Let p be a prime, $p \neq l$, and let G be a* **Q**-*p-elementary group. There exists an integer, m, such that*

$$\{z^{l^m} \mid z \in \psi^h(Det(\mathcal{O}_M[G]^*))\} \subset Det(\mathcal{O}_M[G]^*).$$

Proof By definition (cf. 4.6.28) we may write $G = H \propto (C \times C')$, where H is a p-group, C is a cyclic l-group and C' is a cyclic group whose order is prime to lp. We will prove the proposition by induction on the order of C. When C is trivial we may take $m = 0$, by 4.6.5. Now let us assume that $\#(C) \neq 1$.

Every irreducible character of G is one of the form (cf. Serre, 1977, section 8.2) $V = Ind_J^G(\alpha)$, where $C \times C' \subset J$ and $dim(\alpha) = 1$. By 4.7.17, $\psi^l(V) = \psi^l(Ind_J^G(\alpha)) = Ind_J^G(\alpha^l)$. Let $C'' \subset C'$ denote the subgroup of order l. Hence $C'' \lhd G$ and α^l is trivial when restricted to C''. Therefore $\psi^l(V)$ is a representation of G which is trivial on C''. Let $\pi : G \longrightarrow G/C''$ be the quotient map, so that we have established that $\psi^l(R(G))$ is contained in the image of $\pi^* : R(G/C'') \longrightarrow R(G)$.

Let $z \in \mathcal{O}_M[G]^*$ so that, by Theorem 4.3.10,

$$\frac{\psi^l(Det(F(z)))}{(Det(z))^l} \in Hom_{\Omega_M}(R(G), 1 + l\mathcal{O}_N).$$

Since ψ^h commutes with ψ^l (and with the Frobenius, F),

$$\frac{\psi^l(\psi^h(Det(F(z))))}{(\psi^h(Det(z)))^l} \in Hom_{\Omega_M}(R(G), 1 + l\mathcal{O}_N).$$

If V belongs to $R(G)$ and $\psi^l(V) = \pi^*(V')$ with $V' \in R(G/C'')$ then, by induction, there exists an integer, t, and $z' \in \mathcal{O}_M[G/C'']^*$ such that

$$\{\psi^l(\psi^h(Det(F(z))))\}^{lt}(V)$$

$$= \{\psi^h(Det(z))\}^{lt}(\pi^*(V'))$$

$$= Det(z')(V').$$

Note that, by induction, t may be chosen so as to work for all $z \in \mathcal{O}_M[G]^*$.

We may choose $z'' \in \mathcal{O}_M[G]^*$ such that $\pi(z'') = z'$ so that

$$\{\psi^l(\psi^h(Det(F(z))))\}^{lt}(V)$$

$$= Det(\pi(z''))(V')$$

$$= Det(z')(\pi^*(V'))$$

$$= \psi^l(Det(z''))(V).$$

Using Theorem 4.3.10 once more on

$$\frac{\psi^*(Det(z''))}{(Det(F^{-1}(z'')))^l},$$

we find that, for all $z \in \mathcal{O}_M[G]^*$,

$$(\psi^h(Det(z)))^{l^{t+1}} \in Hom_{\Omega_M}(R(G), 1 + l\mathcal{O}_N).$$

Since $1 + \mathcal{O}_N$ is a pro-l-group, so is $Hom_{\Omega_M}(R(G), 1 + l\mathcal{O}_N)$, so that any finite quotient group is a finite l-group. The group

4.6.35 $$[Det(\mathcal{O}_M[G]^*)Hom_{\Omega_M}(R(G), 1 + l\mathcal{O}_N)]/(Det(\mathcal{O}_M[G]^*))$$

is finite, so that there exists an integer, s, such that $\psi^h(z)^{l^{t+1+s}}$ is trivial in 4.6.35 for all z, which completes the proof of Proposition 4.6.34. □

4.6.36 *Determinants and* $K_1(\mathcal{O}_M[G])$

We are almost ready for the final step in the proof of Theorem 4.6.3. In order to complete that step we shall require a small digression into algebraic K-theory. Further details on this material is to be found in Curtis & Reiner (1987, p. 61 *et seq.*).

Let A be a ring with a unit. We may embed the group, $GL_n(A)$, of invertible matrices in $M_n(A)$ into $GL_{n+1}(A)$ by sending X to

$$\begin{pmatrix} X & 0 \\ 0 & 1 \end{pmatrix}.$$

In this manner we may form the infinite general linear group,

4.6.37 $$GL(A) = \bigcup_n GL_n(A)$$

and thereby define an abelian group, $K_1(A)$, by the formula

4.6.38 $$K_1(A) = GL(A)^{ab} = GL(A)/[GL(A), GL(A)].$$

It is a result of J.H.C. Whitehead that the commutator group,

$$[GL(A), GL(A)],$$

is equal to the group generated by the elementary matrices. In particular, since $\mathcal{O}_M[G]^*$ is equal to $GL_1(\mathcal{O}_M[G])$, the determinantal construction of 4.3.3 extends to give a commutative diagram of the following form:

4.6.39

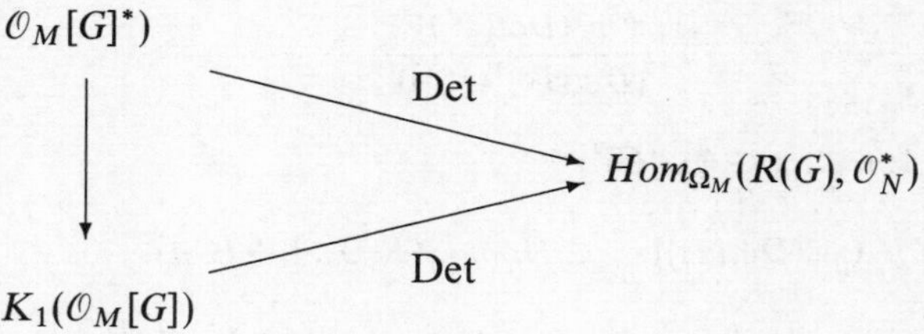

In addition, $K_1(\mathcal{O}_M[G])$ admits restriction and induction maps which make the following diagrams commutative, for all $H \leq G$.

4.6.40

$$\begin{array}{ccc} K_1(\mathcal{O}_M[H]) & \xrightarrow{\text{Det}} & Hom_{\Omega_M}(R(H), \mathcal{O}_N^*) \\ \downarrow{\scriptstyle Ind_H^G} & & \downarrow{\scriptstyle (Res_H^G)^*} \\ K_1(\mathcal{O}_M[G]) & \xrightarrow{\text{Det}} & Hom_{\Omega_M}(R(G), \mathcal{O}_N) \end{array}$$

4.6.41

$$\begin{array}{ccc} K_1(\mathcal{O}_M[G]) & \xrightarrow{Det} & Hom_{\Omega_{M_1}}(R(G), \mathcal{O}_N) \\ \downarrow Res^G_H & & \downarrow (Ind^G_H)^* \\ K_1(\mathcal{O}_M[H]) & \xrightarrow{Det} & Hom_{\Omega_M}(R(H), \mathcal{O}_N) \end{array}$$

In 4.6.40 and 4.6.41 f^* is the map induced by the map, f, of representation rings.

Proposition 4.6.42 *Let $\mathbf{Q}_l \leq M \leq N$ be as in* 4.6.1 *and let $\theta : G \longrightarrow GL_n(M)$ be a representation. If $z \in \mathcal{O}_M[G]^*$ then the map, $(\chi \mapsto Det(z)(\theta \otimes \chi) \in Hom_{\Omega_M}(R(G), \mathcal{O}_N^*)$ lies in the subgroup, $Det(\mathcal{O}_M[G]^*)$.*

Proof The key to this result is the fact that (Curtis & Reiner, 1987; Wall, 1966) $Det(K_1(\mathcal{O}_M[G])) = Det(\mathcal{O}_M[G]^*)$. If $z = \sum \alpha_\gamma \gamma$ then $\theta(z) = \sum \alpha_\gamma \theta(\gamma) \in GL_n(\mathcal{O}_M[G])$ and therefore gives rise to a class, $\theta_*(z) \in K_1(\mathcal{O}_M[G])$. It is clear that the function, $(\chi \mapsto Det(z)(\theta \otimes \chi))$, is equal to $(\chi \mapsto Det(\theta_*(z))(\chi))$, which lies in $Det(K_1(\mathcal{O}_M[G]))$. This completes the proof of 4.6.42. □

4.6.43 *Proof of Theorem* 4.6.3

This proof uses the partial results concerning $\mathbf{Q}$-p-elementary groups (4.6.31–4.6.34) to derive the general case.

Let p be a prime and let G be an arbitrary finite group. By Serre (1977, section 12.6) and Curtis & Reiner (1981) there exists an integer, $n(p)$, which is prime to p and virtual $\mathbf{Q}$-representations

$$\{\theta_i \in R_{\mathbf{Q}}(J_i); J_i \leq G, J_i \ \mathbf{Q}-p-\text{elementary}\},$$

such that

4.6.44 $$n(p) = \textstyle\sum_i Ind^G_{J_i}(\theta_i) \in R(G).$$

If $z \in \mathcal{O}_M[G]^*$ and $T \in R(G)$ we obtain, by Frobenius reciprocity (1.2.39),

$$n(p)T = \sum_i Ind^G_{J_i}(\theta_i Res^G_{J_i}(T)) \in R(G)$$

and

$$(\psi^h(Det(z))(T))^{n(p)} = \prod_i Det(z)(Ind_{J_i}^G(\theta_i Res_{J_i}^G(\psi^h(T))))$$

$$= \prod_i Det(Res_{J_i}^G(z))(\theta_i \psi^h(Res_{J_i}^G(T))),$$

by 4.6.41, where $Res_{J_i}^G(z) \in K_1(\mathcal{O}_M[J_i])$. By 4.6.42 there exist units, $u_i \in \mathcal{O}_M[J_i]^*$, such that, for each i,

$$Det(Res_{J_i}^G(z))(\theta_i \psi^h(Res_{J_i}^G(T))) = \psi^h(Det(u_i))(Res_{J_i}^G(T)).$$

If $p = l$ then, by 4.6.31 applied to J_i, there exists $v_i \in \mathcal{O}_M[J_i]^*$, which satisfies

$$\psi^h(Det(u_i))(Res_{J_i}^G(T))$$

$$= Det(v_i)(Res_{J_i}^G(T))$$

$$= Det(Ind_{J_i}^G(v_i))(T),$$

so that $(\psi^h(Det(z)))^{n(l)} \in Det(\mathcal{O}_M[G]^*)$.

However, if G is a $\mathbf{Q}$-p-elementary group with $p \neq l$ then, by 4.6.34, there exists an m such that $(\psi^h(Det(z)))^{l^m} \in Det(\mathcal{O}_M[G]^*)$. Since $HCF(l^m, n(l)) = 1$ we find that $(\psi^h(Det(z))) \in Det(\mathcal{O}_M[G]^*)$ when G is $\mathbf{Q}$-p-elementary with $p \neq l$.

Finally, let us return to the general case and suppose that $n(l) = p_1^{\alpha_1} \dots p_r^{\alpha_r}$ where $p_1, \dots, p_r$ are distinct primes. Repeating the first step with $\mathbf{Q}$-l-elementary groups replaced successively by $\mathbf{Q}$-p_i-elementary ones we find a set of integers, $n(p_i)$, which satisfy $HCF(n(p_i), p_i) = 1$ and $(\psi^h(Det(z)))^{n(p_i)} \in Det(\mathcal{O}_M[G]^*)$. Therefore $HCF(n(l), n(p_1), \dots, n(p_r)) = 1$ and so $(\psi^h(Det(z))) \in Det(\mathcal{O}_M[G]^*)$, which completes the proof of Theorem 4.6.3. □

4.7 Exercises

4.7.1 If $HCF(k, \#(G)) = 1$ show that

$$\psi^k(Ind_H^G(\rho)) = Ind_H^G(\psi^k(\rho)).$$

4.7.2 Let T be the binary tetrahedral group

$$T = \{a, b, c \mid a^2 = b^2 = c^3, c^6 = 1, bab^{-1} = a^{-1}, aca^{-1} = bc, ac = cb\}.$$

T consists of the 24 elements

$$T = \{a^h b^j c^k ; 0 \leq h < 4, 0 \leq j < 2, 0 \leq k < 3\}.$$

Also $T/\{a^2\} \cong A_4$. Define a two-dimensional, faithful representation, $v : T \longrightarrow SU(2)$, by

$$v(a) = \begin{pmatrix} i & 0 \\ 0 & -i \end{pmatrix},$$

$$v(b) = \begin{pmatrix} 0 & i \\ i & 0 \end{pmatrix}$$

and

$$v(c) = (\sqrt{2})^{-1} \begin{pmatrix} \epsilon^7 & \epsilon^7 \\ \epsilon^5 & \epsilon \end{pmatrix}$$

where $\epsilon = (1+i)/\sqrt{2}, i^2 = -1$.
Verify the following formulae in $R(T)$:
(i) $\psi^2(v) = Ind_{<c^3>}^T(1) - Ind_{<c>}^T(1) - Ind_{<a>}^T(1)$,
(ii) $\psi^4(v) = \psi^2(v) - Ind_{<a^2>}^T(1) + 2Ind_{<a>}^T(1)$
and
(iii) $\psi^6(v) = Ind_{<c>}^T(1) - Ind_{<a>}^T(1)$.

4.7.3 Let K be a local field whose ring of integers is $\mathcal{O}_K$. Let G be a finite group with $\#(G) = n$. Prove that $\mathcal{O}_K[G]$ is a maximal order in $K[G]$ if n is invertible in $\mathcal{O}_K$ (cf. Curtis & Reiner, 1981, p. 582).

4.7.4 Prove, under the action of $\Omega_{\mathbf{Q}}$ on $R(G)$, that complex conjugation fixes every symplectic representation of G.

4.7.5 Explain, to your own satisfaction, how Theorem 4.2.35 is derived from Theorem 4.2.28.

4.7.6 Let $\#(G) = n$ and suppose that $k = 1 + mn$ for some integer, m. By considering $1 + \sigma m \in \mathbf{Z}_p[G]$, when $p | \#(G)$, show by means of 4.2.28 that $S(k) = 0 \in \mathcal{CL}(\mathbf{Z}[G])$.

4.7.7 By means of the Hom-description show that

$$S(kk') = S(k) + S(k') \in \mathcal{CL}(\mathbf{Z}[G]).$$

4.7.8 Prove 4.7.6 and 4.7.7 directly in terms of $\mathbf{Z}[G]$-modules. For example, find a $\mathbf{Z}[G]$-module isomorphism of the form

$$\mathbf{Z}[G] \oplus S(kk') \cong S(k) \oplus S(k').$$

4.7.9 Let C_n denote the cyclic group of order n with generator, x. Show that, if $HCF(k,n) = 1$, then $1 + x + \ldots + x^{k-1} \in \mathbf{Z}_l[C_n]^*$ for all primes, l, which divide n.

4.7.10 Let D_{2n} denote the dihedral group of order $2n$,

$$D_{2n} = \{a, b \mid a^n = 1 = b^2, bab = a^{-1}\}.$$

Let $u_k = 1 + \sum_{i=1}^{k} a^i + \sum_{i=0}^{k-1} a^i b \in \mathbf{Z}[D_{2n}]$.
Assume that $2k+1$ is coprime to n and let l be a prime dividing $2n$.
(i) If $k = tn$ show that $u_k \in \mathbf{Z}_l[D_{2n}]^*$.
(ii) Suppose that $(1+v)u_k = u_{k+r(2k+1)}$ for some $v \in \mathbf{Z}[D_{2n}]$. Prove that

$$(1 + v + a^j b + a^{j+1})u_k = u_{k+(r+1)(2k+1)}$$

for $j = 2k + r(2k+1)$.
(iii) Prove that $u_k \in \mathbf{Z}_l[D_{2n}]^*$.

4.7.11 (*See Martinet,* 1977a, p. 527; *see* 5.2.13.) Let $G = Q_8 = \{x, y \mid x^2 = y^2, y^4 = 1, yxy^{-1} = x^{-1}\}$ be the quaternion group of order eight. Let M be a projective $\mathbf{Z}[G]$-module of rank one (i.e. $\mathbf{Q} \otimes_{\mathbf{Z}} M$ is a free $\mathbf{Q}[G]$-module on one generator). Set $M_+ = \{m \in M \mid x^2(m) = m\}$ and $M_- = \{m \in M \mid x^2(m) = -m\}$.

Hence M_+ is a projective module over the ring $R = \mathbf{Z}[G/\{x^2\}] = \mathbf{Z}[\mathbf{Z}/2 \times \mathbf{Z}/2]$ and hence is free of rank one. Also M_- is a free, rank one module over $S = \mathbf{Z}[i,j,k]$, the integral quaternions.

Let M_+ have an R-basis, Φ, and M_- have an S-basis, Ψ. Prove that
either (a) $\quad \Phi \equiv \Psi$ (mutually $2M$)
or (b) $\quad \Phi \equiv (x + y + xy)\Psi$ (mutually $2M$).

Prove that M is free if and only if (a) holds. Hence prove that $\mathcal{CL}(\mathbf{Z}[G]) \cong \mathbf{Z}/2 \cong T(G)$ generated by the Swan module, $S(3)$.

4.7.12 If $H \leq G$ show that Res_H^G maps $T(G)$ to $T(H)$ and that

$$Res_H^G(S(k)) = S(k) \in T(H).$$

(*Hint:* See Curtis & Reiner, 1987, p. 345.)

4.7.13 Complete the details of the approximation argument to show, in the proof of Theorem 4.5.34, that

$$c(L_0(1-(1-z)\mathcal{O}_M[G])) \geq lc((1-z)J),$$

where J is the Jacobson radical of $\mathcal{O}_M[G]$.

4.7.14 Let p be a prime and let $G = (F_p)^n$, where F_p ($\cong \mathbf{Z}/p$) is the additive group of the field of p elements. Let $x_1, \ldots, x_n$ denote generators for the individual F_p-factors of G. Suppose that $HCF(k,p) = 1$. Consider

$$U = \prod_{[z] \in P(F_p^n)} (1 + \theta(\underline{z}) + \ldots + \theta^{k-1}(\underline{z})) \in \mathbf{Z}_p[G]^*,$$

where $[z] = [z_1, \ldots, z_n]$ runs through representatives of the projective space of G and $\theta(\underline{z}) = x_1^{z_1} \ldots x_n^{z_n}$.

Use U to show that $p^{n-1}T(G) = 0$ if $p \neq 2$ and that $2^{n-2}T(G) = 0$ if $p = 2$.

4.7.15 Generalise the calculation of 4.7.14 to the group

$$G = (F_{p_1} \times \ldots \times F_{p_r})^n,$$

where $\{p_i\}$ are distinct primes.

4.7.16 Let G be a finite abelian l-group and let $\mathbf{Q}_l \subseteq M \subseteq M_1$ be a chain of finite, unramified Galois extensions. Show that the norm

$$N : M_1[G]^* \longrightarrow M[G]^* \cong (M_1[G]^*)^{G(M_1/M)},$$

given by $N(x) = \prod_{g \in G(M_1/M)} g(x)$, is surjective.

4.7.17 Let l be a prime and let $J \leq G$ be finite groups. Assume that the set of all l-primary elements of G forms a normal subgroup which is contained in J. If $\alpha \in R(J)$ show that

$$\psi^l(Ind_J^G(\alpha)) = Ind_J^G(\psi^l(\alpha)) \in R(G).$$

5

A class-group miscellany

Introduction

The determinantal functions play an important role in the Hom-description of the class-group of a group-ring. By applying the determinantal construction to the components of the Explicit Brauer Induction formula for a representation one may construct new types of homomorphisms, which I have christened *restricted determinants*, defined on the representation ring, $R(G)$, and giving rise to elements in the class-group. In Section 1 we describe these new maps and show that their images generate the entire class-group.

In Section 2 we use some homological algebra to derive a new invariant, η, which is a type of 'reduced norm' and which detects elements in the class-group of the quaternion group of order eight, $\mathcal{CL}(\mathbf{Z}[Q_8])$. This class-group is a very popular place in which to test conjectures concerning Galois module structure and I have found this invariant to be a little more convenient than that of Martinet (1977a); see also 4.7.11) since it does not require one to rewrite a projective module as an ideal plus a free module in order to use it. As an example we give a new derivation, in 5.2.33, of the formula for the element of the class-group which is given by a finite abelian group of odd order upon which Q_8 acts.

In Section 3 we derive some new relations between Swan modules, using the geometry of some maps, originally due to Frank Adams, between unit spheres in representations. These relations enable us, in some simple cases, to obtain new upper bounds for the Swan subgroup of G. The motivation for this section is probably more interesting than the results. In the study of the stable homotopy groups of spheres one of the major achievements was the evaluation of the subgroup which is called 'the image of the J-homomorphism'. All that we require to know about this image is that is it a computable subgroup of a very complicated group and in that way is analogous to the Swan subgroup of the class-group of a group-ring. In fact, the analogy goes further. In a famous series of papers Frank Adams (1963, 1965a,b, 1966) showed how to reduce this stable homotopy problem to a conjecture about Adams operations and sphere-bundles. The estimation of lower bounds for the image of the J-homomorphism by the use of Adams operations resembles very closely M.J. Taylor's later method for the determination of the Swan subgroup of a p-group (see 4.4.12). Pursuing this analogy leads one to adapt to the case of the Swan subgroup Adams' method for obtaining upper bounds for the image of the J-homomorphism by means of the algebraic topology of some maps between spheres.

The class-group of a maximal order in the rational group-ring of a finite group also has a Hom-description. In Section 4 we use a modification of the restricted determinants of Section 1 to design a means whereby to detect this class-group in some new Hom-groups which are defined in terms of the cyclic subquotients of G. In Chapter 7, Section 3 these detection homomorphisms are used to give a new proof of a result of David Holland which states that the Fröhlich–Chinburg conjecture is true in the class-group of a maximal order. In this Hom description, as with the Hom description of the 'kernel group', one of the quotients consists of a subgroup of adèlic-valued functions on $R(G)$ whose values satisfy a total positivity condition on symplectic representations. The main new result, which is necessary to make these detection homomorphisms work, is the analysis of the Explicit Brauer Induction homomorphism, a_G, on symplectic representations (see 5.4.42 and 5.4.43). In the proof of this result it was very convenient to use the more geometric version of Explicit Brauer Induction, which was given in Snaith (1988b), to analyse the behaviour of a_G on symplectic lines.

After the calculation of the Swan subgroup of a p-group the next case is the calculation of the Swan subgroup of a nilpotent group. Since a nilpotent group is a product of p-groups this calculation is related to a question of C.T.C. Wall (5.5.3) concerning the Swan subgroup of the product of two groups of coprime order. For nilpotent groups this problem can be reduced to the study of two primes at a time and we determine the answer in the elementary abelian case by the use of determinantal congruences (or, equivalently, the detection maps of Chapter 4). We conjecture that the elementary abelian case suffices to detect the Swan group of an arbitrary nilpotent group and verify this conjecture for several examples which involve small primes. In these examples we make use of some of the relations which were derived in Section 3 in order to reduce the study of the Swan group of a product with a semi-dihedral 2-group to the case of a product with the dihedral group of order eight.

Section 6 studies the elements in the class-group of a cyclic group which are given by p-power roots of unity acted upon by the cyclotomic Galois group. These finite Galois modules are cohomologically trivial and therefore define elements in the class-group of the integral group-ring. These modules not only provide many examples of non-trivial class-group elements but they also figure centrally in the computation, carried out in Chapter 7, of the global Chinburg invariant of a totally real cyclotomic Galois extension of p-power conductor. For these reasons, a discussion of these elements seemed appropriate for inclusion in this miscellany.

Section 7 consists of a collection of exercises, together with research problems which are suggested by our results on the Swan subgroup.

5.1 Restricted determinants

In 4.5.16 we showed how to combine the Explicit Brauer Induction homomorphism of 2.3.15 with the Hom-description of $M\{G\}$ in order to construct a split surjection of the form (see Theorem 4.5.23)

5.1.1
$$\mathscr{A}_G^* : \oplus_{\substack{(H)\\ H \leq G}} M[H^{ab}]^{W_G H} \longrightarrow M\{G\}.$$

In this section we shall briefly describe an analogous construction of maps onto the groups, $\mathscr{CL}(\mathcal{O}_K[G])$ and $D(\mathcal{O}_K[G])$ of 4.2.28 and 4.2.29, where K is a number field.

We maintain the notation of 4.5.16. In particular, if $H \leq G$ then $W_G H = N_G H/H$, where $N_G H$ is the normaliser of H in G.

Let E/K be a large Galois extension of number fields, as in 4.2.8. Recall from 2.3.15 the Explicit Brauer Induction homomorphism

$$a_G : R(G) \longrightarrow R_+(G)$$

given by the formula $a_G(\rho) =$

$$\#(G)^{-1} \sum_{\substack{(H,\phi)\leq(H',\phi') \\ in \mathscr{M}_G}} \#(H)\mu^{\mathscr{M}_G}_{(H,\phi),(H',\phi')} < \phi', Res^G_{H'}(\rho) > (H,\phi)^G,$$

where $(H,\phi)^G$ denotes the G-conjugacy class of a one-dimensional representation, $\phi : H \longrightarrow E^*$. For each $H_1 \leq G$ we may define a homomorphism

5.1.2 $$det_{H_1} : J^*(K[H_1^{ab}])^{W_G H_1} \longrightarrow Hom_{\Omega_K}(R(G), J^*(E))$$

by the formula ($\alpha \in J^*(K[H_1^{ab}])^{W_G H_1}$, $\rho \in R(G)$)

5.1.3 $$det_{H_1}(\alpha)(\rho) = \prod_{(H_1,\phi)^G \in \mathcal{M}_G/G} det(\phi(\alpha))^{\alpha_{(H_1,\phi)^G}} \in J^*(E),$$

where, in 5.1.3, the product is taken over those $(H,\phi)^G$ in

$$a_G(\rho) = \sum \alpha_{(H,\phi)^G}(H,\phi)^G,$$

for which $H = H_1$ (up to conjugation in G).

Note that 5.1.3 is well-defined because α is a $W_G H_1$-invariant idèle and $\phi : H_1 \longrightarrow E^*$ factors through the abelianisation map, $H_1 \longrightarrow H_1^{ab}$. Passing to the quotient in 4.2.28 yields a *restricted determinantal homomorphism*

5.1.4 $$Det_{H_1} : J^*(K[H_1^{ab}])^{W_G H_1} \longrightarrow \mathcal{CL}(\mathcal{O}_K[G])$$

for each subgroup, $H_1 \leq G$.

Now let $\Lambda(G)$, $\mathcal{O}_K[G] \leq \Lambda(G) \leq K[G]$, denote a maximal order. Let $J^*(\Lambda(G)) \subset J^*(K[G])$ denote the subgroup whose elements have all their coordinates lying in the completion, $\Lambda(G)_P$, as P varies through the primes of K. We remark that $\Lambda(G)$ is well-defined up to conjugation by a unit of $K[G]$. Also, when W is abelian then $K[W]$ is isomorphic to a product of number fields by means of an isomorphism which is induced by characters of the form, $\phi : W \longrightarrow E^*$. Since the ring of algebraic integers is a maximal order in a number field this means that when W is abelian ϕ induces a homomorphism, $Det(\phi) : \Lambda(W)^* \longrightarrow \mathcal{O}_E^*$, and an idèlic homomorphism of the form

5.1.5 $$Det_\phi : J^*(\Lambda(W)) \longrightarrow U(\mathcal{O}_E).$$

Therefore the homomorphism of 5.1.2 gives, by means of 4.2.35, a restricted determinantal homomorphism

5.1.6 $$Det_H : J^*(\Lambda(H^{ab}))^{W_G H} \longrightarrow D(\mathcal{O}_K[G]).$$

Remark 5.1.7 (i) The restricted determinants were first introduced in Snaith (1990b); see also Snaith (1990a). These homomorphisms enjoy naturality properties similar to those of $\mathcal{A}_G^*$ in Theorem 4.5.23, which are induced from the naturality property of a_G (see 2.3.8(i)). We leave the derivation of this property to the interested reader and refer to Snaith (1990b) for further details.

(ii) As in the case for Det in 4.6.39, the restricted determinants factor through the adèlic algebraic K-groups, $K_1(J(K[H^{ab}]))^{W_G H}$ and $K_1(J(\Lambda(H^{ab})))^{W_G H}$ where $J(K[H^{ab}])$ is the adèle ring of 4.2.7.

Theorem 5.1.8 *Let G be a finite group and let K be a number field. Then*

$$\text{(i)} \qquad \sum_{H \leq G} Det_H(J^*(K[H^{ab}])^{W_G H}) = \mathcal{CL}(\mathcal{O}_K[G])$$

and

$$\text{(ii)} \qquad \sum_{H \leq G} Det_H(J^*(\Lambda(H^{ab}))^{W_G H}) = D(\mathcal{O}_K[G]).$$

Proof We shall prove only (ii), since part (i) is similar but easier.

Let $\mathcal{C}$ denote a set of conjugacy class representatives of subgroups of G. If $\chi \in R(G)$ and $a_G(\chi) = \sum_i \alpha_i(H_i, \phi_i)^G$ then, for $H \in \mathcal{C}$, define

$$a_{G,H}(\chi) = \sum_{H_i = H} \alpha_i(H_i, \phi_i)^G.$$

We may consider $a_{G,H}(\chi)$ as a linear combination of one-dimensional characters of H and we may form $Ind_H^G(a_{G,H}(\chi)) \in R(G)$. Suppose that $z \in J^*(\Lambda(G))$ then we may consider the function

5.1.9 $\qquad (\chi \longrightarrow Det(z)Ind_H^G(a_{G,H}(\chi))) \in Hom_{\Omega_K}(R(G), U(\mathcal{O}_E)),$

where, as usual, E is a large splitting field for G.

Let P be a prime of K and let Q be a prime of E over P. Let us consider the E_Q-component of 5.1.9 which is given by

$$(\chi \longrightarrow Det(z_P)Ind_H^G(a_{G,H}(\chi))) \in Hom_{\Omega_{K_P}}(R(G), \mathcal{O}_{E_Q}^*).$$

The image of z_P under the composition

5.1.10 $\qquad \Lambda(G)_P^* \longrightarrow K_1(K_P[G]) \xrightarrow{Ind} K_1(K_P[H])^{W_G H}$

$$\longrightarrow K_1(K_P[H^{ab}])^{W_G H} \cong (K_P[H^{ab}]^*)^{W_G H}$$

lies in $(\Lambda(H^{ab})_P)^{W_G H}$. To see this consider the following commutative diagram:

5.1.11

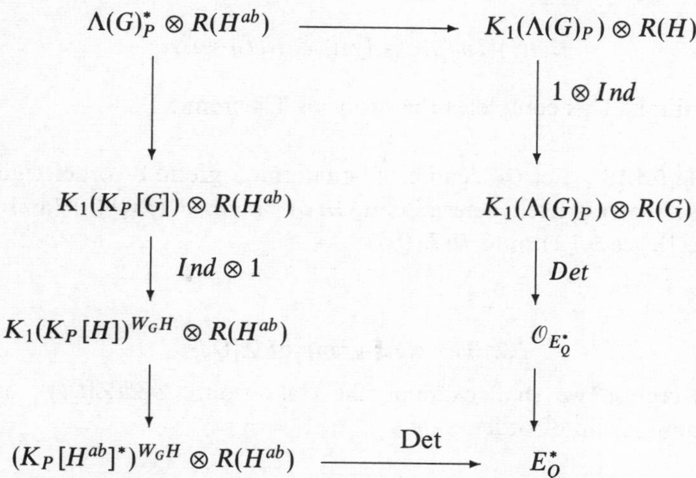

If we start with $z_P \otimes \psi$ in the top left corner of 5.1.11 we see that, for all $\psi \in R(H^{ab})$,

$$Det(w_P \otimes \psi) \in \mathcal{O}^*_{E_Q},$$

where w_P is the image of z_P under 5.1.10. This means that w_P lies in the intersection of $(K_P[H^{ab}]^*)^{W_G H}$ with a maximal $\mathcal{O}_{E_Q}$-order of $E_Q[H^{ab}]$. This is because the characters, ψ, of H^{ab} induce an isomorphism of E-algebras

$$\prod Det_\psi : E[H^{ab}] \longrightarrow \prod_\psi E_Q,$$

which carries a maximal order in $E_Q[H^{ab}]$ onto $\prod_\psi \mathcal{O}_{E_Q}$ (see 5.1.4). Therefore $w_P \in (\Lambda(H^{ab})^*_P)^{W_G H}$.

Now, by the fundamental property of Explicit Brauer Induction,

$$\chi = \sum_{H \in \mathscr{C}} Ind_H^G(a_{G,H}(\chi)) \in R(G)$$

so that

5.1.12 $$Det(z)(\chi) = \prod_{H \in \mathscr{C}} Det(z)\{Ind_H^G(a_{G,H}(\chi))\}.$$

As $z \in J^*(\Lambda(G))$ varies, the functions, $Det(z)$, generate $D(\mathcal{O}_K[G])$ so that the functions

$$\{Det(z)\{Ind_H^G(a_{G,H}(\_))\}; H \in \mathscr{C}, z \in J^*(\Lambda(G))\}$$

also generate $D(\mathcal{O}_K[G])$. However, if $z = (z_P)$ then $w = (w_P) \in J^*(H^{ab})^{W_G H}$, by the previous discussion, and

$$Det(z)\{Ind_H^G(a_{G,H}(\chi))\} = Det_H(w)(\chi),$$

by definition. This completes the proof of Theorem 5.1.8. □

Example 5.1.13 Let Q_8 denote the quaternion group of order eight. In this case the restricted determinant, Det_{Q_8}, maps $J^*(\Lambda(Q_8^{ab}))$ (and even $U(\mathbf{Z}[Q_8^{ab}])$, see 5.7.1) onto $D(\mathbf{Z}[Q_8])$.

5.2 The class-group of $\mathbf{Z}[Q_8]$

In this section we shall examine the class-group, $\mathcal{CL}(\mathbf{Z}[Q_8])$, of the quaternion group of order eight

5.2.1 $$Q_8 = \{x, y \mid x^2 = y^2, y^4 = 1, xyx^{-1} = y^{-1}\}.$$

We shall give an explicit isomorphism

5.2.2 $$n : \mathcal{CL}(\mathbf{Z}[Q_8]) \xrightarrow{\cong} \mathbf{Z}/2$$

(cf. 4.7.11) whose description differs from those to be found in Martinet (1977a) and Taylor (1984, p. 88, proposition 2.21). The map, n, is a type of reduced norm map. It does not seem to have appeared previously in the literature. As an application of 5.2.2 we shall derive a result (5.2.33) which was first obtained in Chinburg (1989, p. 44) from Taylor (1984, p. 88).

Incidentally, our invariant which gives the isomorphism of 5.2.2 is similar in spirit to that of Martinet (1977a), except that 5.2.2 has the advantage of being directly applicable to any finitely generated, projective $\mathbf{Z}[Q_8]$-module, M, whereas to apply the criterion of Martinet (1977a); (see 4.7.11) one must first find an isomorphism of the form

5.2.3 $$M \cong (\mathbf{Z}[Q_8])^r \oplus M_1,$$

where M_1 is an ideal of $\mathbf{Z}[Q_8]$.

Definition 5.2.4 Let M be a finitely generated, projective $\mathbf{Z}[Q_8]$-module. Define M_+ and M_- by the formulae

$$M_+ = \{m \in M \mid x^2(m) = m\}$$

and

$$M_- = \{m \in M \mid x^2(m) = -m\}.$$

Note that M_+ is naturally a module over the ring $\mathbf{Z}[Q_8]/(x^2-1) \cong \mathbf{Z}[\mathbf{Z}/2 \times \mathbf{Z}/2]$, while M_- is a module over $\mathbf{Z}[Q_8]/(x^2+1) \cong \mathbf{H_Z}$, where $\mathbf{H_Z}$ denotes the ring of integral quaternions,

5.2.5 $$\mathbf{H_Z} = \mathbf{Z}[i,j,k]/(i^2 = j^2 = k^2 = -1, ij = k, jk = i, ki = j).$$

Proposition 5.2.6 *Let N be a finitely generated, projective R-module where $R = \mathbf{Z}[\mathbf{Z}/2 \times \mathbf{Z}/2]$ or $\mathbf{H_Z}$, then N is a free R-module.*

Proof See Martinet (1977a). □

Proposition 5.2.7 *Let M be a finitely generated, projective $\mathbf{Z}[Q_8]$-module. Then there is an isomorphism of $\mathbf{Z}[\mathbf{Z}/2 \times \mathbf{Z}/2]$-modules*

$$\psi_M : M/M_- \xrightarrow{\cong} M_+$$

given by $\psi_M(m + M_-) = (1 + x^2)m$.

Proof Clearly ψ_M is injective. In addition, ψ_M is evidently an isomorphism when M is free. If M is a finitely generated, projective module then there exists a complementary module, N, such that $M \oplus N$ is free. Since $\psi_{M\oplus N} = \psi_M \oplus \psi_N$ is an isomorphism one sees that ψ_M is surjective, which completes the proof.

Alternatively, one can show that ψ_M is surjective by observing that the cohomology group

5.2.8 $$H^2(\{x^2\}; M) = M_+/((1 + x^2)M)$$

vanishes, because M is projective over $\mathbf{Z}[\mathbf{Z}/2] = \mathbf{Z}[\{x^2\}]$. □

We are now ready to define the invariant which yields the isomorphism of 5.2.2.

Definition 5.2.9 *The invariant, $n([M])$*

Let M be a finitely generated, projective $\mathbf{Z}[Q_8]$-module. Hence M_+ (and M_-) is a finitely generated, projective module over $\mathbf{Z}[\mathbf{Z}/2 \times \mathbf{Z}/2]$ (and $\mathbf{H_Z}$, respectively) and hence is free.

Therefore we may take a free $\mathbf{H_Z}$-basis

$$\underline{a} = \{a_1, \ldots, a_k\} \in M_-$$

and we may choose, by 5.2.7,

$$\underline{\alpha} = \{\alpha_1, \ldots, \alpha_k\} \in M$$

such that

$$\{\alpha_1 + M_-, \ldots, \alpha_k + M_-\}$$

is a free $\mathbf{Z}[\mathbf{Z}/2 \times \mathbf{Z}/2]$-basis for $M/M_- \cong M_+$.

Now form

$$(1 - x^2)\underline{\alpha} = \{(1 - x^2)\alpha_1, \ldots, (1 - x^2)\alpha_k\} \subset M_-.$$

There is a $k \times k$ matrix, X, with entries in $\mathbf{H_Z}$, such that

5.2.10
$$\begin{pmatrix} (1 - x^2)\alpha_1 \\ \vdots \\ (1 - x^2)\alpha_k \end{pmatrix} = X \begin{pmatrix} a_1 \\ \vdots \\ a_k \end{pmatrix} \in M_-^k.$$

Hence $X \in M_k(\mathbf{H_Z})$, the $k \times k$ integral quaternionic matrices. Let

5.2.11
$$c : M_k(\mathbf{H_Z}) \longrightarrow M_{2k}(\mathbf{Z}[i])$$

denote the complexification map ($i^2 = -1; \mathbf{Z}[i]$, the Gaussian integers) and set

5.2.12
$$n([M]) = det(c(X)) \pmod 4 \in \mathbf{Z}/4.$$

Note that 5.2.12 makes sense since $det(c(X))$ is a positive integer (see 5.7.2).

Theorem 5.2.13 *The construction of Section* 5.2.9 *yields an invariant,*

$$n([M]) \in (\mathbf{Z}/4)^* \cong \mathbf{Z}/2,$$

which depends only on $[M]$, *the class of* M *in* $\mathscr{CL}(\mathbf{Z}[Q_8])$, *and defines an isomorphism*

$$n : \mathscr{CL}(\mathbf{Z}[Q_8]) \xrightarrow{\cong} \mathbf{Z}/2.$$

The proof of Theorem 5.2.13 will be accomplished in the following series of subsidiary results (5.2.14–5.2.18).

Lemma 5.2.14 *If* $X \in M_k(\mathbf{H_Z})$ *then* $det(c(X))$ $(mod\ 4) \in \mathbf{Z}/4$ *depends only upon* X $(mod\ 2M_k(\mathbf{H_Z}))$.

Proof If $\mathbf{H_Z}$ is as in 5.2.5 we may write $X = Y + Wj$, where $Y, W \in M_k(\mathbf{Z}[i])$ so that

5.2.15
$$c(X) = \begin{pmatrix} Y & -W \\ \overline{W} & \overline{Y} \end{pmatrix},$$

where $\overline{W}$ denotes the complex conjugate of W.

Let E_{st} denote the $k \times k$ matrix having a 1 in the (s,t)th entry and zeros elsewhere. Let us examine the effect of changing X by adding $2x$ in the (s,t)th entry, where $x \in \mathbf{Z}[i]$.

Write $M(Y)_{st}$ for the cofactor formed by taking the determinant of the complex matrix obtained from Y by deleting the sth row and the tth column.

Thus

$$\begin{aligned} det(c(X + 2xE_{st})) &= det\begin{pmatrix} Y + 2xE_{st} & -W \\ \overline{W} & \overline{Y} + 2\overline{x}E_{st} \end{pmatrix} \\ &\equiv det(c(X)) + 2x(-1)^{s+t}M(c(X))_{st} \\ &\quad +2\overline{x}(-1)^{k+s+k+t}M(c(X))_{k+s,k+t} \\ &\quad (mod\ 4), \\ &\equiv det(c(X)) + (-1)^{s+t}[2xM(c(X))_{st} \\ &\quad +2\overline{x}M(A)_{s,t}] \\ &\quad (mod\ 4) \end{aligned}$$

where

$$A = \begin{pmatrix} \overline{Y} & \overline{W} \\ -W & Y \end{pmatrix}.$$

Since we are working modulo 4 the signs within the square brackets are irrelevant and therefore we obtain

$$\begin{aligned} &det(c(X + 2xE_{st})) \\ &\equiv det(c(X)) + (-1)^{s+t}2[xM(c(X))_{st} + \overline{xM(c(X))_{st}}] \\ &\equiv det(c(X)) \qquad (mod\ 4). \end{aligned}$$

A similar calculation applies to analyse the case when $X = Y + Wj$ is replaced by $Y + ((W + 2xE_{st})j)$. □

Proposition 5.2.16 *The invariant of* 5.2.9 *yields a well-defined homomorphism*

$$n : \mathscr{CL}(\mathbf{Z}[Q_8]) \longrightarrow (\mathbf{Z}/4)^*.$$

Proof In the notation of 5.2.9, if the $\mathbf{H_Z}$-basis, $\underline{a}$, is changed to $\underline{a'} = V\underline{a}$ where $V \in GL_k(\mathbf{H_Z})$, then X is changed to XY^{-1}. However, $det(c(Y)) \in \mathbf{Z}[i]^*$ is a positive integer, by 5.7.2. Therefore $det(c(Y)) = 1$ and $det(c(XY^{-1})) = det(c(X))det(c(Y))^{-1} = det(c(X))$.

Given a $\mathbf{Z}[\mathbf{Z}/2 \times \mathbf{Z}/2]$-basis,

$$\underline{\alpha} + M_- = \{\alpha_1 + M_-, \ldots, \alpha_k + M_-\},$$

the vector, $\underline{\alpha} = (\alpha_1, \ldots, \alpha_k) \in M^k$ is well-defined only up to the addition of $\underline{\beta} = (\beta_1, \ldots, \beta_k)$ where $\beta_i \in M_-$. However,

$$\begin{aligned}(1 - x^2)(\underline{\alpha} + \underline{\beta}) &= (1 - x^2)\underline{\alpha} + 2\underline{\beta} \\ &= X\underline{a} + 2X'\underline{a}\end{aligned}$$

for some $X' \in M_k(\mathbf{H_Z})$. Since, by 5.2.14, $det(c(X)) \equiv det(c(X + 2X'))$ (mod 4) we see that $n([M])$ is independent of the lifting of $\underline{\alpha} + M_-$ to $\underline{\alpha} \in M^k$.

Now suppose that the $\mathbf{Z}[\mathbf{Z}/2 \times \mathbf{Z}/2]$-basis, $\underline{\alpha} + M_- = \{\alpha_1 + M_-, \ldots, \alpha_k + M_-\}$, is changed to $U\underline{\alpha} + M_-$ for some $U \in GL_k(\mathbf{Z}[\mathbf{Z}/2 \times \mathbf{Z}/2])$. Hence there exists $U' \in M_k(\mathbf{H_Z})$ such that $U\underline{\alpha} + M_-$ lifts to $U\underline{\alpha}$ and

$$\begin{aligned}(1 - x^2)U\underline{\alpha} &= (1 - x^2)U'\underline{\alpha} + M_- \\ &= U'(1 - x^2)\underline{\alpha} \\ &= U'X\underline{a}.\end{aligned}$$

However, $\mathbf{H_Z} \otimes \mathbf{Z}/2 \cong \mathbf{Z}/2[\mathbf{Z}/2 \times \mathbf{Z}/2]$ and therefore the reductions modulo 2 of U and U' coincide. Hence $U' \in M_k(\mathbf{H_Z})$ is a unit (modulo 2) and therefore there exists $V' \in M_k(\mathbf{H_Z})$ such that $1 \equiv U'V'$ (mod $2M_k(\mathbf{H_Z})$). Therefore, by 5.2.14,

$$\begin{aligned}det(c(U'X)) &= det(c(U'))det(c(X)) \\ &\equiv det(c(X)) \qquad (mod\ 4).\end{aligned}$$

Hence we have shown that, given M, the invariant $n([M]) \in \mathbf{Z}/4$ is independent of the choices of bases $\underline{a}$, $\underline{\alpha} + M_-$ and of the lifting $\underline{\alpha}$.

To complete the proof of 5.2.16 it remains to show that

$$n([M \oplus M']) = n([M])n([M']),$$

since $n([\mathbf{Z}[Q_8]]) \equiv 1$. However, $(M \oplus M')_+ = M_+ \oplus M'_+$ and $(M \oplus M')_- = M_- \oplus M'_-$. Therefore bases and liftings for $(M \oplus M')_+$ and $(M \oplus M')_-$ may be obtained by combining bases and liftings for M_+, M_-, M'_+ and M'_-. In this case, if M and M' give rise, respectively, to matrices X and X', in the construction of 5.2.9, then $(M \oplus M')$ will yield the matrix, $X \oplus X'$. Hence

$$\begin{aligned} n([M \oplus M']) &= det(c(X \oplus X')) \\ & det(c(X))det(c(X')) \\ & n([M])n([M']), \end{aligned}$$

which completes the proof. □

Proposition 5.2.17 *In 5.2.16, the homomorphism, n, is surjective.*

Proof Let $M =< 3, \sigma >$, the Swan module of 4.2.43. Here $\sigma = \sum_{g \in Q_8} g \in \mathbf{Z}[Q_8]$. In this case

$$M_- =< 3, \sigma > \cap (1 - x^2)\mathbf{Z}[Q_8] = 3(1 - x^2)\mathbf{Z}[Q_8]$$

and a basis for this $\mathbf{H_Z}$-module is given by $3(1 - x^2)$.

On the other hand, $\sigma - 3(x+y+xy) \in< 3, \sigma >$ is a basis for the quotient $< 3, \sigma > /M_-$. To see this embed $< 3, \sigma > /M_-$ into $\mathbf{Z}[\mathbf{Z}/2 \times \mathbf{Z}/2]$ via the abelianisation map. This induces an isomorphism

$$< 3, \sigma > /M_- \xrightarrow{\cong} < 3, 2\tau >=< 3, \tau > \lhd \mathbf{Z}[\mathbf{Z}/2 \times \mathbf{Z}/2],$$

where $\tau = 1 + x + y + xy \in \mathbf{Z}[\mathbf{Z}/2 \times \mathbf{Z}/2]$.

Since $\tau(3-\tau) = 3\tau - 4\tau = -\tau$ we see that $< 3, \tau >$ is a free $\mathbf{Z}[\mathbf{Z}/2 \times \mathbf{Z}/2]$-module with basis

$$3 - \tau = 3(1 - \tau) + 2\tau,$$

so that $< 3, \sigma > /M_-$ is a free $\mathbf{Z}[\mathbf{Z}/2 \times \mathbf{Z}/2]$-module with basis, $\alpha = \sigma - 3(x + y + xy) \in \mathbf{Z}[Q_8]$.

Hence $(1 - x^2)\alpha = 3(x^2 - 1)(x + y + xy)$ and since

$$det(c(i + j + ij)) = det \begin{pmatrix} i & -(1+i) \\ (1-i) & -i \end{pmatrix}$$

is equal to three we find that $n(< 3, \sigma >) \equiv 3 \pmod 4$. □

5.2.18 *Conclusion of the proof of Theorem 5.2.13*

By virtue of 5.2.3, 5.2.16 and 5.2.17 it will suffice to prove that there are precisely two isomorphism classes of projective $\mathbf{Z}[Q_8]$-modules, M, of rank one. We shall classify these by means of some elementary homological algebra for which the background may be found in Hilton & Stammbach (1971, chapter III) and Snaith (1989b, chapter 2).

Let 1 and τ denote, respectively, the trivial and the non-trivial homomorphisms in $Hom(\{x^2\}, \mathbf{Z}^*)$. Hence we have isomorphisms of $\mathbf{Z}[Q_8]$-modules

5.2.19 $$M_- \cong Ind_{\{x^2\}}^{Q_8}(\tau)\ ,\ M_+ \cong Ind_{\{x^2\}}^{Q_8}(1).$$

Consider the short exact sequence of $\mathbf{Z}[Q_8]$-modules

5.2.20 $$M_- \longrightarrow M \longrightarrow M/M_- \cong M_+.$$

The group

$$Ext^1_{\mathbf{Z}[Q_8]}(Ind_{\{x^2\}}^{Q_8}(1), Ind_{\{x^2\}}^{Q_8}(\tau))$$

classifies, by 5.2.19 and chapter III of Hilton & Stammbach (1971), equivalence classes of sequences such as 5.2.20 under the equivalence relation generated by the relation that $(M_- \longrightarrow A \longrightarrow M_+)$ is equivalent to $(M_- \longrightarrow B \longrightarrow M_+)$ if there is a commutative diagram of the form

5.2.21

$$\begin{array}{ccccc} M_- & \longrightarrow & A & \longrightarrow & M_+ \\ \downarrow{\scriptstyle 1} & & \downarrow & & \downarrow{\scriptstyle 1} \\ M_- & \longrightarrow & B & \longrightarrow & M_+ \end{array}$$

In addition there are 'change of rings' isomorphisms of the form

5.2.22

$$\begin{aligned} Ext^1_{\mathbf{Z}[Q_8]}(Ind_{\{x^2\}}^{Q_8}(1), Ind_{\{x^2\}}^{Q_8}(\tau)) &\cong Ext^1_{\mathbf{Z}[\{x^2\}]}(\mathbf{Z}, Ind_{\{x^2\}}^{Q_8}(\tau)) \\ &\cong H^1(\{x^2\}; Ind_{\{x^2\}}^{Q_8}(\tau)) \\ &\cong H^1(\{x^2\}; \oplus_1^4 \tau) \\ &\cong \mathbf{Z}/2[\mathbf{Z}/2 \times \mathbf{Z}/2] \\ &\cong \mathbf{H}_{\mathbf{Z}} \otimes \mathbf{Z}/2. \end{aligned}$$

Notice that

5.2.23
$$Aut_{Q_8}(Ind_{\{x^2\}}^{Q_8}(\tau)) \cong \mathbf{H}_{\mathbf{Z}}^*$$

and

5.2.24
$$Aut_{Q_8}(Ind_{\{x^2\}}^{Q_8}(1)) \cong \mathbf{Z}[\mathbf{Z}/2 \times \mathbf{Z}/2]^*$$

and that, in terms of 5.2.22, the natural actions of the left sides of 5.2.23 and 5.2.24 correspond to multiplication (reduced modulo 2) by $\mathbf{H}_{\mathbf{Z}}$ and $\mathbf{Z}[\mathbf{Z}/2 \times \mathbf{Z}/2]$ respectively.

We do not wish to classify M up to the equivalence relation generated by 5.2.21 but rather up to isomorphism. Therefore we wish to evaluate

5.2.25
$$\begin{cases} Ext^1_{\mathbf{Z}[Q_8]}(Ind_{\{x^2\}}^{Q_8}(1), Ind_{\{x^2\}}^{Q_8}(\tau))/ \\ (Aut_{Q_8}(Ind_{\{x^2\}}^{Q_8}(\tau)) \times Aut_{Q_8}(Ind_{\{x^2\}}^{Q_8}(1))) \\ \\ \cong \mathbf{Z}/2[\mathbf{Z}/2 \times \mathbf{Z}/2]/(\mathbf{Z}/2 \times \mathbf{Z}/2), \end{cases}$$

where $\mathbf{Z}/2 \times \mathbf{Z}/2$ acts by multiplication. The orbits of this action are readily seen to be

5.2.26
$$\begin{cases} \{0\}, \{1, x, y, xy\}, \\ \{1 + x + y, xy + x + y, x + 1 + xy, y + xy + 1\}, \\ \{1 + x, y + xy\}, \{1 + y, x + xy\}, \\ \{1 + xy, x + y\} \text{ and } \{1 + x + y + xy\}. \end{cases}$$

The proof will be completed if we can show that only the two orbits of order four represent projective modules. In fact, we shall show that only these orbits represent cohomologically trivial modules. For this we need to recall some cohomological facts.

The 'change of rings' isomorphisms in 5.2.22 arise from the pullback diagram obtained from 5.2.20 (see Hilton & Stammbach, 1971; Snaith, 1989b):

5.2.27

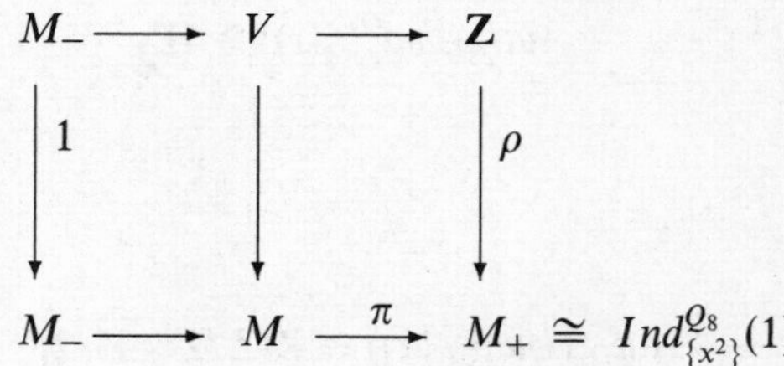

in which

5.2.28 $$\begin{cases} \rho(m) = 1 \otimes m \in \mathbf{Z}[Q_8] \otimes_{\mathbf{Z}[\{x^2\}]} \mathbf{Z} \\ \text{and} \qquad V = \pi^{-1}(\rho(\mathbf{Z})). \end{cases}$$

Also, 5.2.27 is a diagram of $\mathbf{Z}[\{x^2\}]$-module homomorphisms. There results a commutative diagram of coboundary maps and 'change of rings' isomorphisms.

5.2.29

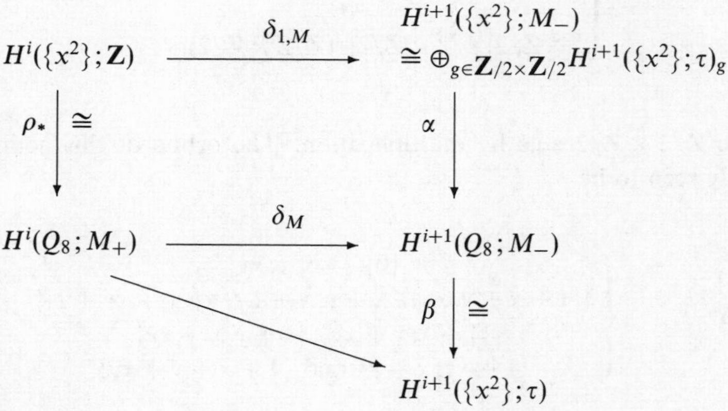

The homomorphism, $\beta\alpha$, in 5.2.29 is simply the map which adds together the four components.

If M is cohomologically trivial then the coboundary

$$\delta_M : H^i(Q_8; M_+) \longrightarrow H^{i+1}(Q_8; M_-)$$

must be an isomorphism for all $i \geq 1$ and surjective for $i = 0$. Identifying this map with $\beta\alpha\delta_{1,M}$ in 5.2.29 it suffices, by the cohomological periodicity of $\{x^2\}$ (Snaith, 1989b, p. 16), to analyse when the homomorphism

$$\beta\alpha\delta_{1,M} : \mathbf{Z} = H^0(\{x^2\};\mathbf{Z}) \longrightarrow H^1(\{x^2\};\tau) \cong \mathbf{Z}/2$$

is onto. However, if one unravels the isomorphisms of 5.2.22, one finds that in

$$\oplus_{g\in\mathbf{Z}/2\times\mathbf{Z}/2} H^1(\{x^2\};\tau)_g \cong \oplus_{g\in\mathbf{Z}/2\times\mathbf{Z}/2}\mathbf{Z}/2,$$

$\delta_{1,M}(1) = \sum(1)_{g_i}$ when M is the module corresponding to the orbit of $\sum_i g_i$ $(g_i \in \mathbf{Z}/2 \times \mathbf{Z}/2)$.

Hence, in the list of 5.2.26, $\beta\alpha\delta_{1,M}(1)$ is zero except for the orbit $\{1, x, y, xy\}$ for which

$$\beta\alpha\delta_{1,M}(1) = \beta\alpha((1)_x) = 1 \in \mathbf{Z}/2$$

and for the orbit $\{1 + x + y, xy + x + y, \ldots\}$ for which

$$\beta\alpha\delta_{1,M}(1) = \beta\alpha((1)_x + (1)_y + (1)_{xy}) = 3 \in \mathbf{Z}/2.$$

This completes the proof of Theorem 5.2.13. □

5.2.30 *Cohomologically trivial modules of finite type*

Let G be a finite group and let M be a $\mathbf{Z}[G]$-module of finite type which is cohomologically trivial. The Grothendieck group (the class-group) of such modules is isomorphic to $\mathcal{CL}(\mathbf{Z}[G])$ (see Tate, 1984, Lemma 8.1, p. 65). Hence the module, M, gives rise to a class, $[M] \in \mathcal{CL}(\mathbf{Z}[G])$. This class is represented by $[B] - [A] \in \mathcal{CL}(\mathbf{Z}[G])$, where

$$0 \longrightarrow A \longrightarrow B \longrightarrow M \longrightarrow 0$$

is any short exact sequence of $\mathbf{Z}[G]$-modules in which A and B are finitely generated and projective.

In particular, any M which is a finite abelian group with a G-action and whose order is prime to $\#(G)$ is cohomologically trivial and gives rise to a class in $\mathcal{CL}(\mathbf{Z}[G])$.

Suppose now that $G = Q_8$, the quaternion group of order eight. We shall examine the class of a finite $\mathbf{Z}[Q_8]$-module of odd order.

Define a homomorphism

5.2.31 $$\chi_+ : (\mathbf{Z}/8)^* \longrightarrow \{\pm 1\} \cong (\mathbf{Z}/4)^*$$

by $\chi_+(-1) = 1, \chi_+(3) = -1$.

Lemma 5.2.32 *Let M be a finite abelian group of odd order upon which Q_8 acts. Define M_-, as in 5.2.4, to be the (-1)-eigengroup of x^2. Then there is an integer, $r_- \geq 0$, such that*

$$\#(M_-) = (r_-)^2.$$

Proof We may decompose M into its p-Sylow subgroups, where p is an odd prime, and then filter each p-Sylow subgroup in such a manner that the associated graded Q_8-module consists of $F_p[Q_8]$-modules, where F_p is the field with p elements. Hence it suffices to assume that M is an F_p-vector space. Therefore we must show that M_- is even-dimensional.

Let $\overline{F}_p$ denote the algebraic closure of F_p. It suffices to show that $M_- \otimes_{F_p} \overline{F}_p$ is even-dimensional over $\overline{F}_p$. However, by Mashke's theorem, $\overline{F}_p[Q_8]$ is semi-simple. In fact,

$$\overline{F}_p[Q_8] \cong M_2(\overline{F}_p) \oplus (\overline{F}_p)^4$$

and the only irreducible component upon which x^2 acts like (-1) is $M_2(\overline{F}_p)$. Since the irreducible $M_2(\overline{F}_p)$-modules are even-dimensional it follows that so also is M_-.

An alternative proof is to be found in the argument preceding 5.2.38. □

Theorem 5.2.33 (*Chinburg*, 1989, section 4.3.6, p. 44) *Let M be a finite abelian group of odd order upon which Q_8 acts. Then, in the notation of* 5.2.30 *and* 5.2.32,

$$[M] = \chi_+(\#(M_+))r_- \in (\mathbf{Z}/4)^* \cong \mathscr{CL}(\mathbf{Z}[Q_8]).$$

Here, as in 5.2.4, *M_+ is the $(+1)$-eigengroup of x^2.*

Proof Choose projective $\mathbf{Z}[Q_8]$-modules, A and B with $\text{rank}(A) = \text{rank}(B) = k$, as in 5.2.30 and consider the following commutative diagram, whose rows are exact:

5.2.34

$$\begin{array}{ccccccccc} 0 & \longrightarrow & A_- & \xrightarrow{f_-} & B_- & \longrightarrow & M_- & \longrightarrow & 0 \\ & & \downarrow & & \downarrow & & \downarrow & & \\ 0 & \longrightarrow & A & \xrightarrow{f} & B & \longrightarrow & M & \longrightarrow & 0 \\ & & \downarrow & & \downarrow & & \downarrow & & \\ 0 & \longrightarrow & A/A_- & \xrightarrow{f_+} & B/B_- & \longrightarrow & M/M_- & \longrightarrow & 0 \end{array}$$

Choose $\mathbf{H_Z}$-bases $\underline{a} = \{a_1, \ldots, a_k\}$ and $\underline{b} = \{b_1, \ldots, b_k\}$ for A_- and B_- respectively. Choose sets of elements $\underline{\alpha} = \{\alpha_1, \ldots, \alpha_k\}$ and $\underline{\beta} = \{\beta_1, \ldots, \beta_k\}$ so that, in the notation of 5.2.9, $\underline{\alpha}+A_-$ and $\underline{\beta}+B_-$ are $\mathbf{Z}[\mathbf{Z}/2\times\mathbf{Z}/2]$-bases for A/A_- and B/B_- respectively. Hence there are matrices

$$X, Y, U, U', V' \in M_k(\mathbf{H_Z})$$

and

$$V \in M_k(\mathbf{Z}[\mathbf{Z}/2 \times \mathbf{Z}/2])$$

such that V and V' coincide modulo 2 and

5.2.35
$$\begin{cases} (1-x^2)\underline{\alpha} & = X\underline{a} \\ (1-x^2)\underline{\beta} & = Y\underline{b} \\ f_-(\underline{a}) & = U\underline{b} \\ \text{and} & \\ f(\underline{\alpha}) & = V\underline{\beta} + U'\underline{b}. \end{cases}$$

The commutative diagram 5.2.34 gives rise to the following calculation:

$$\begin{aligned} XU\underline{b} &= f_-(X\underline{a}) \\ &= f_-((1-x^2)\underline{\alpha}) \\ &= (1-x^2)f(\underline{\alpha}) \\ &= (1-x^2)(V\underline{\beta} + U'\underline{b}) \\ &= V((1-x^2)\underline{\beta}) + U'((1-x^2)\underline{b}) \\ &= V'Y\underline{b} + 2U'\underline{b}. \end{aligned}$$

Therefore

5.2.36 $$XU = V'Y + 2U' \in M_k(\mathbf{H_Z}).$$

Taking complex reduced norms (that is, applying $det(c(-)))$ modulo 4 we obtain

5.2.37 $$det(c(X))det(c(U)) = det(c(V'))det(c(Y)) \in \mathbf{Z}/4.$$

However, if R is the realisation map

$$R : M_k(\mathbf{H_Z}) \longrightarrow M_{4k}(\mathbf{Z}),$$

then

$$\#(M_-) = det(R(U)) = det(c(U))^2$$

since the unreduced norm is the square of the reduced norm in this case (cf. Curtis & Reiner, 1987, section 45, p. 138; Serre, 1977, p. 92). By 5.7.2, $det(c(U))$ is a non-negative integer so that

5.2.38 $$det(c(U)) = r_-.$$

Therefore, from 5.2.37, we see that $det(c(V'))$ is an odd integer and in $\mathcal{CL}(\mathbf{Z}[Q_8]) \cong (\mathbf{Z}/4)^*$ we have

5.2.39 $$[M] = det(c(Y))det(c(X))^{-1} = r_-(det(c(V'))) \in (\mathbf{Z}/4)^*.$$

From 5.2.39 $det(c(V'))$ is an invariant which lies in the class-group and depends only upon M_+. To compute this invariant and thereby complete the proof we shall use induction on $\#(M_+)$.

We shall examine the case in which $\text{rank}(A) = \text{rank}(B) = 1$. In this case we may suppose that

$$V = p + qx + ry + sxy \in \mathbf{Z}[\mathbf{Z}/2 \times \mathbf{Z}/2]$$

and that

$$V' = p + qi + rj + sk \in \mathbf{H_Z}.$$

We have

5.2.40 $$\#(M_+) = det \begin{pmatrix} p & q & r & s \\ q & p & s & r \\ r & s & p & q \\ s & r & q & p \end{pmatrix}$$

$$= (p+q+r+s)(p+q-r-s)(p+r-q-s)(p+s-q-r)$$
$$= [(p+q)^2 - (r+s)^2][(p-q)^2 - (r-s)^2].$$

Hence $p+q+r+s$ is odd and either one or three of p,q,r,s are even. One easily verifies in all the cases that

$$\chi_+(\#(M_+)) = p^2+q^2+r^2+s^2 = det(c(V')) \in (\mathbf{Z}/4)^*.$$

For example, if p is odd and q,r,s even, then $p^2+q^2+r^2+s^2 \equiv 1$ (mod 4) and

$$\begin{aligned}\#(M_+) &\equiv (1-(r+s)^2)(1-(r-s)^2) \quad (mod\ 8)\\ &\equiv 1-2r^2-2s^2 \quad (mod\ 8)\\ &\equiv 1 \quad (mod\ 8)\end{aligned}$$

as claimed.

For the induction step let M be a finite $\mathbf{Z}[Q_8]$-module of odd order with $M_+ \neq \{0\}$. Choose $0 \neq z \in M_+$ and define $\pi : \mathbf{Z}[Q_8] \longrightarrow M$ by $\pi(1) = z$. Since $\#(im(\pi))$ is odd, $ker(\pi)$ will be cohomologically trivial and torsion-free and therefore projective. In the class-group

$$[M] = [im(\pi)] + [M/im(\pi)]$$

and, by induction on the order of the (+1)-eigengroup, we may assume that the theorem has been established for $im(\pi)$ and for $M/im(\pi)$ and therefore holds for M. □

Remark 5.2.41 Theorem 5.2.33 determines the class in $\mathscr{CL}(\mathbf{Z}[Q_8])$ of a finite module of odd order. To give a formula which determines the class corresponding to a cohomologically trivial, finite module whose order is a power of two seems to be much more difficult. However there are many modules of this type. Here is an example, which I learnt from P. Symonds, of such a module representing the non-trivial class in the class-group.

Define a finite $\mathbf{Z}[Q_8]$-module by means of the following exact sequence:

$$0 \longrightarrow < 3, \sigma > \stackrel{h}{\longrightarrow} \mathbf{Z}[Q_8] \longrightarrow M \longrightarrow 0$$

where $h(x) = x(2-\sigma)3^{-1}$ for all $x \in < 3, \sigma >$.

5.3 Relations between Swan modules

Throughout this section we shall assume that G is a finite group in an extension (not necessarily split) of the form

5.3.1

$$\{1\} \longrightarrow J \longrightarrow G \longrightarrow \mathbf{Z}/p \longrightarrow \{1\}$$

where

5.3.2 J is cyclic and p is prime.

We shall use a geometrical technique, which to my knowledge is new but which is very much in the geometrical spirit which motivated R.G. Swan's original introduction of the Swan modules, $S(k)$ (see 4.2.43; Swan, 1960, p. 278).

Theorem 5.3.3 *Let G be as in 5.3.1 and 5.3.2 and let $HCF(k, \#(G)) = 1$. Then*

$$S(k^p) - \sum_g Ind^G_{N_G<g>}(S(k)) = 0 \in D(\mathbf{Z}[G]),$$

where $g \in G - J$ runs over generators for the distinct conjugacy classes of subgroups, $< g >$, of order p and $N_G < g >$ denotes the normaliser of $< g >$ in G.

This result will be proved in 5.3.29–5.3.42. Before proceeding to the proof let us pause to examine some applications.

Example 5.3.4 Let p be an odd prime and let $\mathbf{Z}/p$ act on $\mathbf{Z}/p^n$ $(n \geq 2)$ by $\tau(z) = z^{(1+p)^{p^{n-2}}}$, where $z \in \mathbf{Z}/p^n$ and τ generates $\mathbf{Z}/p$. From Curtis & Reiner (1987, p. 365; 4.4.20):

$$T(\mathbf{Z}/p \propto \mathbf{Z}/p^n) \cong \mathbf{Z}/p^n.$$

If we set $G = \mathbf{Z}/p \propto \mathbf{Z}/p^n$ and $J = \mathbf{Z}/p^n$ we may apply Theorem 5.3.3 to any integer, k, which is not divisible by p. Hence

$$0 = S(k^p) - \sum_g Ind^G_{N_G<g>}(S(k)),$$

where the sum runs over some of the $g \in G - J$ of order p. However, for each of these $N_G < g >= \mathbf{Z}/p \propto \mathbf{Z}/p^{n-1}$ so that if we set $k = t^{p^{n-1}}$ then each $Ind^G_{N_G<g>}(S(k))$ vanishes, by induction on the order of G, and we find that

$$S(t^{p^n}) = 0 \in T(\mathbf{Z}/p \propto \mathbf{Z}/p^n),$$

as expected.

Example 5.3.5 Let $G = Q(8n)$, the generalised quaternion group of order $8n$,

$$Q(8n) = \{X, Y \mid Y^2 = X^{2n}, Y^4 = 1, YXY^{-1} = X^{-1}\}.$$

Setting $J =< X >$, $p = 2$ and observing that there are no elements of order two in $G - J$ we find that, for all n,

5.3.6 $$2T(Q(8n)) = 0.$$

In Bentzen & Madsen (1983) one finds some very difficult and technical calculations of Swan subgroups which are related to 5.3.6. En route to the main theorem of Bentzen & Madsen (1983), one finds a calculation of $T(Q(8p))$, when p is a prime (see Bentzen & Madsen, 1983, theorem 3.5). The preceding application of Theorem 5.3.3 gives a new manner in which to derive an upper bound for $T(Q(8n))$, which is less computational than the method of Bentzen & Madsen (1983).

Notice that the Artin exponent of $Q(8p)$ is equal to four (Bentzen & Madsen, 1983, p. 462), so that 5.3.6 improves upon the estimate given by Theorem 4.4.17.

Let us recall part of Theorem 3.5 of Bentzen & Madsen (1983).

Theorem 5.3.7 *Let p be an odd prime such that either $p \equiv \pm 3$ (mod 8) or $p \equiv 1$ (mod 8) and $ord_p(2)$ is even, where $ord_p(t)$ is the order of t in F_p^*. Then*

$$T(Q(8p)) \cong \mathbf{Z}/2 \oplus \mathbf{Z}/2.$$

Corollary 5.3.8 *Let $Q(8n)$ denote the generalised quaternion group of order $8n$. Suppose that $n = 2^\alpha p_1^{m_1} \ldots p_r^{m_r}$ with $m_i \geq 1$, $\alpha \geq 0$ and $\{p_i; 1 \leq i \leq r\}$ a set of distinct primes such that either $p_i \equiv \pm 3$ (mod 8) or $p_i \equiv 1$ (mod 8) and $ord_{p_i}(2)$ is even, where $ord_p(t)$ is the order of t in F_p^*. Then*

$$T(Q(8n)) \cong (\mathbf{Z}/2)^{r+1}.$$

Proof By 5.3.6, $T(Q(8n))$ is a quotient of

$$\begin{aligned}
&((\mathbf{Z}/2^{\alpha+3})^* \times \textstyle\prod_{i=1}^r (\mathbf{Z}/p_i^{m_i})^*/\{\pm 1\}) \otimes \mathbf{Z}/2 \\
&\cong \mathbf{Z}/2 \times \textstyle\prod_{i=1}^r F_{p_i}^*/(F_{p_i}^*)^2 \\
&\cong (\mathbf{Z}/2)^{r+1}.
\end{aligned}$$

However, inspection of Theorem 3.5 of Bentzen & Madsen (1983) shows that the natural maps to the Swan groups of the subquotients,

$\{Q(8p_i); 1 \leq i \leq r\}$ detect $(\mathbf{Z}/2)^{r+1}$ in $T(Q(8n))$, which completes the proof. □

Before proceeding to the proof of Theorem 5.3.3 we shall need some preparatory results.

Lemma 5.3.9 *Let $\epsilon : \mathbf{Z}[G] \longrightarrow \mathbf{Z}/r$ denote the reduction (modulo r) of the augmentation map. If $HCF(r, \#(G)) = 1$ then*

$$[Ker(\epsilon : \mathbf{Z}[G] \longrightarrow \mathbf{Z}/r)] = -S(r) \in \mathcal{CL}(\mathbf{Z}[G]).$$

Proof By the construction of $S(r)$ given in Curtis & Reiner (1987, section 53, p. 343; 4.2.44) $S(r) = \{r, \sigma\} \lhd \mathbf{Z}[G]$, where $\sigma = \sum_{g \in G} g$. Therefore we have exact sequences of the form

$$0 \longrightarrow Ker(\epsilon) \longrightarrow \mathbf{Z}[G] \longrightarrow \mathbf{Z}/r \longrightarrow 0$$

and

$$0 \longrightarrow (r\mathbf{Z}[G]) \longrightarrow \{r, \sigma\} \stackrel{\pi}{\longrightarrow} \mathbf{Z}/r \longrightarrow 0,$$

where $\pi(r) = 0$ and $\pi(\sigma) = 1$. Since $r(\mathbf{Z}[G]) \cong \mathbf{Z}[G]$, Schanuel's lemma (Swan, 1960, p. 270) implies that

$$Ker(\epsilon) \oplus S(r) \cong \mathbf{Z}[G] \oplus \mathbf{Z}[G],$$

from which the result follows. □

Remark 5.3.10 When G is a p-group the calculation of $T(G)$ is given in 4.4.20 (see also Curtis & Reiner, 1987, section 54). Most of the cases are due to M.J. Taylor, using determinantal congruences and a particular case of the maps, S_T, of 4.4.11. This manner of detection, using ψ^l/l to give a lower bound for $T(G)$, is reminiscent of J.F. Adams' approach to the detection of the image of the J-homomorphism in the stable homotopy groups of spheres (Adams, 1963, 1965a,b, 1966). The more difficult part of the determination of the image of the J-homomorphism was the construction of an upper bound. This led to the celebrated Adams conjecture (Adams, 1963, 1965a,b, 1966) and its ultimate proof by D. Sullivan (1974) and D.G. Quillen (1971). The Adams conjecture concerns the maps between spheres of vector bundles. This motivated the method which we shall now employ to prove Theorem 5.3.3. We shall study maps between unit spheres of representation spaces. This approach will involve a considerable digression into the topology of group actions, for which I must ask the reader's indulgence, before we return to the

Swan subgroup. As to the prerequisites from algebraic topology, the proof will require only the elementary properties of homology theory such as may be found in Greenberg & Harper (1971).

The proof of Theorem 5.3.3 will establish a slightly more general result (see 5.7.5).

5.3.11 Let $X_1, \ldots, X_p$ be topological spaces. The p-fold *join*, $X_1 * \cdots * X_p$, is defined to be the space which is formed in the following manner. If $I = [0, 1]$ is the unit interval consider the set, Z, of $2p$-tuples

$$(t_1, x_1, t_2, x_2, \ldots, t_p, x_p) \in I \times X_1 \times I \times X_2 \times \ldots \times I \times X_p,$$

which satisfy $\sum_{i=1}^{p} t_i^2 = 1$. Impose on Z the relation that

$$(t_1, x_1, \ldots, 0, x_j, \ldots) \sim (t_1, x_1, \ldots, 0, x'_j, \ldots)$$

whenever $t_j = 0$. The quotient space, $Z/\sim$, is $X_1 * \cdots * X_p$.

When $X_1 = X_2 = \ldots = X_p = S^n$, the n-dimensional sphere, then $X_1 * \cdots * X_p$ is homeomorphic to the sphere, $S^{p(n+1)-1}$. For example, such a homeomorphism is given by sending $(t_1, x_1, t_2, x_2, \ldots, t_p, x_p)$ to $(t_1x_1, t_2x_2, \ldots, t_px_p) \in \oplus_{i=1}^{p} R^{n+1}$.

Now let $\phi : J \longrightarrow S^1 \subset C^*$ be a one-dimensional representation, which we will later assume to be a faithful embedding of the cyclic group, J, into the unit circle. Let $S_1^3 = S(\phi \oplus \phi)$ denote the three-dimensional sphere, S^3, upon which J acts via $j(w_1, w_2) = (\phi(j)w_1, \phi(j)w_2)$, where $w_1, w_2 \in C$ and $| w_1 |^2 + | w_2 |^2 = 1$. If $(w_1, w_2) = x$ we shall write $j(x) = \hat{\phi}(j)(x)$ for this action on S_1^3.

Let $z_1, z_2, \ldots, z_p$ be a set of coset representatives for G/J. Therefore there is a homomorphism into the symmetric group

5.3.12 $$\sigma : G \longrightarrow \Sigma_p$$

such that, for $g \in G$,

5.3.13 $$gz_i = z_{\sigma(g)(i)} j(i, g) \qquad (j(i, g) \in J).$$

Let G act on $S_1^3 * \ldots * S_1^3$ (p copies) by the formula

5.3.14 $$\begin{cases} g(t_1, x_1, \ldots, t_p, x_p) \\ = \sigma(g)(t_1, \hat{\phi}(j(1, g))(x_1), \ldots, t_p, \hat{\phi}(j(p, g))(x_p)), \end{cases}$$

where $\sigma(g)$ permutes the p-tuple of pairs $\{(t_i, \hat{\phi}(j(i,g))(x_i))\}$. Clearly $S_1^3 * \ldots * S_1^3$ with this action is just the unit sphere in the representation space afforded by $Ind_J^G(\phi \oplus \phi)$.

Let k be an integer which is coprime to the order of G. Let $k : S_1^3 \longrightarrow S^3(\phi \oplus \phi^k) = S_2^3$, say, denote the map which is the identity on the first circle coordinate and is the kth power map on the second circle coordinate. Define a G-map

5.3.15
$$F_k : S_1^3 * \ldots * S_1^3 \longrightarrow S_2^3 * \ldots * S_2^3$$

by

$$F_k = k * k * \ldots * k$$

(p copies). In view of the previous discussion we have

5.3.16
$$F_k : S(Ind_J^G(\phi \oplus \phi)) \longrightarrow S(Ind_J^G(\phi \oplus \phi^k)),$$

where $S(V)$ denotes the unit sphere in the vector space, V.

Remark 5.3.17 Recall that a continuous map, $f : S^n \longrightarrow S^n$ has a *degree* which is equal to the endomorphism induced by f on the homology group, $H_n(S^n; \mathbf{Z}) \cong \mathbf{Z}$. The kth power map on S^1 has degree k and the degree of the join of a number of such maps is equal to the product of their degrees.

Lemma 5.3.18 *For $H \leq G$ let X^H denote the H-fixed points of X. For all $H \leq G$, F_k restricts to a map of spheres*

$$F_k : S(Ind_J^G(\phi \oplus \phi))^H \longrightarrow S(Ind_J^G(\phi \oplus \phi^k))^H$$

whose degree is a power of k. In particular the degree of F_k in 5.3.16 is $k^{[G:J]}$. Note that $S(V)^H = S(V^H)$ for any G-representation, V.

Proof Since F_k is a G-map it induces maps on H-fixed point sets. Also the degree of F_k is equal to $k^{[G:J]}$, by 5.3.17.

Now let us examine the degree on the H-fixed point sets. By the double coset formula 1.2.40

$$Res_H^G Ind_J^G(\phi \oplus \phi) = \sum_{g \in H \backslash G/J} Ind_{H \cap gJg^{-1}}^H((g^{-1})^*(\phi \oplus \phi)),$$

where $(g^{-1})^*(\phi)(u) = \phi(g^{-1}ug)$.

Also

$$dim(Ind^H_{H\cap gJg^{-1}}((g^{-1})^*(\phi \oplus \phi))^H) = \begin{cases} 0 & \text{if } Res^J_{g^{-1}Hg\cap J}(\phi) \neq 1 \\ 2 & \text{otherwise.} \end{cases}$$

Therefore, since k is coprime to $\#(G)$,

$$dim((Ind^G_J(\phi \oplus \phi))^H) = dim((Ind^G_J(\phi \oplus \phi^k))^H).$$

In addition, there is an injective map

5.3.19 $$\lambda_g : Ind^H_{H\cap gJg^{-1}}((g^{-1})^*(\phi \oplus \phi)) \longrightarrow Ind^G_J(\phi \oplus \phi)$$

given by

5.3.20 $$\lambda_g(h \otimes_{H\cap gJg^{-1}} v) = hg \otimes_J v.$$

The sum of all the λ_g of 5.3.19 for which the H-fixed points are non-trivial yields an isomorphism

5.3.21 $$\oplus_g Ind^H_{H\cap gJg^{-1}}((g^{-1})^*(\phi \oplus \phi))^H \xrightarrow{\cong} Ind^G_J(\phi \oplus \phi)^H.$$

The same is true when the second copy of ϕ is replaced by ϕ^k and then F_k between the H-fixed spheres on the right of 5.3.21 corresponds to the join of the induced maps between H-fixed spheres on the left of 5.3.21. However, a non-zero vector in $Ind^H_{H\cap gJg^{-1}}((g^{-1})^*(\phi \oplus \phi))^H$ is of the form ($v = (z_1, z_2) \in S^3_1$)

$$\left[\sum_{h\in H/(H\cap gHg^{-1})} h\right] \otimes v$$

and the map induced by F_k simply raises the second coordinate of v to the kth power. Hence the degree of F_k on the H-fixed points is a power of k, as required.

Define $D(G, J, \phi, k)$ to be the mapping cone of F_k in 5.3.16. Hence

$$D(G, J, \phi, k) = [S(Ind^G_J(\phi \oplus \phi^k)) \cup (S(Ind^G_J(\phi \oplus \phi)) \times I)]/\sim$$

where $\sim$ is given by

$$(z, 1) \sim (z', 1) \qquad (z, z' \in S(Ind^G_J(\phi \oplus \phi)))$$

and

$$(z, 0) \sim F_k(z).$$

$D(G, J, \phi, k)$ is a finite CW complex with a cellular action induced in the obvious manner from the action on the representation spheres. □

Corollary 5.3.22 *The reduced homology of the fixed point set*

$$\tilde{H}_*(D(G,J,\phi,k)^H;\mathbf{Z}_{(p)})$$

is trivial for all $H \leq G$ and for all localisations, $\mathbf{Z}_{(p)}$, of the integers, $\mathbf{Z}$, at a prime dividing $\#(G)$.

Proof In $\mathbf{Z}_{(p)}$ the integer, k, is invertible. However

$$\tilde{H}_m(S(Ind_J^G(\phi \oplus \phi))^H;\mathbf{Z}_{(p)}) = \tilde{H}_m(S(Ind_J^G(\phi \oplus \phi^k))^H;\mathbf{Z}_{(p)})$$

is zero except when $m = 4pdim(Ind_J^G(\phi)^H) - 1$, in which case it is $\mathbf{Z}_{(p)}$. From the exact sequence

$$\ldots \longrightarrow \tilde{H}_m(S(Ind_J^G(\phi \oplus \phi))^H;\mathbf{Z}_{(p)}) \stackrel{(F_k)_*}{\longrightarrow} \tilde{H}_m(S(Ind_J^G(\phi \oplus \phi^k))^H;\mathbf{Z}_{(p)})$$
$$\longrightarrow \tilde{H}_m(D(G,J,\phi,k)^H;\mathbf{Z}_{(p)}) \longrightarrow \ldots$$

we see, by 5.3.18, that $(F_k)_*$ is invertible (with $\mathbf{Z}_{(p)}$ coefficients) and therefore $\tilde{H}_*(D(G,J,\phi,k)^H;\mathbf{Z}_{(p)}) = 0$, as required. □

5.3.23 *Chain complexes and class-group elements*

We now describe a method, which appears in Oliver (1978), whereby to associate an element of the class-group, $\mathscr{C}\mathscr{L}(\mathbf{Z}[G])$, to a chain complex such as $C_*(D(G,J,\phi,k);\mathbf{Z})$, the cellular chains of $D(G,J,\phi,k)$.

If S is a finite G-set let $\mathbf{Z}(S)$ denote the free $\mathbf{Z}$-module with S as a basis and with the induced action of G. Let $v(G)$ denote the class of all chain complexes of $\mathbf{Z}[G]$-modules of the form

5.3.24 $$0 \longrightarrow \mathbf{Z}(S_n) \stackrel{\delta_n}{\longrightarrow} \mathbf{Z}(S_{n-1}) \stackrel{\delta_{n-1}}{\longrightarrow} \ldots \stackrel{\delta_1}{\longrightarrow} \mathbf{Z}(S_0) \longrightarrow 0$$

such that for all $H \leq G$, and for all i, δ_i maps $\mathbf{Z}(S_i^H)$ into $\mathbf{Z}(S_{i-1}^H)$. An example of such a chain complex is provided by the cellular chains of a finite G–CW-complex, X, with a cellular action of G. In this example the subcomplex $(\mathbf{Z}(S_*^H), \delta_*)$ is provided by the chains of X^H. All chain complexes 5.3.24 of $v(G)$ can be realised geometrically in this manner by some simply connected G-space. Let $v^*(G)$ denote the complexes of $v(G)$ for which $(\mathbf{Z}(S_*^H), \delta_*) \otimes \mathbf{Z}_{(p)}$ is exact for all primes, p, dividing $\#(G)$ and for all p-groups, $H \leq G$. A *resolution* for 5.3.24 is a complex in $v^*(G)$ of the form

5.3.25 $$0 \longrightarrow \mathbf{Z}(\overline{S}_n) \stackrel{\overline{\delta}_n}{\longrightarrow} \mathbf{Z}(\overline{S}_{n-1}) \stackrel{\overline{\delta}_{n-1}}{\longrightarrow} \ldots \stackrel{\overline{\delta}_1}{\longrightarrow} \mathbf{Z}(\overline{S}_0) \longrightarrow 0,$$

such that $\overline{S}_i \geq S_i$ for each i and $\overline{\delta}_i$ extends δ_i and which satisfies the following conditions:

(i) G acts freely on $\overline{S}_i - S_i$ for all i,

(ii) 5.3.25 is exact except at $\mathbf{Z}(\overline{S}_n)$ and $ker(\overline{\delta}_n)$ is a projective module.

In Proposition 4 of Oliver (1978) it is shown that every complex 5.3.24 in $v^*(G)$ admits at least one resolution, as in 5.3.25, and that the class

5.3.26 $$[ker(\overline{\delta}_n)] \in \mathscr{CL}(\mathbf{Z}[G])$$

is a well-defined element, depending only on the chain homotopy class of 5.3.24 in $v^*(G)$. In particular, if X is a connected, finite G–CW-complex, all of whose H-fixed point sets have vanishing reduced $\mathbf{Z}_{(p)}$-homology, we may, by taking 5.3.24 to be the reduced chain complex of X, associate to it in this manner a well-defined *Euler characteristic* which will be denoted by

5.3.27 $$M(X) \in \mathscr{CL}(\mathbf{Z}[G]).$$

Geometrically $M(X)$ is realised by embedding X into a finite G–CW-complex, W, of the same dimension as X such that G acts freely on $W - X$.

We may apply this construction to $D(G, J, \phi, k)$.

Lemma 5.3.28 *In the notation of* 5.3.22 *and* 5.3.23

$$M(D(G,J,\phi,k)) = [Ker(\mathbf{Z}[G] \stackrel{\epsilon}{\longrightarrow} \mathbf{Z}/(k^{[G:J]}))] \in \mathscr{CL}(\mathbf{Z}[G])$$

where $\epsilon(\sum_g \lambda_g g) = \sum_g \lambda_g \pmod{k^{[G:J]}}$.

By 5.3.9, $[Ker(\epsilon)] = -S(k^{[G:J]}) \in \mathscr{CL}(\mathbf{Z}[G])$.

Proof Consider the reduced chain complex of $D(G, J, \phi, k)$:

$$0 \longrightarrow C_n \stackrel{\delta_n}{\longrightarrow} C_{n-1} \stackrel{\delta_{n-1}}{\longrightarrow} \ldots \stackrel{\delta_1}{\longrightarrow} C_0 \stackrel{\eta}{\longrightarrow} \mathbf{Z} \longrightarrow 0,$$

where η is the augmentation which sends every zero-cell to $1 \in \mathbf{Z}$. Since

$$\tilde{H}_i(D(G,J,\phi,k);\mathbf{Z}) = \begin{cases} 0 & \text{if } i \neq n-1, \\ \mathbf{Z}/(k^{[G:J]}) & \text{if } i = n-1, \end{cases}$$

we may build up a resolution in the following manner. We leave C_j unaltered for $j \leq n-1$. In $\mathrm{Ker}(\delta_{n-1})$ we may find an element, x_{n-1}, whose image generates

$$\tilde{H}_{n-1}(D(G,J,\phi,k);\mathbf{Z}) \cong \mathbf{Z}/(k^{[G:J]}).$$

By examining the map F_k explicitly, one can see that x_{n-1} may be taken to be G-fixed. Now we alter C_n to $\overline{C}_n = C_n \oplus \mathbf{Z}[G] < x_n >$ and set $\overline{\delta}_n(x_n) = x_{n-1}$. From this description one readily sees that

$$Ker(\overline{\delta}_n : \overline{C}_n \longrightarrow C_{n-1}) \cong [Ker(\mathbf{Z}[G] \stackrel{\epsilon}{\longrightarrow} \mathbf{Z}/(k^{[G:J]}))],$$

as required. □

Now we shall give the proof of Theorem 5.3.3. While conceptually simple, the proof is technically quite involved and for this reason we shall break the proof into the non-split and the split cases in order that the reader may see some of the details in the (simpler) non-split case first.

5.3.29 *Proof of Theorem* 5.3.3 *in the non-split case*

In this case we shall assume that 5.3.1 is not a split extension, which means that there are no elements $g \in G - J$ of order p. Under these circumstances we must show that $S(k^{[G:J]}) = pS(k) = 0$ in $\mathscr{CL}(\mathbf{Z}[G])$. To do this we will take $\phi : J \longrightarrow \mathbf{C}^*$ to be injective in 5.3.11 and we shall show that

$$M(D(G, J, \phi, k)) = 0 \in \mathscr{CL}(\mathbf{Z}[G]).$$

The result will then follow from 5.3.28. By the discussion of 5.3.23, this relation will follow if we can find a G-embedding of $D(G, J, \phi, k)$ into a connected space W which is acyclic (i.e. $\tilde{H}_*(W; \mathbf{Z}) = 0$) and where G acts freely on $W - D(G, J, \phi, k)$.

Write $X = S_1^3$, $Y = S_2^3$ and let $Z = D(J, J, \phi, k)$ be the mapping cone of the map $k : X \longrightarrow Y$ which was introduced earlier. The projective module associated to Z is equal to

$$M(Z) = -S(k) \in T(J) \subset \mathscr{CL}(\mathbf{Z}[J]),$$

which is trivial if J is cyclic, by 4.2.48. Hence we can find a J-embedding into a simply connected space, D,

$$i : Z \longrightarrow D,$$

where $\tilde{H}_*(D; \mathbf{Z}) = 0$ and J acts freely on $D - i(Z)$. This geometrical realisation is guaranteed by a theorem of C.T.C. Wall (1966) since the space, Z, and the chain complex which D must realise are both simply connected.

Let $* \in i(Z)$ be a J-fixed point (which must be the cone-point since J acts freely on X and Y). Form the following quotient of the p-fold join of D:

5.3.30 $$E = D * D * \ldots * D/\{(t_1, *, t_2, *, \ldots, t_p, *) = *\},$$

on which G acts by the formula of 5.3.14. We may define a G-embedding

5.3.31 $$\pi : D(G, J, \phi, k) \longrightarrow E$$

in the following manner. If $C(X * \ldots * X)$ is the cone on $X * \ldots * X$ then

$$D(G, J, \phi, k) = [C(X * \ldots * X) \cup (Y * \ldots * Y)]/\sim.$$

Define π of 5.3.31 by

$$\pi(t_1, y_1, \ldots, t_p, y_p) = (t_1, i(y_1), \ldots, t_p, i(y_p))$$

and

$$\pi((t_1, x_1, \ldots, t_p, x_p), s) = (t_1, i(s, x_1), \ldots, t_p, i(s, x_p)),$$

where $y_i \in Y, x_i \in X$ and $0 \leq t_i, s \leq 1$. Notice that $\tilde{H}_*(E; \mathbf{Z}) = 0$ because $D * \ldots * D$ is acyclic and E is obtained from this connected, acyclic space by collapsing a (contractible) simplex to a point. Such a collapsing map induces an isomorphism on homology groups.

It remains to verify that if $1 \neq g \in G$ and $\underline{w} = (t_1, w_1, \ldots, t_p, w_p) \notin im(\pi)$ then $g(\underline{w}) \neq \underline{w}$.

We shall divide the possibilities into the following cases:

caseA : $g \in J$ and $\underline{w} \in (i(Z) * \ldots * i(Z))/\sim$,
caseB : $g \notin J$ and $\underline{w} \in (i(Z) * \ldots * i(Z))/\sim$,
caseC : $g \notin J$ and $\underline{w} = (t_1, w_1, \ldots)$ with $w_1 \in D - i(Z)$, and
caseD : $g \in J$ and $\underline{w} = (t_1, w_1, \ldots)$ with $w_1 \in D - i(Z)$.

In case A, because $g \in J \lhd G$, we must examine

$$\begin{aligned} g(\underline{w}) &= (t_1, j(1, g)(w_1), \ldots, t_p, j(p, g)(w_p)) \\ &= \underline{w} \\ &= (t_1, w_1, \ldots, t_p, w_p). \end{aligned}$$

This implies that $j(i, g)(w_i) = w_i$, for each $1 \leq i \leq p$ such that $t_i \neq 0$. However, because ϕ is injective, the action of J on Z is free except for the cone-point, $*$, which is J-fixed. Therefore $\underline{w} = (t_1, *, \ldots, t_p, *) \in im(\pi)$.

In case B, since $g \notin J$ and p is prime, $\sigma(g)$ must be a p-cycle. Let us suppose that $\sigma(g) = (1, 2, \ldots, p)$. Hence we must examine

$$\begin{aligned} g(\underline{w}) &= (t_p, j(p, g)(w_p), \ldots, t_{p-1}, j(p-1, g)(w_{p-1})) \\ &= \underline{w} \\ &= (t_1, w_1, \ldots, t_p, w_p). \end{aligned}$$

This equation implies either that each $w_i \in i(S_2^3)$ and $\underline{w} \in im(\pi)$ or that each w_i lies in the image of the cone on S_1^3. In the latter case each w_i must be at the same level in the cone and therefore, again, $\underline{w} \in im(\pi)$.

In case C let us once more suppose that $\sigma(g) = (1, 2, \ldots, p)$. We may also assume that $t_1 \neq 0$ and that $w_1 \in D - i(Z)$. In this case

$$\begin{aligned} g(t_1, w_1, \ldots, t_p, w_p) &= \sigma(g)(t_1, j(1, g)(w_1), \ldots, t_p, j(p, g)(w_p)) \\ &= (t_1, w_1, \ldots, t_p, w_p), \end{aligned}$$

then

$$t_1 = t_2 = \ldots = t_p$$

and

$$\begin{aligned} w_1 &= j(p, g)(w_p), \\ w_2 &= j(1, g)(w_1), \\ &\vdots \\ w_p &= j(p-1, g)(w_{p-1}). \end{aligned}$$

Substituting each of these equations into the previous one we obtain, in the notation of 5.3.13,

$$\begin{aligned} w_1 &= j(p, g) j(p-1, g) \ldots j(1, g)(w_1) \\ &= z_{\sigma(g)(p)}^{-1} g z_p z_{\sigma(g)(p-1)}^{-1} g z_{p-1} \ldots z_1(w_1) \\ &= (z_1^{-1} g^p z_1)(w_1). \end{aligned}$$

Since $w_1 \in D - i(Z)$ and $g^p \in J$, this implies that $1 = z_1^{-1} g^p z_1$ and therefore that $1 = g^p$, which is the contradiction that completes the proof in case C.

Finally, in case D,

$$g(\underline{w}) = (t_1, g(w_1), \ldots) = (t_1, w_1, \ldots) = \underline{w},$$

so that $g = 1$ because $w_1 \in D - i(Z)$ and J acts freely on $D - i(Z)$. This completes the proof of Theorem 5.3.3 in the non-split case.

5.3.32 *Proof of Theorem* 5.3.3 *in the split case*

We retain the notation of 5.3.29. Also, we shall assume that $G = \mathbf{Z}/p \propto J$ where $\mathbf{Z}/p = <\tau>$. Let $\{\tau_i = \tau^{i-1} \mid 1 \leq i \leq p\}$ serve as a set of coset representatives of G/J. Suppose that

$$\underline{w} = (t_1, w_1, \ldots, t_p, w_p)$$

$(0 \leq t_i \leq 1, \sum_i t_i^2 = 1, w_i \in D)$ denotes a point of $E - im(\pi)$ which is fixed by $g \in G$. By the discussion of 5.3.29, if $g = h\tau$ for some $h \in J$ then $t_1 = t_2 = \ldots = t_p = p^{-1/2}$ and

$$\begin{aligned} w_1 &= j(p, g)(w_p), \\ w_2 &= j(1, g)(w_1), \\ &\vdots \\ w_p &= j(p-1, g)(w_{p-1}), \end{aligned}$$

where $j(i, g) = \tau_{i+1}^{-1} g \tau_i$ and $g = h\tau$ must be of order p. Hence, if $\underline{w}$ is fixed by $g = h\tau$ ($h \in J$) then

5.3.33 $$\underline{w} = (p^{-1/2}, w_1, p^{-1/2}, (\tau_2^{-1} g \tau_1)(w_1), p^{-1/2}, (\tau_3^{-1} g^2 \tau_1)(w_1), \ldots),$$

where $w_1 \in D - i(Z)$. If $g = h\tau^s$ with $2 \leq s \leq p-1$ then $\underline{w}$ has a form similar to 5.3.33 which is determined by its first entry, $w_1 \in D - i(Z)$.

For $g = h\tau$ $(h \in J)$ we obtain an embedding

5.3.34 $$\mu_g : D \xrightarrow{\cong} D_g = \mu_g(D) \subset E$$

given by $\mu_g(w_1) = \underline{w}$, where $\underline{w}$ is given by the formula of 5.3.33. In general, for $g = h\tau^s$ $(h \in J, 2 \leq s \leq p-1)$, we obtain a mapping of the form 5.3.34 which embeds D into E. Specifically, $\mu_{h\tau^s}(w_1)$ is given by $\mu_{h\tau^s}(w_1) = (p^{-1/2}, \alpha_1, \ldots, p^{-1/2}, \alpha_p)$ where $\alpha_{ts+1} = \tau^{-ts}(h\tau^s)^t(w_1)$ $= \tau^{-ts} g^t(w_1)$. Here $g = h\tau^s$ must be of order p.

Lemma 5.3.35 *Let $* \in Z$ denote the (J-fixed) cone-point and let $x_0 = (p^{-1/2}, *, \ldots, p^{-1/2}, *) \in E$. Suppose that $g', g \in G - J$ and that $g^p = 1 = (g')^p$. Then*

(i) $x_0 \in D_g = \mu_g(D)$,

(ii) *either* $D_g \cap D_{g'} = x_0$ *or* $D_g = D_{g'}$,

and

(iii) $D_g = D_{g'}$ *if and only if* $< g > = < g' >$.

In case (iii), $\mu_g(w_1) = \mu_{g'}(w_1)$ *for all* $w_1 \in D$.

Proof Suppose that $g = h\tau^s$ with $h \in J$ then the $(ts+1)$th coordinate is equal to $\tau^{ts}(h\tau^s)^t(*) = *$, since $*$ is J-fixed. This proves part (i).

For part (ii), take $g = h\tau$ and $g = h'\tau^s$ ($1 \leq s \leq p-1$). If $w_1, w_1' \in D - \{*\}$ suppose that $\mu_g(w_1) = \mu_{g'}(w_1')$. By comparison of the first coordinates we see that $w_1 = w_1'$. Comparison of the second coordinates yields $(\tau^{-1} h\tau)(w_1) = \tau^{-ss'}(h'\tau^s)^{s'}(w_1)$, where $ss' \equiv 1 \pmod p$.

Since J acts freely on w_1, $h\tau = (h'\tau^s)^{s'}$ or $< g >=< g' >$. However, in this case, the $(ts+1)$th entry of $\mu_g(w_1)$ equals

$$\tau^{-ts}(h\tau)^{ts} = \tau^{-ts}(h'\tau^s)^{tss'} = \tau^{-ts}(h'\tau^s)^t,$$

which is the $(ts+1)$th entry of $\mu_{g'}(w_1)$. Hence $D_g = D_{g'}$ and $\mu_g = \mu_{g'}$. This proves parts (ii) and (iii). □

Now define

5.3.36 $$W = E - \pi(D(G, J, \phi, k)) \cup (\cup_g D_g),$$

where, in 5.3.36, g runs through representatives of subgroups, $< g >$, of order p in $G - J$.

Lemma 5.3.37 (i) *Let $z \in J$, then the following diagram commutes:*

$$\begin{array}{ccc} D & \xrightarrow{\mu_g} & D_g \\ \downarrow z & & \downarrow z \\ D & \xrightarrow{\mu_{zgz^{-1}}} & D_{zgz^{-1}} \end{array}$$

(ii) *The action of G is free on W in 5.3.36.*

(iii) *Let W be as in* 5.3.36, *then for all $H \leq G$ and for all localisations, $\mathbf{Z}_{(q)}$, of $\mathbf{Z}$ at a prime, q, dividing $\#(G)$*

$$\tilde{H}_*((E - W)^H; \mathbf{Z}_{(q)}) = 0.$$

(iv) *If $z \in N_G < g >$, the normaliser of $< g >$, then $z(D_g) = D_g$.*

Proof For part (i), the $(ts+1)$th coordinate of $\mu_{h\tau^s}(w_1)$ ($h \in J, 1 \leq s \leq p-1$ and $h\tau^s$ of order p) is equal to $\tau^{ts}(h\tau^s)^t(w_1)$. Hence the $(ts+1)$th coordinate of $z(\mu_{h\tau^s}(w_1))$ is equal to

$$\tau^{-ts}z\tau^{ts}\tau^{-ts}(h\tau^s)^t(w_1) = \tau^{-ts}z(h\tau^s)^t(w_1).$$

However, the $(ts+1)$th coordinate of $\mu_{zh\tau^s z^{-1}}(z(w_1))$ is equal to

$$\tau^{-ts}z(h\tau^s)^t z^{-1}z(w_1) = \tau^{-ts}z(h\tau^s)^t(w_1),$$

which proves part (i).

Part (ii) follows from the discussion of 5.3.29 as recapitulated in 5.3.32.

If $Z_g = \mu_g(Z)$ then $\tilde{H}_*(-;\mathbf{Z}_{(q)})$ vanishes on

$$Z_g^H, D_g^H \text{ and } \pi(D(G,J,\phi,k))^H.$$

By 5.3.35, $\cup_g D_g$ is a wedge of the D_gs which intersects $\pi(D(G,J,\phi,k))$ in the wedge of the Z_gs. Hence part (iii) follows from the Mayer–Vietoris homology exact sequence.

Finally, if $z \in N_G < g > \cap J$ then part (iv) follows from part (i). If $z = h'\tau^s \in N_G < g >$ we may, by 5.3.37, assume that $g = h\tau^s$. Then g acts trivially on D_g so that

$$z(D_{h\tau^s}) = h'h^{-1}(D_g) = D_{h'h^{-1}gh(h')^{-1}} = D_{zgz^{-1}} = D_g,$$

which completes the proof of part (iv). □

Corollary 5.3.38 *In the notation of* 5.3.27 *and* 5.3.36

$$0 = M(E-W) \in \mathscr{C}\mathscr{L}(\mathbf{Z}[G]).$$

Proof $E - W$ is embedded in the acyclic space, E, with a free G-action on the complement, W, by 5.3.37 (ii). □

Lemma 5.3.39 *In* 5.3.38

$$\tilde{H}_i(E-W;\mathbf{Z}) = \begin{cases} \mathbf{Z}/k^p & \text{if } i = 4p-1, \\ \oplus_g \mathbf{Z}/k & \text{if } i = 4, \\ 0 & \text{otherwise} \end{cases}$$

where g runs through an indexing set for the distinct D_gs.

Proof This follows from the Mayer–Vietoris sequence for

$$E - W = \pi(D(G,J\phi,k)) \cup (\cup_g D_g)$$

since

$$\pi(D(G,J\phi,k)) \cap (\cup_g D_g) = \cup_g Z_g,$$

$\tilde{H}_*(Z_g;\mathbf{Z})$ is equal to $\mathbf{Z}/k$ in dimension three, $\tilde{H}_*(\pi(D(G,J\phi,k));\mathbf{Z})$ is equal to $\mathbf{Z}/k^p$ in dimension $4p-1$ and each D_g is acyclic. Note that, by 5.3.35, $\cup_g D_g$ and $\cup_g Z_g$ are wedge sums with $*$ as the base-point. □

5.3.40 *Conclusion of the proof of Theorem* 5.3.3

In order to obtain the relation of 5.3.3 in the class-group we will evaluate $M(E-W)$ of 5.3.38. To do this we may assume that $E-W$ is triangulated so that the G-action is simplicial. We must build a resolution (in the sense of 5.3.23) of the augmented chain complex

$$\ldots \longrightarrow C_i(E-W) \longrightarrow C_{i-1}(E-W) \longrightarrow \ldots \longrightarrow C_0(E-W) \longrightarrow \mathbf{Z} \longrightarrow 0,$$

following the recipe of 5.3.23. The process begins in $C_4(E-W)$ where we must kill the generators of the $\mathbf{Z}/k$s which are indexed by the D_{g_i}s, as in 5.3.39.

Fix $g \in G - J$ such that $g^p = 1$ and consider D_g. By the homology calculation of 5.3.39 we may find a 4-chain, σ_g^4, which is a cycle representing the $\mathbf{Z}/k$ in $H_4(E-W)$ which corresponds to D_g and whose cells all lie within D_g. We know from 5.3.35–5.3.37 that $N_G < g >$ acts trivially on the homology class of σ_g^4 while $G/N_G < g >$ acts freely on it, permuting it among the generators of the other $\mathbf{Z}/k$'s which correspond to the $D_{zgz^{-1}}$'s.

In dimension five we insert free modules, $\mathbf{Z}[G] < \delta_g >$, where $\overline{\delta}_5(\delta_g) = \sigma_g^4$, one for each conjugacy class of $g \in G - J$ such that $g^p = 1$.

Suppose that $\overline{\delta}_5(\sum_z \lambda_z z(\delta_g))$ is in $\delta_5(C_5(E-W))$. Passing to homology we find, for each $z \in G - N_G < g >$, that

5.3.41 $$\textstyle\sum_{v \in N_G<g>} \lambda_{zv} \equiv 0 \quad (mod\ k),$$

since, in homology,

$$0 = \overline{\delta}_5\left(\sum \lambda_z z(\delta_g)\right) = \sum_{z \in G/N_G<g>} \left[\sum_{v \in N_G<g>} \lambda_{zv}\right] z(\sigma_g^4).$$

Conversely, if 5.3.41 holds for each $z \in G - N_G < g >$ there is a unique $\alpha \in C_5(E-W)/Ker(\delta_5)$ such that $\overline{\delta}_5(\sum \lambda_z z(\delta_g) - \alpha) = 0$. Hence, since $ker(\delta_5) = im(\delta_6)$, we obtain an exact sequence

5.3.42 $$0 \longrightarrow im(\delta_6) \longrightarrow ker(\overline{\delta}_5) \longrightarrow \oplus_g P_g \longrightarrow 0,$$

where

$$P_g = Ind_{N_G<g>}^G(ker(\mathbf{Z}[N_G < g >] \stackrel{\epsilon}{\longrightarrow} \mathbf{Z}/k)) = -Ind_{N_G<g>}^G(S(k))$$

and the direct sum is taken over representative generators, g, for the conjugacy classes of subgroups of order p which are not contained in J.

The sequence 5.3.42 is exact because the submodule consisting of elements satisfying 5.3.41 for all $z \in G - N_G < g >$ is just $\mathbf{Z}[G] \otimes_{\mathbf{Z}[N_G<g>]} (Ker(\epsilon)) = -Ind_{N_G<g>}^G(S(k))$, by 5.3.9. The sequence of 5.3.42 splits, since

P_g is projective, so that

$$ker(\overline{\delta}_5) = im(\delta_6) \oplus (\oplus_g P_g).$$

It is now simple to complete the resolution. To kill each of the P_gs in dimension five we insert a free module, F_g, in dimension six with $\overline{\delta}_6(F_g) = P_g$. This produces a contribution of $\oplus_g(-P_g)$ in $ker(\overline{\delta}_6)$, where $F_g = P_g \oplus (-P_g)$. Continuing in this manner we obtain a contribution of $\oplus_g(-P_g)$ in the kernel in dimension $4p$, the top dimension. Also, as we get to dimension $4p$ we pick up $-S(k^p)$ from the $\mathbf{Z}/k^p \cong H_{4p-1}(E - W;\mathbf{Z})$, as in 5.3.28, and the top-dimensional kernel of the resolution is equal to

$$\begin{aligned} M(E - W) &= -S(k^p) \oplus (\oplus_g(-P_g)) \\ &= -S(k^p) \oplus (\oplus_g Ind_{N_G<g>}^G)(S(k)), \end{aligned}$$

as required. □

5.4 The class-group of a maximal order

In this section we shall apply the Explicit Brauer Induction map, a_G, in a manner similar to that used in the construction of restricted determinants (5.1.4), to study the class-group of a maximal order in the rational group-ring, $\mathbf{Q}[G]$.

Let $\Lambda(G)$ denote a maximal order of $\mathbf{Q}[G]$, as in 4.2.29. Therefore we have inclusions

5.4.1 $$\mathbf{Z}[G] \subset \Lambda(G) \subset \mathbf{Q}[G].$$

The class-group, $\mathcal{CL}(\Lambda(G))$, admits a Hom-description in terms of the representation ring, $R(G)$, and the group of *fractional ideals* of a large Galois extension, $E/\mathbf{Q}$, which is a splitting field for G, as in 4.2.8. A fractional ideal of E (Lang, 1970, p. 18) is an $\mathcal{O}_E$-submodule of E of the form $c^{-1}I$ where $0 \neq c \in \mathcal{O}_E$ and $I \lhd \mathcal{O}_E$. The set of non-zero fractional ideals forms a group under multiplication and this group is isomorphic to

5.4.2 $$\mathcal{I}(E) = J^*(E)/U(\mathcal{O}_E),$$

where $J^*(E)$ and $U(\mathcal{O}_E)$ are as in 4.2.5 and 4.2.6. A non-zero fractional ideal corresponds to the class of the idèle given by the generators of its completions.

The absolute Galois group of the rational field, $\Omega_{\mathbf{Q}}$, acts on E and hence on $\mathcal{I}(E)$ so that we may consider the group of $\Omega_{\mathbf{Q}}$-equivariant maps

5.4.3 $$Hom_{\Omega_{\mathbf{Q}}}(R(G), \mathcal{I}(E)).$$

Notice that if $E \subset F$ is an inclusion, $\mathcal{I}(E)$ is a subgroup of $\mathcal{I}(F)$ in the obvious manner.

Definition 5.4.4 Let $Fac(G)$ denote the subgroup of 5.4.3 given by

$$Fac(G) =$$

$$\{g \in Hom_{\Omega_{\mathbf{Q}}}(R(G), \mathcal{I}(E)) \mid g(\chi) \in \mathcal{I}(\mathbf{Q}(\chi)) \text{ for all } \chi \in R(G)\},$$

where $\mathbf{Q}(\chi)$ is the smallest Galois extension of $\mathbf{Q}$, the rational numbers, which contains all the character values, $\{Trace(\chi(z)), z \in G\}$ of χ.

The *Fac*-notation is used in Holland (1992) and is derived from its connection with A. Fröhlich's notion of *factorisable* functions. We will not need this notion here, but for further details the interested reader may consult Fröhlich (1988) and Holland (1992).

Definition 5.4.5 A fractional ideal, $c^{-1}I$, of E is *principal* if $I = a\mathcal{O}_E \lhd \mathcal{O}_E$ for some $a \in \mathcal{O}_E$. Define a subgroup,

5.4.6 $$PF^{+}(G) \subset Fac(G)$$

to consist of all $g \in Fac(G)$ such that

(i) $g(\chi)$ is a principal fractional ideal for all $\chi \in R(G)$

and

(ii) for all irreducible, symplectic representations, χ, the fractional ideal, $g(\chi)$, has a generator which is positive under all Archimedean places of E.

Notice that, since χ has a real-valued character, $g(\chi)_{\mathcal{P}}$ will be generated by a real number for all infinite primes, $\mathcal{P}$, so that it makes sense to ask that the generator be *totally positive*, in the above sense. Note also that, on symplectic representations whose complexification is of the form $\rho + \overline{\rho}$ ($\overline{\rho}$ is the complex conjugate of ρ), g is automatically totally positive (Fröhlich, 1983, pp. 22–23). Hence (ii) is equivalent to g being totally positive on *all* symplectic representations of G.

Definition 5.4.7 Let $\Lambda(G)$ be a maximal order in $\mathbf{Q}[G]$. A $\Lambda(G)$-module, M, is said to be *locally freely presented* if there is an exact sequence of $\Lambda(G)$-modules

$$0 \longrightarrow P \longrightarrow N \longrightarrow M \longrightarrow 0$$

in which P and N are locally free $\Lambda(G)$-modules of the same (finite) rank. Hence M is finite.

Let $K_0T(\Lambda(G))$ denote the Grothendieck group of locally freely presented $\Lambda(G)$-modules, taken with respect to exact sequences. Hence if

$$0 \longrightarrow M_1 \longrightarrow M \longrightarrow M_2 \longrightarrow 0$$

is a short exact sequence pf such modules, then

$$[M] = [M_1] + [M_2] \in K_0T(\Lambda(G)).$$

Theorem 5.4.8 (*Fröhlich,* 1984; *Holland,* 1992; *Taylor,* 1984) (i) *There is a natural isomorphism*

$$\mu_G : K_0T(\Lambda(G)) \cong Fac(G).$$

(ii) *There is a commutative diagram in which the horizontal sequence is exact and in which c_G is the canonical Cartan map:*

$$\begin{array}{ccccccccc} 0 & \longrightarrow & PF^+(G) & \longrightarrow & Fac(G) & \longrightarrow & \mathcal{CL}(\Lambda(G)) & \longrightarrow & 0 \\ & & & & \mu_G \downarrow \cong & \nearrow c_G & & & \\ & & & & K_0T(\Lambda(G)) & & & & \end{array}$$

5.4.9 Denote by (H) the G-conjugacy class of a subgroup, $H \leq G$. We have a homomorphism

5.4.10 $$B_G : \sum_{\substack{(H)\\ H\leq G}} R(H^{ab})_{W_GH} \longrightarrow \sum_{\substack{(H)\\ H\leq G}} R(H)_{W_GH} \longrightarrow R(G)$$

defined by

$$B_G(\phi : H^{ab} \longrightarrow E^*) = Ind_H^G(Inf_{H^{ab}}^H(\phi)).$$

We also have a homomorphism

5.4.11 $$A_G : R(G) \longrightarrow \sum_{\substack{(H)\\ H \leq G}} R(H^{ab})_{W_G H},$$

defined by setting the $R(H^{ab})_{W_G H}$-component of $A_G(\chi)$ equal to the sum of all the terms of the form

$$\alpha_{(H,\phi)^G}(H^{ab} \longrightarrow E^*)$$

which appear in the formula (see 2.3.15)

5.4.12 $$a_G(\chi) = \sum_{(J)} \alpha_{(J,\psi)^G}(J,\psi)^G \in R_+(G).$$

Note that, once one has chosen a representative for (J), each $(J,\psi)^G$ determines a homomorphism $J \longrightarrow J^{ab} \longrightarrow E^*$ which is well-defined up to conjugation by elements of $W_G J$.

By Theorem 2.3.9

5.4.13 $$B_G A_G = 1 : R(G) \longrightarrow R(G).$$

Since a_G is $\Omega_{\mathbf{Q}}$-equivariant (see 4.5.18) we immediately obtain the following result:

Proposition 5.4.14 *The homomorphisms of* 5.4.10 *and* 5.4.11 *induce maps*

$$Hom_{\Omega_{\mathbf{Q}}}(R(G), \mathscr{I}(E)) \xrightarrow{B_G^*} \sum_{\substack{(H)\\ H \leq G}} Hom_{\Omega_{\mathbf{Q}}}(R(H^{ab}), \mathscr{I}(E))^{W_G H}$$

and

$$\sum_{\substack{(H)\\ H \leq G}} Hom_{\Omega_{\mathbf{Q}}}(R(H^{ab}), \mathscr{I}(E))^{W_G H} \xrightarrow{A_G^*} Hom_{\Omega_{\mathbf{Q}}}(R(G), \mathscr{I}(E))$$

which satisfy $A_G^* B_G^* = 1$.

Definition 5.4.15 As in 5.4.9, let $H \leq G$ be finite groups and let H^{ab} denote the abelianisation of H. Define $Fac_G(H^{ab})$ to be the following subgroup of $Hom_{\Omega_{\mathbf{Q}}}(R(H^{ab}), \mathscr{I}(E))$:

$$Fac_G(H^{ab}) = \{f \mid f(\phi : H^{ab} \longrightarrow E^*) \in \mathscr{I}(\mathbf{Q}(Ind_H^G(Inf_{H^{ab}}^H(\phi))))\}.$$

Note that $Fac_G(H^{ab})$ is a group of $\Omega_{\mathbf{Q}}$-invariant functions which depends on G, the ambient group, and on $H \leq G$, *not just on* H^{ab}.

Theorem 5.4.16 *The homomorphisms of* 5.4.14 *restrict to give homomorphisms*

$$Fac(G) \xrightarrow{B_G^*} \sum_{\substack{(H)\\ H\leq G}} (Fac_G(H^{ab}))^{W_G H}$$

and

$$\sum_{\substack{(H)\\ H\leq G}} (Fac_G(H^{ab}))^{W_G H} \xrightarrow{A_G^*} Fac(G)$$

which satisfy $\quad A_G^* B_G^* = 1.$

Proof Fix a conjugacy class representative, $H \leq G$, and consider the (H)-component of $B_G^*(f)$ $(f \in Fac(G))$. By definition,

$$B_G^*(f)(\phi : H^{ab} \longrightarrow E^*) = f(Ind_H^G(Inf_{H^{ab}}^H(\phi))),$$

which belongs to $\mathcal{I}(\mathbf{Q}(Ind_H^G(Inf_{H^{ab}}^H(\phi))))$. Hence the (H)-component of $B_G^*(f)$ lies in $Fac_G(H^{ab})$, which completes the verification for B_G^*.

It remains to show that $A_G^*(\{g_{(H)}\})$ lies in $Fac(G)$ if

$$g_{(H)} \in Fac_G(H^{ab})^{W_G H}.$$

Choose a set of representatives, Σ, for the conjugacy classes of subgroups of G. Suppose that $\chi \in R(G)$ and that

5.4.17 $$a_G(\chi) = \textstyle\sum_{H\in\Sigma} a_{G,H}(\chi) \ \in \ R_+(G),$$

where $a_{G,H}(\chi) = \sum_i n_i(H, \phi_i)^G$. Since $a_G(\chi)$ is fixed by $\Omega_{\mathbf{Q}(\chi)}$ the subhomomorphisms, $\{(H, \phi_i)^G\}$, which appear in $a_{G,H}(\chi)$, are permuted by the (abelian) Galois group $G(\mathbf{Q}(\phi_i)/\mathbf{Q}(\chi))$ and the corresponding coefficients, $\{n_i\}$, are constant along orbits. Let

5.4.18 $$stab((H, \phi_i)^G) \leq G(Q(\phi_i)/Q(\chi))$$

denote the stabiliser of $(H, \phi_i)^G$. Also let

5.4.19 $$Z(H, \phi_i) = \{g \in W_G H \mid g^*(\phi_i) = \phi_i : H \longrightarrow E^*\}$$

denote the stabiliser of $(\phi_i : H \longrightarrow E^*)$ in $W_G H$. If $\sigma \in stab((H, \phi_i)^G)$ then there exists $g(\sigma) \in W_G H$ such that

5.4.20 $$\sigma(\phi_i) = g(\sigma)^*(\phi_i) : H \longrightarrow E^*.$$

Therefore there is an injection

5.4.21 $$\gamma_i : stab((H,\phi_i)^G) \longrightarrow W_GH/Z(H,\phi_i),$$

given by

$$\gamma_i(\sigma G(\mathbf{Q}(\phi_i)/\mathbf{Q}(\chi))) = g(\sigma)Z(H,\phi_i).$$

Now consider the subsum of $a_{G,H}(\chi)$ given by the sum of the terms in the orbit of $(H,\phi_i)^G$. This subsum equals

5.4.22 $$n_i\left(\sum_{\tau\in G(\mathbf{Q}(\phi_i)/\mathbf{Q}(\chi))/stab((H,\phi_i)^G)}(H,\tau(\phi_i)^G)\right).$$

To complete the proof it will suffice to show that applying $g_{(H)}$ to 5.4.22 yields a fractional ideal in $\mathcal{I}(\mathbf{Q}(\chi))$.

However, if $w \in G$ then

5.4.23 $$Ind_H^G(\phi_i)(w) = \sum_{\substack{x\in G/H \\ x^{-1}wx\in H}} \phi_i(x^{-1}wx),$$

by 1.2.43. Since

$$H \leq Z(H,\phi_i) \leq W_GH \leq G,$$

we may rewrite 5.4.23 as

5.4.24 $$Ind_H^G(\phi_i)(w) = [Z(H,\phi_i) : H]\sum_{\substack{x\in G/Z(H,\phi_i) \\ x^{-1}wx\in H}} \phi_i(x^{-1}wx).$$

By means of the injection 5.4.21 we may write the coset representatives of $G/Z(H,\phi_i)$ in the form

$$\{x_u\gamma_i(\sigma_v)\},$$

where $1 \leq u \leq u_0$ and σ_v runs through the cosets of $stab((H,\phi_i)^G)$. Hence, in 5.4.24,

5.4.25 $$Ind_H^G(\phi_i)(w) = [Z(H,\phi_i) : H]\sum_{\substack{x_u \\ x_u^{-1}wx_u\in H}} \sum \sigma_v(\phi_i(x_u^{-1}wx_u)).$$

Clearly, 5.4.25 is in the image of the trace from $\mathbf{Q}(\phi_i)$ to K_i where $\mathbf{Q}(\chi) \leq K_i \leq \mathbf{Q}(\phi_i)$ is the fixed field of $stab((H,\phi_i)^G) = G(\mathbf{Q}(\phi_i)/K_i)$. Hence

5.4.26 $$g_{(H)}(H,\phi_i) \in \mathcal{I}(K_i),$$

so that

5.4.27 $$g_{(H)}(n_i(\sum_{\tau\in G(\mathbf{Q}(\phi_i)/\mathbf{Q}(\chi))/stab((H,\phi_i)^G)}(H,\tau(\phi_i)^G)))$$

$$= \prod_{\tau\in G(K_i/\mathbf{Q}(\chi))} \tau(g_{(H)}((H,\phi_i)^G)))^{n_i}.$$

Combining 5.4.26 and 5.4.27 for each i, we see that

5.4.28 $$g_{(H)}(a_{G,H}(\chi)) \in \mathscr{I}(\mathbf{Q}(\chi)),$$

which implies that $A^*_G(\{g_{(H)}\}) \in Fac(G)$, as required. □

5.4.29 Next we must work towards the detection of $PF^+(G)$ of 5.4.6. This will require an analysis of the composite

$$RSp(G) \xrightarrow{c} R(G) \xrightarrow{a_G} R_+(G),$$

where $RSp(G)$ is the Grothendieck group of symplectic representations and c is the (injective) homomorphism which sends a symplectic representation to its underlying complex representation (cf. 4.2.32).

We begin by examining the image of a symplectic representation given by a symplectic line.

Proposition 5.4.30 *Let $v : G \longrightarrow Sp(1)$ be a one-dimensional symplectic representation then*

$$a_G(c(v)) \in R_+(G)$$

is an integral linear combination of elements of the form $\{(H,\phi)^G + (H,\overline{\phi})^G\}$, where $\overline{\phi}$ is the complex conjugate of ϕ, or of the form $(J,\mu)^G$ where $Ind_J^G(\mu)$ is the complexification of a symplectic representation of G.

Proof The representation, $c(v)$, is two-dimensional. For a two-dimensional representation it is easy to show (see Boltje, 1990; Boltje, Snaith & Symonds, 1992) that a_G coincides with the original Explicit Brauer Induction formula of Snaith (1988b), which is described in 2.3.21. Hence

5.4.31 $$a_G(c(v)) = \rho_G(\tau_G(c(v))),$$

where ρ_G and τ_G are as in 2.3.23–2.3.25.

Suppose that, as in 2.3.23,

5.4.32 $$\tau_G(c(v)) = \textstyle\sum_{\alpha\in\mathscr{A}} \chi^{\#}_{\alpha}(H_\alpha \xrightarrow{g_\alpha^{-1}c(v)g_\alpha} NT^2) \in R_+(G, NT^2),$$

then each $g_\alpha^{-1}c(v)g_\alpha$, when described as a complex representation, is the complexification of a symplectic representation of H_α, since it is conjugate to the restriction of $c(v)$ to H_α.

Now consider the homomorphism of 2.3.25:

5.4.33 $$\rho_G : R_+(G, NT^2) \longrightarrow R_+(G).$$

If $g_\alpha^{-1}c(v)g_\alpha(H_\alpha) \leq T^2 \leq NT^2$ then the image is a cyclic group and there is a homomorphism, $\lambda : H_\alpha \longrightarrow S^1$, such that

$$g_\alpha^{-1}c(v)g_\alpha(h) = \begin{pmatrix} \lambda(h) & 0 \\ 0 & \overline{\lambda}(h) \end{pmatrix} \in U(2)$$

for all $h \in H_\alpha$. In this case, by definition of ρ_G in 2.3.25,

$$\rho_G(g_\alpha^{-1}c(v)g_\alpha) = (H_\alpha, \lambda)^G + (H_\alpha, \overline{\lambda})^G.$$

If $g_\alpha^{-1}c(v)g_\alpha(H_\alpha) \not\leq T^2$ then set

$$J = Ker(H_\alpha \longrightarrow NT^2 \longrightarrow NT^2/T^2 \cong \{\pm 1\})$$

and let the restriction of $g_\alpha^{-1}c(v)g_\alpha$ to J have the form

$$g_\alpha^{-1}c(v)g_\alpha(j) = \begin{pmatrix} \phi(j) & 0 \\ 0 & \overline{\phi}(j) \end{pmatrix},$$

for some $\phi : J \longrightarrow S^1$, then

$$\rho_G(g_\alpha^{-1}c(v)g_\alpha) = (J, \phi)^G.$$

However, in this second case (Snaith, 1989b, lemma 3.35(v), p. 234),

$$Ind_J^{H_\alpha}(\phi) = g_\alpha^{-1}c(v)g_\alpha \in R(H_\alpha),$$

which is symplectic. By 5.4.31, this completes the proof of the proposition. □

5.4.34 *Symplectic explicit Brauer induction*

Let G be a finite group and let $RSp(G)$ denote the Grothendieck group of finite-dimensional, symplectic representations of G. Let $RSp_+(G)$ denote the free abelian group on G-conjugacy classes of subhomomorphisms of the form (with $H \leq G$)

5.4.35 $$\phi : H \longrightarrow Sp(1) = S^3.$$

Denote by $(H, \phi)^G$ the class of 5.4.35 in $RSp_+(G)$. Define homomorphisms

$$Res_J^G : RSp_+(G) \longrightarrow RSp_+(J),$$

$$Ind_J^G : RSp_+(J) \longrightarrow RSp_+(G)$$

and

$$bsp_G : RSp_+(G) \longrightarrow RSp(G)$$

in a manner which is analogous to the complex case.

One may define a symplectic Explicit Brauer Induction homomorphism by the method of Symonds (1991, section 6),

5.4.36
$$asp_G : RSp(G) \longrightarrow RSp_+(G).$$

Explicitly, suppose that

5.4.37
$$\rho : G \longrightarrow Sp(n)$$

is a symplectic representation of G. Let G act on the symplectic projective space, $P(\mathbf{H}^n)$, via ρ. Triangulate $P(\mathbf{H}^n)$ in such a manner that G acts simplicially. If σ is a simplex of $G \backslash P(\mathbf{H}^n)$ and $\hat{\sigma}$ is a choice of a simplex above σ in $P(\mathbf{H}^n)$ then the stabiliser, $stab(\hat{\sigma})$, acts via

$$\phi(\hat{\sigma}) : stab(\hat{\sigma}) \longrightarrow Sp(1)$$

on the symplectic line given by any point in the interior of $\hat{\sigma}$. Set

5.4.38
$$asp_G(\rho) = \sum_{\sigma \in G \backslash P(\mathbf{H}^n)} (-1)^{dim(\sigma)} (stab(\hat{\sigma}), \phi(\hat{\sigma}))^G \in RSp_+(G).$$

Remark 5.4.39 The method of Snaith (1988b) also yields a symplectic Explicit Brauer Induction formula

$$TSp_G(\rho) \in RSp_+(G),$$

which is a derivation in the sense of 2.3.28(v). The expressions, $TSp_G(\rho)$ and $asp_G(\rho)$, are related by a formula

$$TSp_G(\rho) = asp_G(\rho) \cdot \epsilon sp_G(\rho),$$

which is analogous to the formula of 2.5.11 Boltje, Snaith & Symonds (1992).

5.4.40 Define a homomorphism

$$c_{+,G} : RSp_+(G) \longrightarrow R_+(G)$$

by the formula

5.4.41

$$c_{+,G}((H,\phi)^G) = Ind_H^G(c(\phi) : H \longrightarrow Sp(1) \longrightarrow U(2)) \in R_+(G).$$

Theorem 5.4.42 *With the notation introduced above*

$$c_{+,G} \cdot asp_G = a_G \cdot c : RSp(G) \longrightarrow R_+(G).$$

Proof Clearly the homomorphism, $a_G \cdot c$, is natural with respect to inclusions of subgroups, $J \leq G$,

$$Res_J^G \cdot a_G \cdot c = a_J \cdot c \cdot Res_J^G.$$

We shall show that $c_{+,G} \cdot asp_G$ is also natural with respect to Res_J^G. Since 5.4.36 commutes with Res_J^G we have only to verify that the same is true for $c_{+,G}$.

If $H, J \leq G$ then there is a double coset formula for the composition

$$Res_J^G Ind_H^G : R_+(H) \longrightarrow R_+(J),$$

whose proof is the same as that of 2.5.7,

$$Res_J^G(Ind_H^G((K,\psi)^H)) = \sum_{w \in J\backslash G/H} Ind_{J\cap wHw^{-1}}^J(w^*(Res_{H\cap w^{-1}Jw}^H((K,\psi)^H))).$$

Hence, if $(H,\phi)^G \in RSp_+(G)$ with $\phi : H \longrightarrow Sp(1)$, then

$$\begin{aligned}
&Res(c_{+,G}((H,\phi)^G)) \\
&= Res_J^G(Ind_H^G(a_H(c(\phi) : H \longrightarrow U(2)))) \\
&= \textstyle\sum_{w\in J\backslash G/H} Ind_{J\cap wHw^{-1}}^J(w^*(Res_{H\cap w^{-1}Jw}^H(a_H(c(\phi))))) \\
&= \textstyle\sum_{w\in J\backslash G/H} Ind_{J\cap wHw^{-1}}^J(w^*(a_{J\cap wHw^{-1}}(c(Res_{H\cap w^{-1}Jw}^H(\phi))))) \\
&= \textstyle\sum_{w\in J\backslash G/H} Ind_{J\cap wHw^{-1}}^J(a_{J\cap wHw^{-1}}(c(Res_{J\cap wHw^{-1}}^{wHw^{-1}}((w^{-1})^*(\phi))))) \\
&= c_{+,J}(Res_J^G((H,\phi)^G)),
\end{aligned}$$

as required.

To complete the proof, as in the proof that a_G is unique if it satisfies 2.2.8, we need only verify that each of the natural transformations, $c_{+,G} \cdot asp_G$ and $a_G \cdot c$, have the same 'leading terms'. That is, we must evaluate the coefficients of the terms of the form $(G,\lambda)^G$ which appear in $c_{+,G} \cdot asp_G(\rho)$ and $a_G \cdot c(\rho)$, where ρ is the symplectic representation of 5.4.37, and for this purpose we may suppose that ρ is irreducible as a symplectic representation. If ρ is a symplectic line then

$$asp_G(\rho) = (G, \rho : G \longrightarrow Sp(1))^G$$

and the coefficient of $(G, \lambda)^G$ in $c_{+,G}(\rho)$ is the coefficient of $(G, \lambda)^G$ in $a_G(c(\rho))$, which equals the Schur inner product, $< \lambda, c(\rho) >$, as required. If ρ is irreducible but not a symplectic line then there are no terms of the form $(G, \phi)^G$ in $asp_G(\rho)$ and no terms of the form $(G, \lambda)^G$ in $c_{+,G}(asp_G(\rho))$. On the other hand, in this case, if $\lambda : G \longrightarrow U(1)$ were a one-dimensional representation for which $< \lambda, c(\rho) >$ was non-zero then $\lambda \oplus \overline{\lambda}$ would be a summand of $c(\rho)$ (even if $\lambda = \overline{\lambda}$) and $\lambda \oplus \overline{\lambda}$ would be the complexification of a symplectic summand of ρ, which is impossible since ρ is irreducible. Hence there are no such terms in $a_G(c(\rho))$ in this case and again the leading terms are equal.

This completes the proof of Theorem 5.4.42. □

Corollary 5.4.43 *Let ρ be a symplectic representation of G, as in 5.4.37. Then*

$$a_G(c(\rho)) \in R_+(G)$$

is an integral linear combination of elements of the form $\{(H, \phi)^G + (H, \overline{\phi})^G\}$, where $\overline{\phi}$ is the complex conjugate of ϕ, and of the form $(J, \mu)^G$ where $Ind_J^G(\mu)$ is the complexification of a symplectic representation.

Proof By Theorem 5.4.42 we have

$$a_G(c(\rho)) = \sum_i n_i Ind_{H_i}^G(a_{H_i}(c(\phi_i) : H_i \longrightarrow Sp(1))),$$

where

$$asp_G(\rho) = \sum_i n_i(H_i, \phi_i)^G \in R_+(G).$$

By 5.4.30 each $a_{H_i}(c(\phi_i))$ is an integral linear combination of terms of the form $\{(K, \psi_i)^{H_i} + (K, \overline{\psi_i})^{H_i}\}$ or of the form $(J, \mu)^{H_i}$ where $Ind_J^{H_i}(\mu)$ is symplectic. In the latter case $Ind_J^G(\mu) = Ind_{H_i}^G(Ind_J^{H_i}(\mu))$ is also symplectic, which completes the proof. □

Definition 5.4.44 In 5.4.16 we detected $Fac(G)$ by means of the groups, $Fac_G(H^{ab})$. Now we define corresponding groups, $PF_G^+(H^{ab})$, by means of which we can detect $PF^+(G)$ of 5.4.6.

For each subgroup, $H \leq G$, define a subgroup (cf. Definition 5.4.5)

5.4.45 $$PF_G^+(H^{ab}) \subset Fac_G(H^{ab})$$

to be the subgroup consisting of all $f \in Fac_G(H^{ab})$ such that

(i) $f(\phi : H^{ab} \longrightarrow E^*)$ is a principal fractional ideal for all $\phi : H^{ab} \longrightarrow E^*$ and

(ii) for all ϕ in (i) such that $Ind_H^G(Inf_{H^{ab}}^H(\phi))$ is symplectic the fractional ideal

$$f(\phi : H^{ab} \longrightarrow E^*) \in \mathscr{I}(\mathbf{Q}(Ind_H^G(Inf_{H^{ab}}^H(\phi))))$$

has a generator which is positive under all Archimedean places of E.

Theorem 5.4.46 *The homomorphisms of* 5.4.16 *induce maps*

$$PF^+(G) \xrightarrow{B_G^*} \oplus_{\substack{(H)\\ H \leq G}} PF_G^+(H^{ab})^{W_G H}$$

and

$$\oplus_{\substack{(H)\\ H \leq G}} PF_G^+(H^{ab})^{W_G H} \xrightarrow{A_G^*} PF^+(G).$$

Proof For $f \in PF^+(G)$, the (H)-component of $B_G^*(f)$ is given by

$$B_G^*(f)(\phi : H^{ab} \longrightarrow E^*)_{(H)} = f(Ind_H^G(Inf_{H^{ab}}^H(\phi))).$$

Thus it is clear that $B_G^*(f)$ takes its values in principal fractional ideals and is totally positive on ϕ whenever $Ind_H^G(Inf_{H^{ab}}^H(\phi))$ is symplectic, which proves the assertion concerning B_G^*.

Let us examine $A_G^*(\{g_{(H)}\})$ where $g_{(H)} \in PF_G^+(H^{ab})^{W_G H}$. In the notation of 5.4.17

$$A_G^*(\{g_{(H)}\})(\chi) = \prod g_{(H)}(a_{G,H}(\chi)).$$

This is a principal fractional ideal and it is totally positive, in the sense of 5.4.4, when χ is symplectic, by virtue of 5.4.43. This is because $g_{(H)}(\phi)g_{(H)}(\overline{\phi})$ automatically has a totally positive generator and $g_{(J)}(\mu)$ has, by assumption, a totally positive generator when $Ind_J^G(\mu)$ is symplectic. □

Definition 5.4.47 For $H \leq G$ define $\mathscr{CL}_G(H^{ab})$ to be the quotient

$$\mathscr{CL}_G(H^{ab}) = \frac{Fac_G(H^{ab})}{PF_G^+(H^{ab})}.$$

Hence $\mathcal{CL}_G(H^{ab})$ depends upon G and upon H, up to conjugacy, not merely H^{ab}. Notice that there is an isomorphism of W_GH-invariants:

5.4.48 $$\mathcal{CL}_G(H^{ab})^{W_GH} \cong \left(\frac{Fac_G(H^{ab})^{W_GH}}{PF_G^+(H^{ab})^{W_GH}}\right).$$

This is seen in the following manner: W_GH acts on $R(H^{ab})$ by permuting the basis of one-dimensional representations, ϕ_i, preserving the character fields, $\mathbf{Q}(Ind_H^G(Inf_{H^{ab}}^H(\phi_i)))$, and the property that $Ind_H^G(Inf_{H^{ab}}^H(\phi_i))$ is symplectic. Therefore $PF_G^+(H^{ab})$ is a direct sum of W_GH-groups of the form $(\mathbf{Z}[W_GH] \otimes_{\mathbf{Z}[U]} \mathbf{Z}) \otimes X$, where X is torsion free and W_GH acts trivially on X. Such W_GH-modules have trivial one-dimensional cohomology since (see Snaith, 1989b)

$$\begin{aligned} & H^1(W_GH;(\mathbf{Z}[W_GH] \otimes_{\mathbf{Z}[U]} \mathbf{Z}) \otimes X) \\ & \cong H^1(W_GH;\mathbf{Z}[W_GH] \otimes_{\mathbf{Z}[U]} \mathbf{Z}) \otimes X \\ & \cong H^1(U;\mathbf{Z}) \otimes X \\ & \cong Hom(U,\mathbf{Z}) \otimes X, \end{aligned}$$

which is trivial, since U is finite. Therefore the short exact sequence of W_GH-modules

$$PF_G^+(H^{ab}) \longrightarrow Fac_G(H^{ab}) \longrightarrow \mathcal{CL}_G(H^{ab})$$

yields a long exact cohomology sequence (Snaith, 1989b, p. 9) which reduces at the left-hand end to the short exact sequence

$$PF_G^+(H^{ab})^{W_GH} \longrightarrow Fac_G(H^{ab})^{W_GH}, \longrightarrow \mathcal{CL}_G(H^{ab})^{W_GH},$$

since the next term is $H^1(W_GH;PF_G^+(H^{ab}))$, which is zero.

Combining 5.4.16 and 5.4.46 we obtain the main result of this section:

Theorem 5.4.49 *The homomorphisms of* 5.4.16 *induce maps*

$$\mathcal{CL}(\Lambda(G)) \xrightarrow{B_G^*} \oplus_{\substack{(H)\\ H\leq G}} \mathcal{CL}_G(H^{ab})^{W_GH}$$

and

$$\oplus_{\substack{(H)\\ H\leq G}} \mathcal{CL}_G(H^{ab})^{W_GH} \xrightarrow{A_G^*} \mathcal{CL}(\Lambda(G))$$

which satisfy $A_G^*B_G^* = 1$.

5.4.50 For each $H \leq G$ and for each $J \leq H^{ab}$ we may define the groups

$$Fac_G(H^{ab}/J), PF^{+}_{G}(H^{ab}/J) \text{ and } \mathscr{CL}_G(H^{ab}/J)$$

in a manner analogous to that of 5.4.15, 5.4.44 and 5.4.47. In addition, if $J \leq J' \leq H^{ab}$, there are inflation maps

$$Inf : Fac_G(H^{ab}/J) \longrightarrow Fac_G(H^{ab}/J')$$

and

$$Inf : PF^{+}_{G}(H^{ab}/J) \longrightarrow PF^{+}_{G}(H^{ab}/J').$$

We may assemble these maps into a homomorphism

$$\lim_{\longleftarrow} Inf : \mathscr{CL}_G(H^{ab}) \longrightarrow \lim_{\substack{\longleftarrow \\ H^{ab}/J cyclic}} \mathscr{CL}_G(H^{ab}/J).$$

Lemma 5.4.51 *The homomorphism of* 5.4.50 *is an isomorphism and induces an isomorphism*

$$\lim_{\longleftarrow} Inf : \mathscr{CL}_G(H^{ab})^{W_G H} \xrightarrow{\cong} \left(\lim_{\substack{\longleftarrow \\ H^{ab}/J cyclic}} \mathscr{CL}_G(H^{ab}/J) \right)^{W_G H}.$$

Proof The homomorphism of 5.4.50 is clearly $W_G H$-equivariant. Hence it suffices to show that it is an isomorphism.

Define a map

$$\psi : \lim_{\substack{\longleftarrow \\ H^{ab}/J cyclic}} \mathscr{CL}_G(H^{ab}/J) \longrightarrow \mathscr{CL}_G(H^{ab})$$

on representing functions in $Fac_G(H^{ab}/J)$ by

$$\psi(\{g_{H^{ab}/J}\})(\phi : H^{ab} \longrightarrow E^*) = g_{H^{ab}/Ker(\phi)}(\phi : H^{ab} \longrightarrow E^*) \in \mathscr{I}(E).$$

This is clearly well-defined and

$$(\lim_{\longleftarrow} Inf)(\psi(\{g_{H^{ab}/J}\}))(\lambda : H^{ab}/J \longrightarrow E^*)$$

$$= \psi(\{g_{H^{ab}/J}\})(H^{ab} \xrightarrow{\pi} H^{ab}/J \xrightarrow{\lambda} E^*)$$

$$= g_{H^{ab}/Ker(\lambda\pi)}(\lambda\pi : H^{ab}/Ker(\lambda\pi) \longrightarrow E^*)$$

but $Ker(\lambda\pi) \geq J$ so that

$$g_{H^{ab}/J}(\lambda) = g_{H^{ab}/Ker(\lambda\pi)}(\lambda\pi)$$

so that $(\lim_{H^{ab}/J} Inf)\psi = 1$.

Conversely,

$$\psi(\lim_{\leftarrow} Inf)(g)(\phi : H^{ab} \longrightarrow E^*)$$

$$= (\lim_{\leftarrow} Inf(g))(\phi : H^{ab}/Ker(\phi) \longrightarrow E^*)$$

$$= g(\phi : H^{ab} \longrightarrow H^{ab}/Ker(\phi) \longrightarrow E^*)$$

so that $\psi(\lim_{H^{ab}/J} Inf) = 1$ also. □

Combining 5.4.49 and 5.4.51 we obtain the following result:

Theorem 5.4.52 *The homomorphisms of* 5.4.16 *induce maps*

$$\mathcal{CL}(\Lambda(G)) \xrightarrow{B_G^*} \oplus_{(H)} \left(\lim_{\substack{\longleftarrow \\ H^{ab}/J cyclic}} \mathcal{CL}_G(H^{ab}/J) \right)^{W_G H}$$

and

$$\oplus_{(H)} \left(\lim_{\substack{\longleftarrow \\ H^{ab}/J cyclic}} \mathcal{CL}_G(H^{ab}/J) \right)^{W_G H} \xrightarrow{A_G^*} \mathcal{CL}(\Lambda(G))$$

which satisfy $A_G^* B_G^* = 1$.

5.5 Swan subgroups for nilpotent groups

In this section we shall study the Swan subgroup, $T(G)$, in the case when G is a nilpotent group. A finite nilpotent group is of the form

5.5.1 $$G \cong G(p_1) \times \ldots \times G(p_r),$$

where $G(p_i)$ is a p_i-group and $p_1, \ldots, p_r$ are distinct primes. The Swan groups, $T(G(p_i))$, are known by 4.4.20 so that the determination of $T(G)$ in 5.5.1 constitutes the next case in the hierarchy of difficulty.

By Curtis & Reiner (1987, section 53.13, p. 347) the restriction maps may be assembled to yield a surjection

5.5.2 $$\prod_{i=1}^{r} Res_{G(p_i)}^{G} : T(G) \longrightarrow \oplus_{i=1}^{r} T(G(p_i)).$$

Therefore our problem is to determine the kernel of 5.5.2. This is a particular case of the following question, which was posed to me by M.J. Taylor and was first raised by C.T.C. Wall (1979, p. 546):

Question 5.5.3 Let G_1 and G_2 be finite groups of coprime order. What is the kernel of

$$Res^G_{G_1} \times Res^G_{G_2} : T(G) \longrightarrow T(G_1) \oplus T(G_2)$$

when $G = G_1 \times G_2$?

5.5.4 In 5.5.1 suppose that $\#(G(p_i)) = p_i^{n_i}$. Recall that, if p is odd,

$$(\mathbf{Z}/p^n)^* \cong F_p^* \times \left(\frac{1+p\mathbf{Z}}{1+p^n\mathbf{Z}}\right) \cong \mathbf{Z}/(p-1) \times \mathbf{Z}/p^{n-1}.$$

When $p = 2$ and $n \geq 2$ we also have

$$(\mathbf{Z}/2^n)^* \cong \{\pm 1\} \times \left(\frac{1+4\mathbf{Z}}{1+2^n\mathbf{Z}}\right) \cong \{\pm 1\} \times \mathbf{Z}/2^{n-2}.$$

Write $A(i)$ for $F_{p_i}^*$ and $B(i)$ for $\mathbf{Z}/p_i^{n_i-1}$ when $p_i \neq 2$ and $A(i) = \{\pm 1\}$, $B(i) = \mathbf{Z}/2^{n_i-2}$ when $p_i = 2$. Let $A(G(p_i))$ denote the Artin exponent of $G(p_i)$ (cf. 4.4.15) so that the map, S, of 4.4.14 factors to give a surjection

5.5.5 $$S : \{(\textstyle\prod_{i=1}^r A(i) \times B(i))/\{\pm 1\}\} \otimes (\mathbf{Z}/(\prod_{j=1}^r A(G(p_j))) \longrightarrow T(G),$$

where G is as in 5.5.1. In some cases, for trivial reasons, 5.5.2 and 5.5.5 combine to show that 5.5.2 is an isomorphism. However, when some p_i divides a $(p_j - 1)$ the image of the factor $A(j)$ under S may yield some 'exotic torsion' of p_i-primary order. Exotic torsion of this type will lie in the kernel of 5.5.2.

The most basic nilpotent groups whose Swan subgroup might admit exotic q-primary torsion are of the form

$$\mathbf{Z}/p \times (\mathbf{Z}/q)^n,$$

where p, q are primes and q divides $(p-1)$. As we shall see, the exotic q-primary torsion in $T(\mathbf{Z}/p \times (\mathbf{Z}/q)^n)$ is the image of the q-primary torsion in F_p^* and, typically, the order of the cyclic group of exotic q-primary torsion in such a group increases with n to a limit. For example, the exotic 2-primary torsion in $T(\mathbf{Z}/3 \times (\mathbf{Z}/2)^n)$ is trivial for $n = 1$ and of order two for $n \geq 2$. In $T(\mathbf{Z}/5 \times (\mathbf{Z}/2)^n)$ the exotic 2-primary torsion is trivial when $n = 1$, is of order two when $n = 2$ and is of order four for $n \geq 3$.

In this section we will determine the exotic q-primary torsion in these examples. Our motivation for studying these elementary examples lies in the following conjecture.

Conjecture 5.5.6 The exotic q-primary torsion (i.e. q-primary torsion in the kernel of 5.5.2) is detected by the family of maps

$$T(G) = T\left(\prod_{i=1}^{r} G(p_i)\right) \longrightarrow T(\mathbf{Z}/p \times (\mathbf{Z}/q)^n),$$

induced by passing to the Swan subgroups of subquotients of G of the form $\mathbf{Z}/p \times (\mathbf{Z}/q)^n$ where p and q are primes and q divides $(p-1)$.

5.5.7 Let p and q be primes such that q divides $(p-1)$. Set $\xi_m = exp(2\pi i/m)$ and let

$$H(p,q;n) = \mathbf{Z}/p \times (\mathbf{Z}/q)^n.$$

Let k be a positive integer such that

5.5.8
$$\begin{cases} k \equiv 1 & (mod\ q^n) \\ \text{and} & \\ k^{q^c} \equiv 1 & (mod\ p) \end{cases}$$

for some $1 \leq c \in Z$.

We wish to analyse when

5.5.9
$$0 = S(k) \in T(H(p,q;n)).$$

Let us establish some notation for the one-dimensional representations of $H(p,q;n)$. For $\mathbf{Z}/p =< x >$, say, let $\phi : \mathbf{Z}/p \longrightarrow \mathbf{Q}(\xi_p)^*$ be given by $\phi(x) = \xi_p$. The one-dimensional representations of $\mathbf{Z}/p$ are then $1, \phi, \phi^2, \ldots, \phi^{p-1}$. For $(\mathbf{Z}/q)^n$ let y_i generate the ith copy of $\mathbf{Z}/q$ and define $\chi_i : (\mathbf{Z}/q)^n \longrightarrow \mathbf{Q}(\xi_q)^*$ to be given by

$$\chi_i(y_j) = \begin{cases} \xi_q & \text{if } i = j, \\ & \\ 1 & \text{if } i \neq j. \end{cases}$$

Write $\chi(\underline{j}) = \chi_1^{j_1} \ldots \chi_n^{j_n}$ so that the irreducible representations of $H(p,q;n)$ are all of the form $\chi(\underline{j}) \otimes \phi^s$ for some $\underline{j} = (j_1, \ldots, j_n)$ and $0 \leq s \leq p-1$.

Using these characters the p-adic group-ring, $\mathbf{Z}_p[H(p,q;n)]$, may be decomposed in the following manner. Let $\mathscr{S}$ be a set of multi-indices, $\underline{j}$, such that the set $\{\chi(\underline{j}) \mid \underline{j} \in \mathscr{S}\}$ is a complete collection of representatives for the $G(\mathbf{Q}(\xi_q)/\mathbf{Q})$-representatives of the $\chi(\underline{j})$s. Define a ring homomorphism

5.5.10 $\prod_{\underline{j}\in\mathcal{S}} \chi(\underline{j}) : \mathbf{Z}_p[(\mathbf{Z}/q)^n] \longrightarrow \mathbf{Z}_p \times \prod_{0\neq\underline{j}\in\mathcal{S}} \mathbf{Z}_p \otimes \mathbf{Z}[\xi_q],$

where $\chi(\underline{j})(\sum m_g g) = \sum m_g \otimes \chi(\underline{j})(g)$. The homomorphism of 5.5.10 is an isomorphism which commutes with the action of F_q^* as $G(\mathbf{Q}(\xi_q)/\mathbf{Q})$ on the right and diagonally on each $\mathbf{Z}/q$-factor as $Aut(\mathbf{Z}/q)$ on the left. Since $\xi_q \in \mathbf{Z}_p^*$ when q divides $p-1$ we have an isomorphism

$$\mathbf{Z}_p \otimes \mathbf{Z}[\xi_q] \xrightarrow{\cong} \prod_1^{q-1} \mathbf{Z}_p$$

given by sending $x \otimes \xi_q$ to $(x\xi_q, x\xi_q^2, \ldots, x\xi_q^{q-1})$. Combining this isomorphism with 5.5.10 yields an isomorphism

5.5.11 $\prod_{\underline{j}} \chi(\underline{j}) : \mathbf{Z}_p[(\mathbf{Z}/q)^n] \longrightarrow \prod_{\underline{j}} \mathbf{Z}_p.$

To see that 5.5.10 and 5.5.11 are isomorphisms we recall that the corresponding map

$$\prod_{\underline{j}\in\mathcal{S}} \chi(\underline{j}) : \mathbf{Q}_p[(\mathbf{Z}/q)^n] \longrightarrow \mathbf{Q}_p \times \prod_{0\neq\underline{j}\in\mathcal{S}} \mathbf{Q}_p \otimes \mathbf{Z}[\xi_q] \cong \prod_{\underline{j}} \mathbf{Q}_p$$

is well-known to be an isomorphism (Lang, 1984, pp. 641–645). The isomorphisms of 5.5.10 and 5.5.11 are given by the restriction of this isomorphism to the maximal $\mathbf{Z}_p$-orders of each side (Curtis & Reiner, 1981, section 27.1, p. 582). Therefore, taking group-rings for $\mathbf{Z}/p$, we obtain an isomorphism of the form

5.5.12 $\prod_{\underline{j}\in\mathcal{S}} \chi(\underline{j}) : \mathbf{Z}_p[H(p,q;n)]^*$

$$\xrightarrow{\cong} \mathbf{Z}_p[\mathbf{Z}/p]^* \times \prod_{0\neq\underline{j}\in\mathcal{S}} \mathbf{Z}_p[\xi_q][\mathbf{Z}/p]^* \cong \prod_{\underline{j}} \mathbf{Z}_p[\mathbf{Z}/p]^*,$$

where $\mathbf{Z}_p[\xi_q]$ denotes $\mathbf{Z}_p \otimes \mathbf{Z}[\xi_q]$.

Similarly, interchanging the roles of p and q, we obtain an isomorphism of the form

5.5.13 $(1,\phi) : \mathbf{Z}_q[H(p,q;n)]^* \xrightarrow{\cong} \mathbf{Z}_q[(\mathbf{Z}/q)^n]^* \times \mathbf{Z}_q[\xi_p][(\mathbf{Z}/q)^n]^*.$

Recall from 4.2.37 that there is an isomorphism of the form

5.5.14 $D(\mathbf{Z}[H(p,q;n)])$

$$\cong \frac{Hom_{\Omega_{\mathbf{Q}}}(R(H(p,q;n)),(\mathbf{Z}_p[\xi_{pq}])^* \times (\mathbf{Z}_q[\xi_{pq}])^*)}{Hom_{\Omega_{\mathbf{Q}}}(R(H(p,q;n)),(\mathbf{Z}[\xi_{pq}])^*) \cdot Det(\mathbf{Z}_p[H(p,q;n)]^* \times \mathbf{Z}_q[H(p,q;n)]^*)}.$$

Proposition 5.5.15 *In the notation of* 5.5.7 *and* 5.5.8 *suppose that there exists a unit, $u \in \mathbf{Z}[\xi_p]^*$ such that*

$$1 = u \in \frac{\mathbf{Z}_q[\xi_p]^*}{(1 + q^n \mathbf{Z}_q[\xi_p])}$$

and $k = u \in F_p^*$, then

$$0 = S(k) \in T(H(p,q;n)).$$

Proof We define local units, $\alpha \in \mathbf{Z}_p[H(p,q;n)]^*$ and $\beta \in \mathbf{Z}_q[H(p,q;n)]^*$, by means of the isomorphisms of 5.5.12 and 5.5.13.

Define β to be equal to

$$\beta = (1 + (k-1)q^{-n}\sigma, 1 + (u-1)q^{-n}\sigma),$$

where $\sigma = \sum_{g \in (\mathbf{Z}/q)^n} g$.

Now consider the pullback square

$$\begin{array}{ccc} \mathbf{Z}_p[\mathbf{Z}/p]^* & \xrightarrow{\phi} & \mathbf{Z}_p[\xi_p]^* \\ \downarrow \epsilon & & \downarrow \\ \mathbf{Z}_p^* & \longrightarrow & F_p^* \end{array}$$

in which ϵ denotes the augmentation. Choose $u_1 \in \mathbf{Z}_p[\mathbf{Z}/p]^*$ such that $\phi(u_1) = u$ and $\epsilon(u_1) = k$ and set α equal to the unit whose 1-component is u_1 and whose $\underline{j}$-component is trivial when $\chi(\underline{j}) = 1$.

Finally, define $f \in Hom_{\Omega_{\mathbf{Q}}}(R(H(p,q;n)), (\mathbf{Z}[\xi_{pq}])^*)$ by $f(1 \otimes \phi) = u$, $f(\chi(\underline{j}) \otimes \phi) = 1$ when $\chi(\underline{j}) \neq 1$ and $f(\chi(\underline{j}) \otimes 1) = 1$. Hence $(Det(\alpha), Det(\beta)) \cdot f^{-1}$ is trivial on all $\chi(\underline{j}) \otimes \phi^s \neq 1$ and $(Det(\alpha), Det(\beta)) \cdot f^{-1}(1) = (k,k)$. By 4.2.47 this is the function that represents $S(k)$. □

Now we come to the main result of this section.

Theorem 5.5.16 *Let k be an integer as in 5.5.8. Then, in the notation of 5.5.7,*

$$0 = S(k) \in T(H(p,q;n))$$

if and only if there exists $u \in \mathbf{Z}[\xi_p]^*$ such that

$$1 = u \in \frac{\mathbf{Z}_q[\xi_p]^*}{(1 + q^n \mathbf{Z}_q[\xi_p])}$$

and $k = u \in F_p^*$.

Proof We must prove the converse of 5.5.15. We shall prove this result (when $q \neq 2$) by means of the determinantal congruences of 4.3.37, which amounts to the analysis of the maps, S_T, of 4.4.11 when $T = \sum_{\underline{j}} \chi(\underline{j})$ and $T = \sum_{\underline{j}} \chi(\underline{j}) \otimes \phi$.

Let us begin by supposing that $q \neq 2$.

If $S(k) = 0$ then there exist local units $\alpha \in \mathbf{Z}_p[H(p,q;n)]^*$ and $\beta \in \mathbf{Z}_q[H(p,q;n)]^*$ together with a global function

$$f \in Hom_{\Omega_{\mathbf{Q}}}(R(H(p,q;n)), (\mathbf{Z}[\xi_{pq}])^*)$$

such that

5.5.17
$$\begin{cases} Det(\alpha)(\chi(\underline{j}) \otimes \phi) = f(\chi(\underline{j}) \otimes \phi) = u(\underline{j}) \in \mathbf{Z}_p[\xi_{pq}]^*, \\ Det(\beta)(\chi(\underline{j}) \otimes \phi) = f(\chi(\underline{j}) \otimes \phi) = u(\underline{j}) \in \mathbf{Z}_q[\xi_{pq}]^*, \\ Det(\alpha)(\chi(\underline{j}) \otimes 1) = f(\chi(\underline{j}) \otimes 1) = a(\underline{j}) \in \mathbf{Z}_p[\xi_q]^* \\ \qquad \text{if } \chi(\underline{j}) \neq 1, \\ Det(\beta)(\chi(\underline{j}) \otimes 1) = f(\chi(\underline{j}) \otimes 1) = a(\underline{j}) \in \mathbf{Z}_q[\xi_q]^* \\ \qquad \text{for all } \chi(\underline{j}), \\ \text{and} \quad Det(\alpha)(1) = kf(1) = ka(\underline{0}) \in \mathbf{Z}_p[\xi_q]^*. \end{cases}$$

This is because there is a q-adic unit, namely $\gamma = 1 + (k-1)q^{-n}\sigma \in \mathbf{Z}_q[(\mathbf{Z}/q)^n]^*$ where $\sigma = \sum_{g \in (\mathbf{Z}/q)^n} g$, such that $\chi(\underline{j})(\gamma) = 1$ if $\chi(\underline{j}) \neq 1$ and $\chi(\underline{0})(\gamma) = k$.

Now consider the representations of $H(p,q;n)$ given by

$$\rho_1 = \sum_{\underline{j}} \chi(\underline{j}) \text{ and } \rho_2 = \sum_{\underline{j}} \chi(\underline{j}) \otimes \phi.$$

We have $\rho_i \equiv 0 \ (mod \ q^n)$ for $i = 1, 2$. Hence, for $i = 1, 2$,

5.5.18
$$Det(\beta)(\psi^q(\rho_i) - q\rho_i) \in 1 + q^{n+1}\mathbf{Z}_q[\xi_{pq}],$$

by 4.3.37. However, $\psi^q(\rho_1) = q^n$ and $\psi^q(\rho_2) = q^n\phi$ so that

$$u(\underline{0})^{q^n-q}\left(\prod_{\underline{0}\neq\underline{j}} u(\underline{j})\right)^{-q} \quad \text{and} \quad a(\underline{0})^{q^n-q}\left(\prod_{\underline{0}\neq\underline{j}} a(\underline{j})\right)^{-q}$$

both lie in $1+q^{n+1}\mathbf{Z}_q[\xi_{pq}]$. However, both these elements are qth powers of $G(\mathbf{Q}(\xi_q)/\mathbf{Q})$-norms and therefore lie in the qth powers of the subgroup $\mathbf{Z}_q[\xi_p]^*$, which contains no non-trivial qth roots. Hence both

$$u(\underline{0})^{q^{n-1}-1}\left(\prod_{\underline{0}\neq\underline{j}} u(\underline{j})\right)^{-1} \quad \text{and} \quad a(\underline{0})^{q^{n-1}-1}\left(\prod_{\underline{0}\neq\underline{j}} a(\underline{j})\right)^{-1}$$

lie in $(1+q^n\mathbf{Z}_q[\xi_p]) \cap \mathbf{Z}[\xi_p]^*$.

On the other hand, at the prime p,

$$Det(\alpha)(\psi^p(\chi(\underline{j})\otimes\phi) - p\chi(\underline{j})\otimes\phi) \text{ and } Det(\alpha)(\psi^p(\chi(\underline{j})) - p\chi(\underline{j}))$$

both lie in $1+p\mathbf{Z}_p[\xi_q]$. Also, $\psi^p(\chi(\underline{j})\otimes\phi) = \chi(\underline{j})$ so that, if $\underline{j}\neq\underline{0}$,

$$\begin{aligned} u(\underline{j}) &\equiv u(\underline{j})^p \ (mod\ p) \\ &\equiv a(\underline{j}) \ (mod\ p) \end{aligned}$$

and

$$\begin{aligned} u(\underline{0}) &\equiv u(\underline{0})^p \ (mod\ p) \\ &\equiv ka(\underline{0}) \ (mod\ p). \end{aligned}$$

Therefore

5.5.19 $\quad u = [a(\underline{0})^{q^{n-1}-1}\prod_{\underline{0}\neq\underline{j}} u(\underline{j})][u(\underline{0})^{q^{n-1}-1}\prod_{\underline{0}\neq\underline{j}} a(\underline{j})]^{-1}$

lies in $(1+q^n\mathbf{Z}_q[\xi_p]) \cap \mathbf{Z}[\xi_p]^*$ and

$$u \equiv k^{q^{n-1}-1} \quad (mod\ p).$$

For $q \neq 2$ the result now follows since k has q-primary order (modulo p).

When $q = 2$ a similar proof would result, by 4.3.37, only in the existence of $u \in \mathbf{Z}[\xi_p]^*$ such that $u^2 \in (1+2^{n+1}\mathbf{Z}_2[\xi_p]) \cap \mathbf{Z}[\xi_p]^*$ and $k = u \in F_p^*$. Therefore we resort to another proof in this case.

We will use an explicit description of generators of $\mathbf{Z}_2[H(p,2;n)]^*$ in order to show that the element, u, of 5.5.19 lies in $\mathbf{Z}[\xi_p]^* \cap (1+2^n\mathbf{Z}_2[\xi_p])$.

By 5.5.13 we may examine $\mathbf{Z}_2[\xi_p][(\mathbf{Z}/2)^n]^*$ and $\mathbf{Z}_2[(\mathbf{Z}/2)^n]^*$ separately. By 5.7.10 generators for $\mathbf{Z}_2[\xi_p][(\mathbf{Z}/2)^n]^*$ may be taken to be either torsion units of the form $\omega \cdot g$, where $\omega \in \mathbf{Z}_2[\xi_p]^*$ is a root of unity and $g \in (\mathbf{Z}/2)^n$, or to be torsion-free units of the form

5.5.20 $$1 + (v-1)2^{-s}(1+L_1)(1+L_2)\ldots(1+L_s),$$

where $v \in 1+2^s\mathbf{Z}_2[\xi_p]$ and, without loss of generality, we may assume that $L_1, \ldots, L_s$ are distinct characters from the set $\{\chi_1, \ldots, \chi_n\}$. If $\beta = (\beta_1, \beta_2)$ in 5.5.13 with β_2 having the form of 5.5.20 then the factor

5.5.21 $$u(\underline{0})^{2^{n-1}-1}(\textstyle\prod_{\underline{0}\neq \underline{j}} u(\underline{j}))^{-1}$$

in 5.5.19 is equal to $v^{2^{n-1}-2^{n-s}}$ which lies in $1+2^n\mathbf{Z}_2[\xi_p]$. On the other hand, if β_2 is a torsion unit, $\omega \cdot g$, then the product in 5.5.21 is equal to $\prod_{\underline{j}} \chi(\underline{j})(g)$, which is equal to 1 since $g^2 = 1$ and $n \geq 2$. A similar discussion applies to $\beta_1 \in \mathbf{Z}_2[(\mathbf{Z}/2)^n]^*$ and shows that the factor

5.5.22 $$a(\underline{0})^{2^{n-1}-1}(\textstyle\prod_{\underline{0}\neq \underline{j}} a(\underline{j}))^{-1}$$

lies in $1+2^n\mathbf{Z}_2$. By 5.5.17, u, in 5.5.19, is a product of units and therefore lies in $\mathbf{Z}[\xi_p]^* \cap (1+2^n\mathbf{Z}_2[\xi_p])$, as required. □

Example 5.5.23 We shall now use Theorem 5.5.16 to verify Conjecture 5.5.6 in some cases of the form $G = G(p) \times G(q)$, where p, q are primes and q divides $p-1$. We will begin with the case when $p = 3$, $q = 2$.

Suppose that $G = G(3) \times G(2)$ with $\#(G(2)) = 2^s$ and $\#(G(3)) = 3^t$ with $1 \leq s, t$. In this case the exotic 2-primary torsion can be at most a copy of F_3^*, which is of order two. By 5.5.16, this copy of F_3^* is detected in $T(\mathbf{Z}/3 \times (\mathbf{Z}/2)^n)$ for any $n \geq 2$. This is because $\mathbf{Z}[\xi_3]^*$ is cyclic of order six, generated by $(-\xi_3)$ and $1, \xi_3, \xi_3^2$ are distinct in the finite field, $F_4 = \mathbf{Z}_2[\xi_3] \otimes \mathbf{Z}/2$, while ± 1 are distinguished by reduction modulo four. Note that every non-abelian 2-group has either a normal subgroup or a quotient of the form $(\mathbf{Z}/2)^n$ with $n \geq 2$ (Gorenstein, 1968, theorem 4.10, p. 199). Table 5.1 summarises the immediate implications of 5.5.16 for the Swan subgroup, $T(G(3) \times G(2))$. In all cases the F_3^* denotes exotic torsion of order two.

Much of Table 5.1 is justified by means of the following observations:

(i) *Artin exponent considerations* (4.4.15, 4.4.17): When $G(3)$ (resp. $G(2)$) is cyclic the Artin exponent of $G = G(3) \times G(2)$ is a power of 2 (resp. 3). See Curtis & Reiner (1987, p. 365) for the Artin exponents of a p-group.

Table 5.1. *Table of* $T(G(3) \times G(2))$ ($\#(G(2)) = 2^s$ *and* $\#(G(3)) = 3^t$)

	$G(3)$*cyclic*	$G(3)$*non-cyclic*
$G(2)$*cyclic*	$\{1\}$	$\mathbf{Z}/3^{t-1}$
$G(2)$ $(s\geq 3)$ *quaternionic*	$F_3^* \times \mathbf{Z}/2$	$F_3^* \times \mathbf{Z}/2 \times \mathbf{Z}/3^{t-1}$
$G(2)$ $(s\geq 4)$ *semi-dihedral*	$F_3^* \times (\mathbf{Z}/2$ *or* $\mathbf{Z}/4)$	$F_3^* \times (\mathbf{Z}/2$ *or* $\mathbf{Z}/4) \times \mathbf{Z}/3^{t-1}$
$G(2)$ $(s\geq 2)$ *dihedral*	$F_3^* \times (\{0\}$ *or* $\mathbf{Z}/2)$	$F_3^* \times (\{0\}$ *or* $\mathbf{Z}/2) \times \mathbf{Z}/3^{t-1}$
$G(2)$ *any other* 2-*group*	$F_3^* \times \mathbf{Z}/2^{s-2}$	$F_3^* \times \mathbf{Z}/2^{s-2} \times \mathbf{Z}/3^{t-1}$

(ii) When $G(2)$ is non-cyclic then G has a quotient of the form $\mathbf{Z}/3 \times (\mathbf{Z}/2)^2$.

The following result concerns the ambiguous cases.

Proposition 5.5.24 *Let SD_n denote the semi-dihedral group of order 2^{n+2} $(n \geq 2)$ and let D_8 denote the dihedral group of order eight. If $T(\mathbf{Z}/3^t \times D_8) \cong F_3^*$ then*

$$T(\mathbf{Z}/3^t \times SD_n) \cong F_3^* \times \mathbf{Z}/2.$$

Proof The semi-dihedral group (Curtis & Reiner, 1987, p. 349; 4.4.20) is given by

$$SD_n = \{a, b \mid a^{2^{n+1}} = 1 = b^2, bab = a^{2^n - 1}\}.$$

We may apply Theorem 5.3.3 to the case when

$$G = \mathbf{Z}/3^t \times SD_n \text{ and } J = \mathbf{Z}/3^t \times < a > \cong \mathbf{Z}/2^{n+1}3^t.$$

The elements of order two in $G - J$ are of the form $\{a^{2j}b\}$ whose normaliser, $N_G < a^{2j}b >$, is isomorphic to $\mathbf{Z}/3^t \times D_8$. Let k be an integer which is coprime to $\#(G)$ and is congruent to 1 modulo 3^t. By Theorem 5.3.3

$$0 = 2S(k) - \sum_g Ind_{\mathbf{Z}/3^t \times D_8}^G (S(k)) \in T(\mathbf{Z}/3^t \times SD_n).$$

However, by hypothesis, $0 = S(k) \in T(\mathbf{Z}/3^t \times D_8)$ so that $2S(k) = 0$. Hence the surjection, S of 4.4.14 becomes

$$S : \mathbf{Z}/2 \times F_3^* \longrightarrow T(\mathbf{Z}/3^t \times SD_n).$$

The $\mathbf{Z}/2$-factor is detected by $T(SD_n) \cong \mathbf{Z}/2$, while the F_3^*-factor is detected by Theorem 5.5.16. □

Example 5.5.25 Consider Theorem 5.5.16 in the case when $p = 5$ and $q = 2$. The units, $\mathbf{Z}[\xi_5]^*$, are generated by $(-\xi_5)$ and $(1 + \xi_5)$. Also

$$(1 + \xi_5)^{15} \equiv 1 - 2(\xi_5 + 3) \ (mod\ 8).$$

From Theorem 5.5.16 ones sees easily that $S(17)$ has order two in $T(\mathbf{Z}/5 \times (\mathbf{Z}/2)^2)$ and has order four in $T(\mathbf{Z}/5 \times (\mathbf{Z}/2)^3)$. In fact, 5.5.5 induces a surjection

$$S : (F_5^* \times \{\pm 1\})/\{\pm 1\} \longrightarrow T(\mathbf{Z}/5 \times (\mathbf{Z}/2)^2),$$

which shows that $S(F_5^*)$ generates $T(\mathbf{Z}/5 \times (\mathbf{Z}/2)^2)$ so that, by 5.5.16,

$$S : F_5^*/(F_5^*)^2 \stackrel{\cong}{\longrightarrow} T(\mathbf{Z}/5 \times (\mathbf{Z}/2)^2).$$

Similarly, there is an isomorphism

$$S : F_5^* \times \mathbf{Z}/2 \stackrel{\cong}{\longrightarrow} T(\mathbf{Z}/5 \times (\mathbf{Z}/2)^3).$$

By Theorem 1.1 of Gorenstein (1968, p. 123), if G is a 2-group all of whose elementary abelian subquotients, $(\mathbf{Z}/2)^k$, have rank at most two ($k \leq 2$), then G has at most two generators. For example, when $\#(G) = 16$ the non-abelian groups of this type are D_{16}, Q_{16}, SD_4 and the semi-direct product of the form $\mathbf{Z}/4 \propto \mathbf{Z}/4$ (Thomas & Wood, 1980, Tables 16/12, 16/14, 16/13 and 16/10 respectively). It is not difficult, by induction on the order, to classify all such non-abelian 2-groups. One method for such a classification is to study central extensions of the type

$$\{1\} \longrightarrow \mathbf{Z}/2 \longrightarrow G \longrightarrow H \longrightarrow \{1\}.$$

In this extension H is again of the same type and the isomorphism classes of possible Gs are in one-one correspondence with the $Aut(H)$-orbits of the cohomology group, $H^2(H; \mathbf{Z}/2)$. The cohomology groups, $H^*(D_{2^n}; \mathbf{Z}/2)$ and $H^*(Q_{2^n}; \mathbf{Z}/2)$, are described in Snaith (1989b, p. 24 and p. 34 respectively). For example, when $n \geq 3$, $H^2(D_{2^n}; \mathbf{Z}/2)$ is an F_2-vector space with basis w, x_1^2 and x_2^2 in the notation of Snaith (1989b, p. 24). There is an outer automorphism, $\Phi : D_{2^n} \longrightarrow D_{2^n}$ such that $\Phi^*(w) = w$, $\Phi^*(x_1^2) = x_1^2$ and $\Phi^*(x_2^2) = x_1^2 + x_2^2$. Hence $H^2(D_{2^n}; \mathbf{Z}/2)$ has six $Aut(D_{2^n})$-orbits which correspond to $\mathbf{Z}/2 \times D_{2^n}$, $D_{2^{n+1}}$, $Q_{2^{n+1}}$, SD_{n-1}, $\mathbf{Z}/4 \propto \mathbf{Z}/2^{n-1}$ and $\mathbf{Z}/2 \propto (\mathbf{Z}/2^{n-1} \times \mathbf{Z}/2)$ (cf. Thomas & Wood, 1980, tables 16/6, 16/12, 16/14, 16/13, 16/10 and 16/8), when $n = 3$. Hence G must be one of $D_{2^{n+1}}$, $Q_{2^{n+1}}$, SD_{n-1} or $\mathbf{Z}/4 \propto \mathbf{Z}/2^{n-1}$.

Now let $G(2)$ and $G(5)$ be a finite 2-group and 5-group respectively. The following table summarises the immediate implications resulting from

Table 5.2. *Table of* $T(G(5) \times G(2))$ ($\#(G(2)) = 2^s$ *and* $\#(G(5)) = 5^t$)

	upper bound	*lower bound*
$G(2)$ *cyclic*	$T(G(5))$	$T(G(5))$
$G(2)$ ($s\geq 3$) *quaternionic*	$F_5^*/(F_5^*)^2 \times \mathbf{Z}/2 \times T(G(5))$	$F_5^*/(F_5^*)^2 \times \mathbf{Z}/2 \times T(G(5))$
$G(2)$ ($s\geq 4$) *semi-dihedral*	$F_5^* \times \mathbf{Z}/4 \times T(G(5))$	$F_5^*/(F_5^*)^2 \times \mathbf{Z}/2 \times T(G(5))$
$G(2)$ ($s\geq 2$) *dihedral*	$F_5^*/(F_5^*)^2 \times \mathbf{Z}/2 \times T(G(5))$	$F_5^*/(F_5^*)^2 \times T(G(5))$
others with rank of subquotients ≤ 2	$F_5^* \times \mathbf{Z}/2^{s-2} \times T(G(5))$	$F_5^*/(F_5^*)^2 \times \mathbf{Z}/2^{s-2} \times T(G(5))$
$G(2)$ *any other 2-group*	$F_5^* \times \mathbf{Z}/2^{s-2} \times T(G(5))$	$F_5^* \times \mathbf{Z}/2^{s-2} \times T(G(5))$

Theorem 5.5.16 and elementary considerations such as Artin exponents. In the table the analogue of 5.5.24 relates the ambiguous dihedral and semi-dihedral cases.

Lemma 5.5.26 *Let p and q be primes such that q properly divides $p-1$. Let f denote the order of q in the multiplicative group, F_p^*. Hence $f \geq 3$. Then*

$$(1-\xi_p)^{q^f-1} \equiv 1 - \sum_{i=1}^{q-1} (q + q/2 + \ldots + q/i)\xi_p^{iq^{f-1}} \quad (mod\ q^2).$$

Proof If $HCF(s, q) = 1$ then $(q^f - sq^e)/(sq^e) = q^{f-e}/s - 1$ is a q-adic unit which is congruent to (-1) (modulo q^2) provided that $e \leq f-2$. Applying this observation to the factors in the binomial coefficients

$$\binom{q^f-1}{i} = \frac{(q^f-1)(q^f-2)\ldots(q^f-i)}{1\cdot 2\cdot\ldots(i-1)\cdot i}$$

shows that the coefficient of ξ_p^i in $(1-\xi_p)^{q^f-1}$ is congruent to one modulo q^2 except when $i = se^{f-1}$ with $1 \leq s \leq q-1$. In the latter case the coefficient is easily seen to be equal to

$$(1-q)(1-q/2)\ldots(1-q/s).$$

Since $q^f - 1$ is divisible by p we have $1 + \xi_p + \ldots + \xi_p^{q^f-1} = 1$ and the result follows immediately. □

Example 5.5.27 Consider Theorem 5.5.16 in the case when $p = 7$ and $q = 3$. The units, $\mathbf{Z}[\xi_7]^*$, are generated by $(-\xi_7)$, $\alpha = (1-\xi_7^2)/(1-\xi_7)$

Table 5.3. *Table of* $T(G(7) \times G(3))(\#(G(3)) = 3^s$ *and* $\#(G(7)) = 7^t)$

	$G(3)$ *cyclic*	$G(3)$ *non-cyclic*
$G(7)$ *cyclic*	$\{1\}$	$\mathbf{Z}/3 \times \mathbf{Z}/3^{s-1}$
$G(7)$ *non-cyclic*	$\mathbf{Z}/7^{t-1}$	$\mathbf{Z}/3 \times \mathbf{Z}/7^{t-1} \times \mathbf{Z}/3^{s-1}$

and $\beta = (1 - \xi_7^3)/(1 - \xi_7)$. When considering 3-primary torsion we may ignore the roots of unity. Also, by 5.5.26 (with $f = 6$),

$$(\alpha^a \beta^b)^{\pm 1} \equiv 1 - 3a\xi_7^3 - 3b\xi_7 + (3a + 3b)\xi_7^5 \quad (mod\ 9),$$

so that $\alpha^a \beta^b$ is congruent to 1 (modulo 9) only when a and b are divisible by 3. In this case the reduction of $\alpha^a \beta^b$ (modulo 7) is equal to $(-1)^{b/3}$. Therefore, since $S(-1)$ is always trivial in the Swan subgroup, 5.5.16 implies that $T(\mathbf{Z}/7 \times (\mathbf{Z}/3)^2)$ contains exotic torsion of order three.

It is straightforward to verify the contents of Table 5.3.

5.6 Cyclic groups

As explained in 7.1.29 a finite abelian group, X, together with a G-action which is cohomologically trivial, yields a class

$$[X] \in \mathcal{CL}(\mathbf{Z}[G]).$$

For such a $\mathbf{Z}[G]$-module there exists an exact sequence of $\mathbf{Z}[G]$-module homomorphisms of the form

$$0 \longrightarrow P_1 \longrightarrow P_0 \longrightarrow X \longrightarrow 0,$$

in which P_0 and P_1 are finitely generated, projective $\mathbf{Z}[G]$-modules. The class, $[X]$, is defined by

$$[X] = [P_0] - [P_1] \in \mathcal{CL}(\mathbf{Z}[G]).$$

In this section, by way of illustration, we shall recall some elementary facts about several examples of cohomologically trivial modules (and their Hom-descriptions) over cyclic groups in which the underlying module is also a cyclic group. These modules are closely related to the modules of roots of unity which will enter into the computation, in Chapter 7, of the global Chinburg invariant in the real cyclotomic case.

Let p be a prime. Let G denote a cyclic p-group of order p^s with

Table 5.4. $H^i(G; M_{a,t})$

	$H^i(G; M_{a,t})$	*Generator, even i*	*Generator, odd i*
p *odd*, $s \leq t$	$\mathbf{Z}/p^{a+1+s-t}$	p^{t-a-1}	p^{t-s}
p *odd*, $t \leq s$	$\mathbf{Z}/p^{a+1}$	p^{t-a-1}	1
$p = 2, s \leq t$, $1 \leq a$	$\mathbf{Z}/2^{a+2+s-t}$	2^{t-a-2}	2^{t-s}
$p = 2, t \leq s$, $1 \leq a$	$\mathbf{Z}/2^{a+2}$	2^{t-a-2}	1
$p = 2, s = t-2$, $a = 0$	0	–	–
$p = 2, t-1 \leq s$, $a = 0$	$\mathbf{Z}/2$	2^{t-1}	1

generator, $g \in G \cong \mathbf{Z}/p^s$. Let $M_{a,t}$ denote the cyclic group, $\mathbf{Z}/p^t$ with a *non-trivial* G-action given, for $x \in M_{a,t}$, by the formulae

5.6.1
$$g(x) = \begin{cases} (1+p)^{p^a} x & \text{if } p \neq 2, t-1-a \leq s, \\ 3^{2^a} x & \text{if } p = 2, t-2-a \leq s. \end{cases}$$

Recall that the Tate cohomology groups, $H^i(G; M_{a,t})$, coincide with the ordinary cohomology groups in positive dimensions and are given by

5.6.2
$$H^i(G; M_{a,t}) = \begin{cases} (M_{a,t})^G / N(M_{a,t}) & \text{if } i \equiv 0 \ (mod\ 2), \\ Ker(N)/((M_{a,t})^G) & \text{if } i \equiv 1 \ (mod\ 2), \end{cases}$$

where $N = 1 + g + \ldots + g^{p^s - 1}$.

The Herbrand difference, $\#(H^0(G; M_{a,t})) - \#(H^1(G; M_{a,t}))$, vanishes for cyclic groups and cyclic modules so that the groups of 5.6.2 are cyclic groups whose order is independent of i. From these observations one sees that the Tate cohomology groups, $H^i(G; M_{a,t})$, are given by Table 5.4 (in which the generators lie in $\mathbf{Z}/p^t \cong M_{a,t}$).

5.6.3 From Table 5.4 we see that $\mathbf{Z}/p^t = M_{a,t}$ is a cohomologically trivial $\mathbf{Z}[\mathbf{Z}/p^s]$-module precisely when p is odd and $a = t - s - 1$ or when $p = 2$ and $a = t - s - 2$. In these cases we obtain a class

$$[M_{a,t}] \in \mathcal{CL}(Z[Z/p^s]).$$

5.6.4 *The Hom-representative of $M_{a,t}$*

Now let us determine the Hom-description, in the sense of 4.2.13, of the class of $M_{a,t}$.

Let $q = (1+p)^{p^a}$ with $a+s+1 = t$ if p is an odd prime and $a+s+2 = t$ if $p = 2$. If $< g >= \mathbf{Z}/p^s$ then $g(x) = qx$ for $x \in M_{a,t} \cong \mathbf{Z}/p^t$. Consider the surjection, $\eta : \mathbf{Z}[\mathbf{Z}/p^s] \longrightarrow M_{a,t}$, given by $\eta(1) = 1$. Hence $Ker(\eta)$ contains p^t and $g - q$. If $\eta(\sum_i \alpha_i g^i) = 0$ then $\sum_i \alpha_i q^i \equiv 0 \ (p^t)$ so that

$$\sum_i \alpha_i g^i = \sum_i \alpha_i (g^i - q^i) + mp^t$$

for some m and therefore this element belongs to $< g-q, p^t > \lhd \mathbf{Z}[\mathbf{Z}/p^s]$.

If l is a prime different from p then we have

$$\mathbf{Z}_l[\mathbf{Z}/p^s] < 1 >= \mathbf{Z}_l \otimes_{\mathbf{Z}} < g - q, p^t > .$$

At the prime p, in $\mathbf{Z}_p \otimes_{\mathbf{Z}} < g - q, p^t >$ we have the identity

$$(g - q)(g^{p^s-1} + qg^{p^s-2} + \ldots + q^{p^s-1}) = 1 - q^{p^s}.$$

However, for all primes p, $1 - q^{p^s} = p^t v$ where $HCF(p, v) = 1$. Therefore

$$\mathbf{Z}_p[\mathbf{Z}/p^s] < g - q >= \mathbf{Z}_p \otimes_{\mathbf{Z}} < g - q, p^t > .$$

Therefore the idèle, $\{a_l\} \in J^*(\mathbf{Q}[\mathbf{Z}/p^s])$ corresponding to this choice of bases is given by

5.6.5
$$a_l = \begin{cases} g - (1+p)^{p^a} & \text{if } l = p, \\ 1 & \text{otherwise.} \end{cases}$$

The following result is immediate from the previous discussion.

Proposition 5.6.6 *Under the conditions of* 5.6.4 *the cohomologically trivial* $\mathbf{Z}[\mathbf{Z}/p^s]$*-module,* $M_{a,t}$*, is represented in the class-group by the function*

$$\chi \mapsto \begin{cases} (\chi(g) - (1+p)^{p^a})^{-1} & \text{if } l = p, \\ 1 & \text{otherwise.} \end{cases}$$

5.6.7 *Some more Hom-representatives*

Let p be any prime. Let u, t be positive integers such that $t \geq u+1$ if p is odd and $t \geq u+2$ if $p = 2$. Let M_p denote the $\mathbf{Z}[\mathbf{Z}/p^u]$-module given by

5.6.8
$$M_p = \begin{cases} \mathbf{Z}/2[\mathbf{Z}/2] & \text{if } p = 2, u = 1, \\ M_{t-u-2,t} & \text{if } p = 2, u > 1, \\ M_{t-u-1,t} & \text{otherwise.} \end{cases}$$

Here $M_{a,t}$ is the module of 5.6.1. Hence M_p is a finite, cohomologically trivial $\mathbf{Z}[\mathbf{Z}/p^u]$-module.

Suppose now that n is any positive integer prime to p so that, if $\mathbf{Z}/n$ acts trivially on M_p, then M_p is a finite, cohomologically trivial $\mathbf{Z}[\mathbf{Z}/np^u]$-module. We shall now generalise the results of 5.6.4–5.6.6 to give the Hom-representative of the element

5.6.9
$$[M_p] \in \mathscr{CL}(\mathbf{Z}[\mathbf{Z}/np^u]).$$

Theorem 5.6.10 *Let x, y be generators of $\mathbf{Z}/n$ and $\mathbf{Z}/p^u$ respectively. The class of M_p in 5.6.9 is represented in terms of the Hom-description by the function which is trivial except at the p-adic places where it is given by*

5.6.11
$$\begin{cases} Det(x+1)^{-1} & \text{if } p = 2, u = 1, \\ Det(xy - 3^{2^{t-u-2}})^{-1} & \text{if } p = 2, u > 1, \\ Det(xy - (1+p)^{p^{t-u-1}})^{-1} & \text{otherwise.} \end{cases}$$

Proof Suppose that $p = 2$ and $u = 1$. The kernel of the map

$$\psi : \mathbf{Z}[\mathbf{Z}/n \times \mathbf{Z}/2] \longrightarrow \mathbf{Z}/2[\mathbf{Z}/2]$$

given by $\psi(x) = 1$ and $\psi(y) = y$ is equal to the ideal, $< 2, x-1 >$. However, in $\mathbf{Z}_2 \otimes_{\mathbf{Z}} < 2, x-1 >$, we have

$$(x+1)(x^{n-1} - x^{n-2} + \ldots + (-1)^{n-1}) = x^n + (-1)^{n-1} = 2$$

so that $\mathbf{Z}_2 \otimes_{\mathbf{Z}} < 2, x-1 >= \mathbf{Z}_2[\mathbf{Z}/n \times \mathbf{Z}/2] < x+1 >$, as required.

In the remaining two cases set $\alpha = (1+p)^{p^a}$ where $a = t-s-1$ if p is odd and $a = t-s-2$ when $p = 2$. The kernel of the map

$$\psi : \mathbf{Z}[\mathbf{Z}/n \times \mathbf{Z}/p^u] \longrightarrow M_p,$$

given by $\psi(x) = 1$ and $\psi(y) = \alpha$, is equal to the ideal $< p^t, y-\alpha, x-1 >$. It is clear that

$$\mathbf{Z}_p \otimes_{\mathbf{Z}} < p^t, y-\alpha, x-1 >= \mathbf{Z}_p \otimes_{\mathbf{Z}} < y-\alpha, x-1 > .$$

However, arguing in the same manner as before, we obtain the identities

$$(xy-\alpha)((xy)^{n-1}+\alpha(xy)^{n-2}+\ldots+\alpha^{n-1}) = y^n - \alpha^n$$

and

$$(xy-\alpha)((xy)^{p^u-1}+\alpha(xy)^{p^u-2}+\ldots+\alpha^{p^u-1}) = x^{p^u} - \alpha^{p^u} = x^{p^u} - 1 + vp^t$$

for some integer, v, such that $HCF(v,p) = 1$, which shows that

$$\mathbf{Z}_p \otimes_{\mathbf{Z}} < p^t, y-\alpha, x-1 >= \mathbf{Z}_p \otimes_{\mathbf{Z}} < xy-\alpha >,$$

as required. □

5.6.12 *The cyclic group of order p^2*

Let p be an odd prime and let ξ_n denote a primitive nth root of unity, as usual. Let g be a generator of the cyclic group of order p^2. For the remainder of this section we shall be concerned with calculations in the kernel subgroup

$$D(\mathbf{Z}[\mathbf{Z}/p^2]) \subset \mathscr{CL}(\mathbf{Z}[\mathbf{Z}/p^2]).$$

We shall show that the class of 5.6.4 (with $t = 3$, $s = 2$) and 5.6.7:

$$[M_{0,3}] = [M_p] = -[< p^3, g-1-p >] \in D(\mathbf{Z}[\mathbf{Z}/p^2])$$

can sometimes be non-zero.

Set

$$\overline{\mathbf{Z}[\xi_p]} = \mathbf{Z}/p \otimes \mathbf{Z}[\xi_p] \cong \mathbf{Z}/p[\pi_1]/(\pi_1^{p-1}),$$

where $\pi_1 = 1 - \xi_p$. From Curtis & Reiner (1987, p. 285) there is an exact sequence of the form

5.6.13 $$\mathbf{Z}[\xi_p]^* \longrightarrow \overline{\mathbf{Z}[\xi_p]}^* \xrightarrow{\delta} D(\mathbf{Z}[\mathbf{Z}/p^2]) \longrightarrow 0,$$

in which the left-hand map is given by reduction modulo p. This exact sequence is derived from the exact K-theory Mayer–Vietoris sequence of a pullback (fibre square) of the following form:

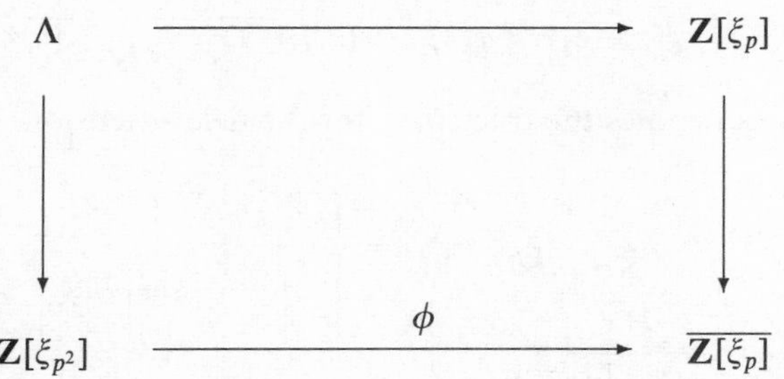

In this diagram the ring, Λ, is given by

$$\Lambda = \mathbf{Z}[g]/(\Phi_p(g)\Phi_{p^2}(g)),$$

where $\Phi_p(t) = 1 + t + \ldots + t^{p-1}$ and $\Phi_{p^2}(t) = 1 + t^p + \ldots + t^{p(p-1)}$ are cyclotomic polynomials. The unlabelled homomorphisms in the diagram are given by sending g to ξ_p or ξ_{p^2} and by reduction modulo p. If $\pi_2 = 1 - \xi_{p^2}$ then we have an isomorphism of the form

$$\mathbf{Z}/p \otimes \mathbf{Z}[\xi_{p^2}] \cong \mathbf{Z}/p[\pi_2]/(\pi_2^{p(p-1)}).$$

The map, ϕ, is given by composing reduction modulo p with the map sending π_2 to π_1.

Lemma 5.6.14 *If $\overline{w}$ denotes the reduction of w modulo p then the homomorphism*

$$\phi : \mathbf{Z}[\xi_{p^2}] \longrightarrow \mathbf{Z}/p \otimes \mathbf{Z}[\xi_{p^2}] \cong \mathbf{Z}/p[\pi_2]/(\pi_2^{p(p-1)})$$

is given by

$$\phi(z) = \overline{N_{\mathbf{Q}(\xi_{p^2})/\mathbf{Q}(\xi_p)}(z)}\ ,$$

where $N_{L/K}$ denotes the norm for the field extension, L/K.

Proof The map, ϕ, is the ring homomorphism which is characterised by $\phi(1) = 1$ and $\phi(\pi_2) = \pi_1$. Therefore it suffices to verify that the map, ψ, defined by

$$\psi(z) = \overline{N_{\mathbf{Q}(\xi_{p^2})/\mathbf{Q}(\xi_p)}(z)}$$

is a ring homomorphism. Since the norm is multiplicative, so is ψ.

However, given z, z_1 it is clear that there exists $u \in \mathbf{Z}[\xi_{p^2}]$ such that

$$\psi(z + z_1) = \psi(z) + \psi(z_1) + \overline{T_{\mathbf{Q}(\xi_{p^2})/\mathbf{Q}(\xi_p)}(u)}\,,$$

where $T_{L/K}$ denotes the trace map for the field extension, L/K. However

$$T_{\mathbf{Q}(\xi_{p^2})/\mathbf{Q}(\xi_p)}(\xi^i_{p^2}) = \begin{cases} p\xi^i_{p^2} & \text{if } i = pj, \\ 0 & \text{otherwise.} \end{cases}$$

Hence $\overline{T_{\mathbf{Q}(\xi_{p^2})/\mathbf{Q}(\xi_p)}(u)} = 0$. □

Now consider the projective $\mathbf{Z}[\mathbf{Z}/p^2]$-module given by the ideal

5.6.15 $$< p^3, g - 1 - p >= I \lhd \mathbf{Z}[\mathbf{Z}/p^2].$$

Clearly, one has a canonical isomorphism of the form

$$\Lambda \otimes_{\mathbf{Z}[\mathbf{Z}/p^2]} I \cong < p^3, g - 1 - p > \lhd \Lambda.$$

This is seen by considering the exact sequence of cohomologically trivial $\mathbf{Z}[\mathbf{Z}/p^2]$-modules

$$0 \longrightarrow I \longrightarrow \mathbf{Z}[\mathbf{Z}/p^2] \longrightarrow \mathbf{Z}/p^3 \longrightarrow 0.$$

Upon tensoring with Λ we obtain an exact sequence of the form

$$Tor^1_{\mathbf{Z}[\mathbf{Z}/p^2]}(\Lambda, \mathbf{Z}/p^3) \longrightarrow \Lambda \otimes_{\mathbf{Z}[\mathbf{Z}/p^2]} I \longrightarrow \Lambda.$$

However, $Tor^1_{\mathbf{Z}[\mathbf{Z}/p^2]}(\Lambda, \mathbf{Z}/p^3)$ is annihilated by p^3, while $\Lambda \otimes_{\mathbf{Z}[\mathbf{Z}/p^2]} I$ is torsion free, since it is a summand of

$$\Lambda \otimes_{\mathbf{Z}[\mathbf{Z}/p^2]} (\mathbf{Z}[\mathbf{Z}/p^2])^r \cong \Lambda^r.$$

Hence the inclusion of I into $\mathbf{Z}[\mathbf{Z}/p^2]$ induces an injective homomorphism

$$\Lambda \otimes_{\mathbf{Z}[\mathbf{Z}/p^2]} I \longrightarrow \Lambda \otimes_{\mathbf{Z}[\mathbf{Z}/p^2]} \mathbf{Z}[\mathbf{Z}/p^2] \cong \Lambda.$$

The image of this injection is clearly the ideal, $< p^3, g - 1 - p >$.

Next observe that there are integers u, v *prime to* p which satisfy

5.6.16 $$\begin{cases} (1 + p)^{p^2} - 1 = p^3 v \\ (1 + p)^p - 1 = p^2 u. \end{cases}$$

Therefore, in Λ, we have the following identities

$$\begin{aligned} p^3 v &= (1+p)^{p^2} - 1 \\ &= (\sum_{j=0}^{p^2-1} g^j (p+1)^{p^2-1-j})(p+1-g) \\ &= (\sum_{j=0}^{p^2-1} g^j (p+1)^{p^2-1-j} - g^j)(p+1-g) \\ &= p(\sum_{j=0}^{p^2-1} g^j [(p+1)^{p^2-1-j} - 1]/p)(p+1-g), \end{aligned}$$

so that, since Λ is torsion-free,

5.6.17 $\quad p^2 v = (\sum_{j=0}^{p^2-2} g^j [(p+1)^{p^2-1-j} - 1]/p)(p+1-g).$

Since $HCF(p^2 v, p^3) = p^2$ we see that

$$\Lambda \otimes_{\mathbf{Z}[\mathbf{Z}/p^2]} I \cong < p^2, g-1-p > \lhd \Lambda.$$

A similar argument shows that there is a canonical isomorphism of the form

$$\mathbf{Z}[\xi_p] \otimes_\Lambda < p^2, g-1-p > \cong < p^2, \xi_p - 1 - p > \lhd \mathbf{Z}[\xi_p]$$

and that, in $\mathbf{Z}[\xi_p]$,

$$pu = \left(\sum_{j=0}^{p-2} \xi_p^j [(p+1)^{p-1-j} - 1]/p \right) (p+1-\xi_p),$$

where u is as in 5.6.16. Hence

$$\mathbf{Z}[\xi_p] \otimes_\Lambda < p^2, g-1-p > \cong < p, \xi_p - 1 - p > = < \pi_1 > \lhd \mathbf{Z}[\xi_p].$$

From 5.6.17 we have, in $\mathbf{Z}[\xi_{p^2}]$,

$$p^2 v = \left(\sum_{j=0}^{p^2-2} \xi_{p^2}^j [(p+1)^{p^2-1-j} - 1]/p \right) (p+1-\xi_{p^2}).$$

In this equation the expression in the first factor is divisible by p in $\mathbf{Z}[\xi_{p^2}]$.To see this, expand $(p+1)^{p^2-1-j}$ by the binomial theorem and observe that every term in the resulting double sum is p-divisible with the exception of the terms

$$-\sum_{j=0}^{p^2-2} \xi_{p^2}^j (j+1).$$

However, in this sum each term with $j = wp - 1$ is clearly p-divisible, while the remaining terms may be grouped into subsums of the form

$$\begin{aligned}&(j+1)\xi_{p^2}^j + (p+j+1)\xi_{p^2}^{p+j} + (2p+j+1)\xi_{p^2}^{p+j} + \ldots \\ &\equiv (j+1)\xi_{p^2}^j(1 + \xi_p + \xi_p^2 + \ldots + \xi_p^{p-1}) \quad (mod\ p) \\ &\equiv 0 \quad (mod\ p).\end{aligned}$$

This calculation shows that $pv \in < p^2, p+1-\xi_{p^2} > \triangleleft \mathbf{Z}[\xi_{p^2}]$ and we obtain a canonical isomorphism of the form

$$\mathbf{Z}[\xi_{p^2}] \otimes_\Lambda < p^2, g-1-p > \cong < p, \xi_{p^2} - 1 - p > = < \pi_2 > \triangleleft \mathbf{Z}[\xi_{p^2}].$$

Now let us recall how the Mayer–Vietoris boundary homomorphism, δ in 5.6.13, is defined. We begin by considering the fibre square of canonical maps.

$$\begin{array}{ccc} \Lambda \otimes_{\mathbf{Z}[\mathbf{Z}/p^2]} I & \longrightarrow & \mathbf{Z}[\xi_p] \otimes_{\mathbf{Z}[\mathbf{Z}/p^2]} I \\ \downarrow & & \downarrow \\ \mathbf{Z}[\xi_{p^2}] \otimes_{\mathbf{Z}[Z/p^2]} I & \longrightarrow & \overline{\mathbf{Z}[\xi_p] \otimes_{\mathbf{Z}[\mathbf{Z}/p^2]} I} \end{array}$$

The previous discussion identifies this fibre square with a diagram

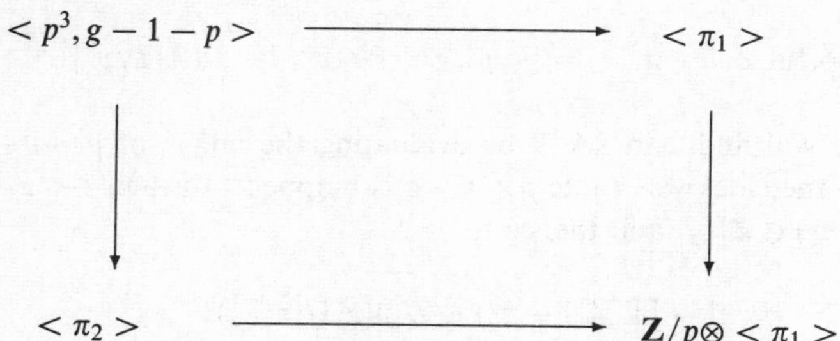

in which the upper horizontal map and both vertical maps are the natural ones. We also have a $\mathbf{Z}[\xi_p]$- and $\mathbf{Z}[\xi_{p^2}]$-module isomorphisms

$$<\pi_1>\cong \mathbf{Z}[\xi_p],$$

$$<\pi_2>\cong \mathbf{Z}[\xi_{p^2}]$$

given by dividing by π_1 or π_2. Hence we may identify the fibre square with one of the form

$$\begin{array}{ccc} <p^3, g-1-p> & \longrightarrow & \mathbf{Z}[\xi_p] \\ \downarrow & & \downarrow \\ \mathbf{Z}[\xi_{p^2}] & \xrightarrow{\mu} & \overline{\mathbf{Z}[\xi_p]} \end{array}$$

in which the unlabelled maps are the natural ones. However, there exists

$$\alpha \in \overline{\mathbf{Z}[\xi_p]}^* \cong \mathbf{Z}/p[\pi_1]/(\pi_1^{p-1})^*$$

such that, for all $z \in \mathbf{Z}[\xi_{p^2}]$,

5.6.18 $$\mu(z) = \alpha\phi(z),$$

where ϕ is as in 5.6.14. The map, δ of 5.6.13, satisfies

5.6.19 $$\delta(\alpha) = [I] \in D(\mathbf{Z}[\mathbf{Z}/p^2]).$$

Lemma 5.6.20 $\delta(1 + \pi_1^{p-2}) = [< p^3, g - 1 - p >] \in D(\mathbf{Z}[\mathbf{Z}/p^2]).$

Proof We will find α in 5.6.19 by evaluating the image of $p + 1 - g$ in $\overline{\mathbf{Z}[\xi_p]}$. Via the clockwise route $p + 1 - g$ is mapped to $p + \pi_1 \in < \pi_1 >$ or to $1 + (p/\pi_1) \in \mathbf{Z}[\xi_p]$ and thence to

5.6.21 $$1 + \prod_{j=2}^{p-1}(1 - \xi_p^j) \in \mathbf{Z}/p[\pi_1]/(\pi_1^{p-1}).$$

However, $1 - \xi_p^j = j\pi_1 + z_j\pi_1^2$ so that 5.6.21 becomes

$$1 + \prod_{j=2}^{p-1}(j\pi_1) = 1 + \pi_1^{p-2} \in \mathbf{Z}/p[\pi_1]/(\pi_1^{p-1}).$$

On the other hand the image of $p + 1 - g$ in $< \pi_2 >$ is equal to $p + \pi_2$, which becomes $1 + (p/\pi_2)$ in $Z[\xi_{p^2}]$. However,

$$N(1 + (p/\pi_2)) = (((p + 1)^p - 1) + \pi_1)\pi_1^{-1} = 1 + pu(p/\pi_1),$$

so that $\phi(1 + (p/\pi_2)) = 1$ and, from 5.6.18,

$$\alpha = 1 + \pi_1^{p-2} \in \mathbf{Z}/p[\pi_1]/(\pi_1^{p-1}),$$

as required. □

Corollary 5.6.22 *If* $p \geq 5$ *is an odd prime then the element*

$$[< p^3, g - 1 - p >] \in D(\mathbf{Z}[\mathbf{Z}/p^2])$$

is non-trivial.

Proof The argument to follow was shown to me by Al Weiss.

By 5.6.13–5.6.20, we must show that

$$\alpha = 1 + \pi_1^{p-2} \in \mathbf{Z}/p[\pi_1]/(\pi_1^{p-1})$$

is not the image of a unit from $\mathbf{Z}[\xi_p]^*$. However, by Kummer's lemma (Curtis & Reiner, 1987, p. 286), we may write any such unit in the form $\xi_p^a\beta$ for some integer, a, and such that $\beta = \overline{\beta}$, the complex conjugate of β. Suppose that

$$\alpha = \xi_p^a\beta.$$

Therefore,

$$\alpha/\overline{\alpha} = (\xi_p^a\beta)/(\xi_p^{-a}\beta) = \xi_p^{2a}.$$

However, $\overline{\pi_1} = -\xi_p^{-1}\pi_1$, which implies that

$$\overline{\alpha} = 1 - \xi_p^2\pi_1^{p-2} = 1 - \pi_1^{p-2} \in \mathbf{Z}/p[\pi_1]/(\pi_1^{p-1}).$$

Therefore,

$$\alpha/\overline{\alpha} = (1 + \pi_1^{p-2})^2 = 1 + 2\pi_1^{p-2} \in \mathbf{Z}/p[\pi_1]/(\pi_1^{p-1}).$$

In particular, since $p \geq 5$, $\alpha/\overline{\alpha}$ is congruent to one modulo π_1^2 while ξ_p^{2a} is congruent to $1 - 2a\xi_p$, which implies that a is divisible by p. However, this is impossible since it would imply that

$$1 = \xi_p^{2a} = \alpha/\overline{\alpha} = 1 + 2\pi_1^{p-2} \in \mathbf{Z}/p[\pi_1]/(\pi_1^{p-1}).$$

This contradiction completes the proof. □

5.7 Exercises

5.7.1 Let Q_8 denote the quaternion group of order eight:

$$Q_8 = \{x, y \mid x^2 = y^2, y^4 = 1, xyx^{-1} = y^{-1}\}.$$

Let $u_2 = x + xy + y \in \mathbf{Z}_2[Q_8^{ab}]^*$. Show, by using 4.2.4, that

$$Det_{Q_8}(u) \in \mathscr{CL}(\mathbf{Z}[Q_8]) = T(Q_8) \cong \mathbf{Z}/2$$

is a generator, where $u \in U(\mathbf{Z}[Q_8^{ab}])$ is equal to u_2 at the prime two and is equal to 1 at all other primes.

5.7.2 Let $\mathbf{H_Z}$ denote the ring of integral quaternions. Suppose that X is a $k \times k$ matrix with entries in $\mathbf{H_Z}$. Show that, in the notation of 5.2.11, $det(c(X))$ is a positive integer.

(*Hint:* See Taylor (1984, p. 9)). To see that $det(c(X))$ is real, and hence an integer, simply write $X = Y + Wj$ where $Y, W \in M_k(\mathbf{Z}[i])$ so that $c(X)$ is equal to

$$\begin{pmatrix} Y & -W \\ \overline{W} & \overline{Y} \end{pmatrix}$$

and observe that this is conjugated to its complex conjugate by the signed permutation matrix

$$\begin{pmatrix} 0 & -1 \\ 1 & 0 \end{pmatrix}$$

in $M_{2k}(\mathbf{Z}[i])$. To see the positivity observe that $det(c(X))$ is the norm of

X when X is a quaternion and use elementary row or column operations to reduce the $k \times k$ case to the 1×1 case.)

5.7.3 In the proof of 5.2.33 prove directly that

$$\chi_+(\#(M_+)) \equiv det(c(V')) \ (mod\ 4).$$

5.7.4 Construct some more $\mathbf{Z}[Q_8]$-modules, similar to that given in 5.2.41, which are cohomologically trivial and whose order is a power of two. Determine their classes in the class-group.

5.7.5 Generalise Theorem 5.3.3 in the following manner. Suppose that there exists an extension of the form

$$\{1\} \longrightarrow J \longrightarrow G \longrightarrow \mathbf{Z}/p \longrightarrow \{1\},$$

where p is a prime. Suppose that J acts freely on spheres, S_1 and S_2, of dimension $n \geq 1$. Suppose, finally, that there exists a J-map, $f : S_1 \longrightarrow S_2$ of degree k with $HCF(k, \#(G)) = 1$.

Prove that, in the notation of Theorem 5.3.3,

$$S(k^p) - \sum_g Ind_{N_G<g>}^G(S(k)) = 0 \in D(\mathbf{Z}[G]).$$

5.7.6 (*Oliver*, 1978, *Proposition* 4) Prove that every chain complex in 5.3.24 has a resolution in the sense of 5.3.25.

5.7.7 Let p and q be odd, positive integers such that $HCF(p, q) = 1$. Let A_p and A_q denote elements of order p and q, respectively, in $\mathbf{Z}/pq$. Define $Q(8, p, q)$ to be the *semi-direct product*

$$Q(8, p, q) = Q(8) \propto \mathbf{Z}/pq,$$

where

$$XA_pX^{-1} = A_p, \quad XA_qX^{-1} = A_q^{-1},$$
$$YA_pY^{-1} = A_p^{-1}, \quad YA_qY^{-1} = A_q.$$

Show that $4T(Q(8, p, q)) = 0$. (*When p and q are prime* it is shown in Bentzen & Madsen (1983) that $2T(Q(8, p, q)) = 0$.)

5.7.8 (*Research problem*) Generalise the result of 5.3.3 to the case of a group, G, which is given by an extension of the form

$$\{1\} \longrightarrow \mathbf{Z}/n \longrightarrow G \longrightarrow (\mathbf{Z}/p)^m \longrightarrow \{1\}.$$

One would expect the relations to take the form of an alternating sum involving $m+1$ terms, rather like an *Euler characteristic*. Some modification of the geometric proof of 5.3.3 might well yield such a generalisation.

5.7.9 (*Research problem*) Generalise the result of 5.3.3 to the case of a group, G, which is given by an extension of the form

$$\{1\} \longrightarrow H \longrightarrow G \longrightarrow \mathbf{Z}/p \longrightarrow \{1\}.$$

More precisely, attempt to prove that the relation

$$S(k^p) - \sum_g Ind_{N_G<g>}^G(S(k)) = 0 \in D(\mathbf{Z}[G]).$$

holds for any integer k for which $HCF(k, \#(G)) = 1$ and $0 = S(k) \in D(\mathbf{Z}[H])$.

It should be admitted that the conjectured relation is suggested merely by the evidence of the metacyclic case (Theorem 5.3.3), the p-group case (Theorem 4.4.20) and its aesthetically pleasant form.

5.7.10 Let $R = \mathbf{Z}_2$ *or* $\mathbf{Z}_2[\xi_p]$ with p an odd prime. Show that the units of $R[(\mathbf{Z}/2)^n]$ are generated by torsion units of the form $\omega \cdot g$ with ω equal to a root of unity in R and $g \in \mathbf{Z}/2$ and by torsion-free units of the form

$$1 + (v-1)2^{-s}(1 + L_1)(1 + L_2)\ldots(1 + L_s),$$

where $L_1, \ldots, L_s$ are distinct characters of $(\mathbf{Z}/2)^n$ and $v \in 1 + 2^sR$. (*Hint:* Let Λ be a ring containing no 2-torsion. Consider the injection

$$\mu : \Lambda[\mathbf{Z}/2] \longrightarrow \Lambda \oplus \Lambda$$

given by $\mu(a + by) = (a + b, a - b)$, where y generates $\mathbf{Z}/2$, and use induction on n. See also Sehgal (1978, p. 54).)

5.7.11 Verify the contents of Table 5.1.

5.7.12 (*Research problem*) Resolve the ambiguities in Table 5.1. (See also 5.5.24.)

5.7.13 Let p be an odd prime whose class number, $h_p = \#(\mathcal{CL}(\mathbf{Z}[\xi_p]))$ is odd. Using a generalisation of 5.5.26 show that the exotic torsion in $T(\mathbf{Z}/p \times (\mathbf{Z}/2)^n)$ is cyclic of order $HCF(2^{n-1}, p-1)$.

5.7.14 Verify, from the definition of 5.6.2, the identities of the cohomology groups which are tabulated in Table 5.4.

5.7.15 Derive the exact sequence of 5.6.13, obtaining a description of the homomorphism, δ, from which to derive the formula of 5.6.19.

5.7.16 Use 5.6.22 to discover some more non-trivial classes among the

$$[M_{a,t}] \in \mathcal{CL}(\mathbf{Z}[\mathbf{Z}/p^s])$$

of 5.6.3.

5.7.17 In the proof of 5.3.29 the vanishing of the Swan group of a cyclic group (see 4.2.48) was used to obtain an embedding of a mapping cone (denoted by Z in 5.3.29) into an acyclic space with a free action on the complement.

Prove the vanishing results of 4.2.48 geometrically, in the spirit of Theorems 5.3.3 and 5.3.29.

6

Complete discrete valuation fields

Introduction

This chapter concerns complete, discrete valuation fields and the conductor invariants of their Galois representations.

In Section 1 we recall the ramification groups, ramification functions, the Artin–Swan conductors and the Artin–Swan representations of a complete, discrete valuation field. The Artin and Swan conductors are integral invariants of Galois representations of complete, discrete valuation fields which are defined when the residue field extension is separable. Their important properties are listed in this section — including invariance under inflation, Galois invariance and inductivity in dimension zero. Each of these conductors is given by the Schur inner product with a corresponding Galois representation whose character is defined in terms of the ramification functions.

In addition, if $G(L/K)$ denotes the Galois group and $g \in G(L/K)$ the ramification functions at $g^{p^{n-1}}$ and at g^{p^n} were conjectured by A. Grothendieck to be congruent modulo p^n. In 6.1.34 we give a proof of these conjectures which is due to S. Sen. In 6.4.2 one sees that these congruences, together with the determinantal congruences of Chapter 4, restrict the possible identities of the ramification subgroups of $G(L/K)$.

Section 1 concludes with the derivation of a formula relating the Artin conductor of V to that of $\psi^p(V)$ when p is the residue characteristic and V is uniformly ramified.

J-P. Serre (1960) asked whether it is possible to extend the theory of the Artin conductor (or, equivalently, the Swan conductor) to the case when the residue field extension is not necessarily separable. This question has been reiterated by several authors, including K. Kato, who gave a definition for a suitable Swan conductor in the case of a one-dimensional representation. In Section 2 we introduce Kato's abelian conductor and establish the formulae which relate its value on a one-dimensional character to the value on the restriction of the character to the inertia group or to the first wild ramification group. We give a number of calculations of the Kato–Swan conductor when the Galois group is isomorphic to $\mathbf{Z}/p \times \mathbf{Z}/p$ and p is the residue characteristic. These examples suffice to show that one cannot construct a non-abelian conductor which generalises all the others, is invariant under inflation and is inductive in dimension zero.

In Section 3 we use Explicit Brauer Induction to construct an integer-valued conductor function which agrees with those of Swan and Kato, when the latter are defined. In addition we show that the obstruction to inductivity in dimension zero for our generalised Swan conductor lies in the cases where the Galois group is $\mathbf{Z}/p^n \times \mathbf{Z}/p$ or $\mathbf{Z}/p^n$ and p is the residue characteristic. This result 6.3.20 is proved by comparing our conductor with the rational-valued, inductive conductor which one may define by applying our construction to the Explicit Brauer Induction homomorphism, d_G, of Chapter 2, Section 4.

Section 4 consists of a collection of exercises concerning conductors.

6.1 Ramification groups and functions

Let L/K be a finite Galois extension of complete, discrete valuation fields. Let $\mathcal{O}_K$, m_K and $\mathcal{O}_K^* = \mathcal{O}_K - m_K$ denote, respectively, the valuation ring of K, the maximal ideal of $\mathcal{O}_K$ and the multiplicative group of units of $\mathcal{O}_K$. Let $G(L/K)$ denote the Galois group and let $\pi_L \in \mathcal{O}_L$ generate $m_L \lhd \mathcal{O}_L$. Let $\overline{L} = \mathcal{O}_L/m_L$ and $\overline{K} = \mathcal{O}_K/m_K$ denote the residue fields of L and K respectively.

We will refer to the case when $\overline{L}/\overline{K}$ is separable as the 'classical case'. *For the moment we will make no such separability assumption.*

Let v_L denote the L-adic valuation, $v_L : L^* \longrightarrow \mathbf{Z}$. Hence $v_L(\pi_L) = 1$.

Following Serre (1979, chapters IV and V), we may define normal subgroups

6.1.1 $$\ldots \lhd G_i \lhd \ldots \lhd G_0 \lhd G_{-1} = G(L/K)$$

by the condition that

6.1.2 $$g \in G_i \longleftrightarrow v_L(g(a) - a) \geq i + 1 \ \text{ for all } \ a \in \mathcal{O}_L.$$

Following Serre (1979, p. 62) and Kato (1989, p. 121) define ramification functions $i_{G(L/K)}$ and $s_{G(L/K)}$ on $G(L/K)$ by the formulae (note that our $s_{G(L/K)}$ is $-S_{G(L/K)}$ of Kato, 1989).

6.1.3

$$\begin{aligned} i_{G(L/K)}(g) &= \inf_{a\in\mathcal{O}_L} v_L(a-g(a)) && \text{if } g \neq 1, \\ s_{G(L/K)}(g) &= \inf_{x\in L^*} v_L(1-g(x)x^{-1}) && \text{if } g \neq 1, \\ i_{G(L/K)}(1) &= -\textstyle\sum_{1\neq g\in G(L/K)} i_{G(L/K)}(g), \\ s_{G(L/K)}(1) &= -\textstyle\sum_{1\neq g\in G(L/K)} s_{G(L/K)}(g). \end{aligned}$$

In the classical case (when $\overline{L}/\overline{K}$ is separable)

6.1.4

$$s_{G(L/K)}(g) = \begin{cases} i_{G(L/K)}(g) - 1 & \text{if } 1 \neq g \in G_0, \\ i_{G(L/K)}(g) & \text{if } 1 \neq g \notin G_0. \end{cases}$$

In fact, in the classical case, $\mathcal{O}_L$ is generated as an $\mathcal{O}_K$-algebra by one element, x, and the infima in 6.1.3 are attained at this element (Serre, 1979, chapter II, section 6, proposition 12 and p. 61). In the general case, if $G_0 = G(L/M) \triangleleft G(L/K)$ then

$$G(M/K) \cong G(L/K)/G_0 \cong G(\overline{M}/\overline{K}),$$

where $\overline{M}/\overline{K}$ is the maximal separable subextension of $\overline{L}/\overline{K}$ and

6.1.5 $$f^{sep}_{L/K} = [\overline{M} : \overline{K}] = [G(L/K) : G_0],$$

where $f^{sep}_{L/K}$ denotes the separable residue degree of L/K.

6.1.6 *The Artin and Swan conductors*

We shall recall briefly the definitions and properties of the classical Artin and Swan representations and their related conductor homomorphisms. For the remainder of this section we shall *assume that $\overline{L}/\overline{K}$ is a separable extension* (the 'classical' case). Under these hypotheses one has complex representations (not merely virtual representations)

6.1.7 $$A_{G(L/K)}, SW_{G(L/K)} \in R(G(L/K)),$$

where $R(G)$ denotes the complex representation ring of G (Serre, 1977, p. 68).

If $V \in R(G)$ let χ_V denote its character function. The characters of $A_{G(L/K)}$ and $SW_{G(L/K)}$ are given, for $g \in G(L/K)$, by the formulae

6.1.8
$$\begin{cases} \chi_{A_{G(L/K)}}(g) & = -f_{L/K} \cdot i_{G(L/K)}(g) \\ \text{and} & \\ \chi_{SW_{G(L/K)}}(g) & = -f_{L/K} \cdot s_{G(L/K)}(g), \end{cases}$$

where $f_{L/K}$ denotes the residue degree of L/K, $f_{L/K} = [\overline{L} : \overline{K}]$.

The representations, A_G and SW_G, are called the Artin representation and the Swan representation, respectively, and, by 6.1.4, they are related by the formula (Serre, 1977, p. 160; Kato, 1989, p. 121)

6.1.9
$$SW_G = A_G + Ind_{G_0}^G(1) - Ind_{\{1\}}^G(1),$$

where $Ind_H^G(V)$ denotes the induced representation of V and $1 \in R(G)$ is the class of the one-dimensional trivial representation.

If $G_i \lhd G(L/K)$ (Serre, 1979, p. 62; 6.1.2) is the ith ramification group (with $G_{-1} = G(L/K)$) then (Serre, 1979, p. 100)

6.1.10
$$\begin{cases} A_{G(L/K)} & = \sum_{i=0}^{\infty} [G_0 : G_i]^{-1} Ind_{G_i}^{G(L/K)}(Ind_{\{1\}}^{G_i}(1) - 1). \\ & \in R(G(L/K)). \end{cases}$$

Notice that $G_i = \{1\}$ for large i so that 6.1.10 is only a finite sum.

From Serre (1979, p. 101), if $G(L/K') \leq G(L/K)$ and $\delta_{K'/K}$ is the discriminant of K'/K then the restriction to $G(L/K')$ is given, in $R(G(L/K'))$, by

6.1.11
$$Res_{G(L/K')}^{G(L/K)}(A_{G(L/K)}) = v_K(\delta_{K'/K}) Ind_{\{1\}}^{G(L/K')}(1) + f_{K'/K} A_{G(L/K')},$$

where v_K is the valuation on K^* and $f_{K'/K}$ denotes the residue degree, $[\overline{K'} : \overline{K}]$.

Remark 6.1.12 Note that, under our current assumptions of separability, $f_{K'/K}$ is also equal to the separable residue degree, $f_{K'/K}^{sep} = [\overline{K'} : \overline{K}]_s$ (Lang, 1984, p. 282). In generalising the conductor homomorphism to the case of non-separable residue field extensions (see 6.3.3) it will become necessary to choose between $f_{K'/K}$ and $f_{K'/K}^{sep}$. We shall choose the latter, where necessary, based on examples (see 6.2.17 and 6.3.20).

Suppose that $N \lhd G$. Define

6.1.13
$$\natural : R(G) \longrightarrow R(G/N)$$

by

$$\natural(V) = C[G/N] \otimes_{C[G]} V \cong V^N,$$

the N-fixed points of V. If $K \le K' \le L$ and K'/K is Galois then (Serre, 1979, p. 101)

6.1.14
$$\natural(A_{G(L/K)}) = A_{G(K'/K)} \quad \in R(G(K'/K)).$$

The *Artin conductor* is the homomorphism

6.1.15
$$f_K : R(G(L/K)) \longrightarrow \mathbf{Z},$$

given by $f_K(\chi) =< A_{G(L/K)}, \chi >_{G(L/K)}$, where $< V, W >_G$ is the Schur inner-product of 1.2.7 ($V, W \in R(G)$).

The *Swan conductor*

6.1.16
$$sw_K : R(G(L/K)) \longrightarrow \mathbf{Z}$$

is defined by $sw_K(\chi) =< sw_{G(L/K)}, \chi >_{G(L/K)}$ and, by 6.1.9,

6.1.17
$$sw_K(\chi) = f_K(\chi)+ < 1, \chi >_{G_0} - dim(\chi).$$

If $N \lhd G$ then the inflation map

6.1.18
$$Inf_{G/N}^G : R(G/N) \longrightarrow R(G)$$

is adjoint to $\natural$ of 6.1.13 in the sense that

$$< V, Inf_{G/N}^G(W) >_G =< \natural(V), W >_{G/N}$$

for $V \in R(G)$ and $W \in R(G/N)$. Therefore, if $K \le K' \le L$ with K'/K Galois then, by 6.1.14,

6.1.19
$$f_K(Inf_{G(K'/K)}^{G(L/K)}(\chi)) = f_K(\chi) \in \mathbf{Z}$$

for all $\chi \in R(G(K'/K))$.

Since Ind_H^G is adjoint to Res_H^G we find for $K \le K' \le L$ and $\chi \in R(G(L/K'))$, by 6.1.11,

6.1.20
$$f_K(Ind_{G(L/K')}^{G(L/K)}(\chi)) = v_K(\delta_{K'/K})dim(\chi) + f_{K'/K} f_{K'}(\chi).$$

By 6.1.17 we easily find, when χ is one-dimensional, that

6.1.21 $$sw_K(Inf^{G(L/K)}_{G(K'/K)}(\chi)) = sw_{K'}(\chi) \in \mathbf{Z}$$

and that

6.1.22

$$f_K(\chi) = \begin{cases} sw_K(\chi) & \text{if } dim(\chi) = 1 \text{ and } \chi = 1 \text{ on } G_0 \\ \text{and} & \\ sw_K(\chi) + 1, & \text{if } dim(\chi) = 1, \text{ otherwise.} \end{cases}$$

Hence, in particular, the trivial character has trivial Artin and Swan conductors.

If $K \leq K' \leq L$ then

6.1.23 $$f_K(Ind^{G(L/K)}_{G(L/K')}(1)) = v_K(\delta_{K'/K}),$$

as a special case of 6.1.20. Also 6.1.20 and 6.1.23 combine to yield

6.1.24

$$\begin{cases} f_K(Ind^{G(L/K)}_{G(L/K')}(\chi - dim(\chi))) = (f_{K'/K}) \cdot f_{K'}(\chi - dim(\chi)) \\ \quad \text{and (see 6.4.3)} \\ sw_K(Ind^{G(L/K)}_{G(L/K')}(\chi - dim(\chi))) = (f_{K'/K}) \cdot sw_{K'}(\chi - dim(\chi)). \end{cases}$$

For any $\chi \in R(G(L/K))$

6.1.25 $$\begin{cases} f_K(\chi) = f_M(Res^{G(L/K)}_{G_0}(\chi)) \\ sw_K(\chi) = sw_M(Res^{G(L/K)}_{G_0}(\chi)) \end{cases}$$

where $G_0 = G(L/M)$. This follows from Frobenius reciprocity (1.2.39) since, by 6.1.10,

6.1.26 $$A_{G(L/K)} = Ind^{G(L/K)}_{G_0}(A_{G(L/M)}).$$

6.1.27 Suppose that the Galois group, $G = G(L/K)$, is given. The disposition of the ramification groups, $\{G_i\}$, cannot be assigned arbitrarily. Similarly there are restrictions on the possible representations in

$R(G)$ that can occur as an Artin representation for some choice of L/K with $G = G(L/K)$. For example, in the classical case some restrictions are given by congruences between values of the ramification functions. These congruences were conjectured by A. Grothendieck and proved by S. Sen (1969, Theorem 1.1). The majority of the rest of this section will be devoted to the study of these congruences. Note that, by 6.1.26, there would be no loss of generality in studying the totally ramified case ($G(L/K) = G_0$) if we were interested in restrictions upon $A_{G(L/K)}$.

Let L be a local field with residue field, $\overline{L}$, of characteristic $p > 0$. Hence L is either a finite extension of the p-adic field, $\mathbf{Q}_p$, or a one-dimensional function field of power series over $\overline{L}$. Hence, in the function field case the constant functions yield a multiplicative map

6.1.28 $$f : \overline{L} \longrightarrow \mathcal{O}_L,$$

such that the reduction of f (modulo m_L) yields the identity map on $\overline{L}$. Similarly, when L is of characteristic zero and $\overline{L}$ is the field with q elements we have the Teichmüller map (Iwasawa, 1986, lemma 2.2, p. 19) which is defined by

$$f(x \ mod \ m_L) = \lim_{n \to \infty} (x^{q^n}).$$

Lemma 6.1.29 *Let L/K be a Galois extension of local fields. There exists a subset, $C \leq \mathcal{O}_L$, such that*

(i) *C is a complete set of representatives for $\mathcal{O}_L/m_L = \overline{L}$*

and

(ii) *if $c \in C$ then $g(c) = c$ for all $g \in G_0 \lhd G(L/K)$.*

Proof Take $C = f(\overline{L})$ where f is an in 6.1.28. □

Lemma 6.1.30 *Let L/K be as in 6.1.27 and let $g \in G_0 \lhd G(L/K)$. If $m = p^e s$ with $HCF(p, s) = 1$ then*

$$v_L(g^m(\pi_L) - \pi_L) = v_L(g^{p^e}(\pi_L) - \pi_L).$$

(Note that, in the notation of 6.1.3, $v_L(g^m(\pi_L) - \pi_L) = i_{G(L/K)}(g^m) = 1 + s_{G(L/K)}(g^m)$ when $g^m \neq 1$.)

Proof Set $h = g^{p^e}$ so that

$$g^m - 1 = (h - 1)(1 + h + \ldots + h^{s-1}).$$

Since $h \in G_0$, $h^i(\pi_L) \equiv \pi_L \pmod{m_L}$ and, for some $\alpha \in \mathcal{O}_L$,

$$\begin{aligned} & v_L(\pi_L + h(\pi_L) + \ldots + h^{s-1}(\pi_L)) \\ & = v_L(s\pi_L + \alpha\pi_L^2) \\ & = v_L(\pi_L) \end{aligned}$$

since $HCF(p,s) = 1$. Both π_L and $s\pi_L + \alpha\pi_L^2$ are choices for a uniformiser (the prime element) of $\mathcal{O}_L$ so that the infimum, $i_{G(L/K)}(-)$ of 6.1.3 is attained at either of these elements. Therefore

$$\begin{aligned} v_L((g^m - 1)\pi_L) &= v_L((h-1)(s\pi_L + \alpha\pi_L^2)) \\ &= v_L((h-1)\pi_L). \end{aligned}$$

□

Lemma 6.1.31 *Let $g \in G_0 \lhd G(L/K)$. For each integer, m, there exists $x_m \in L$ such that*

(i) $v_L(x_m) = m$ and

(ii) $v_L(g(x_m) - x_m) = m + v_L(g^m(\pi_L)\pi_L^{-1} - 1)$.

Proof When $m = 0$ choose x_0 to be any non-zero element of C in 6.1.29.

If $m < 0$, we may set $x_m = (x_{-m})^{-1}$ since

$$v_L(x_m) = v_L((x_{-m})^{-1}) = -v_L(x_{-m}) = -(-m) = m$$

and

$$\begin{aligned} v_L(g(x_m) - x_m) &= v_L((x_{-m} - g(x_{-m}))/x_{-m}g(x_{-m})) \\ &= v_L(x_{-m} - g(x_{-m})) - 2v_L(x_{-m}) \\ &= -m + v_L(g^{-m}(\pi_L)\pi_L^{-1} - 1) + 2m \\ &= m + v_L(g^m(\pi_L)\pi_L^{-1} - 1). \end{aligned}$$

When $m > 0$, set

$$x_m = \prod_{i=0}^{m-1} g^i(\pi_L).$$

Clearly, $v_L(x_m) = \sum_{i=0}^{m-1} v_L(g^i(\pi_L)) = m$ and $g(x_m)x_m^{-1} = g^m(\pi_L)\pi_L^{-1}$ so that $v_L(g(x_m)x_m^{-1} - 1) = v_L(g^m(\pi_L)\pi_L^{-1} - 1)$, as required. □

Corollary 6.1.32 *If $x \in L$ then*

$$x = \sum_{m=v_L(x)}^{\infty} w_m$$

where w_m satisfies 6.1.31(i) *and* (ii).

Proof We may write x in the form of a convergent series

$$x = \sum_{m=v_L(x)}^{\infty} c_m x_m,$$

where $c_j \in C$ in 6.1.29 and x_i is as in 6.1.31. However, since $c_j \in \mathcal{O}_L^*$ and is fixed by g, $w_i = c_i x_i$ also satisfies 6.1.31(i) and (ii). □

Proposition 6.1.33 *In* 6.1.27, *let $g \in G_0 \lhd G(L/K)$ satisfy $g^{p^{n-1}} \neq 1$. Hence, in the notation of* 6.1.3, $s_{G(L/K)}(g) = v_L(g(\pi_L)\pi_L^{-1} - 1)$. *Suppose that, for all $0 < j < n$,*

$$s_{G(L/K)}(g^{p^{j-1}}) \equiv s_{G(L/K)}(g^{p^j}) \ (mod\ p^j),$$

then the integers

$$\{m + s_{G(L/K)}(g^m) \mid v_p(m) < n\} \cup \{s_{G(L/K)}(g^{p^{n-1}})\}$$

are all distinct. Here v_p denotes the p-adic valuation, as in 6.1.30.

Proof Suppose that $m + s_{G(L/K)}(g^m) = j + s_{G(L/K)}(g^j)$. If $v_p(m) = v_p(j)$ then, by 6.1.30, $s_{G(L/K)}(g^m) = s_{G(L/K)}(g^j)$ and therefore $m = j$. On the other hand, $v_p(m-j) = min(v_p(m), v_p(j))$ while the congruences imply that

$$v_p(s_{G(L/K)}(g^j) - s_{G(L/K)}(g^m)) > min(v_p(m), v_p(j))$$

so that $m - j \neq s_{G(L/K)}(g^j) - s_{G(L/K)}(g^m)$.

Finally, if $v_p(m) < n$ then

$$v_p(s_{G(L/K)}(g^{p^{n-1}}) - s_{G(L/K)}(g^m)) > v_p(m),$$

so that $s_{G(L/K)}(g^{p^{n-1}}) \neq m + s_{G(L/K)}(g^m)$. □

We are now ready to establish the congruences of theorem 1.1 of Sen (1969).

Theorem 6.1.34 *Let L/K be a Galois extension of local fields with $char(\overline{L}) = p$. Let $g \in G_0 \lhd G(L/K)$ satisfy $g^{p^n} \neq 1$, then, in the notation of* 6.1.3,

$$s_{G(L/K)}(g^{p^{n-1}}) \equiv s_{G(L/K)}(g^{p^n}) \ (mod\ p^n).$$

Proof By induction on n (replacing g by g^p) we may assume that

$$s_{G(L/K)}(g^{p^{n-1}}) \equiv s_{G(L/K)}(g^{p^n}) \ (mod \ p^{n-1}).$$

Set $s_j = s_{G(L/K)}(g^{p^j})$ then we have to show that $v_p(s_n - s_{n-1}) \geq n$. If this is not so then $v_p(s_n - s_{n-1}) = n-1$. Applying 6.1.31 to g^p with $m = s_{n-1} - s_n$ we obtain $z \in L$ such that $v_L(z) = s_{n-1} - s_n$ and

$$\begin{aligned} v_L(g^p(z) - z) &= s_{n-1} - s_n + s_{G(L/K)}(g^{pm}) \\ &= s_{n-1} - s_n + s_{G(L/K)}(g^{p^n}), \end{aligned}$$

by 6.1.30. Hence $v_L(g^p(z) - z) = s_{n-1} - s_n + s_n = s_{n-1}$.

Set $x = (1 + g + \ldots + g^{p-1})z$. Since $1 + g + \ldots + g^{p-1} \equiv (g-1)^{p-1}$ (mod p) we see that

$$\textbf{6.1.35} \qquad \begin{cases} v_L(x) \geq v_L((g-1)^{p-1}z) > v_L(z) = s_{n-1} - s_n \\ \text{and} \\ v_L((g-1)x) = v_L((g^p - 1)z) = s_{n-1}. \end{cases}$$

Apply 6.1.32 to obtain an expansion $x = \sum_{m=v_L(x)}^{\infty} w_m$ and set $u = (g-1)x = \sum_{m=v_L(x)}^{\infty} u_m$, where $u_m = (g-1)w_m$. If $v_p(m) \geq n$ then $v_L(u_m) \geq v_L(w_m) + s_n$, by 6.1.30. Also, $v_L(x) \leq v_L(w_m)$ so that

$$v_L(u_m) \geq v_L(x) + s_n > v_L(z) + s_n = s_{n-1} = v_L(u),$$

by 6.1.35. Hence

$$\begin{aligned} v_L(u) &= v_L(\textstyle\sum_{v_p(m)<n} u_m + \sum_{v_p(m)\geq n} u_m) \\ &= v_L(\textstyle\sum_{v_p(m)<n} u_m). \end{aligned}$$

However, $v_p(m) < n$ implies, by 6.1.33 and the induction hypothesis, that the integers $v_L(u_m) = m + s_{G(L/K)}(g^m)$ are distinct from one another and from $s_{n-1} = v_L(u)$. Therefore

$$v_L(u) = min_{v_p(m)<n}(v_L(u_m)) \neq v_L(u).$$

This contradiction shows that $v_p(s_n - s_{n-1}) \geq n$ and completes the proof of Theorem 6.1.34. □

Example 6.1.36 Suppose that $char(\overline{L}) = 2$ and $G(L/K) \cong Q_{16}$, the generalised quaternion group of order 16 (see 1.3.7). Hence $Q_{16} = \{x, y \mid x^4 = y^2, y^4 = 1, xyx = y\}$. The congruences of Theorem 6.1.34 show that, for example, if $x^2 \in G_1$ then $x^4 \in G_3$, while if $x \in G_0$ then $x^2 \in G_2$ and $x^4 \in G_6$.

In Serre (1960, section 4, p. 413) (see also Snaith, 1989b, pp. 260 and 261) a 2-adic example is given for which $G(L/Q_2) \cong Q_{16}$ and $i_{G(L/Q_2)}(x^2) = 2$, $i_{G(L/Q_2)}(x^4) = 4$ which shows that $x \notin G_0$ in this example.

6.1.37 *Conductors and Adams operations*

The congruences of Theorem 6.1.34 suggest congruences between the character values of $A_{G(L/K)}$ and $\psi^p(A_{G(L/K)})$, where ψ^p is the Adams operation of 4.1.2. However, the anomalous definition of $i_{G(L/K)}(1)$ seems to obscure the behaviour of such congruences. In 6.4.2 an example is given of the sort of determinantal congruence restrictions which may be derived from 6.1.34.

For similar reasons one cannot expect very good properties for the Artin conductor of $\psi^p(\chi)$. Nevertheless, one may sometimes evaluate $f_K(\psi^p(\chi))$ under conditions which are related to the Explicit Brauer Induction formula, $a_{G(L/K)}(\chi)$. In order to discuss this topic we shall need to recapitulate the upper numbering system for ramification groups.

Let L/K be a finite Galois extension of local fields (i.e. the classical case). If $u \in R$ is a real number and $u \geq -1$ set $G_u = G_i$, where i is the smallest integer such that $i \geq u$. Thus, by 6.1.2,

$$g \in G_u \Longleftrightarrow i_{G(L/K)}(g) \geq u + 1.$$

Set

$$\phi_{L/K}(u) = \phi(u) = \int_0^u \frac{dt}{[G_0 : G_t]}.$$

When $-1 \leq t \leq 0$ we adopt the convention that $[G_0 : G_t]$ equals $[G_{-1} : G_0]^{-1}$ for $t = -1$ and equals $1 = [G_0 : G_0]^{-1}$ for $-1 < t \leq 0$. Hence, for $-1 \leq u \leq 0$, $\phi(u) = u$.

If m is a positive integer and $m \leq u \leq m + 1$, then

$$\phi(u) = (g_0)^{-1}(g_1 + g_2 + \ldots + (u - m)g_{m+1}),$$

where $g_i = \#(G_i)$. Hence $\phi(m) + 1 = (g_0)^{-1}(\sum_{i=0}^{m} g_i)$. Since ϕ is continuous, increasing and piece-wise linear we may define ψ to be its inverse function. The left and right derivatives of ϕ and ψ are easily computed (Serre, 1979, p. 73). Define

$$G^v = G_{\psi(v)} \quad \text{or, equivalently,} \quad G^{\phi(u)} = G_u.$$

Hence $G^{-1} = G(L/K)$, $G^0 = G_0$ and $G^v = \{1\}$ for $v >> 0$. When

$H \lhd G(L/K)$ then $G(L/K)/H = G(F/K)$, $F = L^H$ and $(G(L/K)/H)^v = G^v(L/K)H/H$.

Suppose now that L/K is abelian and that

$$\chi : G(L/K) \longrightarrow \mathbf{C}^*$$

is an injective, one-dimensional representation. By local class field theory we have the Artin reciprocity map

$$K^* \longrightarrow G(K^c/K)_{ab} \longrightarrow G(L/K).$$

Here, as in 4.2.11, K^c is an algebraic closure of K. This map annihilates norms from L^* and factors to yield

$$\omega : K^*/N_{L/K}(L^*) \longrightarrow G(L/K)$$

and (Serre, 1979, corollary 3, p. 228) transforms the filtration

$$\{U_K^n/N_{L/K}(U_L^{\psi(n)})\}$$

into the filtration, $\{G^n(L/K)\}$, where

$$U_L^n = \{u \in \mathcal{O}_L^* \mid u \equiv 1 \ (mod \ m_L^n)\}$$

(Serre, 1979, p. 66). The largest integer, c, such that $G_c(L/K) \neq \{1\}$ is related to $f_K(\chi)$ by Serre (1979, corollary 2, p. 228)

6.1.38 $$f_K(\chi) = \phi(c) + 1.$$

In other words, $(f_K(\chi) - 1)$ is the least integer such that χ is non-trivial on $G_c = G^{\phi(c)} = \omega(U_K^{\phi(c)})$, or $f_K(\chi)$ is the least integer such that χ is trivial on $U_K^{f_K(\chi)})$, when considered (via reciprocity) as the character $\chi\omega : K^* \longrightarrow \mathbf{C}^*$.

Let $e_{L/K}$ be the *ramification index* of L/K so that $[L:K] = f_{L/K}e_{L/K}$. Now suppose that K is a local field over $\mathbf{Q}_p$ and that $f_K(\chi)-1-e_{K/\mathbf{Q}_p} > e_{K/\mathbf{Q}_p}/(p-1)$. In this case the pth power map induces isomorphisms

$$\begin{array}{ccc} U_K^{f_K(\chi)-e_{K/\mathbf{Q}_p}} & \xrightarrow{\cong} & U_K^{f_K(\chi)} \\ \downarrow & & \downarrow \\ U_K^{f_K(\chi)-e_{K/\mathbf{Q}_p}-1} & \xrightarrow{\cong} & U_K^{f_K(\chi)-1} \end{array}$$

by Serre (1979, proposition 9, p. 212). This means that $\psi^p(\chi)$, which is just the p-th power of χ, is non-trivial on $U_K^{f_K(\chi)-e_{K/\mathbf{Q}_p}-1}$ and trivial on $U_K^{f_K(\chi)-e_{K/\mathbf{Q}_p}}$. Therefore we have proved that

Lemma 6.1.39 *If* $f_K(\chi) > 1 + e_{K/\mathbf{Q}_p}(p+1)/p$, *then*

$$f_K(\psi^p(\chi)) = f_K(\chi) - e_{K/\mathbf{Q}_p}.$$

Definition 6.1.40 Let $G(L/K)$ be the Galois group of a finite extension of p-adic local fields. Let $v : G(L/K) \longrightarrow GL_m(\mathbf{C})$ be a representation. Then v will be called *uniformly ramified* if

$$a_{G(L/K)}(v) = \sum_\alpha a_\alpha(G(L/F_\alpha), \chi_\alpha)^{G(L/K)} \in R_+(G(L/K))$$

and, for each α, $f_{F_\alpha}(\chi_\alpha) > 1 + e_{F_\alpha/\mathbf{Q}_p}(p+1)/p$.

Theorem 6.1.41 *If v is uniformly ramified, in the sense of* 6.1.40, *then*

$$f_K(\psi^p(v)) = f_K(v) - e_{K/\mathbf{Q}_p} \cdot dim(v).$$

Proof By 4.1.6, in $R(G(L/K))$, we have

$$v = \sum_\alpha a_\alpha Ind_{G(L/F_\alpha)}^{G(L/K)}(\chi_\alpha)$$

and

$$\psi^p(v) = \sum_\alpha a_\alpha Ind_{G(L/F_\alpha)}^{G(L/K)}(\chi_\alpha^p).$$

Hence, by 6.1.20 and Lemma 6.1.39,

$$\begin{aligned} f_K(\psi^p(v)) &= \textstyle\sum_\alpha a_\alpha(v_K(\delta_{F_\alpha/K}) + f_{F_\alpha/K} f_{F_\alpha}(\chi_\alpha^p)) \\ &= \textstyle\sum_\alpha a_\alpha(v_K(\delta_{F_\alpha/K}) + f_{F_\alpha/K}(f_{F_\alpha}(\chi_\alpha) - e_{F_\alpha/\mathbf{Q}_p})) \\ &= f_K(v) - \textstyle\sum_\alpha a_\alpha \cdot f_{F_\alpha/K} \cdot e_{F_\alpha/\mathbf{Q}_p}. \end{aligned}$$

Now $[F_\alpha : \mathbf{Q}_p] = [F_\alpha : K][K : \mathbf{Q}_p]$ and

$$f_{F_\alpha/\mathbf{Q}_p} = [\overline{F_\alpha} : F_p] = [\overline{F_\alpha} : \overline{K}][\overline{K} : F_p] = f_{F_\alpha/K} f_{K/\mathbf{Q}_p},$$

while $e_{L/K} f_{L/K} = [L : K]$, so that

$$f_{F_\alpha/K} \cdot e_{F_\alpha/\mathbf{Q}_p} = f_{F_\alpha/K} \cdot e_{F_\alpha/K} \cdot e_{K/\mathbf{Q}_p} = [F_\alpha : K] e_{K/\mathbf{Q}_p}.$$

Therefore

$$\begin{aligned} f_K(v) - f_K(\psi^p(v)) &= e_{K/\mathbf{Q}_p}(\textstyle\sum_\alpha a_\alpha[F_\alpha : K]) \\ &= e_{K/\mathbf{Q}_p} \cdot dim(v), \end{aligned}$$

since, by 2.2.42,

$$\begin{aligned} dim(v) &= \textstyle\sum_\alpha a_\alpha dim(Ind_{G(L/F_\alpha)}^{G(L/K)}(\chi_\alpha)) \\ &= \textstyle\sum_\alpha a_\alpha[F_\alpha : K]. \end{aligned}$$

This completes the proof of Theorem 6.1.41. □

6.1.42 Now let us turn to the case of ψ^q when $HCF(p, q) = 1$. Let v be as in 6.1.40 and set

$$\mathscr{A} = \{\alpha \mid \chi_\alpha(G_0(L/F_\alpha)) \neq 1, \chi_\alpha(G_1(L/F_\alpha)) = 1 = \chi_\alpha^q(G_0(L/F_\alpha))\}.$$

We will now calculate $f_K(\psi^q(v))$. By Serre (1979, proposition 1, p. 100; see also 6.1.25) we may assume that L/K is totally ramified. The following result is left to the reader as exercise 6.4.1.

Theorem 6.1.43 *Assume that L/K is totally ramified. With the notation of* 6.1.42

$$f_K(\psi^q(v)) = f_K(v) - \sum_{\alpha \in \mathscr{A}} a_\alpha.$$

6.2 Kato's abelian conductor

In this section we shall recall a generalisation, due to K. Kato (1989) of the Swan conductor. In order to state this definition, which is given in terms of Galois cohomology, we must recall some Galois modules.

6.2.1 Let n be a non-zero integer and let $r \in \mathbf{Z}$. Let K be a complete, discrete valuation field and suppose that K^c is an algebraic closure of K. Let $\mathbf{Z}/n$ denote the Ω_K-module given by the nth roots of unity in K^c, where $\Omega_K = Gal(K^c/K)$ is the absolute Galois group of K. Denote by $\mathbf{Z}/n(r)$ the rth Tate twist of $\mathbf{Z}/n$. Hence $\mathbf{Z}/n(0)$ is the module with the trivial action and for $r > 0$, $\mathbf{Z}/n(r) = (\mathbf{Z}/n)^{\otimes r}$. If char$(K) = p > 0$ we shall also need the module, $W_s\Omega^r_{K,log}[-r]$, given by the logarithmic part of the de Rham–Witt complex, $W_s\Omega^*_K$ (Illusie, 1979, I section 5.7, p. 596).

Define an Ω_K-module, $\underline{\mathbf{Z}/n}(r)$, in the following manner.

6.2.2

$$\underline{\mathbf{Z}/n}(r) = \begin{cases} \mathbf{Z}/n(r) \oplus W_s\Omega^r_{K,log}[-r], & \text{if } char(K) = p, n = p^s m, \\ & s \geq 0 \text{ and } HCF(m,p) = 1, \\ \mathbf{Z}/n(r) & \text{otherwise.} \end{cases}$$

We will write $H^q_n(K)$ for the Galois cohomology group, $H^q(K; \underline{\mathbf{Z}/n}(q-1))$ and $H^q(K)$ for $\lim_{\overrightarrow{n}} H^q_n(K)$. For example,

6.2.3
$$\begin{cases} H^1(K) & \cong Hom_{conts}(\Omega^{ab}_K, \mathbf{Q}/\mathbf{Z}) \\ H^2(K) & \cong Br(K), \end{cases}$$

where $Br(K)$ denotes the Brauer group of K.

6.2.4 Suppose that L/K is a finite Galois extension and that $\chi : G(L/K) \longrightarrow \mathbf{C}^*$ is a one-dimensional representation of the Galois group, $G(L/K)$. Identifying the roots of unity with $\mathbf{Q}/\mathbf{Z}$ we may interpret χ as a continuous homomorphism, $\chi : \Omega_K \longrightarrow \mathbf{Q}/\mathbf{Z}$, and hence obtain

6.2.5
$$\chi \in H^1(K).$$

If M/K is any field extension (not necessarily finite or Galois) denote by

6.2.6
$$\chi_M \in H^1(M)$$

the image of 6.2.5 under the natural map.

If π is the chosen uniformiser in $\mathcal{O}_K$ and T is an indeterminate then, for $n \geq 0$,

6.2.7
$$1 + \pi^{n+1}T \in ((\mathcal{O}_K[T]_{(\pi)})^{(h)})^*,$$

the unit group of the henselianisation, with respect to (π), of the localisation, $\mathcal{O}_K[T]_{(\pi)}$. If M denotes the field of fractions of $\mathcal{O}_K[T]^{(h)}_{(\pi)}$ this implies (Kato, 1989, p. 103) that $1 + \pi^{n+1}T$ represents a class in the cohomology group,

$$H^1_{\acute{e}t}(Spec(\mathcal{O}_K[T]^{(h)}_{(\pi)}); \underline{\mathbf{Z}/s}(1))$$

and hence we obtain

6.2.8 $$1 + \pi^{n+1}T \in \lim_{\overleftarrow{s}} H^1(M; \underline{\mathbf{Z}/s(1)}).$$

We may form the cup-product

6.2.9 $$\{\chi_M, 1 + \pi^{n+1}T\} \in H^2(M).$$

Definition 6.2.10 Let K be a complete, discrete valuation field and let $\chi : G(L/K) \longrightarrow \mathbf{C}^*$ denote a one-dimensional Galois representation, as in 6.2.4. The *Kato–Swan conductor* of χ,

6.2.11 $$sw_K(\chi) \in \mathbf{Z},$$

is defined to be the least integer, $n \geq 0$, such that

$$0 = \{\chi_M, 1 + \pi^{n+1}T\}$$

in 6.2.9.

An important indication of the 'correctness' of this abelian generalisation of the Swan conductor is the following formula of Kato, which often gives sw_K in terms of the ramification function, $s_{G(L/K)}$.

Theorem 6.2.12 *Let L/K be a finite Galois extension of complete, discrete valuation fields. Let $\chi : G(L/K) \longrightarrow \mathbf{C}^*$ be a one-dimensional representation and let $s_{G(L/K)}$ denote the ramification function of* 6.1.3.

If either *$\overline{L}/\overline{K}$ is separable* or *$e_{L/K} = 1$ and $\overline{L}$ is generated over $\overline{K}$ by one element, then*

$$sw_K(\chi) = -(e_{L/K})^{-1} \sum_{g \in G(L/K)} s_{G(L/K)}(g)\chi(g).$$

Proof See Kato (1989, proposition 6.8, p. 12). □

Corollary 6.2.13 *In the classical case the abelian conductor of* 6.2.10 *coincides with the Swan conductor of* 6.1.16.

Proof Let χ be a one-dimensional representation of $G(L/K)$, as in 6.2.10. In the classical case the residue field extension is separable so that 6.2.12 yields

$$\begin{aligned} sw_K(\chi) &= -(e_{L/K})^{-1} \textstyle\sum_{g \in G(L/K)} s_{G(L/K)}(g)\chi(g) \\ &= \#(G(L/K))^{-1} \textstyle\sum_{g \in G(L/K)} (-f_{L/K} s_{G(L/K)}(g))\chi(g) \\ &= < SW_{G(L/K)}, \chi >, \end{aligned}$$

which is the classical definition of the Swan conductor given in 6.1.16. □

For completeness I will sketch a cohomological proof of 6.2.13 which proceeds directly from the definition of 6.2.10.

Proposition 6.2.14 *Let K be a classical local field and suppose that $dim(\chi) = 1$, then the conductors, $sw_K(\chi)$, of* 6.1.16 *and* 6.2.10 *coincide.*

Proof We shall treat only the case when χ is non-trivial on G_0, since the remaining case is simpler. In this case, by 6.1.22, the Artin and Swan conductors of χ differ by one.

If $\{\chi_M, 1 + \pi^{n+1}T\} = 0$ in 6.2.9 then, by Theorem 6.3 of Kato (1989),

$$\{\chi, 1 + \pi^{n+1}\mathcal{O}_K\} \leq H^2(K; \mathbf{Q}/\mathbf{Z}(1)) \cong \mathbf{Q}/\mathbf{Z}$$

vanishes.

Now suppose that $\{\chi, 1 + \pi^{n+1}\mathcal{O}_K\} = 0$ and that $\{\chi_M, 1 + \pi^{m+1}T\} = 0$ with $m > n$. We shall show that $\{\chi_M, 1 + \pi^m T\} = 0$ and hence, by induction, that $\{\chi_M, 1 + \pi^{n+1}T\} = 0$.

Since $m > 0$, we know that $\{\chi_M, 1 + \pi^m T\}$ is p-torsion, by Kato (1989, section 5.5(i)). Now let $\overline{K}$ denote the residue field of K. There is a subgroup, $V_p^2(\mathcal{O}_K[T])$, of $H_p^2(M)$ which is defined in Kato (1989, definition 1.7.1). In addition there is an injective homomorphism, λ_π, and a commutative diagram of the following form (Kato, 1989, section 5.6.1).

6.2.15

$$\begin{array}{ccccc}
H_p^2(\overline{K}[T]) \oplus H_p^1(\overline{K}[T]) & \xrightarrow{\lambda_\pi} & V_p^2(\mathcal{O}_K[T]) & \xrightarrow{j} & H_p^2(M) \\
\downarrow \alpha & & \downarrow \beta & & \\
H_p^2(\overline{K}) \oplus H_p^1(\overline{K}) & \xrightarrow{\lambda'_\pi} & V_p^2(\mathcal{O}_K) & \xrightarrow{\cong} & H_p^2(K)
\end{array}$$

In 6.2.15 $H_p^q(F[T])$ is the qth étale cohomology group of $Spec(F[T])$ with coefficients in $\underline{\mathbf{Z}/p}(q-1)$. The vertical maps are given by the specialisation sending T to 1. Hence α is an isomorphism. The horizontal maps, $\lambda_\pi, \lambda'_\pi$ and j are injective (Kato, 1989, sections 5.6 and 6.3). In addition, by Kato (1989, section 5.1), there exists (a, b) such that

$$\lambda_\pi(a, b) = \{\chi_M, 1 + \pi^m T\}.$$

However,

$$\begin{aligned} \lambda'_\pi(a, b) &= \beta(\{\chi_M, 1 + \pi^m T\}) \\ &= \{\chi, 1 + \pi^m\} \\ &= 0, \end{aligned}$$

so that $(a, b) = 0$ and $\{\chi_M, 1 + \pi^m T\} = 0$, as required.

Finally, from abelian class field theory, we may identify the pairing

$$H^1(K; \mathbf{Q}/\mathbf{Z}(0)) \otimes H^1(K) \longrightarrow H^2(K) \cong \mathbf{Q}/\mathbf{Z}$$

with the evaluation pairing

$$Hom_{conts}(K; \mathbf{Q}/\mathbf{Z}) \otimes \lim_{\substack{\rightarrow \\ n}} K^*/(K^*)^n \longrightarrow \mathbf{Q}/\mathbf{Z}.$$

Therefore the least n for which $\{\chi, 1 + \pi^{n+1}\mathcal{O}_K\} = 0$ is the least n for which $\chi : K^* \longrightarrow \mathbf{C}^*$ vanishes on $U_K^{n+1} = 1 + \pi^{n+1}\mathcal{O}_K$. By Serre (1979, p. 102, proposition 5) the Artin conductor is also equal to the least integer, $n + 1$, such that χ vanishes on U_K^{n+1}. This completes the proof of 6.2.14. □

Remark 6.2.16 In the next section we shall construct a generalisation of the Swan conductor by applying Explicit Brauer Induction to extend the abelian conductor of 6.2.10. For our generalised Swan conductor we will be able establish a number of properties which are possessed by the classical Swan conductor (see 6.3.6 and 6.3.20).

However, we will find that the inductivity in dimension zero (see 6.1.24) is a property which cannot hold in general for any generalisation of the Swan conductor which coincides with 6.2.10 in the abelian case. In order to see this we shall pause to examine several examples in which $G(L/K) \cong \mathbf{Z}/p \times \mathbf{Z}/p$. Local Galois extensions whose group is elementary abelian may all be constructed by the Artin–Schreier theory of Serre (1979, p. 154). Further details on the non-existence of suitable generalisations of the Swan conductor are to be found in Boltje, Cram & Snaith (1993, section 5).

For these examples, which were shown to me by G-M. Cram, we shall calculate the behaviour of any Swan conductor which is invariant under inflation (as in 6.1.21) and coincides with the Swan conductor of 6.2.10 in the abelian case. In examples 6.2.18–6.2.20 we shall see that such a Swan conductor cannot be inductive in dimension zero, in general. Examples 6.2.19 and 6.2.20 have a subextension in common (N_p/K in 6.2.19 is R_p/K in 6.2.20). This feature makes 6.2.19 and 6.2.20 into particularly good illustrations of the sort of relations between values of the Swan conductor which would be implied by inductivity in dimension zero.

Example 6.2.17 Let F_p denote the finite field of order p.

Let $K = F_p((T))((\pi))$, $\mathcal{O}_K = F_p((T))[[\pi]]$, where π is the prime element. To construct Galois extensions of K with group $\mathbf{Z}/p$ we use a modification of Artin–Schreier theory.

Consider the polynomial

$$f_a(X) = X^p - \pi^{p-1}X - a \in K[X].$$

If A is a root of $f_a(X)$ then the other roots are $A + \pi\xi$ where $\xi \in F_p$. Hence either $f_a(X)$ splits completely in K or $K(A)$ is a Galois extension of degree p. Using Galois cohomology one can show that every Galois extension of degree p over K may be obtained in this manner (see, for example, Serre, 1979, Chapter X, section 3, p. 154). Set $M_p = K(\Pi)$ where Π is a root of $f_\pi(X)$. Since $f_\pi(X)$ is an Eisenstein polynomial (Lang, 1984, p. 200) M_p/K is totally ramified — that is, $e_{M_p/K} = p$. Hence $\mathcal{O}_{M_p} = \mathcal{O}_K[\Pi]$, by Serre (1979, chapter I, section 6, proposition 17). Set $M_0 = K(A)$ where A is a root of $f_a(X)$ and $a \in \mathcal{O}_K^*$ is a unit such that a (modulo π) is not a pth power (modulo Π). For example, one may take $a = T$. Hence $f_a(X) \equiv X^p - a \pmod{\pi}$, which is an irreducible polynomial in the residue field, $F_p((T))$. Hence $\mathcal{O}_{M_0} = \mathcal{O}_K[A]$ (see Serre, 1979, chapter I, section 6, proposition 15) and M_0/K is unramified with purely inseparable residue field extension, $\overline{M_0} = F_p(T, \overline{A})$, where $\overline{A}$ is the image of A. For example, if $a = T$ then $\overline{M_0} = F_p(\sqrt[p]{T})$.

Set $L = M_0M_p$. This is a $\mathbf{Z}/p \times \mathbf{Z}/p$-extension of K, since $f_\pi(X)$ remains irreducible over M_0 because π is a prime in M_0. We have $\mathcal{O}_L = \mathcal{O}_{M_0}[\Pi] = \mathcal{O}_K[A, \Pi]$. The Galois group, $G(L/K)$, has generators σ_0, σ_1 with

$$\sigma_0(\Pi) = \Pi, \qquad \sigma_0(A) = A + \pi,$$

$$\sigma_1(\Pi) = \Pi + \pi, \quad \sigma_1(A) = A.$$

Denote the fixed field of $U_i = < \sigma_0^i\sigma_1 >$ by M_i for $i = 0, 1, \ldots, p-1$ so that we have the following diagram of fields:

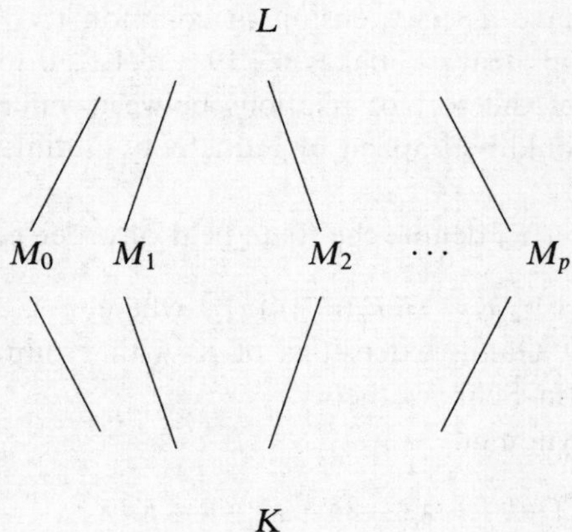

One finds that $M_i = K(A_i)$ with $A_i = A - i\Pi$, since A_i is a root of

$$f_{a_i} = X^p - \pi^{p-1}X - (a - i\pi).$$

Set $U_p =< \sigma_0 >$. Since $a_i = a - i\pi \equiv a \pmod{\pi}$ we see that $\overline{M_0} \cong \overline{M_1} \cong \ldots \cong \overline{M}_{p-1}$.

Now we calculate the functions, i_{U_i} and s_{U_i}, of 6.1.3. We have $U_i = G(L/M_i)$ and $\mathcal{O}_L = \mathcal{O}_{M_i}[\Pi]$ for $i = 0, 1, \ldots, p-1$ so that, for $1 \neq \sigma \in U_i$,

$$\begin{aligned} i_{U_i}(\sigma) &= \inf{}_{x \in \mathcal{O}_L}(v_L(x - \sigma(x))) \\ &= v(\Pi - \sigma(\Pi)) \\ &= v_L(\pi) \\ &= p \end{aligned}$$

and

$$\begin{aligned} s_{U_i}(\sigma) &= \inf{}_{x \in L^*}(v_L(1 - \sigma(x)x^{-1})) \\ &= v_L(1 - \sigma(\Pi)\Pi^{-1}) \\ &= p - 1. \end{aligned}$$

Table 6.1. *Field extensions* (*Example* 6.2.17)

E/F	$e_{E/F}$	$f_{E/F}$	$f^{sep}_{E/F}$
L/M_i $i = 0, \ldots, p-1$	p	1	1
M_i/K $i = 0, \ldots, p-1$	1	p	1
L/M_p	1	p	1
M_p/K	p	1	1

Also, $U_p = G(L/M_p)$ and $\mathcal{O}_L = \mathcal{O}_{M_p}[A]$ so that, for $1 \neq \sigma \in U_p$,

$$\begin{aligned} i_{U_p}(\sigma) &= v(A - \sigma(A)) \\ &= p \end{aligned}$$

and $s_{U_p}(\sigma) = p$.

Next we have $G(L/K)/U_i = G(M_i/K)$ and $\mathcal{O}_{M_i} = \mathcal{O}_K[A_i]$ for $i = 0, 1, \ldots, p-1$ so that, when $\sigma \notin U_i$,

$$\begin{aligned} i_{G(M_i/K)}(\sigma) &= v_{M_i}(A_i - \sigma(A_i)) \\ &= v_{M_i}(\pi) \\ &= 1 \end{aligned}$$

and $s_{G(M_i/K)}(\sigma) = 1$.

Finally, for $G(M_p/K)$, $\mathcal{O}_{M_p} = \mathcal{O}_K[\Pi]$ and, when $\sigma \notin U_p$,

$$\begin{aligned} i_{G(M_p/K)}(\sigma) &= v_{M_p}(\Pi - \sigma(\Pi)) \\ &= v_{M_p}(\pi) \\ &= p \end{aligned}$$

and $s_{G(M_p/K)}(\sigma) = p - 1$.

The table of ramification indices, residue degrees and separable residue degrees of these field extensions are given in Table 6.1

Let $\theta : G(L/M_0) \longrightarrow \mathbf{C}^*$ be given by $\theta(\sigma_1) = \xi_p = exp(2\pi i/p)$. Lift θ to $\hat{\theta} : G(L/K) \longrightarrow \mathbf{C}^*$ given by $\hat{\theta}(\sigma_0^a \sigma_1^b) = \theta(\sigma_1^b) = \xi_p^b$. We have $G(L/K) = U_0 \times U_p$, where $U_p =< \sigma_0 >$. Define $\psi : G(L/K) \longrightarrow \mathbf{C}^*$ by

$\psi(\sigma_0^a\sigma_1^b) = \xi_p^a$ so that

$$Ind_{G(L/M_0)}^{G(L/K)}(\theta - 1) = \sum_{i=0}^{p-1}(\hat{\theta}\psi^i - \psi^i).$$

For $i = 1,\ldots,p-1$ we have $Ker(\psi^i) = U_0$ and

$$\begin{aligned} Ker(\hat{\theta}\psi^i) &= \{\sigma_0^a\sigma_1^b \mid 1 = \xi_p^{b+ia}\} \\ &= \{(\sigma_0^j\sigma_1)^b\} \\ &= U_j, \end{aligned}$$

where $ij \equiv -1 \pmod p$. Also $Ker(\hat{\theta}) = U_p$. Since the ramification indices, residue degrees and separable residue degrees are the same for $U_0,\ldots,U_{p-1}$, we find that

$$sw_K(Ind_{G(L/M_0)}^{G(L/K)}(\theta - 1)) = sw_K(\hat{\theta}).$$

Also, since $f_{M_p/K} = 1$ and $e_{M_p/K} = p$,

$$\begin{aligned} &sw_K(\hat{\theta} : G(M_p/K) \longrightarrow \mathbf{C}^*) \\ &= -p^{-1}\textstyle\sum_{g\in G(M_p/K)} s_{G(M_p/K)}(g)\hat{\theta}(g) \\ &= p^{-1}\{-(p-1)(\textstyle\sum_{g\neq 1}\hat{\theta}(g)) + (p-1)^2\} \\ &= p^{-1}\{(p-1) + (p-1)^2\} \\ &= p-1. \end{aligned}$$

Also, since $f_{L/M_0} = 1$ and $e_{L/M_0} = p$,

$$\begin{aligned} &sw_{M_0}(\theta : G(L/M_0) \longrightarrow \mathbf{C}^*) \\ &= p^{-1}\{-(p-1)(\textstyle\sum_{g\neq 1}\theta(g)) + (p-1)^2\} \\ &= p-1. \end{aligned}$$

Therefore, since $f_{M_0/K}^{sep} = 1$,

$$sw_K(Ind_{G(L/M_0)}^{G(L/K)}(\theta - 1)) = f_{M_0/K}^{sep} sw_{M_0}(\vartheta - 1).$$

Example 6.2.18 As a second example we take the same field extensions and let $\theta_1 : G(L/M_p) \longrightarrow \mathbf{C}^*$ be given by $\theta_1(\sigma_0) = \xi_p$. Lift θ_1 to $\hat{\theta}_1 : G(L/K) \longrightarrow \mathbf{C}^*$ given by $\hat{\theta}_1(\sigma_0^a\sigma_1^b) = \xi_p^a$. Define $\psi_1 : G(L/K) \longrightarrow \mathbf{C}^*$ by $\psi_1(\sigma_0^a\sigma_1^b) = \xi_p^b$. Hence, for $i = 1, \ldots, p-1$ we have $Ker(\psi_1^i) = U_p$ and

$$\begin{aligned} Ker(\hat{\theta}_1\psi_1^i) &= \{\sigma_0^a\sigma_1^b \mid 1 = \xi_p^{a+ib}\} \\ &= \{(\sigma_0^i\sigma_1)^b\} \\ &= U_i. \end{aligned}$$

We have

$$\begin{aligned} & sw_K(Ind_{G(L/M_p)}^{G(L/K)}(\theta_1 - 1)) \\ & = \textstyle\sum_{i=0}^{p-1} sw_K(\hat{\theta}_1\psi_1^i) - \sum_{i=0}^{p-1} sw_K(\psi_1^i) \\ & = psw_K(\hat{\theta}_1) - (p-1)sw_K(\psi_1). \end{aligned}$$

Since $e_{M_0/K} = 1$,

$$\begin{aligned} & sw_K(\hat{\theta}_1 : G(M_0/K) \longrightarrow \mathbf{C}^*) \\ & = \{(-1)(\textstyle\sum_{g\neq 1} \hat{\theta}_1(g)) + (p-1)\} \\ & = p. \end{aligned}$$

Since $e_{M_p/K} = p$ and $f_{M_p/K} = 1$,

$$\begin{aligned} & sw_K(\psi_1 : G(M_p/K) \longrightarrow \mathbf{C}^*) \\ & = p^{-1}\{(-1)(\textstyle\sum_{g\neq 1}(p-1)\psi_1(g)) + (p-1)^2\} \\ & = p - 1. \end{aligned}$$

Hence

$$\begin{aligned} & sw_K(Ind_{G(L/M_p)}^{G(L/K)}(\theta_1 - 1)) \\ & = p^2 - (p-1)^2 \\ & = 2p - 1. \end{aligned}$$

However, since $e_{L/M_p} = 1$,

$$sw_{M_p}(\theta_1 : G(L/M_p) \longrightarrow \mathbf{C}^*)$$

$$= \{(-p)(\textstyle\sum_{g\neq 1} \theta_1(g)) + p(p-1)\}$$

$$= p + p(p-1)$$

$$= p^2.$$

Hence

$$sw_K(Ind_{G(L/M_p)}^{G(L/K)}(\theta_1 - 1)) \neq f_{M_p/K}^{sep} sw_{M_p}(\theta_1 - 1).$$

Example 6.2.19 We will now consider a parametrised version of 6.2.17.

Let $K = F_p((T))((\pi))$ as in 6.2.17. Choose $a \in \mathcal{O}_K^* = F_p((T))[[\pi]]^*$ such that

$$X^p - \overline{a} \in F_p((T))[X] = \overline{K}[X]$$

is irreducible. Choose two integers, $1 \leq s < t$, and assume that s is coprime to p. Define polynomials

$$f_{a,s}(X) = X^p - \pi^{s(p-1)}X - a$$

and

$$f_{a,t}(X) = X^p - \pi^{t(p-1)}X - a$$

with roots $\alpha_{a,s}$ and $\alpha_{a,t}$, respectively.

Set $L = N_0 N_p$, where $N_0 = K(\alpha_{a,t})$ and $N_p = K(\alpha_{a,s})$.

Hence L/K is a Galois extension with group $G(L/K) \cong \mathbf{Z}/p \times \mathbf{Z}/p$ whose generators, $\sigma_{a,s}$ and $\sigma_{a,t}$, are defined by

$$\sigma_{a,s}(\alpha_{a,s}) = \alpha_{a,s} + \pi^s, \quad \sigma_{a,s}(\alpha_{a,t}) = \alpha_{a,t},$$

$$\sigma_{a,t}(\alpha_{a,s}) = \alpha_{a,s}, \qquad \sigma_{a,t}(\alpha_{a,t}) = \alpha_{a,t} + \pi^t.$$

Since $f_{a,s}(X) \equiv f_{a,t}(X) \equiv X^p - \overline{a} \in \overline{K}[X]$ is irreducible, the residue fields of N_0 and N_p are isomorphic and are purely inseparable with

$$[\overline{N}_0 : \overline{K}] = [\overline{N}_p : \overline{K}] = p.$$

The intermediate fields, $K \leq N_i \leq L$, are the fixed fields of $\sigma_{a,s}\sigma_{a,t}^i$ for $0 \leq i \leq p-1$ and of $\sigma_{a,t}$. The fixed field of $\sigma_{a,t}$ is $N_p = L^{\sigma_{a,t}} = K(\alpha_{a,s})$

while $N_i = L^{\sigma_{a,s}\sigma_{a,t}^i} = K(\alpha_{a,t} - i\pi^{t-s}\alpha_{a,s})$ for $0 \le i \le p-1$ since

$$\begin{aligned}
&\sigma_{a,s}\sigma_{a,t}^i(\alpha_{a,t} - i\pi^{t-s}\alpha_{a,s})\\
&= \sigma_{a,s}\sigma_{a,t}^{i-1}(\alpha_{a,t} + \pi^t - i\pi^{t-s}\alpha_{a,s})\\
&\quad\vdots\\
&= \sigma_{a,s}(\alpha_{a,t} + i\pi^t - i\pi^{t-s}\alpha_{a,s})\\
&= \alpha_{a,t} + i\pi^t - i\pi^{t-s}(\alpha_{a,s} + \pi^s)\\
&= \alpha_{a,t} - i\pi^{t-s}\alpha_{a,s}.
\end{aligned}$$

The minimum polynomial of $\alpha_{a,t} - i\pi^{t-s}\alpha_{a,s}$ is

$$f_{a-i\pi^{p(t-s)}a,t}(X) = X^p - \pi^{t(p-1)}X - (a - i\pi^{p(t-s)}a)$$

since

$$\begin{aligned}
&f_{a-i\pi^{p(t-s)}a,t}(\alpha_{a,t} - i\pi^{t-s}\alpha_{a,s})\\
&= \alpha_{a,t}^p - i\pi^{p(t-s)}\alpha_{a,s}^p - \pi^{t(p-1)}\alpha_{a,t}\\
&\qquad + i\pi^{t(p-1)+t-s}\alpha_{a,s} - a + i\pi^{(t-s)p}a\\
&= f_{a,t}(\alpha_{a,t}) - i\pi^{p(t-s)}[\alpha_{a,s}^p - \pi^{tp-s-pt+ps}\alpha_{a,s} - \pi^{p(t-s)-p(t-s)}a]\\
&= (-i\pi^{p(t-s)})f_{a,s}(\alpha_{a,s})\\
&= 0.
\end{aligned}$$

We will now show that $e_{L/N_p} = p$. To accomplish this we will find a prime element, π_L, for L and find its irreducible polynomial over N_p.

Set $\beta = \alpha_{a,t} - \alpha_{a,s}$. Clearly $L = K(\alpha_{a,s}, \beta) = N_p(\beta)$. Also

$$\begin{aligned}
&\beta^p - \pi^{t(p-1)}\beta \\
&= \alpha_{a,t}^p - \alpha_{a,s}^p - \pi^{t(p-1)}\alpha_{a,t} + \pi^{t(p-1)}\alpha_{a,s} \\
&= a - \alpha_{a,s}^p + \pi^{t(p-1)}\alpha_{a,s} \\
&= \pi^{t(p-1)}\alpha_{a,t} - \pi^{s(p-1)}\alpha_{a,s} \\
&= \pi^{s(p-1)}\alpha_{a,s}(\pi^{(t-s)(p-1)} - 1)
\end{aligned}$$

lies in N_p. Also, since $\pi^{(t-s)(p-1)} - 1 \in \mathcal{O}_K^*$,

$$\begin{aligned}
v_{N_p}(\beta^p - \pi^{t(p-1)}\beta) &= v_{N_p}(\pi^{s(p-1)}\alpha_{a,s}) \\
&= v_{N_p}(\pi^{s(p-1)}), \\
&= s(p-1),
\end{aligned}$$

since $\alpha_{a,s} \in \mathcal{O}_{M_p}^*$ and π is prime in N_p. On the other hand, $v_L(\beta^p - \pi^{t(p-1)}\beta) = e_{L/M_p}s(p-1)$. However, $\beta \in \mathcal{O}_L$ so that $v_L(\beta) \geq 0$ and $t > s$ implies that

$$\begin{aligned}
v_{N_p}(\beta^p - \pi^{t(p-1)}\beta) &= e_{L/N_p}s(p-1) \\
&= v_L(\beta^p) \\
&= pv_L(\beta).
\end{aligned}$$

Since $HCF(s,p) = 1$ we must have $e_{L/N_p} = p$ and $v_L(\beta) = s(p-1)$. Now choose integers, r_1 and r_2, such that $r_1 s(p-1) + r_2 p = 1$ and set

$$\pi_L = \beta^{r_1}\pi^{r_2},$$

so that $v_L(\pi_L) = r_1 s(p-1) + r_2 p = 1$ and π_L is a prime element for L.

Now let us record the table of ramification indices, residue degrees and separable residue degrees of these field extensions (see Table 6.2.)

Since $\overline{N}_i = \overline{N}_p$ and $e_{L/N_p} = p$, we must have $1 = f_{L/N_p} = f_{L/N_i}$ for all i. Hence $f_{N_i/K} = p$ for all i and therefore $e_{N_i/K} = 1$ for all i.

Next we compute

$$v_L(\sigma_{a,s}\sigma_{a,t}^i(\pi_L) - \pi_L) = i_{G(L/N_i)}(\sigma_{a,s}\sigma_{a,t}^i)$$

Table 6.2. *Field extensions (Example 6.2.19)*

E/F	$e_{E/F}$	$f_{E/F}$	$f^{sep}_{E/F}$
N_i/K			
$i = 0,\dots,p$	1	p	1
L/N_i			
$i = 0,\dots,p$	p	1	1

for all $0 \le i \le p-1$ and

$$v_L(\sigma_{a,t}(\pi_L) - \pi_L) = i_{G(L/N_p)}(\sigma_{a,t}).$$

Temporarily write σ_i for our chosen generator of $G(L/N_i)$ for $0 \le i \le p$. If $i = 0, 1, \dots, p-1$ then

$$\begin{aligned}\sigma_i(\beta) &= \sigma_{a,s}\sigma^i_{a,t}(\alpha_{a,t} - \alpha_{a,s}) \\ &= \sigma_{a,s}(\alpha_{a,t} + i\pi^t - \alpha_{a,s}) \\ &= \beta + \pi^s(i\pi^{t-s} - 1),\end{aligned}$$

while

$$\begin{aligned}\sigma_p(\beta) &= \sigma_{a,t}(\alpha_{a,t} - \alpha_{a,s}) \\ &= \beta + \pi^t.\end{aligned}$$

Since $\pi_L = \beta^{r_1}\pi^{r_2}$ and $\pi \in K$,

$$\begin{aligned}v_L(\sigma_i(\pi_L) - \pi_L) &= r_2 v_L(\pi) + v_L(\sigma_i(\beta^{r_1}) - \beta^{r_1}) \\ &= pr_2 + v_L(\sigma_i(\beta^{r_1}) - \beta^{r_1}).\end{aligned}$$

When $i = 0, 1, \dots, p-1$, then

$$\begin{aligned}&v_L(\sigma_i(\beta^{r_1}) - \beta^{r_1}) \\ &= v_L\left(\sum_{j=0}^{r_1-1}\begin{pmatrix} r_1 \\ j \end{pmatrix}\beta^j\pi^{s(r_1-j)}(i\pi^{t-s} - 1)^{r_1-j}\right) \\ &= v_L(r_1\beta^{r_1-1}\pi^s(i\pi^{t-s} - 1)) \\ &= (r_1 - 1)s(p-1) + sp \\ &= r_1 s(p-1) + s,\end{aligned}$$

Table 6.3. *Ramification function* (*Example* 6.2.19)

E/F	$s_{G(E/F)}(g)$ $(g \neq 1)$
L/N_i $i = 0,\dots,p-1$	s
L/N_p	$(t-s)p+s$

so that, in these cases,

$$v_L(\sigma_i(\pi_L) - \pi_L) = pr_2 + r_1 s(p-1) + s = 1 + s.$$

When $i = p$,

$$\begin{aligned} & v_L(\sigma_p(\pi_L) - \pi_L) \\ &= pr_2 + v_L\left(\sum_{j=0}^{r_1-1} \binom{r_1}{j} \beta^j \pi^{t(r_1-j)}\right) \\ &= pr_2 + v_L(r_1\beta^{r_1-1}\pi^t) \\ &= pr_2 + (r_1-1)s(p-1) + tp \\ &= (t-s)p + s + 1. \end{aligned}$$

For $1 \neq g \in G(L/N_i)$,

$$s_{G(L/N_i)}(g) = v_L(\sigma_p(\pi_L)\pi_L^{-1} - 1), = i_{G(L/N_i)}(g) - 1,$$

so that we obtain the values given in Table 6.3.

Hence if $\theta : G(L/N_i) \longrightarrow \mathbf{C}^*$ is non-trivial then, by 6.2.12 since $\overline{L} = \overline{N_i}$,

$$\begin{aligned} sw_{N_i}(\theta) &= p^{-1}\{-\textstyle\sum_{1\neq g} s_{G(L/N_i)}(g)(\theta(g) - 1)\} \\ &= p^{-1} s_{G(L/N_i)}(\sigma_i)(p - 1 + 1) \\ &= s_{G(L/N_i)}(\sigma_i) \\ &= \begin{cases} s, & \text{if } i = 0,\dots,p-1, \\ (t-s)p+s & \text{if } i = p. \end{cases} \end{aligned}$$

Now consider $G(N_i/K) \cong G(L/K)/G(L/N_i)$, which is generated by $\sigma_{a,s}$ when $i = 0, 1, \ldots, p-1$. Since N_i/K is unramified,

$$\begin{aligned} s_{G(N_i/K)}(\sigma_{a,s}) &= i_{G(N_i/K)}(\sigma_{a,s}) \\ &= v_{N_i}(\sigma_{a,s}(\alpha_{a,t} - i\pi^{t-s}\alpha_{a,s})) \\ &= v_{N_i}(\alpha_{a,t} - i\pi^{t-s}(\alpha_{a,s} + \pi^s) - \alpha_{a,t} + i\pi^{t-s}\alpha_{a,s}) \\ &= v_{N_i}(i\pi^t) \\ &= t. \end{aligned}$$

When $i = p$, $\sigma_{a,s}$ still generates but

$$\begin{aligned} s_{G(N_p/K)}(\sigma_{a,s}) &= i_{G(N_p/K)}(\sigma_{a,s}) \\ &= v_{N_p}(\sigma_{a,s}(\alpha_{a,s}) - \alpha_{a,s}) \\ &= v_{N_p}(\pi^s) \\ &= s. \end{aligned}$$

Hence if $\theta : G(N_i/K) \longrightarrow \mathbf{C}^*$ is non-trivial then, by 6.2.12,

$$\begin{aligned} sw_K(\theta) &= \{-\textstyle\sum_{1 \neq g} s_{G(N_i/K)}(g)(\theta(g) - 1)\} \\ &= s_{G(N_i/K)}(\sigma_{a,s})(p - 1 + 1) \\ &= \begin{cases} ps, & \text{if } i = p, \\ pt, & \text{if } i = 0, \ldots, p-1. \end{cases} \end{aligned}$$

We record the values of the Swan conductors on non-trivial one-dimensional representations, θ, in Table 6.4

Let $\theta_i : G(L/N_i) \longrightarrow \mathbf{C}^*$ be a non-trivial one-dimensional representation and also denote by θ_i an extension of this homomorphism to $G(L/K)$. Define

$$\psi_i : G(L/K) \longrightarrow G(L/K)/G(L/N_i) \cong G(N_i/K) \longrightarrow \mathbf{C}^*$$

Table 6.4. *Swan conductors (Example 6.2.19)*

E/F	$e_{E/F}$	$f_{E/F} = f^{ins}_{E/F}$	$sw_F(\theta)$
L/N_i			
$i = 0,\dots,p-1$	p	1	s
L/N_p	p	1	$p(t-s)+s$
N_i/K			
$i = 0,\dots,p-1$	1	p	pt
N_p/K	1	p	ps

by the formulae

$$\psi_p(\sigma_{a,s}) = \xi_p, \quad \psi_p(\sigma_{a,t}) = 1 \quad \text{or}$$

$$\psi_i(\sigma_{a,s}) = \xi_p^{-i}, \quad \psi_i(\sigma_{a,t}) = \xi_p \quad \text{if } i = 0,\dots,p-1.$$

Hence

$$Ind^{G(L/K)}_{G(L/N_i)}(\theta_i - 1) = \sum_{j=0}^{p-1} \theta_i\psi_i^j - \sum_{j=0}^{p-1} \psi_i^j.$$

Consider the case when $i = p$. Let us choose $\theta_p(\sigma_{a,s}) = 1$ and $\theta_p(\sigma_{a,t}) = \xi_p$ so that

$$\theta_p\psi_p^j(\sigma^u_{a,t}\sigma^v_{a,s}) = \xi^{u+vj}$$

and $Ker(\theta_p\psi_p^j) =< \sigma_{a,t}^{-j}\sigma_{a,s} >= G(L/N_{-j})$ for $j = 0,\dots,p-1 \in \mathbb{F}_p$. Similarly, $Ker(\psi_p^j) = G(L/N_p)$, so that

$$\begin{aligned} sw_K(Ind^{G(L/K)}_{G(L/N_p)}(\theta_p - 1)) &= p(pt) - (p-1)ps \\ &= p^2(t-s) + ps. \end{aligned}$$

Therefore

$$sw_K(Ind^{G(L/K)}_{G(L/N_p)}(\theta_p - 1)) \neq f^{sep}_{N_p/K} sw_{N_p}(\theta_p - 1) = p(t-s) + s.$$

Now consider simultaneously the cases given by $i = 0, 1, \dots, p-1$. Choose $\theta_i(\sigma_{a,s}) = \xi_p$ and $\theta_i(\sigma_{a,t}) = 1$ so that

$$\theta_i\psi_i^j(\sigma^u_{a,t}\sigma^v_{a,s}) = \xi_p^{v-ijv+uj}.$$

Thus $Ker(\theta_i\psi_i^j)$ runs through the set $\{G(L/N_w) \mid 0 \le w \le p\}$ for $0 \le j \le p$, while $Ker(\psi_i^j)$ runs through the set $\{G(L/N_w) \mid 0 \le w \le p-1\}$ for

$1 \leq j \leq p$. Hence

$$sw_K(Ind_{G(L/N_i)}^{G(L/K)}(\theta_i - 1)) = sw_K(\theta_p : G(N_p/K) \longrightarrow \mathbf{C}^*)$$

$$= ps.$$

Therefore, for $0 \leq i \leq p-1$,

$$sw_K(Ind_{G(L/N_i)}^{G(L/K)}(\theta_i - 1)) \neq f_{N_i/K}^{sep} sw_{N_i}(\theta_i - 1) = s.$$

Example 6.2.20 In this example we shall construct an extension, S/K, by applying the process of 6.2.19 to

$$f_{a,s}(X) = X^p - \pi^{s(p-1)}X - a$$

and

$$f_{b,s}(X) = X^p - \pi^{s(p-1)}X - b,$$

where $a, b \in \mathcal{O}_K^*$ with $v_K(a - b) = 1$. Now set $R_0 = K(\alpha_{b,s})$, $R_p = N_p = K(\alpha_{a,s})$ and $S = K(\alpha_{b,s}, \alpha_{a,s}) = R_0 R_p$ so that $G(S/K) \cong \mathbf{Z}/p \times \mathbf{Z}/p$. In this case the intermediate fields are $K(\alpha_{b,s} - i\alpha_{a,s}) = R_i$ for $i = 0, \ldots, p-1$ *and* $R_p = K(\alpha_{a,s})$. In addition,

$$f_{b-ia,s}(X) = X^p - \pi^{s(p-1)}X - b + ia$$

has $\alpha_{b,s} - i\alpha_{a,s}$ as a root. Among the $f_{b-ia,s}(X)$ there is a special polynomial, namely $f_{b-a,s}(X)$. Since $b - a$ is a prime element this polynomial is Eisenstein. Hence $e_{R_1/K} = p$ and we may choose $\pi_S = \alpha_{b-a,s} = \alpha_{b,s} - \alpha_{a,s}$ for the prime element of S. It follows easily that $\mathcal{O}_K[\alpha_{a,s}, \pi_S] = \mathcal{O}_S$.

Hence $R_0, R_2, \ldots, R_{p-1}, R_p$ are generated by a root of the polynomial $f_{b-ia,s}(X)$ for $i \neq 1$ and the determination of the first two columns of Table 6.5 are identical to the calculations $M_0, M_1, \ldots, M_{p-1}$ in 6.2.17. In the final column of Table 6.5 θ is any non-trivial one-dimensional representation of $G(E/F)$. In the second column it is always true that $f_{E/F} = f_{E/F}^{ins}$ and in the third column it is understood that $1 \neq g \in G(E/F)$.

The verification of the remaining entries in Table 6.5 is left to the reader as an exercise.

Define $\rho_i : G(S/R_i) \longrightarrow \mathbf{C}^*$ by $\rho_i(\sigma_{a,s}\sigma_{b,s}^i) = \xi_p$ if $i = 0, 1, \ldots, p-1$ and $\rho_p(\sigma_{b,s}) = \xi_p$. Define

$$\lambda_i : G(S/K) \longrightarrow G(S/K)/G(S/R_i) \cong G(R_i/K) \longrightarrow \mathbf{C}^*$$

Table 6.5. *Ramification functions and conductors (Example 6.2.20)*

E/F	$e_{E/F}$	$f_{E/F}$	$s_{G(E/F)}(g)$	$i_{G(E/F)}(g)$	$sw_F(\theta)$
S/R_i $i = 0, 2, \dots, p$	p	1	$ps - 1$	ps	$ps - 1$
R_i/K $i = 0, 2, \dots, p$	1	p	s	s	ps
S/R_1	1	p	ps	ps	$p^2 s$
R_1/K	p	1	$ps - 1$	ps	$ps - 1$

by

$$\lambda_p(\sigma_{a,s}) = \xi_p^{-i}, \quad \lambda_p(\sigma_{b,s}) = 1 \quad \text{or}$$

$$\lambda_i(\sigma_{a,s}) = \xi_p^{-i}, \quad \lambda_i(\sigma_{b,s}) = \xi_p \quad \text{if } i = 0, \dots, p-1.$$

Extend ρ_i to $G(S/K)$ by the formulae

$$\rho_p(\sigma_{b,s}) = \xi_p, \quad \rho_p(\sigma_{a,s}) = 1 \quad \text{or}$$

$$\rho_i(\sigma_{a,s}) = \xi_p, \quad \rho_i(\sigma_{b,s}) = 1 \quad \text{if } i = 0, \dots, p-1.$$

Therefore

$$Ind_{G(S/R_i)}^{G(S/K)}(\rho_i - 1) = \sum_{j=0}^{p-1} (\rho_i \lambda_i^j - \lambda_i^j).$$

Hence, if $i \neq 1$,

$$ps - 1 = sw_K(Ind_{G(S/R_i)}^{G(S/K)}(\rho_i - 1)) = f_{R_i/K}^{sep} sw_{R_i}(\rho_i - 1)$$

and, in the remaining case,

$$ps + p - 1 = sw_K(Ind_{G(S/R_1)}^{G(S/K)}(\rho_1 - 1)) \neq f_{R_1/K}^{sep} sw_{R_1}(\rho_1 - 1) = p^2 s.$$

There are a number of non-trivial properties possessed by sw_K which attest to its 'rightness' as a generalisation of the classical Swan conductor in the abelian case. The following result is one such example:

Proposition 6.2.21 *Let M be a henselian discrete valuation field over K such that $\mathcal{O}_K \subset \mathcal{O}_M$ and $m_M = \mathcal{O}_M m_K$. Assume that the residue field, $\overline{M}$, is separable over $\overline{K}$. Let L/K be a finite Galois extension of complete, discrete valuation fields such that $K \leq M \leq L$, then*

$$sw_K(\chi) = sw_M(Res_{G(L/M)}^{G(L/K)}(\chi))$$

for all one-dimensional representations, $\chi : G(L/K) \longrightarrow \mathbf{C}^*$.

Proof See Kato (1989, section 6.2, p. 119). □

Corollary 6.2.22 *Let L/K be a finite Galois extension of complete, discrete valuation fields with Galois group, $G(L/K)$, and inertia group, $G_0(L/K) = G(L/M)$, as in 6.1.1. Let $\chi : G(L/K) \longrightarrow \mathbf{C}^*$ be a one-dimensional representation, then*

$$sw_K(\chi) = sw_M(Res^{G(L/K)}_{G_0(L/K)}(\chi)).$$

Proof In the classical case this result is 6.1.25.

By 6.1.5, M/K is unramified (so that $m_M = \mathcal{O}_M m_K$) and $\overline{M}/\overline{K}$ is the maximal separable subextension of $\overline{L}/\overline{K}$. Also complete, discrete valuation rings are henselian (Milne, 1980, p. 32) so that 6.2.21 applies to yield the result. □

6.2.23 Let L/K be a finite Galois extension of complete, discrete valuation fields. Let T be an indeterminate so that we may form the localisation with respect to π_L, $\mathcal{O}_L[T]_{(\pi_L)}$, of the polynomial ring, $\mathcal{O}_L[T]$.

The field of fractions of the henselianisation of $\mathcal{O}_K[T]_{(\pi_K)}$ was used in the definition of the Swan conductor in 6.2.10. We shall now study the effect of that henselianisation process upon the extension, L/K. We will see that the role of the henselianisations in 6.2.10 is to provide a Galois extension with the same Galois group but with more cohomological structure than the original extension.

Let $\hat{L}$ denote the field of fractions of $\mathcal{O}_L[T]_{(\pi_L)}$. An element of the completed ring is clearly of the form

$$z = \sum_{i=0}^{\infty} a_i T^i \quad (a_i \in \mathcal{O}_L;\ v_L(a_i) \to \infty \ as \ i \to \infty).$$

If $a_0 \in \mathcal{O}_L^* = \mathcal{O}_L - m_L$ then z is invertible. The field $\hat{L}$ consists of Laurent series of the form

$$w = \sum_{i=-n}^{\infty} a_i T^i \quad (a_i \in L;\ v_L(a_i) \to \infty \ as \ i \to \infty).$$

For this ring is clearly a field in which the inverse of w is given by $(a_{-n} \neq 0)$ $w^{-1} = T^n(a_{-n})^{-1}(1 + \sum_{i=-n+1}^{\infty} a_i(a_{-n})^{-1}T^i)^{-1}$. Hence $\hat{L} = \mathcal{O}_L[T]_{(\pi_L)}[(\pi_L T)^{-1}]$.

Proposition 6.2.24 *Let $Aut_{\hat{K}}(\hat{L})$ denote the group of $\hat{K}$-linear field automorphisms of $\hat{L}$. Every $\hat{\phi} \in Aut_{\hat{K}}(\hat{L})$ is of the form*

$$\hat{\phi}\left(\sum_{i=-n}^{\infty} a_i T^i\right) = \sum_{i=-n}^{\infty} \phi(a_i) T^i$$

for some $\phi \in G(L/K)$.

Proof Let $\hat{\phi} : \hat{L} \longrightarrow \hat{L}$ be a $\hat{K}$-algebra isomorphism. Since $T \in \hat{K}$ we have $\hat{\phi}(T) = T$ and therefore $\hat{\phi}$ is determined by $\hat{\phi} : L \longrightarrow \hat{L}$. For $a \in L^*$ suppose that

$$\phi(a) = T^{-j(a)}(\phi_0(a) + \phi_1(a)T + \ldots + \phi_m(a)T^m + \ldots),$$

where $\phi_i(a) \in L$ and $\phi_0(a) \neq 0$. Hence we have a map, $j : L^* \longrightarrow \mathbf{Z}$ such that $j(ab) = j(a) + j(b)$ and $j(K^*) = 0$. Now, clearly, j is trivial on the roots of unity in L^* and on $U_1 = 1 + \pi_L \mathcal{O}_L$ since the former consists of torsion and the latter is divisible by integers which are prime to the residue characteristic. Finally, $j(\pi_L) = 0$ since $\pi_L^{e_{L/K}} \in K^*/(K^* \cap U_L^1) \subset L^*/U_L^1$ so that $e_{L/K} j(\pi_L) = 0$. Hence

$$\phi(a) = \phi_0(a) + \phi_1(a)T + \ldots + \phi_m(a)T^m + \ldots,$$

with $\phi_0 \in G(L/K)$.

We must show that if $\phi_1, \phi_2, \ldots, \phi_{s-1}$ are all identically zero then $\phi_s = 0$, too.

Assume that $char(L) = 0$ and take $x \in L^*$ such that $\phi_s(x) \neq 0$. Let

$$f(T) = t^r + \alpha_1 t^{r-1} + \ldots + \alpha_0 \in K[T]$$

be the irreducible polynomial of x. Hence $f(x) = 0$ and

$$0 = \phi_s(f(x)) = \phi_s(x^r) + \phi_s(\alpha_1 t^{r-1}) + \ldots + \phi_s(\alpha_0).$$

However, since $\phi_s(ab) = \phi_0(a)\phi_s(b) + \phi_s(a)\phi_0(b)$ and $\phi_s(\alpha_i) = 0$, we find that

$$0 = \phi_s(x)\frac{df}{dt}(\phi_0(x)).$$

Therefore $\phi_0(x)$ is a repeated root of $f(T)$, which is impossible.

Now suppose that $char(L) = p$, then $L = L^p$, since L/K is Galois. Therefore the derivation property of ϕ_s implies that it vanishes on L^p, which completes the proof. □

Corollary 6.2.25 *If K is a complete, discrete valuation field let $M(K)$ denote the field of fractions of the henselianisation, $(\mathcal{O}_K[T]_{(\pi_K)})^{(h)}$, which was introduced in 6.2.7. If L/K is a finite Galois extension of complete, discrete valuations fields then $M(L)/M(K)$ is a finite Galois extension and restriction yields a natural isomorphism of Galois groups*

$$G(L/K) \cong G(M(L)/M(K)).$$

Proof By the construction of the henselianisation of a local ring (Milne, 1980, p. 37), the field, $M(L)$, is constructible as the limit of finite Galois extensions of $\hat{L}$ of 6.2.24 for which the residue field extension is trivial. By 6.2.24, we have a split extension of Galois groups

$$H \longrightarrow G(M(L)/\hat{K}) \longrightarrow G(\hat{L}/\hat{K}) \cong G(L/K).$$

Finally, one observes that the $G(L/K)$-fixed points of the integral closure of $\mathcal{O}_{\hat{L}}$ in $M(L)$ is a henselianisation of $\mathcal{O}_{\hat{K}}$ from which the result follows, since $M(K)$ is the field of fractions of this henselianisation. □

Theorem 6.2.26 *Let L/K be a finite Galois extension of complete, discrete valuation fields and suppose that $G(L/K) = G_0(L/K)$ in 6.1.1. Suppose that $G(L/M) \lhd G(L/K)$ is a normal subgroup whose index is prime to the residue characteristic, $0 \neq p = char(\overline{K})$.*

Let $\chi : G(L/K) \longrightarrow \mathbf{C}^*$ be a one-dimensional representation, then

$$[M : K]sw_K(\chi) = sw_M(Res_{G(L/M)}^{G(L/K)}(\chi)),$$

where sw_K is the Swan conductor of 6.2.10.

Proof Set $e = [M : K] = [G(L/K) : G(L/M)]$. Since $\#(G(L/K)) = \#(G_0(L/K)) = e_{L/K} f_{L/K}^{ins}$ and $HCF(e, p) = 1$ we must have $f_{L/K}^{ins} = f_{L/M}^{ins}$ and $e = e_{M/K}$. Therefore there exists $u \in \mathcal{O}_M^*$ such that $\pi_K = \pi_M^e u \in \mathcal{O}_M$. Therefore, if we apply the norm, N, to $1+\pi_M^{ne}\alpha \in 1+\pi_M^{ne}\mathcal{O}_M[T]$ we obtain

$$\begin{aligned} N(1 + \pi_M^{ne}\alpha) &= \textstyle\prod_{g \in G(M/K)}(1 + g(\pi_K^n u^{-n}\alpha)) \\ &= \textstyle\prod_{g \in G(M/K)}(1 + \pi_K^n g(u^{-n}\alpha)), \end{aligned}$$

which lies in $1 + \pi_K^n \mathcal{O}_K[T]$.

Hence we have the inclusion, i, and the norm map which induce homomorphisms

6.2.27 $$\left\{ \frac{(1+\pi_K^n \mathcal{O}_K[T])}{(1+\pi_K^{n+1}\mathcal{O}_K[T])} \stackrel{i}{\longrightarrow} \left(\frac{1+\pi_M^{ne}\mathcal{O}_M[T]}{1+\pi_M^{ne+e}\mathcal{O}_M[T]} \right)^{G(M/K)} \stackrel{N}{\longrightarrow} \frac{(1+\pi_K^n \mathcal{O}_K[T])}{(1+\pi_K^{n+1}\mathcal{O}_K[T])}. \right.$$

The groups in 6.2.27 are annihilated by p so that the formulae, $i(N(z)) = z^e$ and $N(i(w)) = w^e$, imply that i and N are isomorphisms in 6.2.27. In 6.2.27 the left-hand group is isomorphic to $\overline{K}[T]$.

Since $f_{M/K} = f^{ins}_{M/K} f^{sep}_{M/K} = 1$ we have $\overline{K} = \overline{M}$. Setting

$$B_i = \frac{1 + \pi_M^{ne+i}\mathcal{O}_M[T]}{1 + \pi_M^{ne+e}\mathcal{O}_M[T]}$$

we see that B_0 has a composition series of the form

$$0 = B_e \subset B_{e-1} \subset \ldots \subset B_0,$$

with $B_i/B_{i+1} \cong \overline{K}[T]$ for $0 \leq i \leq e-1$. Let $z \in B_1$ be represented by $1 + \pi_M^{ne+1}\alpha \in 1 + \pi_M^{ne+1}\mathcal{O}_M[T]$. Thus

$$\begin{aligned} N(z) &= \textstyle\prod_{g \in G(M/K)}(1 + \pi_K^n g(\pi_M u^{-n}\alpha)) \\ &= 1 + \pi_K^n \beta, \end{aligned}$$

where $\beta \in \mathcal{O}_K[T] \subset \mathcal{O}_M[T]$ vanishes in $\overline{M}[T]$. Since $\overline{K} = \overline{M}$ we see that β vanishes in $\overline{K}[T]$ and $N(z) \in 1 + \pi_K^{n+1}\mathcal{O}_K[T]$. Hence, in 6.2.27, the norm vanishes on elements whose representatives lie in B_1.

Now take $\chi : G(L/K) \longrightarrow \mathbf{C}^*$. In the notation of 6.2.25, suppose that $\{\chi, 1 + \pi_K^n T\} \neq 0$ and that $\{\chi, 1 + \pi_K^{n+1} T\} = 0$. Temporarily let $tors_p(A)$ denote the p-torsion subgroup of A. By Kato (1989, section 5.5(i)), $\{\chi, 1 + \pi_K^n T\} \in tors_p(H^2(M(K)))$. Since $[M : K]$ is prime to p

$$i : tors_p(H^2(M(K))) \longrightarrow tors_p(H^2(M(M)))$$

is injective and therefore

$$0 \neq i(\{\chi, 1 + \pi_K^n T\}) = \{Res^{G(L/K)}_{G(L/M)}(\chi), 1 + \pi_M^{ne} u^n T\} \in H^2(M(M)).$$

Therefore

$$sw_M(Res^{G(L/K)}_{G(L/M)}(\chi)) \geq ne = sw_K(\chi)[M : K].$$

Now we must show that $0 = \{Res^{G(L/K)}_{G(L/M)}(\chi), 1 + \pi_M^{ne+1} T\}$.

Taking the cup-product with $Res^{G(L/K)}_{G(L/M)}(\chi)$ induces a homomorphism of the form

$$B_0 = \frac{1 + \pi_M^{ne}\mathcal{O}_M[T]}{1 + \pi_M^{ne+e}\mathcal{O}_M[T]} \longrightarrow H^2(M(M)).$$

By the previous discussion $1 + \pi_M^{ne+1} T$ is in the kernel of the norm, $N : B_0 \longrightarrow B_0$. Since $HCF(\#(G(L/K)), p) = 1$, $H^1(G(L/K); B_0) = 0$ and therefore there exists a class, $w \in B_0$, such that

$$1 + \pi_M^{ne+1} T = \gamma(w)w^{-1},$$

where $\gamma \in G(L/K)$ is a generator. Therefore, by bimultiplicativity of the cup-product,

$$\begin{aligned}
&\{Res_{G(L/M)}^{G(L/K)}(\chi), 1 + \pi_M^{ne+1} T\} \\
&= \{Res_{G(L/M)}^{G(L/K)}(\chi), \gamma(w)\} - \{Res_{G(L/M)}^{G(L/K)}(\chi), w\} \\
&= \{\gamma(Res_{G(L/M)}^{G(L/K)}(\chi)), \gamma(w)\} - \{Res_{G(L/M)}^{G(L/K)}(\chi), w\} \\
&= \gamma(\{Res_{G(L/M)}^{G(L/K)}(\chi), w\}) - \{Res_{G(L/M)}^{G(L/K)}(\chi), w\}.
\end{aligned}$$

Next we observe that, since the henselianisation, $(K \mapsto M(K))$, preserves the triviality of residue field extensions and preserves Galois groups (by 6.2.25) we have $\overline{M(K)} = \overline{M(M)}$. Therefore, for each n, there is a commutative diagram of natural homomorphisms (cf. 6.2.15) which are defined in Kato (1987, section 3.5.5):

$$\begin{array}{ccc}
H_n^2(\overline{M(K)}) \oplus H_n^1(\overline{M(K)}) & \xrightarrow{\lambda} & H_n^2(M(K)) \\
\downarrow (1,e) & & \downarrow i \\
H_n^2(\overline{M(K)}) \oplus H_n^1(\overline{M(K)}) & \xrightarrow{\lambda'} & H_n^2(M(M))
\end{array}$$

One may see that the left vertical map may be taken to be $(1, e)$ by means of the formula of Kato (1987, section 3.5.5):

$$\lambda(x, y) = i_2(x) + \{i_1(y), \pi_{M(K)}\},$$

where $i_q : H^q(\overline{M(K)}) \longrightarrow H^q(M(K))$ is the natural map.

Now let $n = p^a$ for some (large) integer, $a \geq e$. The extension $M(M)/M(K)$ falls under case I of theorem 3.6 of Kato (1987) and the formula given there shows that $\{Res^{G(L/K)}_{G(L/M)}(\chi), w\} \in im(\lambda')$ and hence $\{Res^{G(L/K)}_{G(L/M)}(\chi), w\} \in im(i)$. Since the image of i is $G(M/K)$-invariant

$$\gamma(\{Res^{G(L/K)}_{G(L/M)}(\chi), w\}) = \{Res^{G(L/K)}_{G(L/M)}(\chi), w\},$$

which completes the proof of Theorem 6.2.26. □

6.3 The non-abelian Swan conductor

Throughout this section let L/K be a finite, Galois extension of complete, discrete valuation fields, as in 6.2.4. In particular, we do *not* assume that the residue field extension is separable. Let $G(L/K)$ denote the Galois group. Under these circumstances we shall construct a non-abelian Swan conductor which generalises the conductors of 6.1.16 and 6.2.11.

Definition 6.3.1 Let G be a finite group. Let $R_+(G)$ denote, as in 2.2.1, the free abelian group on G-conjugacy classes of characters, $\phi : H \longrightarrow \mathbf{C}^*$ where $H \leq G$. Define a natural ring homomorphism

$$\eta_G : R_+(G) \longrightarrow R_+(G)$$

by the formula $\eta_G((H,\phi)^G) = (H,1)^G$. The image of η_G is called the *Burnside ring* of G and the homomorphism, η_G, retracts $R_+(G)$ onto the Burnside ring. Let $\epsilon : R(G) \longrightarrow \mathbf{Z}$ denote the homomorphism which sends a representation to its dimension and set

$$IR(G) = Ker(\epsilon).$$

Theorem 6.3.2 *There is a natural homomorphism*

$$\tilde{a}_G : IR(G) \longrightarrow Ker(\eta_G)$$

defined by

$$\tilde{a}_G(v - dim(v)) = a_G(v) - \eta_G(a_G(v)) \in R_+(G).$$

In addition,

$$b_G(\tilde{a}_G(v - dim(v))) = v - dim(v) \in R(G).$$

Proof This is an immediate consequence of 2.2.42 and 2.3.2.

Definition 6.3.3 (*Definition of sw_K*) We may define an additive homomorphism

$$\hat{sw}_K : Ker(\eta_{G(L/K)}) \longrightarrow \mathbf{Z}$$

by $(K \leq K' \leq L)$

6.3.4
$$\hat{sw}_K((G(L/K'), \phi)^{G(L/K)} - (G(L/K'), 1)^{G(L/K)}) \\ = f^{sep}_{K'/K} \cdot sw_{K'}(\phi)$$

where $sw_{K'}(\phi)$ is as in 6.2.10 and $f^{sep}_{K'/K}$ is the separable residue degree of K'/K. Define

6.3.5
$$sw_K : R(G(L/K)) \longrightarrow \mathbf{Z}$$

to be the homomorphism which is trivial on the trivial representations and on $IR(G(L/K))$ is given by the composition

$$sw_K : IR(G(L/K)) \stackrel{\tilde{a}_{G(L/K)}}{\longrightarrow} Ker(\eta_{G(L/K)}) \stackrel{\hat{sw}_K}{\longrightarrow} \mathbf{Z}.$$

The following is the main result of this section:

Theorem 6.3.6 *Let L/K be a finite, Galois extension of complete, discrete valuation fields with group, $G(L/K)$. The conductor homomorphism, sw_K, defined in* 6.3.3, *possesses the following properties:*

(i) *When $\chi : G(L/K) \longrightarrow \mathbf{C}^*$ is a one-dimensional representation then $sw_K(\chi)$ coincides with the Kato conductor of* 6.2.10.

(ii) *When $\overline{L}/\overline{K}$ is separable (i.e. in the classical case) then sw_K coincides with the classical Swan conductor of* 6.1.16.

(iii) *If $K \leq F \leq L$ is a chain of Galois extensions and $\lambda : G(L/K) \longrightarrow G(F/K)$ is the canonical surjection then*

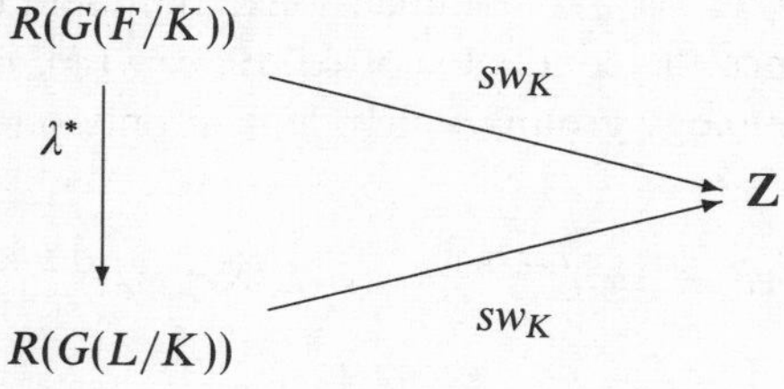

commutes, where $\lambda^* = Inf^{G(L/K)}_{G(F/K)}$.

(iv) *Let $\Omega_{\mathbf{Q}}$ denote the absolute Galois group of the rationals. Hence $\Omega_{\mathbf{Q}}$ acts on $R(G(L/K))$ by means of its action on character values. If $\omega \in \Omega_{\mathbf{Q}}$ and $\chi \in R(G(L/K))$ then*

$$sw_K(\omega(\chi)) = sw_K(\chi).$$

(v) *If* $G(L/M) = G_0 \lhd G(L/K)$ *is as in* 6.1.1 *then*

$$sw_K(\chi) = sw_M(Res^{G(L/K)}_{G_0}(\chi))$$

for all $\chi \in R(G(L/K))$.

(vi) *If* $G(L/K) = G_0(L/K)$ *and* $G(L/M) = G_1(L/K)$ *then, for all* $\chi \in R(G(L/K))$,

$$[M : K] sw_K(\chi) = sw_M(Res^{G(L/K)}_{G(L/M)}(\chi)).$$

Proof If χ is one-dimensional then $sw_K(\chi - 1)$

$$= \hat{sw}_K(\tilde{a}_{G(L/K)}(\chi - 1))$$

$$= \hat{sw}_K((G(L/K), \chi)^{G(L/K)} - (G(L/K), 1)^{G(L/K)}), \quad \text{by 2.3.2(ii)},$$

$$= sw_K(\chi),$$

which proves part (i).

In the classical case, by 6.1.24 and 6.2.13,

$$\hat{sw}_K((G(L/K'), \phi)^{G(L/K)} - (G(L/K'), 1)^{G(L/K)})$$

$$= sw_K(Ind^{G(L/K)}_{G(L/K')}(\phi - 1)),$$

where sw_K is the classical Swan conductor. Hence

$$\hat{sw}_K = sw_K b_{G(L/K)} : Ker(\eta_{G(L/K)}) \longrightarrow \mathbf{Z},$$

which, by 2.3.2, proves part (ii).

Suppose that $\lambda : G(L/K) \longrightarrow G(F/K)$ is the canonical surjection and that $\phi : G(F/K') \longrightarrow \mathbf{C}^*$ is a one-dimensional representation. Then $sw_{K'}(\phi) = sw_{K'}(\lambda\phi)$, since the conductor of 6.2.10, $sw_{K'}(\phi)$, is defined by means of Galois cohomology groups which depend only on K'. Also, in $R_+(G(L/K))$,

$$Inf^{G(L/K)}_{G(F/K)}((G(F/K'), \phi)^{G(F/K)}) = (G(L/K'), \phi\lambda)^{G(L/K)},$$

so that

$$\hat{sw}_K(Inf^{G(L/K)}_{G(F/K)}((G(F/K'), \phi)^{G(F/K)} - (G(F/K'), 1)^{G(F/K)}))$$

$$= \hat{sw}_K((G(L/K'), \phi\lambda)^{G(L/K)} - (G(L/K'), 1)^{G(L/K)}).$$

Part (iii) follows from the naturality (with respect to inflation maps) of $\tilde{a}_G$ in Theorem 6.3.2.

Clearly, the $\Omega_{\mathbf{Q}}$-action does not alter the Swan conductor of 6.2.10, $sw_K(\phi)$. Part (iv) follows from the fact that, in Theorem 2.3.2, a_G commutes with the $\Omega_{\mathbf{Q}}$-action. This is seen by appealing to the fact that a_G is uniquely characterised by the two properties of 2.3.2. Hence $a_G(-) = \omega^{-1}(a_G(\omega(-)))$ for $\omega \in \Omega_{\mathbf{Q}}$, since both homomorphisms fulfil the characterisation.

To prove (v) we first note that, by naturality with respect to restriction (in the sense of 2.3.2(i)) of a_G, $\tilde{a}_G$ and η_G, it suffices to show that

$$\hat{sw}_K((G(L/F),\phi)^{G(L/K)} - (G(L/F),1)^{G(L/K)})$$

$$= \hat{sw}_M(Res_{G_0}^{G(L/K)}((G(L/F),\phi)^{G(L/K)} - (G(L/F),1)^{G(L/K)})),$$

where $G_0 = G(L/M)$. By 2.2.3

$$Res_{G_0}^{G(L/K)}((G(L/F),\phi)^{G(L/K)} - (G(L/F),1)^{G(L/K)})$$

$$= \sum\nolimits_{z\in G_0\backslash G(L/K)/G(L/F)}(G_0 \cap zG(L/F)z^{-1},(z^{-1})^*(\phi))^{G_0} \\ -(G_0 \cap zG(L/F)z^{-1},1)^{G_0},$$

where $(z^{-1})^*(\phi) = \phi(z^{-1} - z)$. By definition 6.3.3,

$$\hat{sw}_M((G_0 \cap zG(L/F)z^{-1},(z^{-1})^*(\phi))^{G_0} - (G_0 \cap zG(L/F)z^{-1},1)^{G_0})$$

$$= \hat{sw}_M((G(L/Mz(F)),(z^{-1})^*(\phi))^{G_0} - (G_0 \cap G(L/Mz(F)),1)^{G_0})$$

$$= f^{sep}_{Mz(F)/M} \cdot sw_{Mz(F)}(Res^{G(L/z(F))}_{G(L/Mz(F))}((z^{-1})^*(\phi)))$$

$$= f^{sep}_{MF/M} sw_{MF}(Res^{G(L/F)}_{G(L/MF)}(\phi)),$$

since it is clear from the definition of $sw_K(\chi)$ that

$$sw_{Mz(F)}(Res^{G(L/z(F))}_{G(L/Mz(F))}((z^{-1})^*(\phi))) = sw_{MF}(Res^{G(L/F)}_{G(L/MF)}(\phi)).$$

Now, by Kato (1989, lemma 6.2); (see also 6.2.22), since $\overline{M}/\overline{K}$ is separable,

$$\begin{aligned} sw_F(\phi : G(L/F) \longrightarrow \mathbf{C}^*) &= sw_{MF}(Res^{G(L/F)}_{G(L/F)\cap G_0}(\phi)) \\ &= sw_{MF}(Res^{G(L/F)}_{G(L/MF)}(\phi)). \end{aligned}$$

Hence

$$\hat{sw}_K((G(L/F),\phi)^{G(L/K)} - (G(L/F),1)^{G(L/K)})$$

$$= f^{sep}_{F/K} \cdot sw_F(\phi : G(L/F) \longrightarrow \mathbf{C}^*)$$

$$= f^{sep}_{F/K} \cdot sw_{MF}(Res^{G(L/F)}_{G(L/MF)}(\phi)).$$

Finally,

$$\begin{aligned}
\#(G_0\backslash G(L/K)/G(L/F)) &= [G(L/K) : G(L/MF)] \\
&= [G(L/K) : G_0]([G(L/MF) : G_0])^{-1} \\
&= f^{sep}_{L/K}([G(L/F) : G_0 \cap G(L/F)])^{-1} \\
&= f^{sep}_{L/K}(f^{sep}_{L/F})^{-1} \qquad \text{(by 6.1.5)} \\
&= f^{sep}_{F/K},
\end{aligned}$$

so that, by the preceding discussion,

$$\hat{sw}_M(Res^{G(L/K)}_{G_0}((G(L/F),\phi)^{G(L/K)} - (G(L/F),1)^{G(L/K)}))$$

$$= f^{sep}_{F/K}(sw_{MF}(Res^{G(L/F)}_{G(L/MF)}(\phi))),$$

which completes the proof of part (v).

As in part (v), in order to prove (vi) it suffices by naturality to show that

$$\hat{sw}_K((G(L/F),\phi)^{G(L/K)} - (G(L/F),1)^{G(L/K)})[M : K]$$

$$= \hat{sw}_M(Res^{G(L/K)}_{G_1}((G(L/F),\phi)^{G(L/K)} - (G(L/F),1)^{G(L/K)})).$$

By a computation similar to that used in the proof of part (v), the latter expression is equal to

$$\#(G_1\backslash G(L/K)/G(L/F))sw_{MF}(Res_{G_1(L/F)}^{G(L/F)}(\phi))$$

$$= [G_0(L/K) : G_1(L/K)]([G_0(L/MF) : G_1(L/MF)])^{-1} \cdot sw_{MF}(Res_{G_1(L/F)}^{G(L/F)}(\phi))$$

$$= [M : K]sw_F(\phi) \qquad \text{by 6.2.26}$$

$$= [M : K]f_{F/K}^{sep} sw_F(\phi)$$

$$= [M : K]\hat{sw}_K((G(L/F), \phi)^{G(L/K)} - (G(L/F), 1)^{G(L/K)}),$$

as required. This completes the proof of Theorem 6.3.6. □

Definition 6.3.7 Let L/K be a finite Galois extension of complete, discrete valuation fields.

We shall say that the Swan conductor, sw_K, of 6.3.3 is *inductive in dimension zero* on $R(G(L/K))$ if, for all $G(L/F) \leq G(L/K)$ and $\chi \in R(G(L/F))$,

$$sw_K(Ind_{G(L/F)}^{G(L/K)}(\chi - dim(\chi))) = f_{F/K}^{sep} sw_F(\chi - dim(\chi)).$$

6.3.8 In the examples of 6.2.18, 6.2.19 and 6.2.20 we saw that it is not possible in general to construct a conductor function which extends the abelian conductor of 6.2.10, is inflative in the sense of 6.3.6(iii) and which is inductive in dimension zero, in the sense of 6.3.7. Nevertheless, the remainder of this section will be devoted to proving (Theorem 6.3.20) that the obstruction to inductivity in dimension zero for sw_K lies with p-groups which are of the form $\mathbf{Z}/p^n$ or $\mathbf{Z}/p^n \times \mathbf{Z}/p$, where p is the residue characteristic. That is, we shall show that sw_K is inductive in dimension zero on $R(G(L/K))$ if and only if sw_F is inductive in dimension zero on $R(G(L/F))$ for each p-subgroup, $G(L/F)$, of the above form.

We will begin by reducing the question to the case of p-groups.

Proposition 6.3.9 *Let L/K be a finite Galois extension of complete, discrete valuation fields and suppose that $G_0 = G(L/M) \leq G(L/K)$. If sw_M is inductive in dimension zero on $R(G(L/M))$, in the sense of* 6.3.7, *then sw_K is inductive in dimension zero on $R(G(L/K))$.*

Proof If $x \in IR(G(L/F))$ then (if $G_0 = G(L/M) = G_0(L/M)$)

$$sw_K(Ind_{G(L/F)}^{G(L/K)}(x))$$

$$= sw_M(Res_{G(L/M)}^{G(L/K)} Ind_{G(L/F)}^{G(L/K)}(x)) \qquad \text{by } 6.3.6(v)$$

$$= \sum_{z \in G_0 \backslash G(L/K)/G(L/F)} sw_M(Ind_{G_0 \cap G(L/z(F))}^{G_0}((z^{-1})^*(Res_{G(L/MF)}^{G(L/F)}(x))))$$

$$= \sum_{z \in G_0 \backslash G(L/K)/G(L/F)} sw_{Mz(F)}((z^{-1})^*(Res_{G_0(L/F)}^{G(L/F)}(x)))$$

$$= [G(L/K) : G_0G(L/F)] sw_{MF}(Res_{G_0(L/F)}^{G(L/F)}(x))$$

$$= f_{F/K}^{sep} \cdot sw_F(x),$$

by hypothesis, since the equality $[G(L/K) : G_0G(L/F)] = f_{F/K}^{sep}$ was established in the course of the proof of 6.3.6(v). □

Proposition 6.3.10 *Let L/K be a finite Galois extension of complete, discrete valuation fields and suppose that $G_1 = G(L/M) \leq G(L/K) = G_0(L/K)$. If sw_M is inductive in dimension zero on $R(G(L/M))$, in the sense of* 6.3.7, *then sw_K is inductive in dimension zero on $R(G(L/K))$.*

Proof If $x \in IR(G(L/F))$ then (if $G_1 = G(L/M)$)

$$[M : K] sw_K(Ind_{G(L/F)}^{G(L/K)}(x))$$

$$= sw_M(Res_{G(L/M)}^{G(L/K)} Ind_{G(L/F)}^{G(L/K)}(x)) \qquad \text{by } 6.3.6(vi)$$

$$= \sum_{z \in G_1 \backslash G(L/K)/G(L/F)} sw_M(Ind_{G_1 \cap G(L/z(F))}^{G_1}((z^{-1})^*(Res_{G(L/MF)}^{G(L/F)}(x))))$$

$$= \sum_{z \in G_1 \backslash G(L/K)/G(L/F)} sw_{Mz(F)}((z^{-1})^*(Res_{G_1(L/F)}^{G(L/F)}(x)))$$

$$= [G(L/K) : G_1G(L/F)] sw_{MF}(Res_{G_1(L/F)}^{G(L/F)}(x))$$

$$= [M : K][G_0(L/MF) : G_1(L/MF)]^{-1} sw_{MF}(Res_{G_1(L/F)}^{G(L/F)}(x))$$

$$= [M : K] sw_F(x) \qquad \text{by } 6.3.6(\text{vi})$$

$$= [M : K] f_{F/K}^{sep} \cdot sw_F(x),$$

by hypothesis, since the equality

$$[G(L/K) : G_1G(L/F)] = [M : K][G_0(L/MF) : G_1(L/MF)]^{-1}$$

was established in the course of the proof of 6.3.6(vi). □

6.3.11 In order, as described in 6.3.8, to reduce the question of the inductivity of sw_K to abelian p-groups of small rank we shall construct an inductive, rational-valued conductor

6.3.12 $$SW_{G(L/K)} : R(G(L/K)) \longrightarrow \mathbf{Q}$$

using the homomorphism, $d_{G(L/K)}$, of 2.4.1–2.4.12 to replace $\tilde{a}_{G(L/K)}$ in the construction of sw_K in 6.3.5.

It is important to notice that $SW_{G(L/K)}$, constructed in this manner, will automatically be inductive in degree zero, in the sense of 6.3.7, but will *not necessarily be inflative* in the sense of 6.3.6(iii).

Lemma 6.3.13 *Let d_G denote the homomorphism of* 2.4.1–2.4.12

$$d_G : R(G) \longrightarrow R_+(G) \otimes \mathbf{Q}.$$

Then, in the notation of 6.3.1,

$$d_G(IR(G)) \subseteq Ker(\eta_G) \otimes \mathbf{Q}.$$

Proof This follows from the formula of 2.4.14(ii) since, if $dim(\chi) = 0$ and $K \leq H$ are cyclic, $Ind_K^H(Res_K^G(\chi))$ may be expressed as $\sum_i(\phi_i - \psi_i)$ where $\phi_i, \psi_i : H \longrightarrow \mathbf{C}^*$ are one-dimensional representations. □

Definition 6.3.14 *Definition of $SW_{G(L/K)}$* Let L/K be a finite, Galois extension of complete, discrete valuation fields. Define a rational-valued homomorphism

$$SW_{G(L/K)} : R(G(L/K)) \longrightarrow \mathbf{Q}$$

to be the homomorphism which annihilates trivial representations and on $IR(G(L/K))$ is equal to the composition

$$R(G(L/K)) \stackrel{d_{G(L/K)}}{\longrightarrow} Ker(\eta_{G(L/K)}) \otimes \mathbf{Q} \stackrel{\hat{sw}_K \otimes 1}{\longrightarrow} \mathbf{Q},$$

where $\hat{sw}_K$ is the homomorphism of 6.3.4. This definition makes sense by virtue of 6.3.13.

Proposition 6.3.15 *In 6.3.14, $SW_{G(L/K)}$ possesses the following properties:*

(i) *Let $\Omega_{\mathbf{Q}}$ denote the absolute Galois group of the rationals. Hence $\Omega_{\mathbf{Q}}$ acts on $R(G(L/K))$ by means of its action on character values. If $\omega \in \Omega_{\mathbf{Q}}$ and $\chi \in R(G(L/K))$, then*

$$SW_{G(L/K)}(\omega(\chi)) = SW_{G(L/K)}(\chi).$$

(ii) *If $G(L/M) = G_0 \lhd G(L/K)$ is as in 6.1.1, then*

$$SW_{G(L/K)}(\chi) = SW_{G(L/M)}(Res_{G_0}^{G(L/K)}(\chi))$$

for all $\chi \in R(G(L/K))$.

(iii) *If $G(L/K) = G_0(L/K)$ and $G(L/M) = G_1(L/K)$, then, for all $\chi \in R(G(L/K))$,*

$$[M : K]SW_{G(L/K)}(\chi) = SW_{G(L/M)}(Res_{G(L/M)}^{G(L/K)}(\chi)).$$

(iv) *$SW_{G(L/K)}$ is inductive in dimension zero, in the sense of 6.3.7.*

Proof The proofs of (i)–(iii) are the same as those for the corresponding results ((iv)–(vi) respectively) of 6.3.6. For part (iv) it suffices, by 2.4.1(i), to observe that for $G(L/F) \le G(L/K)$

$$f_{F/K}^{sep}\hat{sw}_F((G(L/N),\phi)^{G(L/F)} - (G(L/N),1)^{G(L/F)})$$

$$= f_{F/K}^{sep} f_{N/F}^{sep} sw_N(\phi)$$

$$= f_{N/K}^{sep} sw_N(\phi)$$

$$= \hat{sw}_K((G(L/N),\phi)^{G(L/K)} - (G(L/N),1)^{G(L/K)})$$

$$= \hat{sw}_K(Ind_{G(L/F)}^{G(L/K)}((G(L/N),\phi)^{G(L/F)} - (G(L/N),1)^{G(L/F)})).$$

□

6.3.16 Let $p_1, \ldots, p_r$ be distinct primes and let

$$C = \mathbf{Z}/p_1^{n_1} \times \ldots \times \mathbf{Z}/p_r^{n_r}$$

be a cyclic group. Write $C^0 \le C$ for the subgroup

$$C = \mathbf{Z}/p_1^{n_1-1} \times \ldots \times \mathbf{Z}/p_r^{n_r-1},$$

with the convention that $\{1\}^0 = \{1\}$. Recall that the *Euler totient function*, $\phi(n)$, is the multiplicative function on the positive integers which is characterised by the formula $\phi(p^m) = p^{m-1}(p-1)$ on prime powers (Hunter, 1964). Hence

$$\phi(n) = \#\{m \mid 1 \leq m \leq n, HCF(n,m) = 1\}.$$

Lemma 6.3.17 *Let G be any finite group. With the notation of* 6.3.16

(i) $d_G(1)$

$$= \#(G)^{-1} \sum_{\substack{(C,\psi) \\ C cyclic, Res^C_{C_0}(\psi)=1}} \#(C^0)\phi(\#(Ker(\psi)/C^0))\mu_{Ker(\psi),C}(C,\psi)^G$$

and

(ii) *for all* $\mu : G \longrightarrow \mathbf{C}^*$

$$d_G(\mu) = \#(G)^{-1} \sum_{\substack{(C,\psi) \\ C cyclic, Res^C_{C_0}(\psi)=1}} \#(C^0)\phi(\#(Ker(\psi)/C^0))$$

$$\mu_{Ker(\psi),C}(C, \psi Res^G_C(\mu))^G.$$

Proof Clearly part (ii) follows from part (i), by 2.4.1(ii). Part (i) follows from the formula of 2.4.14(ii) upon noticing that $\mu_{K,H} = 0$ unless $[H : K]$ is square-free when one collects the remaining terms, using the familiar identity (Hunter, 1964)

$$\sum_{H^0 \leq K \leq Ker(\psi)} \#(K/H^0)\mu_{K,Ker(\psi)} = \phi(\#(Ker(\psi)/H^0)).$$

□

In the following result it is essential that G be a p-group. It is for this reason that we have reduced the study of sw_K to the case when $G(L/K) = G_1(L/K)$ by means of 6.3.6(v) and (vi).

Proposition 6.3.18 *Let G be a finite p-group. Let* $N \lhd G$ *be a normal subgroup and let* $\mu : G \longrightarrow \mathbf{C}^*$ *be a one-dimensional representation which factors through* G/N. *Then* $d_G(\mu - 1)$ *is a* **Q**-*linear combination of pairs,* $(H,\psi)^G$ *with* $Res^H_{H \cap N}(\psi) = 1$.

Proof By the formula of 6.3.17 we have $d_G(\mu - 1) =$

$$\#(G)^{-1} \sum_{\substack{(C,\psi) \\ C cyclic, Res^C_{C_0}(\psi)=1}} \#(C^0)\phi(\#(Ker(\psi)/C^0))\mu_{Ker(\psi),C}X(C,\psi),$$

where $X(C,\psi) = [(C, \psi Res^G_C(\mu))^G - (C,\psi)^G]$.

Now consider the pair $(C, \psi)^G$. If $C \cap N = C$ then $\mu\psi = \psi$ on C, since μ is trivial on N. If $C \cap N \neq C$ then $C \cap N \leq C^0$, since C is a cyclic p-group, and hence $\mu\psi$ and ψ are both trivial on $C \cap N$. This completes the proof of 6.3.18. □

Proposition 6.3.19 *Let L/K be a finite, Galois extension of complete, discrete valuation fields. Suppose that, for all one-dimensional representations of subgroups $\psi : G(L/F) \longrightarrow \mathbf{C}^*$*

$$SW_{G(L/F)}(\psi - 1) = sw_F(\psi - 1),$$

then

$$SW_{G(L/K)} = sw_K : R(G(L/K)) \longrightarrow \mathbf{Z} \subset \mathbf{Q}.$$

Proof Let $\chi \in R(G(L/K))$ and suppose that

$$a_{G(L/K)}(\chi) = \sum_i a_i (G(L/F_i), \phi_i)^{G(L/K)}.$$

Therefore
$sw_K(\chi - dim(\chi))$

$$= \sum_i a_i \hat{sw}_K((G(L/F_i), \phi_i)^{G(L/K)} - (G(L/F_i), 1)^{G(L/K)})$$

$$= \sum_i a_i f^{sep}_{F_i/K} sw_{F_i}(\phi_i)$$

$$= \sum_i a_i f^{sep}_{F_i/K} sw_{F_i}(\phi_i - 1)$$

$$= \sum_i a_i f^{sep}_{F_i/K} SW_{G(L/F_i)}(\phi_i - 1)$$

$$= \sum_i a_i SW_{G(L/K)}(Ind^{G(L/K)}_{G(L/F_i)}(\phi_i - 1))$$

by 6.3.15(iv)

$$= SW_{G(L/K)}(b_{G(L/K)}(\tilde{a}_{G(L/K)}(\chi - dim(\chi))))$$

by 6.3.2

$$= SW_{G(L/K)}(\chi - dim(\chi)),$$

as required. □

Theorem 6.3.20 *Let L/K be a finite, Galois extension of complete, discrete valuation fields. Let $p = char(\overline{K})$ and suppose that $G(L/K)$ is a p-group. Suppose that we have subgroups, $G(L/M) \lhd G(L/F) \leq G(L/K)$, which satisfy the following conditions:*

(i) *If $G(M/F) \cong G(L/F)/G(L/M) \cong \mathbf{Z}/p^n$ then*

$$sw_F(Ind_{G(M/N)}^{G(M/F)}(\theta - 1)) = f_{N/F}^{sep} sw_N(\theta - 1),$$

where $\mathbf{Z}/p^{n-1} \cong G(M/N) = G(M/F)^0 \leq G(M/F)$ and $\theta : G(M/N) \longrightarrow \mathbf{C}^*$ is a faithful one-dimensional representation.

(ii) *If $G(M/F) \cong G(L/F)/G(L/M) \cong \mathbf{Z}/p^n \times \mathbf{Z}/p$ then*

$$sw_F(Ind_{G(M/N)}^{G(M/F)}(\theta - 1)) = f_{N/F}^{sep} sw_N(\theta - 1),$$

where $\mathbf{Z}/p^n = G(M/N) \leq G(M/F)$ and $\theta : G(M/N) \longrightarrow \mathbf{C}^*$ is a faithful one-dimensional representation.

Then, for all $G(L/R) \leq G(L/K)$ and $\lambda : G(L/R) \longrightarrow \mathbf{C}^*$,

$$SW_{G(L/R)}(\lambda - 1) = sw_R(\lambda - 1).$$

Corollary 6.3.21 *Under the hypotheses of* 6.3.20, *sw_K is inductive in dimension zero on $R(G(L/K))$ in the sense of* 6.3.7.

Proof This follows from 6.3.19 and 6.3.20. □

6.3.22 *Proof of Theorem* 6.3.20

Suppose that $G(L/F) \lhd G(L/K)$ is a cyclic, normal subgroup of order p. Suppose also that $\lambda : G(L/K) \longrightarrow \mathbf{C}^*$ is trivial on $G(L/F)$. We will begin by showing that

6.3.23 $$SW_{G(L/K)}(\lambda - 1) = SW_{G(F/K)}(\overline{\lambda} - 1),$$

where, in 6.3.23, $\overline{\lambda} : G(F/K) \cong G(L/K)/G(L/F) \longrightarrow \mathbf{C}^*$ is induced by λ.

By 6.3.14 and 6.3.17,

6.3.24

$$\#(G(L/K))SW_{G(L/K)}(\lambda - 1)$$

$$= \sum_{G(L/T)} \sum_{\psi} \#(G(L/T)^0)\phi(\#(Ker(\psi)/G(L/T)^0))$$

$$\mu_{Ker(\psi),G(L/T)} f^{sep}_{T/K} sw_T(\lambda\psi - \psi)$$

$$= \sum_{G(L/T)} \#(G(L/T)^0) f^{sep}_{T/K} [(p-1) sw_T(Res^{G(L/K)}_{G(L/T)}(\lambda - 1))$$

$$- \sum_{\psi} sw_T(\lambda\psi - \psi)]$$

where, in 6.3.24, the first sum is taken over cyclic subgroups, $G(L/T)$, such that $\{1\} \neq G(L/T) \leq G(L/K)$ and over $\psi : G(L/T)/G(L/T)^0 \longrightarrow \mathbf{C}^*$, and the second sum is taken over cyclic subgroups, $G(L/T)$, and faithful representations, $\psi : G(L/T)/G(L/T)^0 \cong \mathbf{Z}/p \longrightarrow \mathbf{C}^*$.

Similarly,

6.3.25

$$\#(G(L/K))SW_{G(F/K)}(\overline{\lambda} - 1)$$

$$= p \sum_{G(F/S)} \#(G(F/S)^0) f^{sep}_{S/K} [(p-1) sw_S(Res^{G(L/K)}_{G(L/S)}(\lambda - 1))$$

$$- \sum_{\psi} sw_S(\lambda\psi - \psi)],$$

where the sum is taken over $G(L/F) \leq G(L/S) \leq G(L/K)$ such that $G(F/S)$ is non-trivial, cyclic and $\psi : G(L/S)/G(L/S)^0 \longrightarrow \mathbf{C}^*$ is a faithful one-dimensional representation.

Hence, with the summation as in 6.3.24 and 6.3.25, we must prove that

6.3.26

$$\sum_{G(L/T)} \#(G(L/T)^0) f^{sep}_{T/K} sw_T((Res^{G(L/K)}_{G(L/T)}(\lambda - 1))(p - Ind^{G(L/T)}_{G(L/T)^0}(1)))$$

$$= \sum_{G(F/S)} p\#(G(F/S)^0) f^{sep}_{S/K} sw_S((Res^{G(L/K)}_{G(L/S)}(\lambda - 1))(p - Ind^{G(F/S)}_{G(F/S)^0}(1))).$$

In order to verify the identity of 6.3.26, consider the following map between the indexing sets of the two sums (note that the term with $T = F$

vanishes in 6.3.24):

$$\{\{1\} \neq G(L/T) \leq G(L/K) \mid G(L/T) \text{ cyclic, } T \neq F\}$$

$$\downarrow \Gamma$$

$$\{G(L/F) \leq G(L/S) \leq G(L/K) \mid F \neq S, G(F/S) \text{ cyclic}\},$$

given by the formula

6.3.27 $$\Gamma(G(L/T)) = G(L/T)G(L/F) = G(L/T \cap F).$$

We wish to show that Γ is surjective. We claim that

6.3.28 $$\begin{cases} \text{either} \quad (a) \quad G(L/S) \quad \text{is} \quad \text{cyclic} \\ \text{or} \quad (b) \quad G(L/S) = G(L/F) \times C \end{cases}$$

for some cyclic subgroup, C. For if $G(L/S)$ is not cyclic then the cyclic group, $G(L/S)/G(L/F) \cong G(F/S)$ acts trivially on $G(L/F)$, being a p-group, so that $G(L/S)$ is abelian and must be isomorphic to $\mathbf{Z}/p \times \mathbf{Z}/p^n$ for some n. An examination of the lattice of subgroups of $\mathbf{Z}/p \times \mathbf{Z}/p^n$ shows that each subgroup of order p has a complement and therefore $G(L/S) \cong G(L/F) \times C$, as claimed. This means that Γ is surjective.

Now we will prove that (noting that $p\#(G(F/S)^0) = \#(G(L/S))p^{-1}$)

6.3.29 $$\frac{\#(G(L/S))}{p\#(G(L/T)^0)} sw_S(Res^{G(L/K)}_{G(L/S)}(\lambda - 1)(p - Ind^{G(L/S)}_{G(L/U)}(1)))$$

$$= \sum_{\Gamma(G(L/T))=G(L/S)} f^{sep}_{T/K} sw_T(Res^{G(L/K)}_{G(L/T)}(\lambda - 1)(p - Ind^{G(L/T)}_{G(L/T)^0}(1))),$$

where $G(L/U)$ is the pre-image of $G(F/S)^0$ under $G(L/S) \longrightarrow G(F/S)$. Note that $S = F \cap T$ under the map of 6.3.27 and, as we shall explain, $\#(G(L/T)^0)$ is constant for all T such that $\Gamma(G(L/T)) = G(L/S)$, so that dividing by $\#(G(L/T)^0)$ makes sense in 6.3.29.

In case (a) of 6.3.28 $G(L/S)$ is cyclic and $\Gamma^{-1}(G(L/S))$ consists only of $G(L/S)$ so that $T = S$, $\#(G(L/S)) = p\#(G(L/T)^0)$, $G(L/U) = G(L/T)^0$ and both sides of 6.3.29 coincide.

In case (b) of 6.3.28, $\Gamma^{-1}(G(L/S))$ consists of the p complements of $G(L/F)$ in $G(L/S)$, $G(L/E_1), \ldots, G(L/E_p)$, say. Hence $G(L/E_i) \cong \mathbf{Z}/p^n$ for $1 \leq i \leq p$. Therefore, by part (ii) of the hypothesis, the sum over $G(L/T)$ of 6.3.29 becomes (where $G(L/U) \cap G(L/E_i) = G(L/UE_i)$ is independent of i and

$$(\#(G(L/E_i)))^{-1}\#(G(L/S))p^{-1} = p^{1-n}p^{n+1}p^{-1} = p$$

for all i)

$$\sum_{i=1}^{p} f_{E_i/K}^{sep} sw_{E_i}((Res_{G(L/E_i)}^{G(L/K)}(\lambda-1))(p - Ind_{G(L/UE_i)}^{G(L/E_i)}(1)))$$

$$= \sum_{i=1}^{p} f_{S/K}^{sep} sw_S((Res_{G(L/S)}^{G(L/K)}(\lambda-1))(pInd_{G(L/E_i)}^{G(L/S)}(1) - Ind_{G(L/UE_i)}^{G(L/S)}(1)))$$

$$= pf_{S/K}^{sep} sw_S((Res_{G(L/S)}^{G(L/K)}(\lambda-1))(\sum_{i=1}^{p} Ind_{G(L/E_i)}^{G(L/S)}(1) - Ind_{G(L/UE_p)}^{G(L/S)}(1))).$$

Now $G(L/S)/G(L/UE_i) \cong \mathbf{Z}/p \times \mathbf{Z}/p$ and

$$\sum_{i=1}^{p} Ind_{G(L/E_i)}^{G(L/S)}(1) = p + \sum_{\psi} \psi \in R(G(L/S)),$$

where $\psi : G(L/S)/G(L/UE_i) \cong G(UE_i/S) \longrightarrow \mathbf{C}^*$ runs over all characters whose kernels belong to the set $\{G(L/E_i)\}$. On the other hand,

$$Ind_{G(L/UE_p)}^{G(L/S)}(1) = \sum_{\rho} \rho \in R(G(L/S)),$$

where ρ runs through all $\rho : G(UE_i/S) \longrightarrow \mathbf{C}^*$. Hence the difference between these two expressions equals

$$p - Ind_{G(L/U)}^{G(L/S)}(1) \in R(G(L/S))$$

and we find that the $G(L/T)$-sum in 6.3.29 reduces to

$$pf_{S/K}^{sep} sw_S(Res_{G(L/S)}^{G(L/K)}(\lambda-1)(p - Ind_{G(L/U)}^{G(L/S)}(1))),$$

as required to verify 6.3.29 and 6.3.26, and thereby to establish 6.3.23.

Since sw_K is invariant under inflation, in the sense of 6.3.23, we may filter the kernel of $\lambda : G(L/K) \longrightarrow \mathbf{C}^*$ by subgroups

$$G(L/F_0) \lhd G(L/F_1) \lhd \ldots \lhd G(L/F_t) = Ker(\lambda) \lhd G(L/K)$$

in such a manner that $[F_{i+1} : F_i] = p$ for all i. By induction on the order of $G(L/K)$ we may therefore reduce to showing that

6.3.30 $$SW_{G(L/K)}(\lambda - 1) = sw_K(\lambda - 1)$$

in the case when $G(L/K)$ is a cyclic p-group. If $\#(G(L/K))$ is equal to 1 or p then 6.3.30 is immediate from the definition. Also we may suppose that $\lambda : G(L/K) \longrightarrow \mathbf{C}^*$ is faithful. Suppose that $G(L/K) \cong \mathbf{Z}/p^n$ with

$n \geq 2$ and let $\overline{\lambda} : G(L/F) \cong \mathbf{Z}/p^{n-1} \longrightarrow \mathbf{C}^*$ be a faithful representation. Hence

$$Ind_{G(L/F)}^{G(L/K)}(\overline{\lambda} - 1) = \sum_{i=1}^{p} \omega_i(\lambda) - \sum_{i=1}^{p} \psi_i,$$

where $\omega_1, \ldots, \omega_p \in \Omega_{\mathbf{Q}}$ and $\psi_1, \ldots, \psi_p : G(L/K) \longrightarrow \mathbf{C}^*$ are trivial on $G(L/F)$. Therefore, by 6.3.6(iv), 6.3.15(i), 6.3.20(i) and induction,

$$\begin{aligned}
&psw_K(\lambda - 1) - \textstyle\sum_{i=1}^{p} SW_{G(L/K)}(\psi_i - 1) \\
&= psw_K(\lambda - 1) - \textstyle\sum_{i=1}^{p} sw_K(\psi_i - 1) \\
&= sw_K(Ind_{G(L/F)}^{G(L/K)}(\overline{\lambda} - 1)) \\
&= f_{F/K}^{sep} sw_F(\overline{\lambda} - 1) \\
&= f_{F/K}^{sep} SW_{G(L/F)}(\overline{\lambda} - 1) \\
&= SW_{G(L/K)}(Ind_{G(L/F)}^{G(L/K)}(\overline{\lambda} - 1)) \\
&= pSW_{G(L/K)}(\lambda - 1) - \textstyle\sum_{i=1}^{p} SW_{G(L/K)}(\psi_i - 1),
\end{aligned}$$

which proves 6.3.30 and completes the proof of Theorem 6.3.20. □

6.4 Exercises

6.4.1 Prove the formula of Theorem 6.1.43.

6.4.2 Let Q_{2^m} ($m \geq 3$) denote the generalised quaternion group of order 2^m:

$$Q_{2^m} = \{x, y \mid x^{2^{m-2}} = y^2, y^4 = 1, xyx = y\}.$$

Suppose that L/K is a finite Galois extension of 2-adic local fields with group $G(L/K) = G_0(L/K) \cong Q_{2^m}$. Let $A_{G(L/K)} \in R(Q_{2^m})$ denote the Artin representation of 6.1.7.

(i) Prove that, in the sense of 4.3.31,

$$A_{G(L/K)} - \psi^2(A_{G(L/K)}) \equiv 0 \quad (mod\ 2).$$

(ii) For $z \in Z_2[Q_{2^m}]^*$ prove that, in the sense of 4.3.38,

$$Det(z)(3\psi^2(A_{G(L/K)}) - \psi^4(A_{G(L/K)}) - 2A_{G(L/K)}) \in \{\pm 1\} + 4\mathcal{O}_N,$$

where N/K is a finite Galois extension which is large enough to contain all the 2^{m-1}th roots of unity.

6.4.3 In 6.1.24 prove that the Artin conductor, f_K, is inductive in dimension zero if and only if the same is true of the Swan conductor, sw_K. That is, show that

$$f_K(Ind_{G(L/K')}^{G(L/K)}(\chi - dim(\chi))) = (f_{K'/K}) \cdot f_{K'}(\chi - dim(\chi))$$

holds for all $\chi \in R(G(L/K'))$ if and only if

$$sw_K(Ind_{G(L/K')}^{G(L/K)}(\chi - dim(\chi))) = (f_{K'/K}) \cdot sw_{K'}(\chi - dim(\chi)).$$

6.4.4 Let $s_{G(L/K)}$ and $i_{G(L/K)}$ be the ramification functions of 6.1.3.

(i) In the classical case, find an element at which these infima are attained.

(ii) Verify the formula of 6.1.4.

6.4.5 Complete the verification of the table of ramification functions and conductors in 6.2.20 (Table 6.5) and verify the other conductor calculations which are omitted in that example.

6.4.6 (*Research problem*) Let L/K be a finite, Galois extension of complete, discrete valuation fields. If $K \leq F \leq L$ is a subextension, find a formula for

$$sw_K(Ind_{G(L/F)}^{G(L/K)}(1)) \in \mathbf{Z},$$

where sw_K is as in 6.3.3.

6.4.7 (*Research problem*) Let p be a prime. Let L/K be a finite, Galois extension of complete, discrete valuation fields in characteristic p. Is the conductor, sw_K, of 6.3.3 inductive in dimension zero modulo $(p-1)$?

A weaker question would be the following: is sw_K inductive in dimension zero modulo $(p-1)$ when K is a higher dimensional local field, in the sense of Kato (1987)?

7

Galois module structure

Introduction

In Section 1 we consider a finite Galois extension of local fields, L/K. The theory of local class formations asserts the existence of a *fundamental class* which gives a canonical generator for $H^2(G(L/K); L^*)$. To each generator of this cyclic group and to each cohomologically trivial subgroup, $U \leq L^*$, such that L^*/U is a finitely generated $\mathbf{Z}[G(L/K)]$-module we show how to assign an Euler characteristic in the class-group. When applied to the canonical generator this gives the local Chinburg invariant, $\Omega(L/K, U)$, which lies in the class-group, $\mathcal{CL}(\mathbf{Z}[G(L/K)])$. We develop sufficient cohomological technology to be able to describe $\Omega(L/K, U)$ canonically. This is accomplished by means of a description of the fundamental class, which is due to J-P. Serre, together with a homological construction which is given in terms of a K-division algebra containing L as a maximal subfield. We use this canonical description to give a new proof of the theorem of Chinburg which states that $\Omega(L/K, U_L^1)$ is trivial when L/K is tamely ramified.

In Section 2 we define the global Chinburg invariant, $\Omega(L/K, 2)$ which lies in $\mathcal{CL}(\mathbf{Z}[G(L/K)])$, where L/K is a Galois extension of number fields. This was originally described as another type of Euler characteristic construction. Here we describe $\Omega(L/K, 2)$ in terms of local Chinburg invariants at the wildly ramified primes in a manner which is due to S. Kim. In addition we show how to use the Artin root numbers of the symplectic representations of $G(L/K)$ to construct a 2-torsion element, $W_{L/K} \in \mathcal{CL}(\mathbf{Z}[G(L/K)])$. In the tamely ramified case the Chinburg invariant becomes simply the projective module given by the ring of integers of L, which was conjectured by Fröhlich to be equal to the analytic Cassou-Noguès–Fröhlich class, $W_{L/K}$. This conjecture was proved by M.J. Taylor. The Fröhlich–Chinburg conjecture asserts that $\Omega(L/K,2)$ and $W_{L/K}$ should be equal in general.

To date, the best general result concerning the Fröhlich–Chinburg conjecture is David Holland's result, which shows that the conjecture holds when mapped to the class-group of a maximal order of the rational group-ring. In Section 3 Holland's result is proved in a new manner by means of the detection machinery which was developed in Chapter 5, Section 4.

Section 4 contains a derivation of the Hom-description for the global Chinburg invariant, $\Omega(\mathbf{Q}(\xi_{p^{s+1}})/\mathbf{Q},2)$, of the totally real cyclotomic extension of p-power conductor when p is an odd, regular prime. These Chinburg invariants had not been considered previously and the appearance of the p-adic L-function in their Hom-description is particularly satisfying and probably significant. Using naturality under passage to quotient groups the Chinburg invariant, $\Omega(F_{s+1}/\mathbf{Q},2)$, is shown to vanish, where $F_{s+1}/\mathbf{Q}$ is the intermediate extension having Galois group of order p^s. This provides further examples in which the Fröhlich–Chinburg conjecture of 7.2.14 is true. Unlike the material of Section 3, these calculations are not derived by the use of Explicit Brauer Induction, but they seem particularly appropriate for inclusion into this chapter.

Section Five consists of a collection of exercises concerning class formations, cohomology and the Fröhlich–Chinburg conjecture.

7.1 Local Chinburg invariants

Let G be a finite group and suppose that A,B are $\mathbf{Z}[G]$-modules. We will begin by recalling some homological algebra. Suppose that

7.1.1 $$\ldots \longrightarrow P_n \stackrel{d}{\longrightarrow} \ldots P_1 \stackrel{d}{\longrightarrow} P_0 \stackrel{\epsilon}{\longrightarrow} A \longrightarrow 0$$

is a *projective* $\mathbf{Z}[G]$*-resolution* of A. Hence each P_i is a projective $\mathbf{Z}[G]$-module and the $\mathbf{Z}[G]$-homomorphisms of 7.1.1 satisfy

$$dd = 0, \;\; \epsilon d = 0$$

and 7.1.1 is *exact* (i.e. the kernel of each map equals the image of its predecessor). The group, $Ext^i_{\mathbf{Z}[G]}(A,B)$, is defined to be the ith homology group of the chain complex

7.1.2 $$\ldots \stackrel{d^*}{\longleftarrow} Hom_{\mathbf{Z}[G]}(P_n,B) \stackrel{d^*}{\longleftarrow} \ldots \stackrel{d^*}{\longleftarrow} Hom_{\mathbf{Z}[G]}(P_0,B) \longleftarrow 0.$$

Hence

7.1.3
$$Ext^i_{\mathbf{Z}[G]}(A,B) = \frac{ker(d^* : Hom_{\mathbf{Z}[G]}(P_i,B) \longrightarrow Hom_{\mathbf{Z}[G]}(P_{i+1},B))}{im(d^* : Hom_{\mathbf{Z}[G]}(P_{i-1},B) \longrightarrow Hom_{\mathbf{Z}[G]}(P_i,B))},$$

where we adopt the convention that $P_i = 0$ if $i < 0$. Up to a canonical isomorphism, this definition is independent of the choice of resolution in 7.1.1.

The ith *cohomology group*, $H^i(G;B)$, is defined (Snaith, 1989b, p.2) to be given by

7.1.4
$$H^i(G;B) = Ext^i_{\mathbf{Z}[G]}(\mathbf{Z},B),$$

where G acts trivially on $\mathbf{Z}$, the integers.

A canonical projective resolution of $\mathbf{Z}$ is given by the *bar resolution* (Snaith, 1989b, p. 3)

7.1.5
$$\ldots \xrightarrow{d_2} B_2G \xrightarrow{d_1} B_1G \xrightarrow{d_0} B_0G \xrightarrow{\epsilon} 0.$$

In 7.1.5 B_nG is the free left $\mathbf{Z}[G]$-module on G^n. If $(g_1,\ldots,g_n) \in G^n$ we write $[g_1 \mid g_2 \mid \ldots \mid g_n]$ for the corresponding $\mathbf{Z}[G]$-basis element of B_n. We write $[\;]$ for the basis element of B_0G. The $\mathbf{Z}[G]$-homomorphisms of 7.1.5 are given by

$$\epsilon(g_1[\;]) = 1, \qquad \text{and}$$

$$\begin{aligned} d_n([g_1 \mid g_2 \mid \ldots \mid g_{n+1}]) &= g_1[g_2 \mid \ldots \mid g_{n+1}] + \\ &\quad \textstyle\sum_{i=1}^n (-1)^i [g_1 \mid \ldots \mid g_ig_{i+1} \mid \ldots \mid g_{n+1}] + \\ &\quad (-1)^{n+1}[g_1 \mid g_2 \mid \ldots \mid g_n]. \end{aligned}$$

One may show that 7.1.5 is exact by constructing a contracting homotopy

$$0 \longrightarrow \mathbf{Z} \xrightarrow{\eta} B_0G \xrightarrow{s_0} B_1G \xrightarrow{s_1} \ldots$$

given by the formulae

$$\eta(1) = [\;], \qquad \text{and}$$

$$s_n(g_1[g_2 \mid \ldots \mid g_n]) = [g_1 \mid g_2 \mid \ldots \mid g_n] \text{ for } n \geq 0.$$

One readily verifies the following identities

$$1 = \epsilon\eta,$$

$$1 = \eta\epsilon + d_0 s_0 \qquad \text{and}$$

$$1 = s_{n-1}d_{n-1} + d_n s_n \qquad \text{for } n \geq 1.$$

In this section we will mainly be interested in $H^2(G; B)$. For future reference we will record the description of $H^2(G; B)$ which the bar resolution provides.

Proposition 7.1.6 *Let B be a left $\mathbf{Z}[G]$-module (written in additive notation). Then*

$$H^2(G; B) \cong Z_2/B_2,$$

where

$$Z_2 = \left\{ f : G \times G \longrightarrow B \mid \begin{array}{l} g_1 f(g_2, g_3) + f(g_1, g_2 g_3) \\ = f(g_1 g_2, g_3) + f(g_1, g_2) \\ \textit{for all } g_1, g_2, g_3 \in G \end{array} \right\}$$

and

$$B_2 = \left\{ f : G \times G \longrightarrow B \mid \begin{array}{l} f(g_1, g_2) = g_1 h(g_2) + h(g_1) - h(g_1 g_2) \\ \textit{for all } g_1, g_2 \in G \textit{ and} \\ \textit{for some } h : G \longrightarrow B \end{array} \right\}.$$

(Z_2 and B_2 are called the groups of 2-cocycles *and* 2-coboundaries, *respectively.)*

Now suppose that L/K is a finite Galois extension of local fields with group $G(L/K)$. The multiplicative group, L^*, is a $\mathbf{Z}[G(L/K)]$-module and there is a canonical isomorphism (Serre, 1979, pp. 166 and 195):

7.1.7 $$inv : H^2(G(L/K); L^*) \xrightarrow{\cong} \mathbf{Z}/[L : K] \subset \mathbf{Q}/\mathbf{Z}.$$

An explicit description of the generator, $inv^{-1}(-[L : K]^{-1})$ may be given by means of the fact (cf. 5.2.20 and 5.2.21) that $H^2(G(L/K); L^*)$ classifies extensions of the form

$$0 \longrightarrow L^* \longrightarrow A \longrightarrow B \longrightarrow \mathbf{Z} \longrightarrow 0$$

and generators of $H^2(G(L/K); L^*)$ classify those extensions for which A and B are *cohomologically trivial* $\mathbf{Z}[G]$-modules (i.e. $H^i(H; M) = 0$ for all $i > 0$ and $H \leq G$ when $M = A, B$).

The following construction is taken from Serre (1979, p. 202).

Let W/K be the maximal unramified subextension of L/K. Let K_{nr}/K denote the maximal unramified extension so that $G_0(K_{nr}/K) = \{1\}$ and there are isomorphisms

$$G(K_{nr}/K) \cong G(\overline{K_{nr}}/\overline{K}) \cong G(F_q^c/F_q),$$

where F_q is the finite field of order q and F_q^c is its algebraic closure. Let $F \in G(K_{nr}/K)$ correspond to the Frobenius element (see 4.3.7) given by the qth power map on F_q^c. Set $L_0 = K_{nr}L$. Consider the sequence

7.1.8 $$0 \longrightarrow L^* \xrightarrow{i} (L \otimes_K K_{nr})^* \xrightarrow{(1\otimes F)/1} (L \otimes_K K_{nr})^* \xrightarrow{w} \mathbf{Z} \longrightarrow 0.$$

In 7.1.8, if $\alpha \in L, \beta \in K_{nr}$,

7.1.9 $$(1 \otimes F)(\alpha \otimes \beta) = \alpha \otimes F(\beta).$$

If $d = [W : K]$, there is an isomorphism of K-algebras

7.1.10 $$\lambda : L \otimes_K K_{nr} \longrightarrow \oplus_{i=1}^d L_0$$

given by the formula

$$\lambda(\alpha \otimes \beta) = (F^{d-1}(\beta)\alpha, F^{d-2}(\beta)\alpha, \ldots, \beta\alpha).$$

The map, w in 7.1.8, is given by

7.1.11 $$w = (\textstyle\sum_{i=1}^d v_{L_0}) \cdot \lambda : (L \otimes_K K_{nr})^* \longrightarrow \oplus_{i=1}^d L_0^* \longrightarrow \mathbf{Z}.$$

Now let us consider the Galois action in terms of 7.1.10. We have a map, given by restriction of the action of g to W (denoted by $(g \mid W)$)

7.1.12 $$\begin{cases} h_1 : G(K_{nr}/K) \times G(L/K) \longrightarrow G(W/K) \\ h_1(F^i, z) = (F^i \mid W)(z \mid W)^{-1}. \end{cases}$$

If $(F^i, z) \in Ker(h_1)$ then we may define an element of $G(L_0/K)$ which is equal to F^i on K_{nr} and to z on L. This induces an isomorphism

7.1.13 $$Ker(h_1) \xrightarrow{\cong} G(L_0/K).$$

Define an element

7.1.14
$$\begin{cases} F_0 \in G(L_0/K) \\ \text{by } (F_0 \mid K_{nr}) = F^d \text{ and } (F_0 \mid L) = 1. \end{cases}$$

Lemma 7.1.15 *Define*

$$(1 \otimes F) : \oplus_{i=1}^d L_0 \longrightarrow \oplus_{i=1}^d L_0$$

by $(1 \otimes F)(x_1, \ldots, x_d) = (F_0(x_d), x_1, \ldots, x_{d-1})$, *then the following diagram commutes:*

$$\begin{array}{ccc} L \otimes_K K_{nr} & \xrightarrow{(1 \otimes F)} & L \otimes_K K_{nr} \\ \downarrow \lambda & & \downarrow \lambda \\ \oplus_{i=1}^d L_0 & \xrightarrow{(1 \otimes F)} & \oplus_{i=1}^d L_0 \end{array}$$

Proof If $\alpha \in L$ and $\beta \in K_{nr}$ then

$$\begin{aligned} \lambda(1 \otimes F)(\alpha \otimes \beta) &= \lambda(\alpha \otimes F(\beta)) \\ &= (F^d(\beta)\alpha, F^{d-1}(\beta)\alpha, \ldots, F(\beta)\alpha), \end{aligned}$$

while

$$\begin{aligned} (1 \otimes F)(\lambda(\alpha \otimes \beta)) &= (1 \otimes F)(F^{d-1}(\beta)\alpha, F^{d-2}(\beta)\alpha, \ldots, \beta\alpha) \\ &= (F_0(\beta\alpha), F^{d-1}(\beta)\alpha, \ldots, F(\beta)\alpha) \end{aligned}$$

and the result follows since, by definition,

$$F_0(\beta\alpha) = F_0(\beta)\alpha = F^d(\beta)\alpha.$$

□

Lemma 7.1.16 *The sequence of* 7.1.8 *is exact.*

Proof Let $\pi_L \in \mathcal{O}_L$ denote the uniformiser of L. Hence there exists $u \in (L \otimes_K K_{nr})^*$ such that $\lambda(u) = (\pi_L, 1, 1, \ldots, 1)$ and therefore $w(u) = v_{L_0}(\pi_L) = 1$, by 7.1.11, so that w is onto. By means of 7.1.15, $(1 \otimes F)/1$ may be identified with the map

$$(1 \otimes F)/1 : \oplus_{i=1}^d L_0^* \longrightarrow \oplus_{i=1}^d L_0^*$$

given by the formula

7.1.17
$$(1 \otimes F)/1(x_1, \ldots, x_d) =$$
$$(F_0(x_d)x_1^{-1}, x_1x_2^{-1}, x_2x_3^{-1}, \ldots, x_{d-1}x_d^{-1}).$$

Hence the kernel of $(1 \otimes F)/1$ consists of d-tuples of the form $(y, y, \ldots, y)$, where $y \in L_0^*$ and $F_0(y) = y$ or, equivalently, $y \in L^*$. Since $\lambda(i(y)) = \lambda(y \otimes 1) = (y, y, \ldots, y)$ we see that $Ker((1 \otimes F)/1) = Im(i)$.

Finally, suppose that $\alpha \in (L \otimes_K K_{nr})^*$ satisfies $w(\alpha) = 0$. Let $\lambda(\alpha) = (x_1, \ldots, x_d)$ with $x_i \in L_0$. By 7.1.11, $\sum_{i=1}^d v_{L_0}(x_i) = 0$ so that $x = \prod_{i=1}^d x_i \in \mathcal{O}_{L_0}^*$, the units of $\mathcal{O}_{L_0}$. We claim that there exists $u \in \mathcal{O}_{L_0}^*$ such that $F_0(u)u^{-1} = x$. Let us assume this fact for the moment. Consider the equation

$$\begin{aligned} &(F_0(y_d)y_1^{-1}, y_1y_2^{-1}, \ldots, y_{d-1}y_d^{-1}) \\ &= (1 \otimes F)/1(y_1, \ldots, y_d) \qquad \text{by 7.1.17} \\ &= (x_1, \ldots, x_d) \end{aligned}$$

We may solve this by choosing $y_d = u$ and then selecting $y_1, y_2, \ldots, y_{d-1}$ to satisfy $y_iy_{i+1}^{-1} = x_{i+1}$ for $1 \leq i \leq d-1$. With these choices the first coordinate becomes

$$\begin{aligned} F_0(y_d)y_1^{-1} &= F_0(u)y_1^{-1} \\ &= u(\textstyle\prod_{i=1}^d x_i)y_1^{-1} \\ &= y_dx_1y_1y_2^{-1}y_2y_3^{-1} \ldots y_{d-1}y_d^{-1}y_1^{-1} \\ &= x_1, \end{aligned}$$

as required. Therefore, to complete the proof, we must construct $u \in \mathcal{O}_{L_0}^*$.

There is a finite intermediate Galois extension, E/K, such that $K \subset W \subset L \subset E \subset L_0$ with $x \in \mathcal{O}_E^*$. Let $\overline{E}$ denote the residue field of E and let $\overline{x} \in \overline{E}^*$ denote the image of x. We have $\overline{W} = \overline{L} = F_q$, $\overline{L_0} = \overline{K_{nr}} = F_q^c$ and F_0 induces the Frobenius on $\overline{E}/F_q$. Since $q^d - 1$ is prime to q we may take a (q^d-1)th root of $\overline{x}$ in $\overline{E}$, by making E sufficiently large. Therefore choose $\overline{u_1} \in \overline{E}^*$ such that

$$\overline{u_1}^{q^d-1} = F_0(\overline{u_1})(\overline{u_1})^{-1} = \overline{x}.$$

We may lift $\overline{u_1}$ to $u_1 \in \mathcal{O}_E^*$ to obtain an equation of the form

$$F_0(u_1)u_1^{-1} = x + \pi x_1$$

where $x_1 \in \mathcal{O}_E$ and $\pi = \pi_{L_0} = \pi_L$ denotes the prime element. Now suppose that we have found $x_m \in \mathcal{O}_E$ and $u_m \in \mathcal{O}_E^*$ such that, for $m \geq 1$,

$$F_0(u_m)u_m^{-1} = x + \pi^m x_m = x(1 + \pi^m x_m x^{-1}).$$

Consider the congruences ($v \in \mathcal{O}_E$)

$$\begin{aligned} &F_0(1+\pi^m v)(1+\pi^m v)^{-1} \\ &\equiv (1+\pi^m F_0(v))(1-\pi^m v) \quad (mod\ \pi^{m+1}\mathcal{O}_E) \\ &\equiv 1 + \pi^m(F(v) - v) \qquad (mod\ \pi^{m+1}\mathcal{O}_E). \end{aligned}$$

By 7.5.2, enlarging E if necessary, we can find $v \in \mathcal{O}_E$ such that

$$F(v) - v \equiv x_m x^{-1} \ (mod\ \pi\mathcal{O}_E).$$

Replacing u_m by $u_{m+1} = u_m(1 + \pi^m v) \in \mathcal{O}_E^*$ we obtain $x_{m+1} \in \mathcal{O}_E$ such that

$$F_0(u_{m+1})u_{m+1}^{-1} = x + \pi^{m+1}x_{m+1}.$$

The sequence, $\{u_m\}$, converges π-adically to $u \in \mathcal{O}_{L_0}^*$ such that

$$F_0(u)u^{-1} = x,$$

which completes the proof of 7.1.16. □

7.1.18 Given a short exact sequence of $\mathbf{Z}[G]$-modules

$$0 \longrightarrow A \xrightarrow{\alpha} B \xrightarrow{\beta} C \longrightarrow 0$$

there is a natural coboundary map (Snaith, 1989b, p. 9)

$$\delta : H^i(G; C) \longrightarrow H^{i+1}(G; A)$$

which fits into a long exact sequence of the form

$$\ldots \longrightarrow H^i(G;A) \xrightarrow{\alpha_*} H^i(G;B) \xrightarrow{\beta_*} H^i(G;C) \xrightarrow{\delta} H^{i+1}(G;A) \longrightarrow \ldots.$$

Therefore, when B is cohomologically trivial, δ is an isomorphism for $i \geq 1$. In terms of the bar resolution δ is describable in the following manner. Let $[g] \in H^i(G; C)$ be represented by a $\mathbf{Z}[G]$-homomorphism

$$g : B_i G \longrightarrow C$$

such that $gd_i = 0$. Since B_iG is a free $\mathbf{Z}[G]$-module we may choose a $\mathbf{Z}[G]$-homomorphism

$$h : B_iG \longrightarrow B$$

such that $\beta h = g$. The homomorphism

$$hd_i : B_{i+1}G \longrightarrow B_iG \longrightarrow B$$

becomes trivial when composed with β, since $\beta hd_i = gd_i = 0$. Hence we obtain a $\mathbf{Z}[G]$-homomorphism

$$H : B_{i+1}G \longrightarrow A = Ker(\beta)$$

such that $\alpha H = hd_i$. Also H is an $(i+1)$-cocycle, since $\alpha H d_{i+1} = hd_i d_{i+1} = 0$ and α is injective. We define $\delta[g]$ by the formula

7.1.19 $$\delta[g] = [H] \in H^{i+1}(G; A).$$

In particular, when we have an exact sequence of the form of 7.1.8, we may derive two coboundary maps

$$\delta : H^0(G; \mathbf{Z}) \longrightarrow H^1(G; Ker(w)) = H^1(G; Im((1 \otimes F)/1))$$

and

$$\delta : H^1(G; Im((1 \otimes F)/1)) \longrightarrow H^2(G; L^*).$$

There is an isomorphism $H^0(G; \mathbf{Z}) \cong \mathbf{Z}$ in which $n \in \mathbf{Z}$ is represented by the homomorphism, $g_n : B_0G \longrightarrow \mathbf{Z}$, given by $g_n([\ ]) = n$. Therefore 7.1.8 gives rise to a class

$$\delta(\delta(1)) = \delta(\delta([g_1])) \in H^2(G(L/K); L^*) \cong \mathbf{Z}/\#(G).$$

Theorem 7.1.20 (*Serre*, 1979, p. 202) *In the exact sequence of* 7.1.8, *the module* $(L \otimes_K K_{nr})^*$ *is cohomologically trivial and the equivalence class of the extension of* 7.1.8 *corresponds to* $\delta(\delta(1)) \in H^2(G(L/K); L^*)$. *Furthermore, in* 7.1.7,

$$\delta(\delta(1)) = -inv^{-1}([L : K]^{-1}).$$

Proof By the theory of the classification of 2-extensions by means of cohomology 7.1.8 corresponds to $\delta(\delta(1))$ (see Hilton & Stammbach, 1971).

The remainder of the proof is left to the reader as exercise 7.5.1. □

7.1.21 Theorem 7.1.20 permits us to give an explicit description of the 2-cocycle in 7.1.6 which represents

$$inv^{-1}(-[L:K]^{-1}) \in H^2(G(L/K); L^*).$$

Following 7.1.18 we begin with $g_1 : B_0G(L/K) \longrightarrow \mathbf{Z}$ defined by $g_1([\ ]) = 1$ and lift it to the homomorphism

$$h : B_0G(L/K) \longrightarrow \oplus_{i=1}^d L_0^*$$

given by $h([\ ]) = (\pi_L, 1, 1, \ldots, 1)$. Now define

$$i = hd_0 : B_1G(L/K) \longrightarrow \oplus_{i=1}^d L_0^*$$

so that, for $z \in G(L/K)$,

7.1.22
$$\begin{cases} i([z]) & = z(h([\ ]))(h([\ ]))^{-1} \\ & = z(\pi_L, 1, 1, \ldots, 1)(\pi_L^{-1}, 1, 1, \ldots, 1). \end{cases}$$

Now choose $j : B_1G(L/K) \longrightarrow \oplus_{i=1}^d L_0^*$ such that

$$(1 \otimes F)(j)/j = i : B_1G(L/K) \longrightarrow \oplus_{i=1}^d L_0^*$$

and set

$$f : G(L/K) \times G(L/K) \longrightarrow L^*$$

to be any 2-cocycle such that, for $z_1, z_2 \in G(L/K)$,

7.1.23
$$\begin{cases} f(z_1, z_2) & = j(d_1[z_1 \mid z_2]) \\ & = z_1(j([z_2]))j([z_1z_2])^{-1}j([z_1]) \in L^*, \end{cases}$$

where L^* is embedded diagonally into $\oplus_{i=1}^d L_0^*$.

There is a second manner in which to describe the extensions which are classified by $H^2(G(L/K); L^*)$. This uses the fact that $H^2(G(L/K); L^*)$ also classifies central simple K-algebras of index $n = [L : K]$ (Reiner, 1975, pp. 242–3). Before describing this we will need a simple cohomological result.

Lemma 7.1.24 *Let G be a finite group and let $IG \lhd \mathbf{Z}[G]$ denote the augmentation ideal. The coboundary*

$$\delta : \mathbf{Z} \cong H^0(G; \mathbf{Z}) \longrightarrow H^1(G; IG)$$

associated to the short exact sequence

$$0 \longrightarrow IG \longrightarrow \mathbf{Z}[G] \stackrel{\epsilon}{\longrightarrow} \mathbf{Z} \longrightarrow 0$$

is given by

$$\delta(1) = [w],$$

where $w : B_1G \longrightarrow IG$ *is given by* $w([g]) = g - 1$ *for* $g \in G$.

Proof Lift $g_1 : B_0G \longrightarrow \mathbf{Z}$ to $g_1 : B_0G \longrightarrow \mathbf{Z}[G]$ given by $h([\ ]) = 1 \in \mathbf{Z}[G]$. Hence $hd_0([g]) = h(g[\ ] - [\ ]) = g - 1$ is a representative for $\delta(1)$, by 7.1.19. □

7.1.25 *Central simple algebras*

Let L/K be a finite Galois extension of local fields with Galois group, $G(L/K)$, and $n = [L : K]$. Given a 2-cocycle

$$f : G(L/K) \times G(L/K) \longrightarrow L^*$$

we may construct a central simple K-algebra of index n from f (Reiner, 1975, p. 242). Let V_f denote the L-vector space on a basis $\{u_g \mid g \in G(L/K)\}$ where $u_1 = 1$. Endow V_f with the K-algebra structure defined by the relations

$$u_g x u_g^{-1} = g(x) \quad (x \in L^*)$$

and

$$f(g_1, g_2)u_{g_1g_2} = u_{g_1}u_{g_2} \quad (g_1, g_2 \in G(L/K)).$$

With this structure V_f becomes a central simple K-algebra which depends, up to isomorphism, only on the cohomology class,

$$[f] \in H^2(G(L/K); L^*)$$

in 7.1.6.

In addition, the invariant

$$inv([f]) \in \mathbf{Z}/\#(G(L/K))$$

is called the *Hasse invariant* of V_f. The V_f which are *division algebras* are precisely those whose Hasse invariants lie in $(\mathbf{Z}/\#(G(L/K)))^*$ (Reiner, 1975, sections 14.6 and 31.1).

Suppose now that $D = V_f$ is such a division algebra. By Reiner (1975, p. 240, section 28.10) there is an embedding, $L \subset D$, as a maximal subfield. This embedding is unique up to inner automorphisms of D.

Let $N_D L^*$ denote the normaliser of L^* in $D^* = D - \{0\}$. Conjugation by $z \in N_D L^*$ induces a K-automorphism of L and hence we obtain a homomorphism

$$\phi : N_D L^* \longrightarrow G(L/K),$$

which is surjective, by the Skölem–Noether theorem (Reiner, 1975, p. 103). The kernel of ϕ contains L^*. However, if $z \in Ker(\phi)$ then $L(z)$ is a subfield of D. Since L is a maximal subfield of D we must have $Ker(\phi) = L^*$ and therefore

7.1.26 $$L^* \longrightarrow N_D L^* \stackrel{\phi}{\longrightarrow} G(L/K)$$

is exact. In fact, $N_D L^* =< L^*; u_g, g \in G(L/K) >$ and $\phi(u_g) = g$.

By Hilton & Stammbach (1971, p. 198) 7.1.26 yields a short exact sequence of (left) $\mathbf{Z}[G]$-modules of the form

7.1.27 $$L^* \stackrel{\chi}{\longrightarrow} \mathbf{Z}[G] \otimes_{\mathbf{Z}[N_D L^*]} I N_D L^* \stackrel{\nu}{\longrightarrow} I G(L/K)$$

given by

$$\chi(z) = 1 \otimes (z - 1) \qquad (z \in L^*)$$

$$\nu(g \otimes (m - 1)) = g\phi(m) - g \quad (g \in G(L/K), m \in N_D L^*).$$

Now let us calculate the image of $[w] \in H^1(G(L/K); IG(L/K))$ of 7.1.24 under the coboundary map

$$\delta : H^1(G(L/K); IG(L/K)) \longrightarrow H^2(G(L/K); L^*)$$

associated to 7.1.27. We begin by lifting w to the homomorphism

$$v \in Hom_{G(L/K)}(B_1 G(L/K), \mathbf{Z}[G] \otimes_{\mathbf{Z}[N_D L^*]} I N_D L^*)$$

given by $v([g]) = 1 \otimes (u_g - 1)$. Hence $vd_1 : B_2G(L/K) \longrightarrow L^*$ is given, in $\mathbf{Z}[G] \otimes_{\mathbf{Z}[N_DL^*]} IN_DL^*$, by

$$\begin{aligned} vd_1([g_1 \mid g_2]) &= g_1 \otimes (u_{g_2} - 1) - 1 \otimes (u_{g_1g_2} - 1) + 1 \otimes (u_{g_1} - 1) \\ &= 1 \otimes u_{g_1}(u_{g_2} - 1) - 1 \otimes (u_{g_1g_2} - 1) + 1 \otimes (u_{g_1} - 1) \\ &= 1 \otimes (u_{g_1}u_{g_2} - u_{g_1g_2}) \\ &= 1 \otimes (f(g_1, g_2) - 1)u_{g_1g_2} \\ &= g_1g_2 \otimes (u_{g_1g_2})^{-1}(f(g_1, g_2) - 1)u_{g_1g_2} \\ &= g_1g_2 \otimes (g_1g_2)^{-1}(f(g_1, g_2) - 1) \\ &= 1 \otimes (f(g_1, g_2) - 1), \end{aligned}$$

which is the image of $f(g_1, g_2) \in L^*$. Hence

$$\delta([w]) = [f] \in H^2(G(L/K); L^*).$$

Since $inv([f]) \in (\mathbf{Z}/[L : K])^*$ the theory of local class formations implies that the homomorphisms

$$H^i(G(L/K); \mathbf{Z}) \xrightarrow{\delta} H^{i+1}(G(L/K); IG(L/K)) \xrightarrow{\delta} H^{i+2}(G(L/K); L^*)$$

are isomorphisms for all $i \geq 1$ and therefore that

$$H^i(G(L/K); \mathbf{Z}[G] \otimes_{\mathbf{Z}[N_DL^*]} IN_DL^*) = 0$$

for all $i \geq 1$. Furthermore, this discussion together with that of 7.1.18 yields the following result:

Proposition 7.1.28 *Let L/K be a finite Galois extension of local fields with $n = [L : K]$. Let D denote the division algebra of index n over K and with Hasse invariant, $s \in (\mathbf{Z}/n)^*$. Then*

(i) *$\mathbf{Z}[G(L/K)] \otimes_{\mathbf{Z}[N_DL^*]} IN_DL^*$ is cohomologically trivial and*

(ii) *the 2-extension*

$$L^* \longrightarrow \mathbf{Z}[G(L/K)] \otimes_{\mathbf{Z}[N_DL^*]} IN_DL^* \longrightarrow \mathbf{Z}[G(L/K)] \xrightarrow{\epsilon} \mathbf{Z},$$

obtained by splicing together the sequences of 7.1.24 *and* 7.1.27, *is represented by $inv^{-1}(s) \in H^2(G(L/K); L^*)$.*

7.1.29 $\Omega(L/K, U)$

We are now in a position to describe the *local Chinburg invariants* (cf. Chinburg, 1985).

Suppose that

$$L^* \longrightarrow A \longrightarrow B \longrightarrow \mathbf{Z}$$

is a 2-extension of $\mathbf{Z}[G(L/K)]$-modules in which A and B are cohomologically trivial and L/K is a finite extension of p-adic local fields, as in 7.1.7. The module, A, cannot be chosen to be projective since it will contain torsion elements and, furthermore, it is not finitely generated. The module, B, may be chosen to be torsion-free and finitely generated and, being cohomologically trivial, will therefore be projective. In 7.1.28 $B = \mathbf{Z}[G(L/K)]$, for example.

Suppose also that this extension is classified by

$$inv^{-1}([L : K]^{-1}) \in H^2(G(L/K); L^*).$$

If $U \subset L^*$ is a $\mathbf{Z}[G(L/K)]$-submodule which is cohomologically trivial then the induced map

$$H^2(G(L/K); L^*) \longrightarrow H^2(G(L/K); L^*/U)$$

is an isomorphism and consequently the sequence

$$L^*/U \longrightarrow A/U \longrightarrow B \longrightarrow \mathbf{Z}$$

corresponds to $inv^{-1}([L : K]^{-1}) \in H^2(G(L/K); L^*/U) \cong \mathbf{Z}/n$.

Suppose, in addition, that L^*/U is a finitely generated $\mathbf{Z}[G(L/K)]$-module, then we may assume, by 7.1.28, that A/U and B are finitely generated (and hence projective, if torsion free) $\mathbf{Z}[G(L/K)]$-modules. In any case, a finitely generated, cohomologically trivial $\mathbf{Z}[G(L/K)]$-module, X, has a finitely generated projective resolution of the form

$$0 \longrightarrow P_1 \longrightarrow P_0 \longrightarrow X \longrightarrow 0$$

and therefore defines a class

$$[X] = [P_0] - [P_1] \in \mathcal{CL}(\mathbf{Z}[G(L/K)]).$$

Therefore we have classes $[A/U]$ and $[B]$ in $\mathcal{CL}(\mathbf{Z}[G(L/K)])$, and we may define a class

7.1.30 $$\Omega(L/K, U) = [A/U] - [B] \in \mathcal{CL}(\mathbf{Z}[G(L/K)]).$$

Proposition 7.1.31 *$\Omega(L/K, U)$ in 7.1.30 is independent of the choice of A and B in the 2-extension*

$$L^* \longrightarrow A \longrightarrow B \longrightarrow \mathbf{Z}.$$

Proof Let

$$L^* \longrightarrow A' \longrightarrow B' \longrightarrow \mathbf{Z}$$

be another 2-extension, with B' finitely generated and A', B' cohomologically trivial, which represents $inv^{-1}([L : K]^{-1}) \in H^2(G(L/K); L^*)$. Therefore there is a commutative diagram resulting from the equivalence of these two 2-extensions:

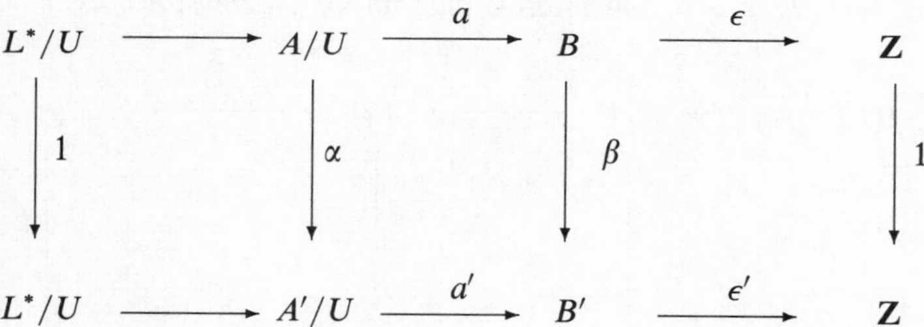

We may form a chain complex of $\mathbf{Z}[G(L/K)]$-homomorphisms

$$0 \longrightarrow A/U \stackrel{(a,\alpha)}{\longrightarrow} B \oplus A'/U \stackrel{(\beta,-a')}{\longrightarrow} B' \longrightarrow 0.$$

It suffices to show that this sequence is exact, since exactness implies the relation

$$[A/U]+[B'] = [A/U \oplus B'] = [B \oplus A'/U] = [B]+[A'/U] \in \mathcal{CL}(\mathbf{Z}[G(L/K)]).$$

However, at the left, if $(a, \alpha)(z) = (0, 0)$ then $z \in L^*/U \subset A/U$ and then $0 = \alpha(z) = z$. Also, at the right, if $b' \in B'$ we may choose $b \in B$ such that $\epsilon(b) = \epsilon'(b')$ and then $b' - \beta(b) \in ker(\epsilon') = im(a')$. Finally, if $b \in B$ and $x \in A'/U$ satisfy $\beta(b) = a'(x)$ then $\epsilon(b) = \epsilon'(a'(x)) = 0$ so that $b \in im(a)$ and $b = a(y)$. Therefore

$$\begin{aligned} a'(x - \alpha(y)) &= \beta(b) - a'(\alpha(y)) \\ &= \beta(b) - \beta(a(y)) \\ &= 0 \end{aligned}$$

so that $x = \alpha(y) + w$ for some $w \in L^*/U$ and

$$\begin{aligned}(a,\alpha)(y+w) &= (a(y)+a(w), \alpha(y)+\alpha(w)) \\ &= (a(y), \alpha(y)+w) \\ &= (b,x),\end{aligned}$$

which establishes exactness in the middle. □

Proposition 7.1.32 *Let k be an integer such that*

$$HCF(k, \#(G(L/K))) = 1.$$

Suppose that there is a commutative diagram of 2-extensions, as in 7.1.29,

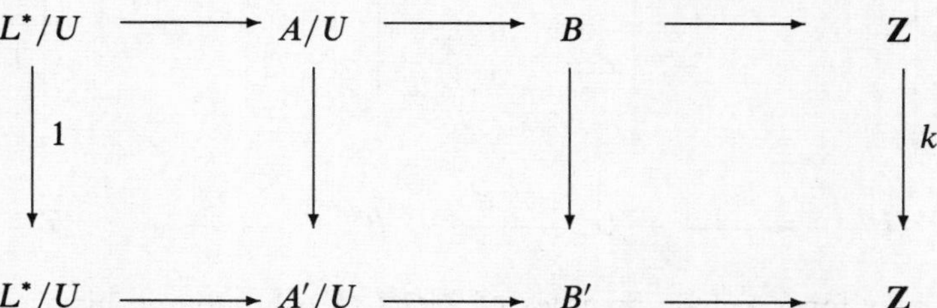

where the right-hand map is multiplication by k. Then

$$[A/U] - [B] = [A'/U] - [B'] + S(k) \in \mathcal{CL}(\mathbf{Z}[G(L/K)]),$$

where $S(k)$ is the Swan module of 4.2.44.

Proof In this case the diagram chase in the proof of 7.1.31 yields an exact sequence

$$0 \longrightarrow A/U \xrightarrow{(a,\alpha)} B \oplus A'/U \xrightarrow{(\beta,-a')} B' \longrightarrow \mathbf{Z}/k \longrightarrow 0.$$

Each of these modules is cohomologically trivial so that

$$[A/U] - [A'/U] - [B] + [B'] - [\mathbf{Z}/k]$$

vanishes in the class-group. However, by 5.3.9,

$$S(k) = [\mathbf{Z}[G(L/K)]] + [\mathbf{Z}/k] = [\mathbf{Z}/k] \in \mathcal{CL}(\mathbf{Z}[G(L/K)])$$

and the result follows.

Alternatively, if the reader prefers to work only with projective modules the result may be proved by the use of Schanuel's lemma (Swan, 1960, p. 270) to splice the above sequence together with

$$0 \longrightarrow k\mathbf{Z}[G(L/K)] \longrightarrow S(k) \longrightarrow \mathbf{Z}/k \longrightarrow 0$$

to eliminate the $\mathbf{Z}/k$'s. □

Corollary 7.1.33 *If $L^* \longrightarrow A' \longrightarrow B' \longrightarrow \mathbf{Z}$ is a 2-extension representing $inv^{-1}(k/[L:K]) \in H^2(G(L/K); L^*)$ and $U \subset L^*$ is as in 7.1.29 then*

$$[A'/U] - [B'] = \Omega(L/K, U) - S(k) \in \mathcal{CL}(\mathbf{Z}[G(L/K)]).$$

Proof By the classification of 2-extensions (Hilton & Stammbach, 1971, p. 206) one knows that there exists a diagram, as in 7.1.32, in which the upper row represents the generator, $inv^{-1}([L:K]^{-1})$. □

Corollary 7.1.34 *Let $f : G(L/K) \times G(L/K) \longrightarrow L^*$ be a 2-cocycle representing the 2-extension of 7.1.8. Let D denote the division algebra, V_f, of 7.1.25, then*

$$\Omega(L/K, U) = [(\mathbf{Z}[G(L/K)] \otimes_{\mathbf{Z}[N_D L^*]} I N_D L^*)/U] \in \mathcal{CL}(\mathbf{Z}[G(L/K)]).$$

Proof By 7.1.20 and 7.1.33 the module in question represents

$$\Omega(L/K, U) - S(-1)$$

but $S(-1) \cong \mathbf{Z}[G(L/K)]$ and therefore represents the trivial class in $\mathcal{CL}(\mathbf{Z}[G(L/K)])$. □

Proposition 7.1.35 *Let L/K be a tamely ramified, finite Galois extension of p-adic local fields. Then, for $i \geq 1$,*

$$H^i(G(L/K); U_L^1) = 0,$$

where $U_L^n = 1 + \pi_L^n \mathcal{O}_L \subset \mathcal{O}_L^*$ denotes the group of units of level n.

Proof By 7.5.2,

$$H^i(G(\overline{L}/\overline{K}); \overline{L})) = \begin{cases} 0 & \text{if } i \geq 1, \\ \overline{K} & \text{if } i = 0. \end{cases}$$

However, if $G_0 \lhd G(L/K)$ denotes the inertia group then G_0 is cyclic of order prime to p and there is an extension of the form

$$G_0 \longrightarrow G(L/K) \longrightarrow G(\overline{L}/\overline{K}).$$

There is also an isomorphism of $G(L/K)$-modules of the form

$$U_L^n/U_L^{n+1} \cong \overline{L},$$

where $G(L/K)$ acts via $G(\overline{L}/\overline{K})$. There is a spectral sequence (Snaith, 1989b, p. 23) of the form

$$E_2^{s,t} = H^s(G(\overline{L}/\overline{K}); H^t(G_0; \overline{L}) \Longrightarrow H^{s+t}(G(L/K); \overline{L}).$$

However, since $\overline{L}$ consists of p-torsion and $HCF(\#(G_0), p) = 1$, $H^s(G_0; \overline{L}) = 0$ if $s > 0$, by Snaith (1989b, p. 12) and Hilton & Stammbach (p. 228).

Therefore $E_2^{s,t} = 0$ except when $(s,t) = (0,0)$. This implies that $H^j(G(L/K); \overline{L}) = 0$ if $j > 0$. To see this we need to know very little about spectral sequences. In fact we need only that $H^j(G(L/K); \overline{L})$ has a (finite) filtration whose sth composition factor is $E_\infty^{s,t}$ and that the latter is computed from $E_2^{s,t} = 0$ by successive operations of taking homology subquotients. Hence all the composition factors are clearly zero when $j > 0$.

From the long exact cohomology sequences of the short exact sequences

$$0 \longrightarrow U_L^n/U_L^{n+1} \longrightarrow U_L^1/U_L^{n+1} \longrightarrow U_L^1/U_L^n \longrightarrow 0$$

we see, by induction, that $H^j(G(L/K); U_L^1/U_L^m) = 0$ whenever $j > 0$ and $m \geq 1$. Finally, if $j \geq 1$,

$$H^j(G(L/K); U_L^1) \cong \lim_{\overleftarrow{m}} H^j(G(L/K); U_L^1/U_L^m) = 0.$$

□

Proposition 7.1.36 *Suppose in* 7.1.35 *that L/K is unramified. Thus $G(L/K)$ ($\cong G(\overline{L}/\overline{K})$) is cyclic of order d, generated by the Frobenius, F. There is an exact sequence of* $\mathbf{Z}[G(L/K)]$*-modules of the form*

$$0 \longrightarrow \mathbf{Z}[G(L/K)] \xrightarrow{F-q} \mathbf{Z}[G(L/K)] \xrightarrow{\eta} \overline{L}^* \longrightarrow 0,$$

where $\overline{K} = F_q$.

Proof Let ξ be a generator for the cyclic group, $\overline{L}^*$. Define η by $\eta(1) = \xi$ so that η is clearly surjective. Now suppose that $\sum_{i=1}^d a_i F^i \in ker(\eta)$ so that

$$\xi^{\sum_{i=1}^d a_i q^i} = 1$$

in $\overline{L}^*$. Hence

$$\sum_{i=1}^{d} a_i q^i \equiv 0 \quad (mod\ (q^d - 1)).$$

For each $i \geq 0$,

$$F^i - q^i = (F - q)(F^{i-1} + qF^{i-2} + \ldots + q^{i-1}) \in (F - q)\mathbf{Z}[G(L/K)]$$

and in particular $(1 - q^d) \in (F - q)\mathbf{Z}[G(L/K)]$. Hence

$$\begin{aligned} \sum_{i=1}^d a_i F^i &\equiv \sum_{i=1}^d a_i (F^i - q^i) \quad (mod\ (F - q)\mathbf{Z}[G(L/K)]) \\ &\equiv 0 \qquad\qquad\qquad\qquad (mod\ (F - q)\mathbf{Z}[G(L/K)]). \end{aligned}$$

□

Corollary 7.1.37 *In* 7.1.36

$$0 = [\overline{L}^*] \in \mathcal{CL}(\mathbf{Z}[G(L/K)]).$$

Proposition 7.1.38 *Let L/K be a finite, unramified Galois extension of p-adic local fields then*

$$0 = \Omega(L/K, U_L^1) \in \mathcal{CL}(\mathbf{Z}[G(L/K)]).$$

Proof By 4.2.48 the Swan modules, $S(k)$, are trivial in $\mathcal{CL}(\mathbf{Z}[G(L/K)])$. Hence, by 7.1.33, we may choose any 2-cocycle,

$$f : G(L/K) \times G(L/K) \longrightarrow L^*,$$

whose class in $H^2(G(L/K); L^*)$ represents a division algebra, D, and then

$$\Omega(L/K, U_L^1) = [(\mathbf{Z}[G(L/K)] \otimes_{\mathbf{Z}[N_D L^*]} IN_D L^*)/U_L^1] \in \mathcal{CL}(\mathbf{Z}[G(L/K)]).$$

As a $\mathbf{Z}[G(L/K)]$-module

$$L^*/U_L^1 \cong \overline{L}^* \times \mathbf{Z},$$

where $G(L/K)$ acts trivially on $\mathbf{Z}$. From 7.1.27 there is a short exact sequence of the form

$$0 \longrightarrow \overline{L}^* \times \mathbf{Z} \longrightarrow \mathbf{Z}[G(L/K)] \otimes_{\mathbf{Z}[N]} IN \longrightarrow IG(L/K) \longrightarrow 0,$$

where $N = N_D L^*$ and the middle module represents $\Omega(L/K, U_L^1)$ in the class-group. By 7.1.37, we may divide out by $\overline{L}^*$ to obtain

$$0 \longrightarrow \mathbf{Z} \longrightarrow \mathbf{Z}[G(L/K)] \otimes_{\mathbf{Z}[N_1]} IN_1 \longrightarrow IG(L/K) \longrightarrow 0,$$

where $N_1 = N/\overline{L}^*$. We must evaluate the middle module of this exact sequence.

However, this sequence is obtained from the extension

7.1.39 $$\mathbf{Z} \longrightarrow N_1 \longrightarrow G(L/K) \cong \mathbf{Z}/d$$

in the same manner as was used to obtain 7.1.27 from 7.1.26. We also have cohomology isomorphisms

$$\begin{aligned} H^2(G(L/K); L^*) &\cong H^2(G(L/K); L^*/U_L^1) \\ &\cong H^2(G(L/K); \overline{L}^* \times \mathbf{Z}) \\ &\cong H^2(G(L/K); \overline{L}^*) \oplus H^2(G(L/K); \mathbf{Z}) \\ &\cong H^2(G(L/K); \mathbf{Z}) \\ &\cong \mathbf{Z}/d \end{aligned}$$

so that the class of the central extension of 7.1.39 coincides with the class of $[f]$ in $H^2(G(L/K); L^*)$ and therefore is equal to an element of $(\mathbf{Z}/d)^*$. This means that $N_1 \cong \mathbf{Z}$ and that 7.1.39 is equal to

$$\mathbf{Z} \xrightarrow{d} \mathbf{Z} \xrightarrow{\pi} \mathbf{Z}/d.$$

In this case it will suffice to construct an isomorphism of the form

$$\psi : \mathbf{Z}[\mathbf{Z}/d] \longrightarrow \mathbf{Z}[\mathbf{Z}/d] \otimes_{\mathbf{Z}[\mathbf{Z}]} I\mathbf{Z}.$$

Let z generate N_1. Define ψ by the formula

$$\psi(\alpha) = \alpha \otimes (z - 1) \quad (\alpha \in \mathbf{Z}[\mathbf{Z}/d]).$$

Therefore the composition

$$\mathbf{Z}[\mathbf{Z}/d] \xrightarrow{\psi} \mathbf{Z}[\mathbf{Z}/d] \otimes_{\mathbf{Z}[\mathbf{Z}]} I\mathbf{Z} \longrightarrow I\mathbf{Z}/d$$

is given by sending α to $\alpha(z-1)$. The kernel of this surjective map is $< 1+F+\ldots+F^{d-1} >$. However,

$$\begin{aligned}\psi(1+F+\ldots+F^{d-1}) &= (1+F+\ldots+F^{d-1})\otimes(z-1)\\ &= 1\otimes(1+z+\ldots+z^{d-1})(z-1)\\ &= 1\otimes(z^d-1),\end{aligned}$$

which is the generator of the kernel of $\mathbf{Z}[\mathbf{Z}/d]\otimes_{\mathbf{Z}[\mathbf{Z}]} I\mathbf{Z} \longrightarrow I\mathbf{Z}/d$, thereby showing that ψ is an isomorphism. □

7.1.40 *The tamely ramified case*

The majority of the remainder of this section will be devoted to showing that $\Omega(L/K, U_L^1)$ vanishes when L/K is tamely ramified. This result (Theorem 7.1.56) was first proved by Chinburg (1985, section VI). Our proof differs from that of Chinburg (1985) and is based upon 7.1.34. Of course, Theorem 7.1.56 generalises 7.1.38, which I have included because it seemed worth while to start with a simple example.

Let us begin by setting up a presentation for the Galois group of a tame local extension. Let $a \in G_0$ be a generator and let $b \in G(L/K)$ be an element whose image in $G(\overline{L}/\overline{K})$ is the inverse Frobenius, F^{-1}. Assume that $\overline{K} = F_q$ and that $\overline{L} = F_{q^d}$. Then

7.1.41 $$G(L/K) = \{a, b \mid a^e = 1, b^d = a^c, b^{-1}ab = a^q\}.$$

Let $f : G(L/K) \times G(L/K) \longrightarrow L^*$ be a 2-cocycle whose class in $H^2(G(L/K); L^*)$ represents the division algebra, D, with Hasse invariant equal to $1/de = [L:K]^{-1}$. The calculation of 7.1.25 amounts to showing, in the classification of group extensions by $H^2(G(L/K); L^*)$, that $[f]$ also corresponds to the extension of 7.1.26:

$$L^* \longrightarrow N_D L^* \longrightarrow G(L/K).$$

For the remainder of this section set $N = N_D L^*/U_L^1$. Let ξ generate $\overline{L}^*$ and let π denote the image of π_L in

$$L^*/U_L^1 \cong \overline{L}^* \times \mathbf{Z} = < \xi, \pi > .$$

Note that this module is *not* a sum of $\overline{L}^*$ and $\mathbf{Z}$, as $\mathbf{Z}[G(L/K)]$-modules, unless L/K is unramified.

We have a short exact sequence of $\mathbf{Z}[G(L/K)]$-modules

7.1.42 $$0 \longrightarrow \overline{L}^* \times \mathbf{Z} \longrightarrow \mathbf{Z}[G(L/K)] \otimes_{\mathbf{Z}[N]} IN \xrightarrow{\lambda} IG(L/K) \longrightarrow 0$$

and we must evaluate the class of the middle module in $\mathscr{CL}(\mathbf{Z}[G(L/K)])$. For this purpose define a homomorphism

$$\psi : \mathbf{Z}[G(L/K)]z_1 \oplus \mathbf{Z}[G(L/K)]z_2 \longrightarrow \mathbf{Z}[G(L/K)] \otimes_{\mathbf{Z}[N]} IN$$

by the formulae:

7.1.43 $$\begin{cases} \psi(z_1) = 1 \otimes (u_a - 1) \\ \psi(z_2) = 1 \otimes (u_b - 1) \end{cases}$$

in the notation of 7.1.25.

Define elements R, S and T by the formulae:

7.1.44 $$\begin{cases} R = (1 + a + \ldots + a^{e-1})z_1, \\ S = (b(1 + a + \ldots + a^{q-1}) - 1)z_1 - (a-1)z_2, \\ T = (1 + b + \ldots + b^{d-1})z_2 - (1 + a + \ldots + a^{c-1})z_1. \end{cases}$$

Lemma 7.1.45 *The elements R,S and T lie in ker($\lambda\psi$).*

Proof We have

$$\begin{aligned} \lambda\psi(R) &= \lambda((1 + a + \ldots + a^{e-1}) \otimes (u_a - 1)) \\ &= (1 + a + \ldots + a^{e-1})(a-1) \\ &= a^e - 1 \\ &= 0, \end{aligned}$$

$$\begin{aligned}
\lambda\psi(S) &= \lambda((b(1+a+\ldots+a^{q-1})-1)\otimes(u_a-1)\\
&\qquad -(a-1)\otimes(u_b-1))\\
&= (b(1+a+\ldots+a^{q-1})-1)(a-1)-(a-1)-(a-1)(b-1)\\
&= b(a^q-1)-a+1-ab+b+a-1\\
&= ba^q-b-ab+b\\
&= 0
\end{aligned}$$

and

$$\begin{aligned}
\lambda\psi(T) &= \lambda((1+b+\ldots+b^{d-1})\otimes(u_b-1)\\
&\qquad -(1+a+\ldots+a^{c-1})\otimes(u_a-1))\\
&= (1+b+\ldots+b^{d-1})(b-1)-(1+a+\ldots+a^{c-1})(a-1)\\
&= b^d-1-(a^c-1)\\
&= 0,
\end{aligned}$$

as required. □

Lemma 7.1.46 *If $z \in ker(\psi)$ then*

$$z = \beta R + \gamma S + mT$$

for some $\beta, \gamma \in \mathbf{Z}[G(L/K)]$ and $m \in \mathbf{Z}$.

Proof Write $z = \beta_1 z_1 + \gamma_1 z_2$ and write γ_1 in the form

$$\gamma_1 = \sum_{j=0}^{d-1} \gamma_{1,j} b^j$$

with $\gamma_{1,j} \in \mathbf{Z}[G_0]$. The image of $\lambda\psi(z)$ in $IG(\overline{L}/\overline{K})$ is equal to

$$0 = \sum_{j=0}^{d-1} \epsilon(\gamma_{1,j}) b^j (b-1)$$

where $\epsilon : \mathbf{Z}[G_0] \longrightarrow \mathbf{Z}$ is the augmentation homomorphism. Hence there exists an integer, m, such that

$$\epsilon(\gamma_{1,j}) = m$$

for all $j = 0, \ldots, d$. Subtracting mT from z, we may suppose that $\epsilon(\gamma_{1,j}) = 0$ for all $j = 0, \ldots, d$. Therefore

$$\gamma_{1,j} b^j = b^j \gamma'_{1,j}(a-1)$$

and we may add

$$\left(\sum_{j=0}^{d-1} b^j \gamma'_{1,j} \right) S$$

to z to give an element of the form $\beta_1 z_1$. Clearly, for an element of this form to be in $\ker(\psi)$, β_1 must be equal to a multiple of $(1+a+\ldots+a^{e-1})$, which completes the proof. □

Next we must evaluate $\psi(R)$, $\psi(S)$ and $\psi(T)$ in terms of the 2-cocycle, $f : G(L/K) \times G(L/K) \longrightarrow \overline{L}^* \times \mathbf{Z}$.

Lemma 7.1.47

$$\psi(R) = \prod_{j=0}^{e-1} f(a^j, a) \in \overline{L}^* \times \mathbf{Z}.$$

Proof By definition

$$\begin{aligned} \psi(R) &= (1 + a + \ldots + a^{e-1}) \otimes (u_a - 1) \\ &= \textstyle\sum_{j=0}^{e-1} 1 \otimes (u_{a^j} u_a - u_{a^j}) \\ &= \textstyle\sum_{j=0}^{e-1} 1 \otimes (f(a^j, a) u_{a^{j+1}} - 1) - 1 \otimes (u_{a^j} - 1). \end{aligned}$$

However, as in 7.1.25, if $y \in \overline{L}^* \times \mathbf{Z}$ and $g \in G(L/K)$ then

$$\begin{aligned} 1 \otimes (y u_g - u_g) &= g \otimes u_g^{-1}(y-1) u_g \\ &= g \otimes (g^{-1}(y) - 1) \\ &= 1 \otimes (y-1) \end{aligned}$$

so that

$$\begin{aligned}\psi(R) &= \textstyle\sum_{j=0}^{e-1} 1 \otimes (f(a^j, a) - 1) + 1 \otimes (u_{a^{j+1}} - 1) - 1 \otimes (u_{a^j} - 1) \\ &= \textstyle\sum_{j=0}^{e-1} 1 \otimes (f(a^j, a) - 1) \\ &= 1 \otimes (\textstyle\prod_{j=0}^{e-1} f(a^j, a) - 1),\end{aligned}$$

which is the image of $\prod_{j=0}^{e-1} f(a^j, a) \in \overline{L}^* \times \mathbf{Z}$. □

Lemma 7.1.48

$$\psi(S) = \left(\prod_{j=0}^{q-1} f(ba^j, a)\right) f(a, b)^{-1}.$$

Proof By definition

$$\begin{aligned}\psi(S) &= \textstyle\sum_{j=0}^{q-1} ba^j \otimes (u_a - 1) - 1 \otimes (u_a - 1) - (a - 1) \otimes (u_b - 1) \\ &= \textstyle\sum_{j=0}^{q-1} 1 \otimes (f(ba^j, a)u_{ba^{j+1}} - u_{ba^j}) - 1 \otimes (u_a - 1) \\ &\quad -1 \otimes (f(a, b)u_{ab} - 1) + 1 \otimes (u_b - 1) + 1 \otimes (u_a - 1) \\ &= \textstyle\sum_{j=0}^{q-1} 1 \otimes (f(ba^j, a)u_{ba^{j+1}} - u_{ba^{j+1}}) \\ &\quad + \textstyle\sum_{j=0}^{q-1} 1 \otimes (u_{ba^{j+1}} - u_{ba^j}) \\ &\quad -1 \otimes (f(a, b)u_{ab} - u_{ab}) - 1 \otimes (u_{ab} - 1) + 1 \otimes (u_b - 1) \\ &= 1 \otimes ((\textstyle\prod_{j=0}^{q-1} f(ba^j, a)) - 1) + 1 \otimes (u_{ba^q} - u_b) \\ &\quad -1 \otimes (f(a, b) - 1) - 1 \otimes (u_{ab} - u_b) \\ &= 1 \otimes ((\textstyle\prod_{j=0}^{q-1} f(ba^j, a))f(a, b)^{-1} - 1),\end{aligned}$$

as required. □

Lemma 7.1.49

$$\psi(T) = \left(\prod_{j=0}^{d-1} f(b^j, b)\right) \left(\prod_{s=0}^{c-1} f(a^s, a)\right)^{-1} \in \overline{L}^* \times \mathbf{Z}.$$

Proof By definition

$$\begin{aligned}
\psi(T) &= \textstyle\sum_{j=0}^{d-1} b^j \otimes (u_b - 1) - \sum_{s=0}^{c-1} a^s \otimes (u_a - 1) \\
&= \textstyle\sum_{j=0}^{d-1} 1 \otimes (f(b^j, b)u_{b^{j+1}} - u_{b^j}) \\
&\quad - \textstyle\sum_{s=0}^{c-1} 1 \otimes (f(a^s, a)u_{a^{s+1}} - u_{a^s}) \\
&= \textstyle\sum_{j=0}^{d-1} 1 \otimes (f(b^j, b)u_{b^{j+1}} - u_{b^{j+1}}) \\
&\quad + \textstyle\sum_{j=0}^{d-1} 1 \otimes (u_{b^{j+1}} - u_{b^j}) \\
&\quad - \textstyle\sum_{s=0}^{c-1} 1 \otimes (f(a^s, a)u_{a^{s+1}} - u_{a^{s+1}}) \\
&\quad - \textstyle\sum_{s=0}^{c-1} 1 \otimes (u_{a^{s+1}} - u_{a^s}) \\
&= \textstyle\sum_{j=0}^{d-1} 1 \otimes (f(b^j, b) - 1) + 1 \otimes (u_{b^d} - 1) \\
&\quad - \textstyle\sum_{s=0}^{c-1} 1 \otimes (f(a^s, a) - 1) - 1 \otimes (u_{a^c} - 1)
\end{aligned}$$

and $b^d = a^c$, which yields the required formula. □

7.1.50 Now we must evaluate $f(z_1, z_2) \in \overline{L}^* \times \mathbf{Z}$ using the method of 7.1.21.

Suppose that

$$\frac{a(\pi)}{a} = u \in \overline{L}^* \subset \overline{L}^* \times \mathbf{Z}.$$

Thus u is a primitive eth root of unity in $\overline{L}^*$ and $a(u) = u$ since $a \in G_0$. There exists $\gamma \in K_{nr}^*$ such that

$$u = \gamma^{q^d - 1} = \frac{F_0(\gamma)}{\gamma} \in L^*$$

and

$$\frac{F_0(\gamma^s)}{\gamma^s} = u^s = \frac{a^s(\pi)}{\pi}.$$

Hence, in the notation of 7.1.22, if $s \geq 0$

$$\begin{aligned} i(a^s) &= (a^s(\pi)\pi^{-1}, 1, \ldots, 1) \in \oplus_1^d L_0^* \\ &= (F_0(\gamma^s)\gamma^{-s}, \gamma^s\gamma^{-s}, \ldots, \gamma^s\gamma^{-s}) \\ &= (1 \otimes F)/1(\gamma^s, \gamma^s, \ldots, \gamma^s). \end{aligned}$$

Here we have used the fact that $a \in Ker(h_1)$ in 7.1.13 so that a acts component by component on $\oplus_1^d L_0^*$.

Therefore

$$j([a^s]) = \begin{cases} (\gamma^s, \gamma^s, \ldots, \gamma^s) & \text{if } s \neq 0 \\ (1, 1, \ldots, 1) & \text{if } s = 0. \end{cases}$$

Therefore

$$\begin{aligned} f(a^s, a) &= \frac{a^s(j(a))j(a^s)}{j(a^{s+1})} \\ &= \frac{j(a)j(a^s)}{j(a^{s+1})} \\ &= \begin{cases} 1 & \text{if } s \neq e-1, \\ (\gamma^e, \gamma^e, \ldots, \gamma^e) & \text{if } s = e-1. \end{cases} \end{aligned}$$

From 7.1.47 we obtain the following result:

Lemma 7.1.51 $\psi(R) = \gamma^e \in \overline{L}^* \subset \overline{L}^* \times \mathbf{Z}$.

Proof The embedding of L^* into $(L \otimes_K K_{nr})^* \cong \oplus_1^d L_0^*$ sends γ^e to $(\gamma^e, \gamma^e, \ldots, \gamma^e)$ since $(\gamma^e)^{q^d-1} = u^e 1$ implies that γ is a root of unity which lies in $W \subset L$.

7.1.52 Now suppose that

$$\frac{b(\pi)}{\pi} = u_1 \in L^*$$

then, as in 7.1.50, there exists $\gamma_1 \in K_{nr}^*$ such that

$$u_1 = \gamma_1^{q^d-1} = \frac{F_0(\gamma_1)}{\gamma_1}.$$

In the notation of 7.1.12 and 7.1.13, $b = (F, 1) \cdot (F^{-1}, b)$ and $(F^{-1}, b) \in Ker(h_1)$. Hence

$$\begin{aligned} b(\pi, 1, 1, \ldots, 1) &= (F, 1)(F^{-1}, b)(\pi, 1, \ldots, 1) \\ &= (F, 1)(b(\pi), 1, 1, \ldots, 1) \\ &= (1, b(\pi), 1, \ldots, 1), \end{aligned}$$

by 7.1.15. Therefore

$$i(b) = (\pi^{-1}, b(\pi), 1, 1, \ldots, 1),$$

while

$$((1 \otimes F)/1)(b(\pi)\gamma_1, \gamma_1, \ldots, \gamma_1, \gamma_1)$$

$$= (F_0(\gamma_1), b(\pi)\gamma_1, \ldots, \gamma_1)(b(\pi)^{-1}\gamma_1^{-1}, \ldots, \gamma_1^{-1})$$

$$= (u_1 b(\pi)^{-1}, b(\pi), 1, 1, \ldots, 1)$$

$$= (\pi^{-1}, b(\pi), 1, \ldots, 1),$$

so that

$$j(b) = (b(\pi)\gamma_1, \gamma_1, \ldots, \gamma_1).$$

Similarly,

$$j(ab) = (ab(\pi)\gamma\gamma_1, \gamma\gamma_1, \ldots, \gamma\gamma_1)$$

and

7.1.53 $$f(a, b) = [a(b(\pi)\gamma_1, \gamma_1, \ldots, \gamma_1)] \frac{(\gamma, \gamma, \ldots, \gamma)}{(ab(\pi)\gamma\gamma_1, \gamma\gamma_1, \ldots, \gamma\gamma_1)} = 1,$$

since a acts trivially on K_{nr}^*. □

Lemma 7.1.54

$$\psi(S) = 1 \in \overline{L}^* \times \mathbf{Z}.$$

Proof Combining 7.1.48 and 7.1.53 we find that

$$\begin{aligned} \psi(S) &= \textstyle\prod_{j=0}^{q-1} f(ba^j, a) \\ &= \textstyle\prod_{j=0}^{q-1} \frac{ba^j(j(a)) \cdot j(ba^j)}{j(ba^{j+1})} \\ &= (\textstyle\prod_{j=0}^{q-1} b(j(a)))j(ba^q)^{-1} j(b), \end{aligned}$$

since a acts trivially on $j(a) = (\gamma, \gamma, \ldots, \gamma) \in \oplus_1^d K_{nr}^*$. However, $j(b) = (b(\pi)\gamma_1, \gamma_1, \ldots, \gamma_1)$ and $j(ba^q) = j(ab) = (ab(\pi)\gamma\gamma_1, \gamma\gamma_1, \ldots, \gamma\gamma_1)$ by the preceding calculation. Also

$$\begin{aligned} b(j(a))^q &= b(\gamma^q, \gamma^q, \ldots, \gamma^q) \\ &= (F, 1)(F^{-1}, b)(\gamma^q, \gamma^q, \ldots, \gamma^q) \\ &= (F, 1)(\gamma, \gamma, \ldots, \gamma) \\ &= (F_0(\gamma), \gamma, \gamma, \ldots, \gamma) \\ &= (u\gamma, \gamma, \gamma, \ldots, \gamma). \end{aligned}$$

Therefore

$$\begin{aligned} \psi(S) &= \tfrac{(u\gamma, \gamma, \gamma, \ldots, \gamma)(b(\pi)\gamma_1, \gamma_1, \ldots, \gamma_1)}{(ab(\pi)\gamma\gamma_1, \gamma\gamma_1, \ldots, \gamma\gamma_1)} \\ &= (ub(\pi)ab(\pi)^{-1}, 1, 1, \ldots, 1) \\ &= 1, \end{aligned}$$

since $ab(\pi) = a(u_1\pi) = u_1 u\pi = ub(\pi)$. □

Lemma 7.1.55 *For some* α,

$$\psi(T) = (\xi^\alpha, \pi) \in \overline{L}^* \times \mathbf{Z}.$$

Proof We will write $x \sim y$ if $x, y \in \overline{L}^* \times \mathbf{Z}$ have the same second coordinate.

By 7.1.49,

$$\begin{aligned} \psi(T) &\sim \textstyle\prod_{j=0}^{d-1} f(b^j, b) \\ &= (\textstyle\prod_{j=0}^{d-1} b^j(b(\pi)\gamma_1, \gamma_1, \ldots, \gamma_1))j(b^d)^{-1} \\ &\sim (\textstyle\prod_{j=0}^{d-1} b^j(b(\pi), 1, \ldots, 1))j(a^c)^{-1} \\ &\sim (\textstyle\prod_{j=0}^{d-1}(F^j, 1)(F^{-j}, b^j)(b(\pi), 1, \ldots, 1)). \end{aligned}$$

However

$$\begin{aligned}(F^j,1)(F^{-j},b^j)(b(\pi),1,\ldots,1) &\sim (F^j,1)(b^{j+1}(\pi),1,\ldots,1)\\ &= (1,1,\ldots,b^{j+1}(\pi),1,1,\ldots,1)\\ &\sim (1,1,\ldots,\pi,1,1,\ldots,1),\end{aligned}$$

where π is in the jth coordinate. Therefore

$$\psi(T) \sim (\pi,\pi,\ldots,\pi),$$

which is the image of $\pi \in L^*$ in $\oplus_1^d L_0^*$, as required. □

Theorem 7.1.56 *Let L/K be a tamely ramified, finite Galois extension of p-adic local fields. Then*

$$0 = \Omega(L/K, U_L^1) \in \mathscr{CL}(\mathbf{Z}[G(L/K)]).$$

Proof Consider the map, ψ, of 7.1.43. The composition of ψ with the projection, λ, onto $IG(L/K)$ is clearly surjective, because $a-1$ and $b-1$ generate $IG(L/K)$. Let us examine the kernel of ψ, which we will prove to be a free $\mathbf{Z}[G(L/K)]$-module. By 7.1.46 the kernel consists of elements of the form $z = \beta R + \delta S + mT$. By 7.1.51, 7.1.54 and 7.1.55

$$\psi(z) = \beta(\gamma^e, 0) + (\xi^{\alpha m}, m) \in \overline{L}^* \times \mathbf{Z}.$$

Since u is a primitive eth root of unity and $u = \gamma^{q^d-1}$ then γ^e is a generator of $\overline{L}^*$, which implies that ψ is surjective. In addition, since $\mathbf{Z}[G] < R >\cong \mathbf{Z}[G(\overline{L}/\overline{K})] < R >$, 7.1.36 implies that βR lies in $\mathbf{Z}[G(\overline{L}/\overline{K})](b^{-1}-q) < R >$. However

$$\begin{aligned}(-b)(b^{-1}-q)R &= (-b)(b^{-1}-q)(1+a+\ldots+a^{e-1})z_1\\ &= (qb-1)(1+a+\ldots+a^{e-1})z_1\\ &= (1+a+\ldots+a^{e-1})S.\end{aligned}$$

Hence we see that the kernel of ψ is isomorphic to $\mathbf{Z}[G(L/K)] < S >$ which is a free module of rank one. Hence we have a short exact sequence of the form

$$0 \longrightarrow \mathbf{Z}[G(L/K)] \longrightarrow \oplus_1^2 \mathbf{Z}[G(L/K)] \xrightarrow{\psi} \mathbf{Z}[G(L/K)] \otimes_{\mathbf{Z}[N]} IN \longrightarrow 0,$$

which implies that the class of $\mathbf{Z}[G(L/K)] \otimes_{\mathbf{Z}[N]} IN$ is trivial in $\mathscr{CL}(\mathbf{Z}[G(L/K)])$. □

We conclude this section with a result concerning the local Chinburg invariant, $\Omega(L/K, U)$ of 7.1.29, which is related to the exercise 7.5.1.

Theorem 7.1.57 *Let L/K be a finite Galois extension of local fields and let $d = [\overline{L} : \overline{K}]$ denote the residue degree. Let*

7.1.58 $$0 \longrightarrow L^* \stackrel{i}{\longrightarrow} \oplus_1^d L_0^* \stackrel{(1\otimes F)/1}{\longrightarrow} \oplus_1^d L_0^* \stackrel{w}{\longrightarrow} \mathbf{Z} \longrightarrow 0$$

denote the 2-extension of 7.1.8. Suppose that $J = G(L/M) \lhd G(L/K)$ is a normal subgroup. Then taking J-fixed points of 7.1.58 yields a 2-extension of $\mathbf{Z}[G(M/K)]$-modules which is equivalent to

$$0 \longrightarrow M^* \stackrel{i}{\longrightarrow} \oplus_1^r M_0^* \stackrel{(1\otimes F)/1}{\longrightarrow} \oplus_1^r M_0^* \stackrel{w}{\longrightarrow} \mathbf{Z} \longrightarrow 0,$$

($r = [\overline{M} : \overline{K}]$) the 2-extension of 7.1.8 associated to M/K.

Proof Firstly we remark that taking the J-fixed points of 7.1.58 yields an exact sequence, because $\oplus_1^d L_0^*$ is cohomologically trivial.

Since the inertia group, $I = G_0(L/M) \lhd G(L/M)$, is the kernel of $G(L/M) \longrightarrow G(\overline{L}/\overline{M}) \longrightarrow G(\overline{L}/\overline{K})$ it is a characteristic subgroup and hence is normal in $G(L/K)$. Therefore we may compute the J-invariants of a $\mathbf{Z}[G(L/K)]$-module, X, by means of the isomorphism

$$X^J \cong (X^I)^{J/I}.$$

Therefore we may reduce the computation to the following two cases:

Case A: $J = I$, $\overline{K} = F_q$, $\overline{M} = \overline{L} = F_{q^d}$ or

Case B: $I = \{1\}$, $\overline{K} = F_q$, $\overline{M} = F_{q^r}$, $\overline{L} = F_{q^d}$ and $d = rs$.

In case A, $J = I \subset G(L_0/K)$ and the J-invariants of 7.1.58 are given by

7.1.59

$$0 \longrightarrow M^* = (L^J)^* \stackrel{i}{\longrightarrow} \oplus_1^d (L_0^J)^* \stackrel{(1\otimes F)/1}{\longrightarrow} \oplus_1^d (L_0^J)^* \stackrel{v}{\longrightarrow} \mathbf{Z} \longrightarrow 0.$$

However, 7.1.59 is clearly the 2-extension of 7.1.8 for M/K, since $L_0^J = (K_{nr}L)^J = K_{nr}(L^J) = K_{nr}M = M_0$.

In case B it suffices (Hilton & Stammbach, 1971, p. 148) to construct a commutative diagram of $G(L/K)/J$-extensions of the form

7.1.60

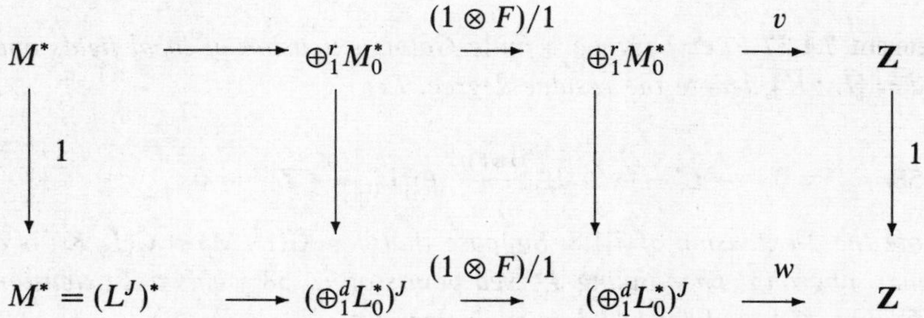

$$\begin{array}{ccccccc} M^* & \longrightarrow & \oplus_1^r M_0^* & \xrightarrow{(1\otimes F)/1} & \oplus_1^r M_0^* & \xrightarrow{v} & \mathbf{Z} \\ \downarrow 1 & & \downarrow & & \downarrow & & \downarrow 1 \\ M^* = (L^J)^* & \longrightarrow & (\oplus_1^d L_0^*)^J & \xrightarrow{(1\otimes F)/1} & (\oplus_1^d L_0^*)^J & \xrightarrow{w} & \mathbf{Z} \end{array}$$

Let $b \in J$ map to $F^{-r} \in G(\overline{L}/\overline{M})$ then b acts on $\oplus_1^d L_0^*$ as $b = (F^r, 1)(F^{-r}, b)$ where $x = (F^{-r}, b) \in G(L_0/K) \subset G(K_{nr}/K) \times G(L/K)$. Hence, if $z_i \in L_0^*$ then

$$\begin{aligned} b(z_1, \ldots, z_d) &= (F^r, 1)(x(z_1), \ldots, x(z_d)) \\ &= (F_0(x(z_{d-r+1})), \ldots, F_0(x(z_d)), x(z_1), \ldots, x(z_{d-r})). \end{aligned}$$

Therefore $b(z_1, \ldots, z_d) = (z_1, \ldots, z_d)$ if and only if

7.1.61
$$\begin{cases} z_1 = F_0(x(z_{d-r+1})) \\ z_2 = F_0(x(z_{d-r+2})) \\ \vdots \\ z_r = F_0(x(z_d)) \\ z_{r+1} = x(z_1) \\ \vdots \\ z_d = x(z_{d-r}). \end{cases}$$

Therefore the coordinates, $z_{d-r+1}, \ldots, z_d$, determine all the other $\{z_i\}$ and we obtain a diagram of the form of 7.1.60 by choosing

$$\phi : \oplus_1^r M_0^* \longrightarrow \oplus_1^d L_0^*$$

to have the form

7.1.62 $$\phi(z_{d-r+1},\ldots,z_d) = (\ldots,z_{d-r+1},\ldots,z_d).$$

The valuation of L_0^* induces that of M_0^* in such a way as to make the right-hand square commute. It is also evident that the left-hand square commutes once we have verified that the central square is commutative. However, by 7.1.61,

$$(1 \otimes F)(z_{d-r+1},\ldots,z_d) = (\ldots,z_{d-r},z_{d-r+1},\ldots,z_{d-1})$$

and

$$\begin{aligned} z_{d-r} &= x(z_{d-2r}) \\ &= \vdots \\ &= F_0(x^{s-1}(z_d)), \end{aligned}$$

which ensures that $(1 \otimes F)\phi = \phi(1 \otimes F)$ on $(z_{d-r+1},\ldots,z_d) \in \oplus_1^r M_0^*$, since $F_0 x^{s-1}$ is equal to $F^{d-r(s-1)} = F^r$ on K_{nr} and is equal to $b^{-1} = 1$ on M. By definition 7.1.14, this is the manner in which F_0 acts on M_0. □

7.2 The global Chinburg invariant

Throughout this section let L/K be a finite Galois extension of number fields. In Chinburg (1985) three class-group invariants, $\Omega(L/K,i) \in \mathcal{CL}(\mathbf{Z}[G(L/K)])$, were constructed in a cohomological manner which is analogous to the construction of the local invariant of 7.1.29. In this section we shall examine $\Omega(L/K,2)$ and we shall give a description of this class, which is originally due to S. Kim (1991); (see also Kim, 1992) and is also used in S.M.J. Wilson (1990).

For each finite prime, $P \lhd \mathcal{O}_K$, choose a prime lying over it, $Q \lhd \mathcal{O}_L$. Hence $G(L/K)$ contains a subgroup, called the *decomposition group* of Q, which is isomorphic to $G(L_Q/K_P)$. We shall say that P is *tame* if L_Q/K_P is tamely ramified (i.e. $G_1(L_Q/K_P) = 1$) and that P is *wild* otherwise. By a theorem of E. Noether, if P is tame then $\mathcal{O}_{L_Q}$ is a free $\mathcal{O}_{K_P}[G(L_Q/K_P)]$-module of rank one. Therefore we may choose an adèle

7.2.1 $$(a_P) = a \in \textstyle\prod_P \mathcal{O}_{L_Q}$$

(the product being taken over finite primes of K with Q being the chosen prime over P) such that

7.2.2 $$\begin{cases} \text{(i) } a_P \in \mathcal{O}_{L_Q} \text{ and } K_P[G(L_Q/K_P)]a_P = L_Q \text{ for each } P \\ \text{(ii) } \mathcal{O}_{K_P}[G(L_Q/K_P)]a_P = \mathcal{O}_{L_Q} \text{ for each tame } P. \end{cases}$$

By 4.2.17 there are isomorphisms

$$L \otimes_K K_P \cong \prod_{R|P} L_R \cong Ind_{G(L_Q/K_P)}^{G(L/K)}(L_Q)$$

and

$$\mathcal{O}_L \otimes_{\mathcal{O}_K} \mathcal{O}_{K_P} \cong \prod_{R|P} \mathcal{O}_{L_R} \cong Ind_{G(L_Q/K_P)}^{G(L/K)}(\mathcal{O}_{L_Q})$$

so that

7.2.3 $$\begin{cases} \text{(i) } K_P[G(L/K)]a_P \cong \prod_{R|P} L_R \text{ for each } P \\ \text{(ii) } \mathcal{O}_{K_P}[G(L/K)]a_P \cong \prod_{R|P} \mathcal{O}_{L_R} \text{ for each tame } P. \end{cases}$$

We will abbreviate $\mathcal{O}_{K_P}[G(L_Q/K_P)]a_P$ to X_Q and set

$$X = \mathcal{O}_K[G(L/K)]a.$$

This is to be interpreted as meaning that X is the intersection of L with the product of its P-completions, X_P, where both are considered as subgroups of the adèles. Hence X is a locally free $\mathcal{O}_K[G(L/K)]$-module whose P-completion is

7.2.4 $$X_P = \mathcal{O}_{K_P}[G(L/K)]a_P = Ind_{G(L_Q/K_P)}^{G(L/K)}(X_Q).$$

7.2.5 In addition we shall assume henceforth (by replacing X by mX for a suitable integer, $m \in \mathbf{Z}$, if necessary) that the Q-adic exponential defines an isomorphism

$$exp : X_Q \xrightarrow{\cong} 1 + X_Q \subset \mathcal{O}_{L_Q}^*$$

for all wild L_Q/K_P.

Definition 7.2.6 Since X is locally free, it is cohomologically trivial and so also is X_Q for each Q. Hence X defines a class in $\mathcal{CL}(\mathbf{Z}[G(L/K)])$ and we may define

$$\Omega(L/K, 2) = [X] + \sum_{P wild} Ind_{G(L_Q/K_P)}^{G(L/K)}(\Omega(L_Q/K_P, 1 + X_Q))$$

in $\mathcal{CL}(\mathbf{Z}[G(L/K)])$. Here $\Omega(L_Q/K_P, 1 + X_Q)$ is the local Chinburg invariant of 7.1.29. The assumption of 7.2.5 ensures that $1 + X_Q$ is cohomologically trivial.

Proposition 7.2.7 *In* 7.2.6 $\Omega(L/K,2)$ *is independent of the choice of* X, *satisfying the conditions of* 7.2.2(i),(ii) *and* 7.2.5.

Proof Let P be a wild prime of K and suppose that X_Q and Y_Q are two choices for the lattice in 7.2.2 such that $X_Q \subset Y_Q$. Hence we have, in the notation of 7.1.34, a commutative diagram of the following form:

$$\begin{array}{ccccc}
L_Q^*/(1+X_Q) & \to & (\mathbf{Z}[G(L_Q/K_P)] \otimes_{\mathbf{Z}[N_D L^*]} IN_D L^*)/(1+X_Q) & \to & IG(L_Q/K_P)) \\
\downarrow & & \downarrow \beta_P & & \downarrow \cong 1 \\
L_Q^*/(1+Y_Q) & \to & (\mathbf{Z}[G(L_Q/K_P)] \otimes_{\mathbf{Z}[N_D L^*]} IN_D L^*)/(1+Y_Q) & \to & IG(L_Q/K_P)
\end{array}$$

The vertical maps are surjective and the horizontal sequences are short exact so that there is an isomorphism

7.2.8 $$Ker(\beta_P) \cong (1+Y_Q)/(1+X_Q) \cong Y_Q/X_Q.$$

The $\mathbf{Z}[G(L_Q/K_P)]$-modules of 7.2.8 are finite and cohomologically trivial, by 7.2.2(ii).

Similarly Y/X is a finite $\mathbf{Z}[G(L/K)]$-module, which is therefore the sum of its P-completions $(Y/X)_P \cong Y_P/X_P$ (taken only over the wild P, since $X_Q = Y_Q$ when L_Q/K_P is tame, by 7.2.2(ii)). Hence, in $\mathcal{CL}(\mathbf{Z}[G(L/K)])$, we have

$$\begin{aligned}
[Y]-[X] &= [Y/X] \\
&= \textstyle\sum_{P wild}[Y_P/X_P] \\
&= \textstyle\sum_{P wild}[Y_P]-[X_P] \\
&= \textstyle\sum_{P wild} Ind_{\mathbf{Z}[G(L_Q/K_P)]}^{\mathbf{Z}[G(L/K)]}([Y_Q]-[X_Q]), && \text{by 7.2.4} \\
&= \textstyle\sum_{P wild} Ind_{\mathbf{Z}[G(L_Q/K_P)]}^{\mathbf{Z}[G(L/K)]}(Ker(\beta_P)), && \text{by 7.2.8} \\
&= \textstyle\sum_{P wild} Ind_{\mathbf{Z}[G(L_Q/K_P)]}^{\mathbf{Z}[G(L/K)]}(\Omega(L_Q/K_P, 1+Y_Q) \\
&\qquad -\Omega(L_Q/K_P, 1+X_Q)).
\end{aligned}$$

This proves the result when $X \subset Y$. Given two general lattices, X and Y, we may find a third lattice, W, which lies in $X \cap Y$ and may apply the preceding argument to compare the effect of choosing X or Y with that of choosing W. □

Remark 7.2.9 When L_Q/K_P is tame for all P we say that L/K is *tame*. Observe that

$$\Omega(L/K, 2) = [\mathcal{O}_L] \in \mathcal{CL}(\mathbf{Z}[G(L/K)])$$

when L/K is tame.

7.2.10 *The analytic class, $W_{L/K}$*

Let L/K be a finite, Galois extension of number fields and let $\rho : G(L/K) \longrightarrow GL(V)$ be a finite-dimensional, complex representation. To this data is attached the *extended Artin L-function*

$$\Lambda_K(s, \rho)$$

(cf. Snaith, 1989b, p. 253), which is a meromorphic function of the complex variable, s, satisfying the following properties:

(i) $\Lambda_K(s, \rho_1 \oplus \rho_2) = \Lambda_K(s, \rho_1)\Lambda_K(s, \rho_2)$.

(ii) If $K \subset L \subset N$ is a chain of finite Galois extensions and $G(N/K) \longrightarrow G(L/K)$ is the canonical map then

$$\Lambda_K(s, Inf_{G(L/K)}^{G(N/K)}(\rho)) = \Lambda_K(s, \rho).$$

(iii) If F is an intermediate field of L/K and $\psi : G(L/F) \longrightarrow GL(W)$ is a representation then

$$\Lambda_K(s, Ind_{G(L/F)}^{G(L/K)}(\psi)) = \Lambda_F(s, \psi).$$

(iv) If $\overline{\rho}$ denotes the complex conjugation of ρ then

$$\Lambda_K(1 - s, \rho) = W_K(\rho)\Lambda_K(s, \overline{\rho}),$$

where $W_K(\rho)$ is a complex number of modulus one.

The invariant, $W_K(\rho)$, is called the *Artin root number* of ρ. By properties (i)–(iv)

$$W_K : R(G(L/K)) \longrightarrow \mathbf{C}^*$$

is a homomorphism into the unit circle. Furthermore, if $\rho = \overline{\rho}$ then $W_K(\rho) \in \{\pm 1\}$ and if ρ is an orthogonal representation then $W_K(\rho) = 1$ (Tate, 1977, p. 130; Snaith, 1989b, p. 289, (3.9) (proof)). Therefore, if

ρ is an irreducible complex representation derived from the underlying complex vector space of a symplectic representation then $W_K(\rho) \in \{\pm 1\}$.

In terms of these invariants of analytic origins we may define a class

7.2.11 $$W_{L/K} \in \mathcal{CL}(\mathbf{Z}[G(L/K)])$$

in the following manner. We shall use the Hom-description of 4.2.28. Let $E/\mathbf{Q}$ be a large Galois extension of the rational numbers, as in 4.2.8. The absolute Galois group, $\Omega_{\mathbf{Q}} = G(\mathbf{Q}^c/\mathbf{Q})$, acts transitively on the infinite places of E. Let v_∞ denote one of these places. Define a homomorphism

$$W'_{L/K} \in Hom_{\Omega_{\mathbf{Q}}}(R(G(L/K)), J^*(E))$$

by the following formula for the vth coordinate of $W'_{L/K}(\rho)$, where ρ is any irreducible representation:

$$W'_{L/K}(\rho)_v = \begin{cases} 1 & \text{if } v \text{ is finite} \\ 1 & \text{if } \rho \text{ is not symplectic} \\ \alpha^{-1}(W_K(\alpha(\rho))) & \text{if } \rho \text{ symplectic, } v = \alpha^{-1}(v_\infty), \\ & \qquad \alpha \in \Omega_{\mathbf{Q}}. \end{cases}$$

By construction, $W'_{L/K}(\rho)$ is $\Omega_{\mathbf{Q}}$-equivariant and therefore represents a class, $W_{L/K}$, in 7.2.11. However, we must show that the construction of $W_{L/K}$ is independent of the choice of v_∞. In the tamely ramified case it is possible to define $W_{L/K}$ without making a choice of v_∞ and the construction of $W_{L/K}$ was introduced in this form by Ph. Cassou-Noguès (1978) and by A. Fröhlich. The generalisation of 7.2.11 is due to Chinburg (1989, p. 18).

Proposition 7.2.12 *The class, $W_{L/K}$, of* 7.2.11 *is independent of the choice of v_∞.*

Proof If ρ is a representation of $G(L/K)$ then, for each prime $P \lhd \mathcal{O}_K$ with $Q \lhd \mathcal{O}_L$ above it, we may form the *conductor ideal* (Martinet, 1977b, p. 14)

$$I_P(\rho) = P^{f_{K_P}(\rho_P)} \lhd \mathcal{O}_{K_P},$$

where $\rho_P = Res^{G(L/K)}_{G(L_Q/K_P)}(\rho)$ and f_{K_P} is the Artin conductor of 6.1.15. The *Artin conductor ideal*, $f(\rho)$, is the product of these ideals and its *absolute norm*, $Nf(\rho)$, is the integer given by the product

7.2.13 $$Nf(\rho) = \prod_P \#(\mathcal{O}_{K_P}/I_P(\rho)).$$

Write $\surd(Nf(\rho))$ for the positive square-root of 7.2.13.

Since $\surd(Nf(\rho))$ is a positive real number the elements

$$\alpha^{-1}(\surd(Nf(\alpha(\rho)))) = \alpha^{-1}(\surd(Nf(\rho))) \in E^*_{\alpha^{-1}(v_\infty)}$$

are positive reals at all infinite places $\alpha^{-1}(v_\infty)$. Hence we may define

$$h \in Hom_{\Omega_{\mathbf{Q}}}(R(G(L/K)), J^*(E))$$

by replacing $\alpha^{-1}(W_K(\alpha(\rho)))$ with $\alpha^{-1}(\surd(Nf(\alpha(\rho))))$ in the definition of $W'_{L/K}(\rho)_v$. The resulting class is equal to one at all finite places and is totally positive on symplectic representations at all infinite places. By Taylor (1984, p. 9); (see also Chinburg, 1989, p. 19, (2.6.1) (proof)) h is a determinant and represents the trivial element in $\mathcal{CL}(\mathbf{Z}[G(L/K)])$. Hence $W_{L/K}$ may be represented by $W'_{L/K} \cdot h^{-1}$. However, when ρ is irreducible and symplectic,

$$\frac{\alpha^{-1}(W_K(\alpha(\rho)))}{\alpha^{-1}(\surd(Nf(\alpha(\rho))))} = \frac{W_K(\rho)}{\surd(Nf(\rho))}$$

(Chinburg, 1983, p. 327) so that the $\alpha^{-1}(v_\infty)$-component of this representative for $W_{L/K}$ is independent of α and hence does not depend on the choice of v_∞. □

7.2.14 *The Fröhlich-Chinburg Conjecture*

Let L/K be a finite Galois extension of number fields. In this case Chinburg (Cassou-Noguès *et al.*, 1991; Chinburg, 1989; Chinburg, 1985) has conjectured that

$$\Omega(L/K, 2) = W_{L/K} \in \mathcal{CL}(\mathbf{Z}[G(L/K)]).$$

This conjecture has a considerable amount of supporting evidence. In the first place it generalises, by 7.2.9, a conjecture of Fröhlich which states, if L/K is tamely ramified, that

$$[\mathcal{O}_L] = W_{L/K} \in \mathcal{CL}(\mathbf{Z}[G(L/K)]).$$

This case was proved by M.J. Taylor (1981) and is also described in Fröhlich (1983). Furthermore, several wildly ramified examples have been proved (Chinburg, 1989; Kim, 1991). Also, M. Rogers has proved that $\Omega(L/K, 2) - W_{L/K}$ is in the kernel of the map from $\mathcal{CL}(\mathbf{Z}[G(L/K)])$ to $G_0(\mathbf{Z}[G(L/K)])$, the Grothendieck group of all finitely-generated $\mathbf{Z}[G(L/K)]$-modules. In fact (Chinburg, 1983, p. 327), $D(\mathbf{Z}[G(L/K)])$

is contained in this kernel and D. Holland (1992) has shown the stronger result that

$$\Omega(L/K,2) - W_{L/K} \in D(\mathbf{Z}[G(L/K)]).$$

This amounts to the evaluation of $\Omega(L/K,2)$ in $\mathcal{CL}(\Lambda(G(L/K)))$, the class-group of a maximal order. Using the detection mechanism which was developed in 5.4.52 we will give a different proof of Holland's theorem in the next section.

The mixture of the algebraic and the analytic aspects of L/K in the Fröhlich–Chinburg conjecture is especially attractive and number theory owes a great debt to Ali Fröhlich for his discovery of this deep connection and for his considerable contributions to its successful progress.

7.3 The Chinburg invariant modulo $D(\mathbf{Z}[G])$

In this section we will prove a result of D. Holland (1992) which states that the Fröhlich–Chinburg conjecture is true modulo $D(\mathbf{Z}[G(L/K)])$. This result was mentioned in 7.2.14. As in the previous section, let L/K be a finite Galois extension of number fields with group $G(L/K)$. Let $E/\mathbf{Q}$ be a large Galois extension of the rationals, as in 4.2.8, and let $\Lambda(G(L/K))$ denote a maximal order of $\mathbf{Q}[G(L/K)]$, as in 5.4.1. Extension of scalars induces a homomorphism of class-groups

7.3.1 $$\mathcal{CL}(\mathbf{Z}[G(L/K)]) \longrightarrow \mathcal{CL}(\Lambda(G(L/K))),$$

whose kernel is $D(\mathbf{Z}[G(L/K)])$. Since we wish to evaluate the image

7.3.2 $$\Omega(L/K,2) \in \mathcal{CL}(\Lambda(G(L/K)))$$

we must discuss means whereby we can detect elements in $\mathcal{CL}(\Lambda(G))$, where G is any finite group. In our calculations we will be interested in related computations which take place in the Grothendieck group of locally freely presented $\Lambda(G)$-modules, $K_0T(\Lambda(G))$, of 5.4.7. Recall from 5.4.8 that there is a natural isomorphism

7.3.3 $$\mu_G : Fac(G) \xrightarrow{\cong} K_0T(\Lambda(G))$$

and that the kernel of the natural surjection

$$c_G : K_0T(\Lambda(G)) \longrightarrow \mathcal{CL}(\Lambda(G))$$

is isomorphic to $PF^+(G)$ of 5.4.6. Therefore, with this type of application in mind, we shall begin by studying invariants of functions which lie in the group, $Hom_{\Omega_{\mathbf{Q}}}(R(G), \mathcal{I}(E))$ or in the subgroup, $Fac(G)$, of 5.4.4.

7.3.4 We begin by observing that a function

$$g \in Hom_{\Omega_{\mathbf{Q}}}(R(G), \mathcal{I}(E))$$

is determined by its *local components* in the following manner. For each integral prime, p, choose a prime, $P \lhd \mathcal{O}_E$, which divides p. A non-zero fractional ideal is determined by the class of the idèle given by the generators of its completions. The action of $\Omega_{\mathbf{Q}}$ permutes the primes, P over p, with stabiliser $\Omega_{\mathbf{Q}_p}$. Hence we see that g is determined by its local components

$$g_p \in Hom_{\Omega_{\mathbf{Q}_p}}(R(G), \mathcal{I}(E_P)).$$

Henceforth we shall usually work in terms of these local components.

Definition 7.3.5 Let G be a finite group and let p be an integral prime. Define $\hat{S}^p(G)$ to be the set of all pairs

$$(C, \phi)^G \in R_+(G)$$

where C is cyclic and $HCF(p, \#(im(\phi))) = 1$. Hence, if C_p is the Sylow p-subgroup of C, we may write $C \cong C_p \times C'$ and ϕ factorises through the projection onto C',

$$\phi : C \longrightarrow C' \longrightarrow E^*.$$

Each such ϕ determines an irreducible idempotent, e_ϕ, of $\mathbf{Z}_p[C']$ (cf. 5.5.12) since representatives of the Galois orbits of homomorphisms, $\psi : C' \longrightarrow E^*$, induce an isomorphism of the form

7.3.6 $$\prod \psi : \mathbf{Z}_p[C'] \xrightarrow{\cong} \prod_\psi \mathcal{O}_{\mathbf{Q}_p(\psi)}.$$

In 7.3.6 the product is taken over representatives of the $\Omega_{\mathbf{Q}}$-orbits of the $\{\psi\}$.

In terms of characters the idempotent, e_ϕ, induces an irreducible idempotent of $\mathbf{Q}_p[C']$ whose character as a $\mathbf{Q}_p$-representation is equal to

$$\chi(e_\phi) = \sum_\omega \omega(\phi),$$

where ω runs through $\Omega_{\mathbf{Q}}/stab(\phi)$. That is, $\chi(e_\phi)$ is the sum of all the distinct $\Omega_{\mathbf{Q}_p}$-conjugates of ϕ.

Define $S^p(G)$ to be the set of pairs (C, e) where $C \leq G$ is a cyclic subgroup and e is an irreducible idempotent of $\mathbf{Z}_p[C']$. All such idempotents are of the form, $e = e_\phi$, for some ϕ. We will often abbreviate $\chi(e_\phi)$ to $\chi(e)$ in the remainder of this section.

Let $f^p : S^p(G) \longrightarrow \mathscr{I}(\mathbf{Q}_p)$ be any map. We say that

7.3.7
$$g \in Hom_{\Omega_{\mathbf{Q}_p}}(R(G), \mathscr{I}(E_P))$$

is the *canonical factorisation* of f^p if, for all $(C, e) \in S^p(G)$,

7.3.8
$$g(Ind_C^G(Inf_{C'}^C(\chi(e))) = f^p((C, e)).$$

If f^* is a family of maps of the form

7.3.9
$$f^* = \{f^p : S^p(G) \longrightarrow \mathscr{I}(\mathbf{Q}_p)\}$$

we say that

$$g \in Hom_{\Omega_{\mathbf{Q}}}(R(G), \mathscr{I}(E))$$

is the *canonical factorisation* of f^* in 7.3.9 if, for each p, the p-component of g is the canonical factorisation of f^p in the sense of 7.3.8.

Definition 7.3.10 Now we shall introduce the functions, f^* as in 7.3.9, which are associated to $\mathbf{Z}[G]$-modules. Our treatment follows Holland (1992, section 2) but the reader should also compare that of Burns (1991) and Fröhlich (1988). See also Cassou-Noguès *et al.* (1991, p. 95).

Let M_1 and M_2 be $\mathbf{Z}[G]$-modules, which need not be finitely generated. Let

$$i : M_1 \longrightarrow M_2$$

be an injective $\mathbf{Z}[G]$-homomorphism with *finite* cokernel, $coker(i)$. Define $f_i^* = \{f_i^p ; p \text{ } prime\}$ in the following manner. If $(C, e) \in S^p(G)$ we may form the $\mathbf{Z}_p[C']$-modules $(i = 1, 2)$

$$M_{i,p}^{C_p} = (M_i \otimes \mathbf{Z}_p)^{C_p} \quad \text{and} \quad M_{i,p}^{C_p} e$$

where e is the idempotent of 7.3.5. Define $f_i^p(C, e)$ to be the order ideal given by the $\mathbf{Z}_p$-order

7.3.11
$$f_i^p(C, e) = \#(coker(i^{C_p})_p \cdot e) = \#((M_{2,p}^{C_p} \cdot e)/(M_{1,p}^{C_p} \cdot e)) \in \mathscr{I}(\mathbf{Q}_p).$$

(In general, if M is a finitely generated, torsion $\mathcal{O}_E$-module its *order ideal* in $\mathcal{I}(E)$ is given by the unique ideal, $J \lhd \mathcal{O}_E$, such that $\mathcal{O}_E/J$ and M have the same $\mathcal{O}_E$-composition factors.)

Following Holland (1992) we shall alternatively write f_i^* as f_{M_1,M_2}^* when there is no ambiguity about the map, $i : M_1 \longrightarrow M_2$. In addition, if $M_1 = 0$ and $M_2 = N$ is finite we will abbreviate $f_{0,N}^* = f_i^*$ to f_N^*. Should it be necessary to indicate the dependence upon G we will denote f_i^* by $f_{G,i}^*$ and f_N^* by $f_{G,N}^*$.

Theorem 7.3.12 *Let G be a finite group then the family of maps, $f_i^* = \{f_i^p ; p \text{ prime}\}$ of 7.3.11 possesses the following properties:*

(i) *Let $j : M_1 \longrightarrow M_2$ and $i : M_2 \longrightarrow M_3$ be injective $\mathbf{Z}[G]$-homomorphisms with finite cokernels, then*

$$f_{ij}^* = f_i^* f_j^* = \{f_i^p f_j^p ; p \ \text{prime}\}.$$

(ii) *For $H \leq G$, let $N_G H$ denote the normaliser of H in G. If, for each $H \leq G$, there is an $\mathbf{Z}[N_G H]$-isomorphism of the form*

$$coker(i^H) \cong coker(j^H),$$

then

$$f_i^* = f_j^*.$$

(iii) *Suppose that, for each $H \leq G$, there is a $\mathbf{Z}[N_G H]$-isomorphism of the form*

$$coker(i^H) \cong (coker(i))^H,$$

then

$$f_i^* = f_{coker(i)}^*.$$

In particular, this is true for the injection $i : M_2 \longrightarrow M_3$ if $H^1(H; M_2) = 0$ for all $H \leq G$.

(iv) *Suppose that there is a commutative diagram of $\mathbf{Z}[G]$-modules*

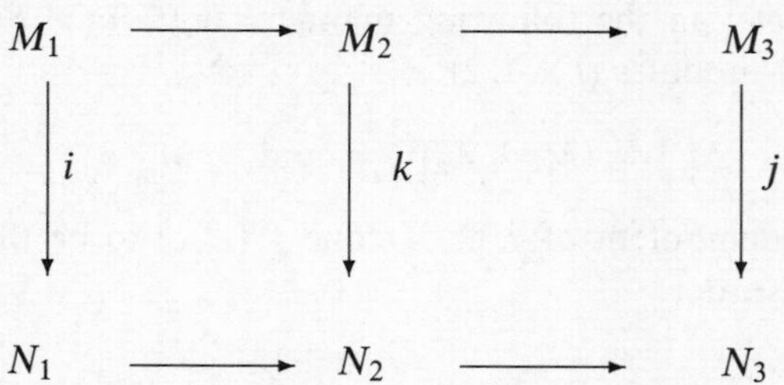

and that, for each $H \leq G$, the induced sequence

$$0 \longrightarrow coker(i^H) \longrightarrow coker(k^H) \longrightarrow coker(j^H) \longrightarrow 0$$

is an exact sequence of $\mathbf{Z}[N_G H]$-modules. Then

$$f_k^* = f_i^* f_j^*.$$

Proof We begin by observing that, if $(C,e) \in S^p(G)$, then $C \leq N_G C_p$ and hence $e \in \mathbf{Z}_p[N_G C_p]$. Also $(- \otimes_{\mathbf{Z}} \mathbf{Z}_p)$ is an exact functor.

Therefore, for part (i), we have inclusions $M_{1,p}^{C_p} \longrightarrow M_{1,p}^{C_p}$ and $M_{2,p}^{C_p} \longrightarrow M_{3,p}^{C_p}$. Consequently there are inclusions $M_{1,p}^{C_p} e \longrightarrow M_{1,p}^{C_p} e$ and $M_{2,p}^{C_p} e \longrightarrow M_{3,p}^{C_p} e$, from which the result follows since

$$\begin{aligned} f_{ij}^p(C,e) &= \#(M_{3,p}^{C_p} e / M_{1,p}^{C_p} e) \\ &= \#(M_{3,p}^{C_p} e / M_{2,p}^{C_p} e) \cdot \#(M_{2,p}^{C_p} e / M_{1,p}^{C_p} e) \\ &= f_i^p(C,e) f_j^p(C,e). \end{aligned}$$

Assume that $i : M \longrightarrow N$ and $j : M' \longrightarrow N'$ satisfy

$$(M_p')^{C_p} / (N_p')^{C_p} \cong (M_p)^{C_p} / (N_p)^{C_p}$$

for all $(C,e) \in S^p(G)$. Since taking idempotents is an exact functor, applying e to this isomorphism yields an isomorphism of the form

$$(M_p')^{C_p} e / (N_p')^{C_p} e \cong (M_p)^{C_p} e / (N_p)^{C_p} e,$$

which proves part (ii).

If $i : M_2 \longrightarrow M_3$ is an injective $\mathbf{Z}[G]$-homomorphism then $f_{coker(i)}^*$ is given by

$$f_{coker(i)}^p(C,e) = \#((coker(i)_p^{C_p})e),$$

which, under the conditions of part (iii), is equal to

$$\#(M_{3,p}^{C_p} e / M_{2,p}^{C_p} e) = f_i^p(C,e).$$

In particular, if $H^1(H; M_2) = 0$ then the exact cohomology sequence of

$$M_2 \longrightarrow M_3 \longrightarrow coker(i)$$

starts with

$$0 \longrightarrow M_2^H \longrightarrow M_3^H \longrightarrow coker(i)^H \longrightarrow H^1(H; M_2) = 0$$

and we obtain an isomorphism between $coker(i)^H$ and $coker(i^H)$, as required.

Applying e to the cokernel exact sequence in part (iv) yields an exact sequence of finite abelian p-groups

$$0 \longrightarrow coker(i^{C_p})_p e \longrightarrow coker(k^{C_p})_p e \longrightarrow coker(j^{C_p})_p e \longrightarrow 0$$

for each $(C, e) \in S^p(G)$. The order of the central group, $f_k^p(C, e)$, is therefore equal to the product of the orders of the other two groups, which is $f_i^p(C, e) f_j^p(C, e)$, as required. □

Theorem 7.3.13 *Let G be a finite group and let $f^* = \{f^p : S^p(G) \longrightarrow \mathscr{I}(\mathbf{Q}_p)\}$ be as in 7.3.9. Then, if the canonical factorisation of f^* exists in 7.3.5, it is unique.*

Proof Choose a prime, p, and consider the canonical factorisation of $f^p : S^p(G) \longrightarrow \mathscr{I}(\mathbf{Q}_p)$. The difference between two canonical factorisations of f^p will be a canonical factorisation of the constant function whose value is the ideal, $\mathbf{Z}_p$. We must show that any canonical factorisation

$$g \in Hom_{\Omega_{\mathbf{Q}_p}}(R(G), \mathscr{I}(E_P))$$

of the constant map is also equal to the constant function whose value is the ideal, $\mathbf{Z}_p$.

Since $Hom_{\Omega_{\mathbf{Q}_p}}(R(G), \mathscr{I}(E_P))$ is torsion free, Artin's induction theorem (see 2.1.3) implies that the natural maps, induced by Ind_C^G with C cyclic, yield an injective map

$$Hom_{\Omega_{\mathbf{Q}_p}}(R(G), \mathscr{I}(E_P)) \longrightarrow \oplus_{\substack{C \\ cyclic}} Hom_{\Omega_{\mathbf{Q}_p}}(R(C), \mathscr{I}(E_P)).$$

Therefore we must show that

7.3.14 $$\mathbf{Z}_p = g(Ind_C^G(\lambda)) \in \mathscr{I}(E_P)$$

for all $\lambda : C \longrightarrow E^*$, where C is cyclic. If $C = C_p \times C'$, as in 7.3.5, we may write λ as the product of $\phi : C' \longrightarrow E^*$ and $\psi : C_p \longrightarrow E^*$. By 7.3.8, we know that 7.3.14 is true when ψ is trivial. This is because, if $e = e_\phi, \chi(e) = \chi(e_\phi)$ and ϕ are as in 7.3.5, then

$$\begin{aligned} \mathbf{Z}_p &= g(Ind_C^G(Inf_{C'}^C(\chi(e)))) \\ &= \textstyle\sum_\omega g(Ind_C^G(Inf_{C'}^C(\omega(\phi)))) \\ &= g(Ind_C^G(Inf_{C'}^C(\phi)))^t, \end{aligned}$$

where t is the number of distinct homomorphisms in the $\Omega_{\mathbf{Q}}$-orbit of ϕ.

Fix $(C, \phi) \in \hat{S}^p(G)$, as in 7.3.5, and let $H \leq C_p$ be a subgroup. Since $H \times C'$ is cyclic we have

$$\begin{aligned} \mathbf{Z}_p &= g(Ind_{H\times C'}^G(Inf_{C'}^{H\times C'}(\phi))) \\ &= g(Ind_{C_p\times C'}^G(Ind_H^{C_p}(1) \otimes \phi)) \\ &= \prod_\alpha g(Ind_{C_p\times C'}^G(\psi_\alpha \otimes \phi)), \end{aligned}$$

where $Ind_H^{C_p}(1) = \sum_\alpha \psi_\alpha$ for one-dimensional representations,

$$\psi_\alpha : C_p \longrightarrow C_p/H \longrightarrow E^*.$$

By induction on the order of the image of ψ_α we may suppose that

$$\mathbf{Z}_p = g(Ind_{C_p\times C'}^G(\psi_\beta \otimes \phi))$$

for all ψ_β except those $\{\psi_\alpha\}$ for which $\#(im(\psi_\alpha))$ is maximal. However, these ψ_α comprise a Galois orbit consisting of $\psi_{\alpha_1}, \ldots, \psi_{\alpha_m}$, say. Hence

$$\begin{aligned} \mathbf{Z}_p &= \prod_{s=1}^m g(Ind_{C_p\times C'}^G(\psi_{\alpha_s} \otimes \phi)) \\ &= g(Ind_{C_p\times C'}^G(\psi_{\alpha_1} \otimes \phi))^m \end{aligned}$$

so that $\mathbf{Z}_p = g(Ind_{C_p\times C'}^G(\psi_\alpha \otimes \phi))$ for all the ψ_α's. This completes the induction step, which starts with $\psi = 1$, to show that $\mathbf{Z}_p = g(Ind_C^G(\psi\otimes\phi))$ for all C, ϕ and ψ. $\square$

7.3.15 Let $K_0T(\mathbf{Z}[G])$ denote the Grothendieck group of finite, locally freely presented $\mathbf{Z}[G]$-modules, whose definition is analogous to that of $K_0T(\Lambda(G))$ in 5.4.7. As in 5.4.8 there is a Cartan map

$$c_G : K_0T(\mathbf{Z}[G]) \longrightarrow \mathcal{CL}(\mathbf{Z}[G]).$$

In addition, we have the Hom-description of 4.2.28

$$Det : \mathcal{CL}(\mathbf{Z}[G]) \xrightarrow{\cong} \frac{Hom_{\Omega_{\mathbf{Q}}}(R(G), J^*(E))}{Hom_{\Omega_{\mathbf{Q}}}(R(G), E^*) \cdot Det(U(\mathbf{Z}[G]))}.$$

There is a similar Hom-description (Taylor, 1984, p. 10)

$$K_0T(\mathbf{Z}[G]) \cong \frac{Hom_{\Omega_{\mathbf{Q}}}(R(G), J^*(E))}{Det(U(\mathbf{Z}[G]))}$$

and c_G may be identified with the canonical quotient map. Therefore a finite, cohomologically trivial $\mathbf{Z}[G]$-module, T, defines a class

7.3.16 $$[T] \in K_0T(\mathbf{Z}[G]),$$

which is represented by a homomorphism

7.3.17 $$h_T \in Hom_{\Omega_{\mathbf{Q}}}(R(G), J^*(E)).$$

In addition, there is a natural ideal map

7.3.18 $$I : J^*(E) \longrightarrow \mathscr{I}(E),$$

whose kernel is $U(E)$.

The following result relates canonical factorisations with representatives of the image of $[T]$.

Theorem 7.3.19 *Let $[T] \in K_0T(\mathbf{Z}[G])$ be the class of a finite, cohomologically trivial module, which is represented by h_T in* 7.3.17. *Let $f_T^* = \{f_T^p : S^p(G) \longrightarrow \mathscr{I}(\mathbf{Q}_p)\}$ be the family of functions defined in* 7.3.10. *Then*

$$I \cdot h_T \in Hom_{\Omega_{\mathbf{Q}}}(R(G), \mathscr{I}(E))$$

lies in $Fac(G)$ of 5.4.4 *and is the canonical factorisation of f_T^*.*

Proof As in 5.4.7, $K_0T(\mathbf{Z}[G])$ is generated by the classes, $[T]$, modulo relations which come from short exact sequences of such $\{T\}$. However, a short exact sequence of finite, cohomologically trivial $\mathbf{Z}[G]$-modules remains exact over $\mathbf{Z}[N_GH]$ upon taking H-fixed points. Hence, by 7.3.12(iv), $(T \mapsto f_T^*)$ factors through $K_0T(\mathbf{Z}[G])$. Since $K_0T(\mathbf{Z}[G])$ is generated by classes of the form $[\mathbf{Z}[G]/J]$ where $J \lhd Z[G]$ is a locally free ideal, we may assume that $T = \mathbf{Z}[G]/J$. In this case $J = \mathbf{Z}[G]\alpha$, for some idèle $\alpha \in J^*(\mathbf{Q}[G])$ and, by 4.2.28, $h_T = Det(\alpha)$. By naturality of the isomorphism of 5.4.8(i), $g = I \cdot Det(\alpha) \in Fac(G)$. Furthermore, if p is a prime, the local components of g are given by

$$g_p = I \cdot Det(\alpha_p)$$

by Fröhlich (1983, II, lemma 2.1). Hence we may restrict our attention to the local components

$$g_p \in Hom_{\Omega_{\mathbf{Q}_p}}(R(G), \mathscr{I}(E_P)).$$

For $(C, e) \in S^p(G)$ we must show that

7.3.20 $$Det(\alpha_p)(Ind_C^G(Inf_{C'}^C(\chi(e)))) = \#(T_p^{C_p}e).$$

Since $Res_C^G = Det(\alpha)(Ind_C^G(-))$ represents $Res_C^G([T]) \in K_0T(\mathbf{Z}[G])$ (cf. 4.4.18) we may assume that $G = C$. Furthermore, by 7.5.6, $Det(\alpha)(Inf_{C'}^C(-))$ represents $[T^{C_p}] \in K_0T(\mathbf{Z}[C'])$ so that we may assume that $HCF(p, \#(C)) = 1$. In this case $\mathbf{Z}_p[C]e \subset \mathbf{Q}_p[C]e$ which is the $\mathbf{Q}_p$-representation whose $\mathbf{Q}_p$-character is

$$\chi(e) = \chi(e_\phi) = \sum_{\Omega_p \ orbits} \omega(\phi),$$

as in 7.3.5. Hence the left side of 7.3.20 becomes

$$Det(\alpha)(\chi(e)),$$

which is the order of the finite cokernel of the map induced by α:

$$\alpha : \mathbf{Z}_p[C]e \longrightarrow \mathbf{Z}_p[C]e$$

(cf. 5.2.32(proof)). Finally, the order of this cokernel is

$$\#(\mathbf{Z}_p[C]e/\alpha\mathbf{Z}_p[C]e) = \#(T_p^{C_p}e),$$

as required. □

Proposition 7.3.21 *Let $N \lhd G$ be a normal subgroup and let $i : V \longrightarrow W$ be an injective homomorphism of $\mathbf{Z}[G/N]$-modules such that $W/i(V)$ is finite. If $f^*_{G/N,i}$ has canonical factorisation*

$$\{g_p\} = g \in Hom_{\Omega_{\mathbf{Q}}}(R(G/N), \mathscr{I}(E))$$

then $(Z \mapsto g(Z^N))$ is the canonical factorisation of $f^*_{G,i}$.

Proof Once more we will work with the local components.

Let $(C, e) \in S^p(G)$ then we must show that

7.3.22 $$g_p((Ind_C^G(Inf_{C'}^C(\chi(e))))^N) = f^p_{G,i}(C, e),$$

where $\chi(e)$ is as in 7.3.5. Since

$$Inf_{C'}^C(\chi(e)) = 1 \otimes \chi(e) : C \cong C_p \times C' \longrightarrow GL_t(\mathbf{Q}_p)$$

we have, by 2.5.14,

$$(Ind_C^G(Inf_{C'}^C(\chi(e))))^N = Ind_{C/C\cap N}^{G/N}((1 \otimes \chi)^{C\cap N}).$$

Hence

$$g_p((Ind_C^G(Inf_{C'}^C(\chi(e))))^N)$$

$$= g_p(Ind_{C/C\cap N}^{G/N}((1 \otimes \chi)^{C\cap N}))$$

$$= \begin{cases} \mathbf{Z}_p, & \text{if } ker(\chi(e)) \not\subseteq C' \cap N, \\ f_{G/N,i}^p((C/C\cap N, \overline{e})), & \text{otherwise,} \end{cases}$$

where $\overline{e} \in \mathbf{Z}_p[C/C\cap N]$ is the image of the idempotent, e.

However, if $X = W$ or V, then $X_p^{C_p} = X_p^{C_p/C_p\cap N}$ since N acts trivially on X. Therefore

$$X_p^{C_p}e = X_p^{C_p/C_p\cap N}\overline{e},$$

which is trivial unless ϕ (or equivalently, $\chi(e) = \chi(e_\phi)$) vanishes on $C' \cap N$. These observations easily imply 7.3.22. □

Proposition 7.3.23 *Let $H \leq G$ be a subgroup and let $i : V \longrightarrow W$ be an injective $\mathbf{Z}[H]$-module homomorphism. If $f_{H,i}^*$ has canonical factorisation*

$$\{g_p\} = g \in Hom_{\Omega_{\mathbf{Q}}}(R(H), \mathscr{I}(E))$$

then $(Z \mapsto g(Res_H^G(Z))$ is the canonical factorisation of $f_{G,j}^*$, where j is the tensor product map

$$j = \mathbf{Z}[G] \otimes_{\mathbf{Z}[H]} i : \mathbf{Z}[G] \otimes_{\mathbf{Z}[H]} V \longrightarrow \mathbf{Z}[G] \otimes_{\mathbf{Z}[H]} W.$$

Proof Let $(C, e) \in S^p(G)$ then we must show that

7.3.24 $$g_p(Res_H^G(Ind_C^G Inf_{C'}^C(\chi(e)))) = f_{G,j}^p(C, e).$$

By the double coset formula of 1.2.40, the left side of 7.3.24 is equal to

$$\prod_{z\in H\backslash G/C} g_p(Ind_{H\cap zCz^{-1}}^H(z^{-1})^*(Res_{z^{-1}Hz\cap C}^C(Inf_{C'}^C(\chi(e)))))$$

$$= \prod_{z\in H\backslash G/C} g_p(Ind_{H\cap zCz^{-1}}^H(Inf_{H\cap zC'z^{-1}}^{H\cap zCz^{-1}}((z^{-1})^*(\chi(e))))).$$

Let $z(e) = zez^{-1} \in Z_p[zCz^{-1}]$ and let e_z denote the unique indecomposable idempotent of $\mathbf{Z}_p[(H \cap (zCz^{-1}))']$ such that

$$e_z(z(e)) = z(e).$$

Here, as in 7.3.5,

$$H \cap (zCz^{-1}) = (H \cap (zCz^{-1}))_p \times (H \cap (zCz^{-1}))'.$$

Let $\chi(e_z)$ denote the character of the $\mathbf{Q}_p$-representation,

$$\mathbf{Q}_p[(H \cap (zCz^{-1}))']e_z.$$

Let m_z denote the degree of the 'abstract' field extension

$$\mathbf{Q}_p[(zCz^{-1})']z(e)/\mathbf{Q}_p[(H \cap (zCz^{-1}))']e_z.$$

In other words,

$$m_z = \frac{dim_{\mathbf{Q}_p}(\mathbf{Q}_p[(zCz^{-1})']z(e))}{dim_{\mathbf{Q}_p}(\mathbf{Q}_p[(H \cap (zCz^{-1}))']e_z)},$$

the ratio of the $\mathbf{Q}_p$-dimensions of the local fields which correspond to the factors in the idempotent decompositions of 7.3.6.

Hence

$$Res^{zCz^{-1}}_{H \cap zCz^{-1}}(1 \otimes (z^{-1})^*(\chi(e))) = m_z(1 \otimes \chi(e_z))$$

and we find that the left side of 7.3.24 becomes

$$\begin{aligned}&\textstyle\prod_{z \in H\backslash G/C} g_p(Ind^H_{H \cap zCz^{-1}}(1 \otimes \chi(e_z)))^{m_z}\\ &= \textstyle\prod_{z \in H\backslash G/C}(f^p_{H,i}(H \cap zCz^{-1}, e_z))^{m_z}.\end{aligned}$$

Similarly, if $X = V$ or W, there is a $\mathbf{Z}_p$-isomorphism of the form

$$\begin{aligned}&(Ind^G_H(X_p))^{C_p}e\\ &\cong \oplus_{z \in H\backslash G/C}\mathbf{Z}_p[C']z(e) \otimes_{\mathbf{Z}_p[(H \cap zCz^{-1})']e_z} (X_p^{(H \cap zCz^{-1})_p})e_z\end{aligned}$$

from which we deduce that

$$f^p_{G,j}(C, e) = \prod_{z \in H\backslash G/C} (f^p_{H,i}(H \cap zCz^{-1}, e_z))^{m_z},$$

as required. □

Example 7.3.25 Let us consider the case in which L/K is a wildly ramified Galois extension of p-adic local fields with *cyclic* Galois group, $G(L/K)$, generated by x of order n.

Consider the following commutative diagram:

7.3.26

$$\begin{array}{ccccccccc} 0 & \longrightarrow & \mathbf{Z} & \longrightarrow & \mathbf{Z}[G(L/K)] & \stackrel{\beta}{\longrightarrow} & IG(L/K) & \longrightarrow & 0 \\ & & \downarrow \gamma & & \downarrow i & & \downarrow 1 & & \\ 0 & \longrightarrow & L^*/(1+X) & \longrightarrow & \mathbf{Z}[G(L/K)] \otimes_{\mathbf{Z}[N]} IN & \longrightarrow & IG(L/K) & \longrightarrow & 0 \end{array}$$

In 7.3.26 $\beta(z) = (x-1)z$ and $i(z) = z \otimes (u_x - 1)$. The lower sequence of 7.3.26 is that of 7.2.7 (see also 7.1.28).

The map, γ, may be assumed to be injective. In fact, we may arrange that

7.3.27 $$\gamma(1) = \pi_K \in K^*/(1 + X^{G(L/K)}) \cong (L^*/(1+X))^{G(L/K)}.$$

We may establish 7.3.27 in the following manner. By the calculation of 7.1.47, a generator for $ker(\beta)$ of 7.3.26 is given by $\sum_{i=0}^{n-1} x^i \otimes (u_x - 1)$ and

$$\gamma(1) = \sum_{i=0}^{n-1} x^i \otimes (u_x - 1) = \prod_{i=0}^{n-1} f(x^i, x) \in L^*/(1+X),$$

where $f : G(L/K) \times G(L/K) \longrightarrow L^*$ is the 2-cocycle of 7.1.34. We may alter f by any boundary map, $d(h)$, where $h : G(L/K) \longrightarrow L^*$ is any map. If $h(x) = \pi_L$ then changing f to $f \cdot d(h)$ changes $\gamma(1)$ by a factor which is equal to

$$\prod_{i=0}^{n-1} \frac{x^i(h(x))h(x^i)}{h(x^{i+1})} = N_{L/K}(\pi_L),$$

where $N_{L/K} : L^* \longrightarrow K^*$ denotes the *norm map*. Up to units in $\mathcal{O}_K^*$, $N_{L/K}(\pi_L)$ is equal to $\pi_K^{f_{L/K}}$ where $f_{L/K}$ is the residue degree. Therefore, since π_K is of infinite order in $L^*/(1+X)$, we may change f so as to ensure that $\gamma(1)$ has infinite order. In addition, if $f_{L/K} = 1$, we may ensure that $\gamma(1)$ is represented by π_K, in 7.3.27.

Now suppose that $f_{L/K} \neq 1$. Local class field theory yields an isomorphism

7.3.28 $$G(L/K) \cong K^*/(N_{L/K}(L^*))$$

By Serre (1979, lemma 4, p. 178), $\prod_{i=0}^{n-1} f(x^i, x) \in L^*$ lies in K^* and corresponds to $x \in G(L/K)$ under the isomorphism of 7.3.28. The valuation of K induces a surjective homomorphism, with kernel isomorphic to $G_0(L/K)$,

$$v_K : K^*/(N_{L/K}(L^*)) \longrightarrow \mathbf{Z}/f_{L/K}.$$

Since v_K is surjective, we see that $\gamma(1)$ may also be taken to be represented by π_K in the case when $f_{L/K} \neq 1$.

Now set $T = coker(i)$ in 7.3.26. Hence T is a finite $\mathbf{Z}[G(L/K)]$-module which is cohomologically trivial. Since $[T] \in K_0T(\mathbf{Z}[G(L/K)])$ maps to $\Omega(L/K, 1+X) \in \mathscr{CL}(\mathbf{Z}[G(L/K)])$ under the Cartan map, we wish to calculate f_T^*. By 7.3.12

7.3.29 $$f_T^* = f_i^* = f_\gamma^*.$$

Consider the following commutative diagram.

7.3.30

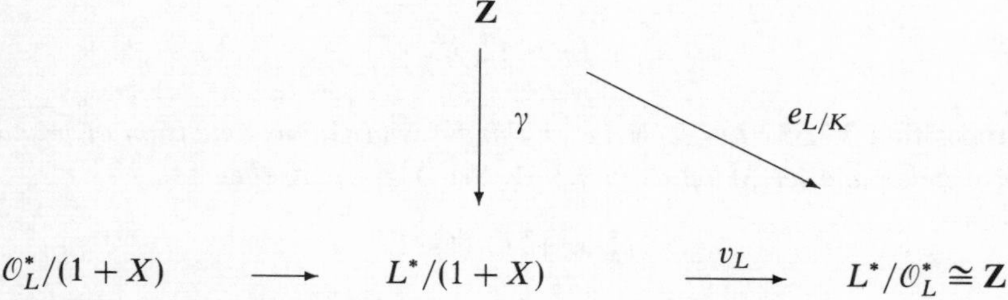

Note that $1 + X \subset \mathcal{O}_L^*$ since L/K is wild. Set

7.3.31 $$M = \mathcal{O}_L^*/(1+X)$$

and let $(C, e) \in S^p(G(L/K))$. Therefore

7.3.32

$$f_\gamma^p(C, e) = \#([(L^*/1+X)_p^{C_p} e]/\mathbf{Z}_p e)$$

$$= \begin{cases} \#(M_p^{C_p} e) & \text{if } \chi(e) \neq 1, \\ \#(M_p^{C_p} e) \cdot e_{L^C/K} & \text{if } \chi(e) = 1. \end{cases}$$

The first case of 7.3.32 follows from the fact that $e \neq 1$ annihilates trivial modules such as $\mathbf{Z}_p$. When $e = 1$ we have $W^{C_p}e = W_p^C$ and

$$v_L((L^*/1+X)_p^C) = v_L(((L^C)^*/1+X^C)_p) = e_{L/L^C}\mathbf{Z},$$

which yields the second case of 7.3.32 since $e_{L^C/K}e_{L/L^C} = e_{L/K}$.

Definition 7.3.33 Let L/K be any finite, wild Galois extension of p-adic local fields. Define a family of functions (cf. 7.3.9)

$$\mathscr{E}^* = \{\mathscr{E}^p : S^p(G(L/K)) \longrightarrow \mathscr{I}(E)\}$$

by the formula

$$\mathscr{E}^p(C,e) = \begin{cases} e_{L^C/K} & \text{if } \chi(e) = 1, \\ 1 & \text{if } \chi(e) \neq 1. \end{cases}$$

Hence, with this notation, we have proved the following result:

Lemma 7.3.34 *When L/K is cyclic in* 7.3.33 *then, in* 7.3.25,

$$f_T^* = f_M^*\mathscr{E}^*.$$

Proposition 7.3.35 *Let L/K be any finite, wild Galois extension of p-adic local fields and let M be as in* 7.3.31. *Set $A = \mathcal{O}_L/X$ then*

$$f_M^* = f_A^* f_{\overline{L}^*}^* (f_{\overline{L}}^*)^{-1}.$$

Proof Let $R_L = ideal\{\pi_L\} \lhd \mathcal{O}_L$ and let $m : 1+X \longrightarrow \mathcal{O}_L^*$, $a : X \longrightarrow \mathcal{O}_L$ denote the inclusions. By the hypothesis of 7.2.5 we may assume that exp induces an isomorphism, if $X \subseteq R_L^t$,

7.3.36 $$R_L^t/X \cong (1+R_L^t)/(1+X).$$

For this value of t, define inclusions $a_0 : R_L \longrightarrow \mathcal{O}_L$, $a_s : R_L^{s+1} \longrightarrow R_L^s$ for $s = 1,\ldots,t-1$ and $a_{t+1} : X \longrightarrow R_L^t$. Similarly define inclusions $m_0 : 1+R_L \longrightarrow \mathcal{O}_L^*$, $m_s : 1+R_L^{s+1} \longrightarrow 1+R_L^s$ for $s = 1,\ldots,t-1$ and $m_{t+1} : 1+X \longrightarrow 1+R_L^t$.

By 7.3.12(i)

$$f_A^* = f_a^* = \prod_{s=0}^{t} f_{a_s}^*$$

and

$$f^*_M = f^*_m = \prod_{s=0}^{t} f^*_{m_s}.$$

For each $H \leq G$,

$$\mathcal{O}_L^H = \mathcal{O}_{L^H}, \qquad (\mathcal{O}_L^*)^H = \mathcal{O}^*_{L^H}$$

$$(\overline{L^H})^* = (\overline{L}^*)^H, \quad \overline{L^H} = (\overline{L})^H$$

so that, by 7.3.12(iii), for $0 \leq s \leq t-1$

$$f^*_{a_s} = f^*_{coker(a_s)}, \; f^*_{m_s} = f^*_{coker(m_s)}.$$

Since X is cohomologically trivial

$$(R^t_L/X)^H \cong (R^t_L)^H/X^H = (R^t_{L^H})/X^H$$

and

$$(1 + R^t_L/1 + X)^H \cong (1 + R^t_{L^H})/(1 + X^H).$$

Hence, by 7.3.12(iii),

$$f^*_{a_t} = f^*_{coker(a_t)}, \; f^*_{m_t} = f^*_{coker(m_t)}.$$

By 7.3.36 and the exponential isomorphisms

$$R^s_L/R^{s+1}_L \cong (1 + R^s_L)/(1 + R^{s+1}_L)$$

for $1 \leq s$, we find that

$$\begin{aligned} f^*_A &= \textstyle\prod_{s=0}^t f^*_{coker(a_s)} \\ &= f^*_{coker(a_0)}(\textstyle\prod_{s=1}^t f^*_{coker(m_s)}) \\ &= f^*_{\overline{L}} f^*_M (f^*_{coker(m_0)})^{-1} \\ &= f^*_{\overline{L}} f^*_M (f^*_{\overline{L}^*})^{-1}, \end{aligned}$$

as required. □

7.3.37 Suppose that L/K is as in 7.3.35 with $\overline{L} = F_{q^n}$, $\overline{K} = F_q$. Let p be a prime and $(C, e_\phi) \in S^p(G(F_{q^n}/F_q))$. Here $\phi : C' \longrightarrow E^*$ has image of order prime to p. As in 7.3.5, let $\chi(e_\phi) : C' \longrightarrow GL_v(\mathbf{Z}_p)$ (where $v = dim_{\mathbf{Q}_p}(\chi(e_\phi)))$ be the $\mathbf{Q}_p$-representation associated to ϕ.

Suppose that $C = G(F_{q^n}/F_{q^d})$ and that $n = p^\alpha t d$ with $HCF(p,t) = 1$. Hence

$$C_p = G(F_{q^n}/F_{q^{dt}}), \ \ C' = G(F_{q^n}/F_{q^{dp^\alpha}})$$

and, by 7.1.36, we have a short exact sequence of $C/C_p = G(F_{q^{dt}}/F_{q^d})$-modules

7.3.38

$$0 \longrightarrow \mathbf{Z}_p[G(F_{q^n}/F_q)]^{C_p} \xrightarrow{q-F} \mathbf{Z}_p[G(F_{q^n}/F_q)]^{C_p}$$

$$\longrightarrow (F^*_{q^n})_p^{C_p} \cong (F^*_{q^{dt}})_p \longrightarrow 0.$$

The Frobenius, F, generates the cyclic group, $G(F_{q^n}/F_q)$, so that there is an isomorphism

$$\lambda : \mathbf{Z}_p[G(F_{q^{dt}}/F_q)] \xrightarrow{\cong} \mathbf{Z}_p[G(F_{q^n}/F_q)]^{C_p}$$

given by

$$\lambda\left(\sum_{i=0}^{dt-1} \alpha_i F^i\right) = \left(\sum_{i=0}^{dt-1} \alpha_i F^i\right)(1 + F^{dt} + F^{2dt} + \ldots + F^{(p^\alpha-1)dt}).$$

Hence 7.3.38 may be identified with the sequence

7.3.39 $$\mathbf{Z}_p[G(F_{q^{dt}}/F_q)] \xrightarrow{q-F} \mathbf{Z}_p[G(F_{q^{dt}}/F_q)] \longrightarrow (F^*_{q^{dt}})_p$$

of $\mathbf{Z}_p[C/C_p]$-modules. The generator of $C/C_p \subset G(F_{q^{dt}}/F_q)$ is F^d. Hence, as a $\mathbf{Z}_p[C/C_p]$-module,

7.3.40 $$\mathbf{Z}_p[G(F_{q^{dt}}/F_q)] \cong \oplus_{j=0}^{d-1} \mathbf{Z}_p[C/C_p] < F^j > .$$

Therefore, applying the idempotent, e_ϕ, to 7.3.40 yields the d-fold sum of the $\mathbf{Z}_p[C/C_p]$-module given by $\chi(e_\phi) : C/C_p \longrightarrow GL_v\mathbf{Z}_p$. However, $q - F$ cyclically permutes the summands, $\mathbf{Z}_p[C/C_p] < F^j >$. In other words the injection

$$\mathbf{Z}_p[G(F_{q^{dt}}/F_q)]e_\phi \xrightarrow{q-F} \mathbf{Z}_p[G(F_{q^{dt}}/F_q)]e_\phi$$

may be identified with

$$Ind_{G(F_{q^{dt}}/F_{q^d})}^{G(F_{q^{dt}}/F_q)}(\mathbf{Z}_p^v) \xrightarrow{q-F} Ind_{G(F_{q^{dt}}/F_{q^d})}^{G(F_{q^{dt}}/F_q)}(\mathbf{Z}_p^v),$$

where $C/C_p \cong G(F_{q^{dt}}/F_q)$ acts on $\mathbf{Z}_p^v$ via $\chi(e_\phi)$. As in 5.2.33(proof) the order of the cokernel of $q-F$ is given by the determinant of $q-F$ on this induced module. In other words,

$$f_{\overline{L}^*}^*(C, e_\phi) = Det(q-F)(Ind_C^{G(\overline{L}/\overline{K})}(Inf_{C'}^C(\chi(e_\phi)))).$$

In order to compute $f_{\overline{L}}^*$ we may proceed in a similar manner, using the sequence

7.3.41

$$0 \longrightarrow (\mathbf{Z}_p[G(F_{q^n}/F_q)])^{f_{K/\mathbf{Q}_p}} \xrightarrow{p} (\mathbf{Z}_p[G(F_{q^n}/F_q)])^{f_{K/\mathbf{Q}_p}} \longrightarrow \overline{L} \longrightarrow 0.$$

This sequence is exact because, by the Normal Basis Theorem, $\overline{L}$ is isomorphic to $\overline{K}[G(F_{q^n}/F_q)]$. In 7.3.41, $f_{K/\mathbf{Q}_p}$ is the residue degree so that $q = p^{f_{K/\mathbf{Q}_p}}$. Hence

$$f_{\overline{L}}^*(C, e_\phi) = Det(q)(Ind_C^{G(\overline{L}/\overline{K})}(Inf_{C'}^C(\chi(e_\phi))))$$

and

7.3.42 $$f_{\overline{L}^*}^*(f_{\overline{L}}^*)^{-1}(C, e_\phi) = Det(1-q^{-1}F)(Ind_C^{G(\overline{L}/\overline{K})}(Inf_{C'}^C(\chi(e_\phi)))).$$

7.3.43 Now let us consider the general case (i.e. not necessarily cyclic). Let L/K be a finite, wildly ramified Galois extension of p-adic local fields with Galois group, $G(L/K)$. We wish to evaluate the image in $\mathcal{CL}(\Lambda(G(L/K)))$ of the local Chinburg invariant, $\Omega(L/K, 1+X)$. For this purpose we shall use the following two commutative diagrams:

7.3.44

$$\begin{array}{ccccc}
K_0T(\mathbf{Z}[G(L/K)]) & \longrightarrow & K_0T(\Lambda(G(L/K))) & \cong & Fac(G(L/K)) \\
\downarrow & & \downarrow & & \\
\mathcal{CL}(\mathbf{Z}[G(L/K)]) & \longrightarrow & \mathcal{CL}(\Lambda(G(L/K))) & &
\end{array}$$

7.3.45

$$\begin{array}{ccccc}
K_0T(\mathbf{Z}[G(L/K)]) & \xrightarrow{\text{Res}} & K_0T(\mathbf{Z}[G(L/F)]) & \longrightarrow & K_0T(\mathbf{Z}[G(M/F)]) \\
\downarrow & & \downarrow & & \downarrow \\
Fac(G(L/K)) & \xrightarrow{\text{Res}} & Fac(G(L/F)) & \longrightarrow & Fac(G(M/F)) \\
 & & & & \cup \\
 & & & & Fac_{G(L/K)}(G(M/F))
\end{array}$$

In 7.3.45, we have $G(L/F) \leq G(L/K)$ and $G(L/M) \lhd G(L/F)$ with cyclic quotient (via the canonical isomorphism)

$$G(L/F)/G(L/M) \cong G(M/F).$$

By 7.3.19, the vertical maps in 7.3.45 are described in terms of the canonical factorisations of 7.3.9.

In order to calculate the image of $\Omega(L/K, 1 + X)$ in the group, $Fac_{G(L/K)}(G(M/F))$ of 5.4.15, we will lift $\Omega(L/K, 1 + X)$ to a finite, cohomologically trivial module, T, which possesses a class in

$$K_0T(\mathbf{Z}[G(L/K)])$$

We will then chase $[T]$ anti-clockwise around 7.3.45 and evaluate the result in $Fac(G(M/F))$. Upon doing so we shall find that it lies in the subgroup, $Fac_{G(L/K)}(G(M/F))$. Unfortunately, during this process we will express the images of $[T]$ in terms of modules which do not define elements of $\mathcal{CL}(\mathbf{Z}[G(L/F)])$ or $\mathcal{CL}(\mathbf{Z}[G(M/F)])$ so that we cannot chase $[T]$ around 7.3.44.

We will begin by considering the following commutative diagram, which is similar to that studied in the cyclic case in 7.3.26. In particular, the horizontal rows of this diagram are *short exact*:

7.3.46

$$\begin{array}{ccccc} \mathbf{Z} & \xrightarrow{\sigma(L/K)} & \mathbf{Z}[G(L/K)] & \xrightarrow{\beta} & (n-\sigma(L/K))\mathbf{Z}[G(L/K)] \\ \downarrow \lambda & & \downarrow j_1 & & \downarrow j_2 \\ L^*/(1+X) & \longrightarrow & \mathbf{Z}[G(L/K)] \otimes_{\mathbf{Z}[N]} IN & \longrightarrow & IG(L/K) \end{array}$$

In 7.3.46, $\sigma(L/K) = \sum_{g \in G(L/K)} g \in \mathbf{Z}[G(L/K)]$ and $n = [L : K]$. The upper left-hand map is given by multiplication by $\sigma(L/K)$. The map, j_2, is the inclusion and $j_1(z) = \sum_{g \in G(L/K)} z \otimes (u_1 - u_g)$. Hence the map, λ, is determined by j_1 and, as a consequence of 7.3.27, we shall see in 7.3.49 that λ (and hence j_1) may be chosen to be injective.

Let $G(L/F) \leq G(L/K)$ be a subgroup of index $d = [L : F]$. Choose coset representatives $x_1, \ldots, x_d$ for $G(L/K)/G(L/F)$. As a $\mathbf{Z}[G(L/F)]$-module $\mathbf{Z}[G(L/K)]$ is free on the $\{x_i\}$ while $IG(L/K)$ is isomorphic to $IG(L/F) \oplus (\oplus_{i=2}^{d} \mathbf{Z}[G(L/F)])$. In fact, as a diagram of $\mathbf{Z}[G(L/F)]$-modules 7.3.46 becomes the following diagram (see 7.5.8).

7.3.47

$$\begin{array}{ccccc} \mathbf{Z} & \xrightarrow{\sigma} & \oplus_{i=1}^{d} \mathbf{Z}[G(L/F)] & \longrightarrow & \begin{array}{c}(r-\sigma(L/F))\mathbf{Z}[G(L/F)] \\ \oplus(\mathbf{Z}[G(L/F)]^{d-1})\end{array} \\ \downarrow \lambda & & \downarrow \psi_1 & & \downarrow \phi_1 \\ L^*/(1+X) & \longrightarrow & \mathbf{Z}[G(L/K)] \otimes_{\mathbf{Z}[N]} IN & \longrightarrow & IG(L/F) \oplus (\mathbf{Z}[G(L/F)]^{d-1}) \end{array}$$

where $\sigma = (\sigma(L/F), 0, \ldots, 0)$, $dr = n$ and

7.3.48

$$\begin{cases} \phi_1(b_1, 0, \ldots, 0) & = (d^2 b_1, db_1, \ldots, db_1) \\ \phi_1(0, \ldots, o, b_i, 0, \ldots, 0) & = (d(r - \sigma(L/F))b_i, -\sigma(L/F)b_i, \ldots, \\ & \quad (n - \sigma(L/F))b_i, -\sigma(L/F)b_i, \ldots, \\ & \quad -\sigma(L/F)b_i). \end{cases}$$

In 7.3.48 the term $(n-\sigma(L/F))b_i$ $(2 \le i)$ is in the ith coordinate.

Lemma 7.3.49 *In 7.3.46 and 7.3.47 the map, λ, may be assumed to be injective.*

Proof It suffices to show that λ is injective in 7.3.47 for one choice of L/F.

Let $G(L/F) \le G(L/K)$ be a non-trivial cyclic subgroup with generator, x, of prime order, r. By 7.5.4 the restriction map

$$\begin{aligned} &H^2(G(L/K); L^*) \cong H^2(G(L/K); L^*/(1+X)) \\ &\longrightarrow H^2(G(L/F); L^*/(1+X)) \cong \mathbf{Z}/r \end{aligned}$$

is onto. Hence, if we glue the lower sequence of 7.3.47 to

$$IG(L/F) \oplus F \xrightarrow{\rho\oplus 1} \mathbf{Z}[G(L/F)] \oplus F \xrightarrow{(\epsilon,0)} \mathbf{Z},$$

where $F = \oplus_{i=2}^{d} \mathbf{Z}[G(L/F)]$ and ρ is the inclusion, we obtain a 2-extension

$$L^*/(1+X) \longrightarrow \mathbf{Z}[G(L/K)] \otimes_{\mathbf{Z}[N]} IN \longrightarrow \mathbf{Z}[G(L/F)] \oplus F \longrightarrow \mathbf{Z},$$

which represents a generator of $H^2(G(L/F); L^*/(1+X))$. Choose a split injective $\mathbf{Z}[G(L/F)]$-homomorphism, $s : F \longrightarrow \mathbf{Z}[G(L/K)] \otimes_{\mathbf{Z}[N]} IN$, which splits F from the lower sequence in 7.3.47. Dividing out by F yields a 2-extension

$$L^*/(1+X) \longrightarrow (\mathbf{Z}[G(L/K)] \otimes_{\mathbf{Z}[N]} IN)/s(F) \longrightarrow \mathbf{Z}[G(L/F)] \longrightarrow \mathbf{Z},$$

which also represents a generator of $H^2(G(L/F); L^*/(1+X))$.

We have a commutative diagram

$$\begin{array}{ccccc} \mathbf{Z} & \xrightarrow{(\sigma(L/F),0,\ldots,0)} & & & \oplus_{i=1}^{d} \mathbf{Z}[G(L/F)] \\ \downarrow \lambda & & & & \downarrow \lambda' \\ L^*/(1+X) & \longrightarrow & (\mathbf{Z}[G(L/K)] \otimes_{\mathbf{Z}[N]} IN)/s(F) & \xrightarrow{\pi} & IG(L/F) \end{array}$$

in which

$$\pi(\lambda'(1,0,0,\ldots,0)) = r - \sigma(L/F) = \sum_{v=1}^{r-1} (1-x^v) \in IG(L/F).$$

We may choose the representing 2-cocycle, $f : G(L/K) \times G(L/K) \longrightarrow L^*$, so that on $G(L/F)$ it makes the map, γ, injective in 7.3.26, as in 7.3.25. With this choice one sees that

$$\begin{aligned} \lambda(1) &= \textstyle\sum_{v=1}^{r-1} v\gamma(1) \\ &= (r(r-1)/2)\gamma(1), \end{aligned}$$

which is an element of infinite order in $L^*/(1+X)$, by 7.3.30. □

7.3.50 Suppose that $G(L/M) \lhd G(L/F)$ is a normal subgroup such that $G(L/F)/G(L/M) \cong G(M/F)$ is cyclic of order $m = [M : F]$. Hence, in the notation of 7.3.47, $m[L : M] = r = [L : F]$. Set $c = [L : M]$ so that $mc = r$.

We wish to consider the diagram of $\mathbf{Z}[G(M/F)]$-modules which is obtained by taking the $G(L/M)$-invariants of 7.3.47. Notice that the upper sequence of 7.3.47 remains short exact upon applying $(-)^{G(L/M)} = H^0(G(L/M); -)$ since $H^1(G(L/M); \mathbf{Z}) \cong Hom(G(L/M), \mathbf{Z}) = 0$. Also $H^1(G(L/M); L^*) = 0$, by Hilbert's Theorem 90 (Serre, 1979, p. 150), while

$$H^2(G(L/M); 1+X) \cong H^2(G(L/M); X) = 0.$$

Therefore $H^1(G(L/M); L^*/(1+X))$ is trivial and the $G(L/M)$-invariants of the lower row of 7.3.47 also form a short exact sequence. By 7.5.9, the $G(L/M)$-invariants of 7.3.47 form the following commutative diagram in which the horizontal rows are short exact while the vertical maps are injective with finite cokernels.

7.3.51

$$\begin{array}{ccccc} \mathbf{Z} & \stackrel{\sigma'}{\longrightarrow} & \oplus_{i=1}^{d} \mathbf{Z}[G(M/F)] & \stackrel{\delta}{\longrightarrow} & \begin{array}{c} c(m-\sigma(M/F))\mathbf{Z}[G(M/F)] \\ \oplus(\mathbf{Z}[G(M/F)]^{d-1}) \end{array} \\ \downarrow \lambda & & \downarrow \psi_2 & & \downarrow \phi_2 \\ \frac{M^*}{(1+X(M))} & \longrightarrow & (\mathbf{Z}[G(L/K)] \otimes_{\mathbf{Z}[N]} IN)^{G(L/M)} & \longrightarrow & \begin{array}{c} IG(M/F) \\ \oplus(\mathbf{Z}[G(M/F)]^{d-1}) \end{array} \end{array}$$

In 7.3.51, $\sigma' = (\sigma(M/F), 0, \ldots, 0)$, $\delta = (c(m - \sigma(M/F)), 1, \ldots, 1)$, $X(M) = X^{G(L/M)}$ and

7.3.52

$$\begin{cases} \phi_2(z_1, 0, \ldots, 0) & = (d^2 z_1, dz_1, \ldots, dz_1) \\ \phi_2(0, \ldots, o, z_i, 0, \ldots, 0) & = c(d(m - \sigma(M/F))z_i, -\sigma(L/F)z_i, \ldots, \\ & \quad (dm - \sigma(M/F))z_i, -\sigma(L/F)z_i, \ldots, \\ & \quad -\sigma(L/F)z_i). \end{cases}$$

In 7.3.52, $(dm - \sigma(M/F))z_i$ lies in the ith coordinate.

The diagram of 7.3.51 may be modified by an isomorphism which leaves λ and ϕ_2 unchanged and puts the lower short exact sequence into a more amenable form.

There are isomorphisms

$$Ext^1_{\mathbf{Z}[G(M/F)]}(IG(M/F) \oplus (\oplus_{i=2}^d \mathbf{Z}[G(M/F)]), M^*/(1 + X(M)))$$

$$\cong Ext^1_{\mathbf{Z}[G(M/F)]}(IG(M/F), M^*/(1 + X(M)))$$

$$\cong H^2(G(M/F); M^*/(1 + X(M)))$$

$$\cong \mathbf{Z}/[M : F],$$

where $m = [M : F]$, in the notation of 7.3.50. By 7.1.57, the lower 2-extension corresponds to the generator, $inv^{-1}([M : F]^{-1})$. Hence, by the theory of the cohomological classification of 2-extensions (Hilton & Stammbach, 1971), we may alter ψ_2 to obtain a new diagram which is isomorphic to 7.3.51 and has the following form. The details are left to the interested reader as exercise 7.5.10.

7.3.53

$$\begin{array}{ccccc} \mathbf{Z} & \xrightarrow{\sigma'} & \oplus_{i=1}^d \mathbf{Z}[G(M/F)] & \xrightarrow{\delta} & \begin{array}{c} c(m - \sigma(M/F))\mathbf{Z}[G(M/F)] \\ \oplus(\mathbf{Z}[G(M/F)]^{d-1}) \end{array} \\ \downarrow \lambda & & \downarrow \psi_2' & & \downarrow \phi_2 \\ \frac{M^*}{(1+X(M))} & \xrightarrow{i} & W \oplus (\mathbf{Z}[G(M/F)]^{d-1}) & \xrightarrow{\mu \oplus 1} & \begin{array}{c} IG(M/F) \\ \oplus(\mathbf{Z}[G(M/F)]^{d-1}) \end{array} \end{array}$$

where $W = \mathbf{Z}[G(M/F)] \otimes_{\mathbf{Z}[N_1]} IN_1$. In 7.3.53 the 2-extension

$$M^*/(1+X(M)) \xrightarrow{i} \mathbf{Z}[G(M/F)] \otimes_{\mathbf{Z}[N_1]} IN_1 \xrightarrow{\mu} IG(M/F)$$

is obtained from the lower short exact sequence of 7.3.26 upon replacing L, K, N by M, F, N_1.

Before embarking upon the first main result of this section, the evaluation of $\Omega(L/K, 1+X)$ in $\mathscr{CL}(\Lambda(G(L/K)))$, we require two more straightforward technical results.

Lemma 7.3.54 *Let $G(M/F)$ be a cyclic subquotient of $G(L/K)$, as in 7.3.49 and 7.3.50. Define a function*

$$r : S^*(G(M/F)) \longrightarrow \mathscr{I}(E)$$

by

$$r(G(M/E), e) = \begin{cases} 1 & \text{if } 1 \neq \chi(e), \\ (p - \text{part of } [F:K]) & \text{if } 1 = \chi(e) \end{cases}$$

for $(G(M/E), e) \in S^p(G(M/F))$.

(i) *Then r has a canonical factorisation*

$$r_1 \in Hom_{\Omega_{\mathbf{Q}}}(R(G(M/F)), \mathscr{I}(E))$$

given by

$$r_1(V) = ideal < [F:K]^{dim(V^{G(M/F)})} > .$$

(ii) *In* (i)

$$r_1 \in PF^+_{G(L/K)}(G(M/F)) \subset Fac(G(M/F)).$$

Proof Part (i) follows from the fact that, if $C = G(M/E)$,

$$dim((Ind_C^{G(M/F)}(Inf_{C'}^C(V)))^{G(M/F)}) =< V, 1 >_{C'} .$$

When $V = \chi(e)$, an irreducible $\mathbf{Q}_p$-representation, this dimension is zero except when $\chi(e) = 1$.

Part (ii) follows from the fact that r_1 takes values in $\mathscr{I}(\mathbf{Q})$. □

Lemma 7.3.55 *Let $G(L/M) \lhd G(L/F) \leq G(L/K)$ be as in 7.3.49 and 7.3.50. Consider the following short exact sequence of $\mathbf{Z}[G(\overline{M}/\overline{F})]$-modules:*

$$0 \longrightarrow IG(\overline{M}/\overline{F}) \xrightarrow{j} \mathbf{Z}[G(\overline{M}/\overline{F})]/(\sigma(\overline{M}/\overline{F})) \longrightarrow \mathbf{Z}/(f_{M/F}) \longrightarrow 0.$$

(i) *For* $(G(\overline{M}/\overline{E}) \in S^p(G(\overline{M}/\overline{F}))$

$$f^p_{G(\overline{M}/\overline{F}),j}(G(\overline{M}/\overline{E}), e) = \begin{cases} f_{E/F} & \text{if } 1 = \chi(e), \\ 1 & \text{if } 1 \neq \chi(e). \end{cases}$$

(ii) *The canonical factorisation of* $f^*_{G(M/F),j}$ *exists and lies in*

$$PF^+_{G(L/K)}(G(M/F)).$$

Proof If $1 \neq \chi(e)$ then e annihilates trivial modules and $IG(\overline{M}/\overline{F})_p e \cong \mathbf{Z}_p[G(\overline{M}/\overline{F})]e$ so that $f^p_{G(\overline{M}/\overline{F}),j}(G(\overline{M}/\overline{E}), e) = 1$ in this case.

If $1 = \chi(e)$ then $f^p_{G(\overline{M}/\overline{F}),j}(G(\overline{M}/\overline{E}), e)$ is equal to the p-part of the order of the cokernel of

$$IG(\overline{M}/\overline{F})^{G(\overline{M}/\overline{E})} \cong IG(\overline{E}/\overline{F})$$

$$\longrightarrow (\mathbf{Z}[G(\overline{M}/\overline{F})]/(\sigma(\overline{M}/\overline{F})))^{G(\overline{M}/\overline{E})} \cong \mathbf{Z}[G(\overline{E}/\overline{F})]/(\sigma(\overline{E}/\overline{F})),$$

which establishes part (i).

For part (ii) we first note that, by 7.3.21, $f^*_{G(M/F),j}$ has the canonical factorisation

$$(V \mapsto s(V^{G_0(M/F)})),$$

where s is as in 7.5.13.

Suppose that V is an irreducible representation of $G(M/F)$, then either $V = V^{G_0(M/F)}$ or $0 = V^{G_0(M/F)}$. In the second case $s(V) = 1$, which is certainly a principal fractional ideal of $\mathbf{Q}(Ind^{G(L/K)}_{G(L/F)}(Inf^{G(L/F)}_{G(M/F)}(V)))$ satisfying the total positivity condition of 5.4.45(ii). If $V = V^{G_0(M/F)}$ then V is a one-dimensional representation which factorises through $G(\overline{M}/\overline{F})$. Hence $Ind^{G(L/K)}_{G(L/F)}(Inf^{G(L/F)}_{G(M/F)}(V))$ factorises through $G(\overline{L}/\overline{K})$ and, since $G(\overline{L}/\overline{K})$ is cyclic,

$$\mathbf{Q}(Ind^{G(L/K)}_{G(L/F)}(Inf^{G(L/F)}_{G(M/F)}(V))) = \mathbf{Q}(V).$$

Therefore, in both cases, $s(V)$ is a principal fractional ideal of $\mathbf{Q}(Ind^{G(L/K)}_{G(L/F)}(Inf^{G(L/F)}_{G(M/F)}(V)))$.

It remains to verify the total positivity condition of 5.4.45(ii). Dividing into cases as before we see that we need only consider the case when $V = V^{G_0(M/F)}$ is one-dimensional. However, in this case

$$Ind^{G(L/K)}_{G(L/F)}(Inf^{G(L/F)}_{G(M/F)}(V)) = Inf^{G(L/K)}_{G(\overline{L}/\overline{K})}(Ind^{G(\overline{L}/\overline{K})}_{G(\overline{M}/\overline{F})}(V))$$

is a sum of one-dimensional representations and is symplectic if and only if $Ind_{G(\overline{M}/\overline{F})}^{G(\overline{L}/\overline{K})}(V)$ is symplectic.

We may assume that V is a faithful representation of $G(\overline{M}/\overline{F})$.

In general, if $\{1\} \neq \mathbf{Z}/a \leq \mathbf{Z}/ab$ and $\theta : \mathbf{Z}/a \longrightarrow \mathbf{C}^*$ is a faithful, one-dimensional representation, we may extend θ to $\theta : \mathbf{Z}/ab \longrightarrow \mathbf{C}^*$ and then

$$Ind_{\mathbf{Z}/a}^{\mathbf{Z}/ab}(\theta) = \theta(1 + \theta^a + \ldots + \theta^{ab-a}).$$

For this to be symplectic we must have $b = 2\beta$ and

$$\sum_{i=0}^{b-1} \theta^{ia+1} = \sum_{i=0}^{b-1} \theta^{-ia-1} \in R(\mathbf{Z}/2a\beta).$$

This condition can only happen if $a = 2$ but then $Ind_{\mathbf{Z}/a}^{\mathbf{Z}/ab}(\theta)$ cannot be symplectic because it is the sum of symplectic lines plus $1 + \theta^{a\beta}$, which is not symplectic. This discussion shows that, when V factorises through $G(\overline{M}/\overline{F})$, $Ind_{G(L/F)}^{G(L/K)}(Inf_{G(M/F)}^{G(L/F)}(V))$ can only be symplectic if $V = W \oplus \overline{W}$ for some $W \in R(G(\overline{M}/\overline{F}))$ and in this case the total positivity of $s(V)$ is automatic (cf. Theorem 5.4.46(proof)). $\square$

Theorem 7.3.56 *Let L/K be a finite, wildly ramified Galois extension of p-adic local fields with Galois group, $G(L/K)$. In the notation of* 7.3.43 *and* 7.3.44, *the image of the local Chinburg invariant, $\Omega(L/K, 1+X)$, in the class-group of the maximal order, $\Lambda(G(L/K))$, is given by*

$$\Omega(L/K, 1+X) = [\mathcal{O}_L/X] \in \mathcal{CL}(\Lambda(G(L/K))),$$

where $[\mathcal{O}_L/X]$ denotes the class represented by the canonical factorisation of $f^*_{\mathcal{O}_L/X}$.

Proof We begin with $[T] \in K_0T(\mathbf{Z}[G(L/K)])$ where $T = coker(j_1)$ in 7.3.46. By 7.3.19, f^*_T has a canonical factorisation which gives the image of $[T]$ in $Fac(G(L/K))$. By 5.4.52, it suffices to find the image of $[T]$ in

$$\frac{Fac_{G(L/K)}(G(M/F))}{PF^+_{G(L/K)}(G(M/F))}$$

for each $K \leq F \leq M \leq L$ as in 7.3.50. By the naturality in G of 7.3.19, the image of $[T]$ in $Fac(G(M/F))$ is represented by $coker(\psi_2)$ of 7.3.51. In other words, we must evaluate the canonical factorisation of $f^*_{G(M/F),\psi_2}$ in $Fac(G(M/F))$. However, by 7.3.12(iv) applied to 7.3.51,

7.3.57 $$f^*_{\psi_2} = f^*_\lambda f^*_{\phi_2}.$$

Suppose now that the composition

$$\mathbf{Z} \xrightarrow{\lambda} L^*/(1+X) \xrightarrow{v_L} \mathbf{Z}$$

is multiplication by $ve_{L/K}$. One finds easily, by 7.3.12(i), that $f^*_\lambda = f^*_v f^*_\gamma$, where f^*_γ is as in 7.3.26. Therefore, by 7.3.34 and 7.3.35

7.3.58 $$f^*_\lambda = f^*_v f^*_{\mathcal{O}_M/(X(M))} f^*_{\overline{M}^*} (f^*_{\overline{M}})^{-1} \mathscr{E}^* \in Fac(G(M/F)).$$

By 7.5.11, f^*_v has a canonical factorisation which lies in

$$PF^+_{G(L/K)}(G(M/F)),$$

since $g(\chi) \in \mathscr{I}(\mathbf{Q})$ in 7.5.11.

Now let us turn our attention to $f^*_{\phi_2}\mathscr{E}^*$.

If $(G(M/E), e) \in S^p(G(M/F))$ and $\chi(e) = 1$ then $f^p_{\phi_2}(G(M/E), e)\mathscr{E}^p(G(M/E), e)$ is equal to the p-part of

$$\# \left(\frac{IG(L/K)^{G(L/E)}}{(([L:K] - \sigma(L/K))\mathbf{Z}[G(L/K)])^{G(L/E)}} \right) e_{E/F}$$

$$= [L:E]([L:K]^{[E:K]-2})e_{E/F},$$

by 7.3.10 and 7.3.12(ii). If $\chi(e) \neq 1$ then $\mathscr{E}^p(G(M/E), e) = 1$ and

$$f^p_{\phi_2}(G(M/E), e)\mathscr{E}^p(G(M/E), e)$$

$$= \# \left(\frac{IG(L/K)_p e}{([L:K] - \sigma(L/K))\mathbf{Z}_p[G(L/K)]e} \right)$$

$$= \# \left(\frac{\mathbf{Z}_p[G(L/K)]e}{[L:K]\mathbf{Z}_p[G(L/K)]e} \right)$$

since e annihilates trivial modules.

Consider the multiplication map (see 7.5.12)

$$[L:K] : \mathbf{Z}[G(L/K)] \longrightarrow \mathbf{Z}[G(L/K)].$$

The above calculation shows that, in $Fac(G(M/F))$,

$$\{f^p_{\phi_2}\mathscr{E}^p(f^p_{[L:K]})^{-1}\}(G(M/E), e)$$

is trivial if $1 \neq \chi(e)$ and otherwise is given by the p-part of

$$e_{E/F}[L:E]([L:K]^{[E:K]-2})([L:K]^{-[E:K]})$$

$$= e_{E/F}[L:E][L:K]^{-2} f_{E/F} f_{E/F}^{-1}$$

$$= [L:E][E:F][F:K][F:K]^{-1} f_{E/F}^{-1}[L:K]^{-2}$$

$$= [F:K]^{-1}[L:K]^{-1} f_{E/F}^{-1}.$$

By 7.3.54, 7.3.55 and 7.5.11, this function has a canonical factorisation which lies in $PF^{+}_{G(L/K)}(G(M/F))$. Therefore, by 7.3.57 and 7.3.58, we have shown that $[T]$ is represented in

$$Fac(G(M/F))/PF^{+}_{G(L/K)}(G(M/F))$$

by the canonical factorisation of

$$f^{*}_{\mathcal{O}_M/(X_M)} f^{*}_{\overline{M}^{\cdot}} (f^{*}_{\overline{M}})^{-1}.$$

However, $f^{*}_{\overline{M}^{\cdot}} (f^{*}_{\overline{M}})^{-1}$ has a canonical factorisation which lies in

$$PF^{+}_{G(L/K)}(G(M/F)).$$

This last fact is seen by observing that 7.3.21 and 7.3.42 imply that the canonical factorisation of $f^{*}_{\overline{M}^{\cdot}} (f^{*}_{\overline{M}})^{-1}$ is equal to the image of (the p-part of) the function on $R(G(L/K))$ given by sending V to

$$Det\left(1 - \frac{(F \mid V^{G_0(L/K)})}{\#(\overline{K})}\right),$$

which is the value at $s = 1$ of an Euler factor (above p) from the Artin L-function (Martinet, 1977b). Since this function depends only on $V^{G_0(L/K)}$ — as a representation of the cyclic quotient, $G(\overline{L}/\overline{K})$ — one sees that this function is in $PF(G(L/K))$ by the argument which was used to establish 7.3.55(ii). By 5.4.46 the image, $f^{*}_{\overline{M}^{\cdot}} (f^{*}_{\overline{M}})^{-1} \in Fac(G(M/F))$, lies in $PF^{+}_{G(L/K)}(G(M/F))$, as required. This completes the proof of Theorem 7.3.56. □

7.3.59 We shall now state a result which relates the analytic class, $W_{L/K}$, to the canonical factorisation of modules. This result is proved using the theory of norm resolvents to represent classes in Hom-descriptions and it is given in Holland (1992, section 4); (see also Wilson, 1990). We shall not give a proof of this result here, since our Explicit Brauer Induction techniques have nothing to add to this step in the programme.

Let L/K be a finite Galois extension of number fields with Galois group $G(L/K)$ and let $X = \mathcal{O}_K[G(L/K)]a$ be as in 7.2.3. Let $b \in X$ be a free generator for L as a $K[G(L/K)]$-module.

Theorem 7.3.60 (i) *The function, $f^*_{\mathcal{O}_L/b\mathcal{O}_K[G(L/K)]}$, has a canonical factorisation which represents a class*

$$[\mathcal{O}_L/b\mathcal{O}_K[G(L/K)]] \in \mathcal{CL}(\Lambda(G(L/K))).$$

(ii) *If $W_{L/K} \in \mathcal{CL}(\mathbf{Z}[G(L/K)])$ is the class defined in* 7.2.11 *then*

$$W_{L/K} = [\mathcal{O}_L/b\mathcal{O}_K[G(L/K)]] \in \mathcal{CL}(\Lambda(G(L/K))).$$

Theorem 7.3.61 *Let L/K be a finite Galois extension of number fields with Galois group, $G(L/K)$. Let*

$$\Omega(L/K,2), W_{L/K} \in \mathcal{CL}(\mathbf{Z}[G(L/K)])$$

be as in 7.2.6 *and* 7.2.11, *respectively. Then*

$$\Omega(L/K,2) = W_{L/K} \in \mathcal{CL}(\Lambda(G(L/K))).$$

Proof Let X be as in 7.2.3 and 7.2.4. By definition

$$\begin{aligned}&\Omega(L/K,2)\\ &= [X] + \textstyle\sum_{P \ wild} Ind_{G(L_Q/K_P)}^{G(L/K)}(\Omega(L_Q/K_P, 1+X_Q))\\ &= [X] + \textstyle\sum_{P \ wild} Ind_{G(L_Q/K_P)}^{G(L/K)}[\mathcal{O}_{L_Q}/(\mathcal{O}_{K_P}[G(L_Q/K_P)]a_P)]\end{aligned}$$

in $\mathcal{CL}(\Lambda(G(L/K)))$. (In 7.2.3 $\mathcal{O}_{K_P}[G(L_Q/K_P)]a_P$ was denoted by $X_Q \subset \mathcal{O}_{L_Q}$, while in 7.3.51 it was denoted by $X(L_Q)$.)

Now let $b \in X$ be a free generator for L as a $K[G(L/K)]$-module so that

7.3.62 $$[X] = [X/b\mathcal{O}_K[G(L/K)]] \in \mathcal{CL}(\Lambda(G(L/K))).$$

By 7.2.2–7.2.4, for each wild prime P,

7.3.63 $$Ind_{G(L_Q/K_P)}^{G(L/K)}[\mathcal{O}_{L_Q}/(\mathcal{O}_{K_P}[G(L_Q/K_P)]a_P)] \cong (\mathcal{O}_L/X) \otimes_{\mathcal{O}_K} \mathcal{O}_{K_P}.$$

Since $\mathcal{O}_L/X$ is trivial at tame primes we may combine our formula for $\Omega(L/K, 2)$ together with 7.3.62 and 7.3.63 to obtain

$$\Omega(L/K, 2)$$

$$= [X/b\mathcal{O}_K[G(L/K)]] + [\mathcal{O}_L/X]$$

$$= [\mathcal{O}_L/b\mathcal{O}_K[G(L/K)]]$$

in $\mathcal{CL}(\Lambda(G(L/K)))$, which completes the proof of Theorem 7.3.61. □

7.4 Real cyclotomic Galois module structure

7.4.1 *Based exact sequences*

We shall begin this section by establishing some elementary properties of 'torsion elements' (reminiscent of Reidermeister torsion and Whitehead torsion) which lie in $K_1(\mathbf{Q}_p[G])$ and are associated to exact sequences of locally free, based modules. This passage from exact sequences of based modules to K-theory will be applied below to analyse the Hom-description of the global Chinburg invariant.

Let Λ be a ring and suppose that

$$\Pi: \qquad \{0\} \longrightarrow A \stackrel{\lambda}{\longrightarrow} B \stackrel{\pi}{\longrightarrow} C \longrightarrow \{0\}$$

is a short exact sequence of Λ-modules with *preferred bases*

$$\underline{x}_A(*) = \{x_A(1), \ldots, x_A(t)\},$$

$$\underline{x}_B(*) = \{x_B(1), \ldots, x_B(s)\}$$

and

$$\underline{x}_C(*) = \{x_C(1), \ldots, x_C(s-t)\}$$

for A, B and C, respectively.

This information yields an element, $[\Pi] \in K_1(\Lambda)$, in the following manner. Choose a Λ-homomorphism, ϕ, from the free Λ-module on $\{y_C(i); 1 \leq i \leq s-t\}$

$$\phi : \Lambda < y_C(1), \ldots, y_C(s-t) > \longrightarrow B$$

such that $\phi(y_C(i)) = x_C(i)$ for each $1 \leq i \leq s-t$. Hence we have

Lemma 7.4.2 *The homomorphism*

$$\lambda + \phi : A \oplus \Lambda < y_C(*) > \longrightarrow B$$

is an isomorphism.

Therefore there exists an invertible matrix, $U \in GL_s(\Lambda)$, such that

$$(\lambda + \phi) \begin{pmatrix} x_A(1) \\ \vdots \\ x_A(t) \\ \\ y_C(1) \\ \vdots \\ y_C(s-t) \end{pmatrix} = U \begin{pmatrix} x_B(1) \\ \vdots \\ \\ \\ \\ \\ x_B(s) \end{pmatrix}.$$

Definition 7.4.3 *Define* $[\Pi] = [U] \in K_1(\Lambda) = GL(\Lambda)_{ab}$.

Proposition 7.4.4 $[U]$ *does not depend on the choice of* ϕ.

Proof If ϕ' is a second choice of homomorphism then we have a commutative diagram of the following form, in which $\mu = (\lambda + \phi')^{-1}(\lambda + \phi)$.

$$\begin{array}{ccccc} A & \longrightarrow & A \oplus \Lambda < y_C(*) > & \longrightarrow & C \\ \downarrow 1 & & \downarrow \mu & & \downarrow 1 \\ A & \longrightarrow & A \oplus \Lambda < y_C(*) > & \longrightarrow & C \end{array}$$

Therefore the difference in the two values of $[U]$, which result from the two choices, is represented by the matrix of μ with respect to the basis $\{\underline{x}_A(*), \underline{y}_C(*)\}$. This matrix has the form

$$\begin{pmatrix} I_t & Y \\ 0 & I_{s-t} \end{pmatrix},$$

which represents zero in $K_1(\Lambda)$, as required. □

Now suppose that $\Lambda = \mathbf{Z}_p[G]$, a p-adic group-ring. Suppose that V is a finitely generated, free $\mathbf{Z}_p[G]$-module and that X is a projective

$\mathbf{Z}[G]$-module together with an embedding, $k : X_p = X \otimes_{\mathbf{Z}} \mathbf{Z}_p \longrightarrow V$. In addition, let us suppose that V/X_p is a finite p-group. Suppose also that we have a finitely generated, projective $\mathbf{Z}[G]$-resolution of the form

$$\{0\} \longrightarrow P_1 \stackrel{i}{\longrightarrow} P_0 \stackrel{\pi}{\longrightarrow} V/X_p \longrightarrow \{0\}.$$

Let $x_{1,p}(*) = \{x_{1,p}(1), \ldots, x_{1,p}(t)\}$ be a basis for $P_1 \otimes_{\mathbf{Z}} \mathbf{Z}_p$ over $\mathbf{Z}_p[G]$ and let $x_{0,p}(*) = \{x_{0,p}(1), \ldots, x_{0,p}(t)\}$ be a basis for $P_0 \otimes_{\mathbf{Z}} \mathbf{Z}_p$. Also let $x_{1,0}(*) = \{x_{1,0}(1), \ldots, x_{1,0}(t)\}$ be a basis for $P_1 \otimes_{\mathbf{Z}} \mathbf{Q}$ over the rational group-ring, $\mathbf{Q}[G]$, and let $x_{0,0}(*) = i(x_{1,0}(*))$ so that

$$\{x_{0,0}(1), \ldots, x_{0,0}(t)\}$$

is a $\mathbf{Q}[G]$-basis for $P_0 \otimes_{\mathbf{Z}} \mathbf{Q}$. Let

$$V_1, V_0 \in GL_t(\mathbf{Q}_p[G])$$

be the matrices given by $(i = 0, 1)$

$$V_i \begin{pmatrix} x_{i,0}(1) \\ \vdots \\ x_{i,0}(t) \end{pmatrix} = \begin{pmatrix} x_{i,p}(1) \\ \vdots \\ x_{i,p}(t) \end{pmatrix}.$$

Therefore

$$\begin{aligned} i(x_{1,p}(*)) &= i(V_1 x_{1,0}(*)) \\ &= V_1 i(x_{1,0}(*)) \\ &= V_1 x_{0,0}(*) \\ &= V_1 V_0^{-1} x_{0,p}(*) \end{aligned}$$

so that the matrix of $i : P_1 \otimes_{\mathbf{Z}} \mathbf{Z}_p \longrightarrow P_0 \otimes_{\mathbf{Z}} \mathbf{Z}_p$ is equal to

$$V_1 V_0^{-1} \in GL_t(\mathbf{Q}_p[G]).$$

Notice that $V_1 V_0^{-1}$ has entries which lie in $\mathbf{Z}_p[G]$ and that the class, $[V_i] \in K_1(\mathbf{Q}_p[G])$ is the p-component of the K_1-idèle which is associated to the based module, P_i, in the Hom-description construction of 4.2.13 and 4.6.36.

From the projective $\mathbf{Z}[G]$-resolution of V/X_p we may construct an exact sequence of $\mathbf{Z}_p[G]$-modules of the form

$$\{0\} \longrightarrow P_1 \otimes_{\mathbf{Z}} \mathbf{Z}_p \stackrel{(i,j)}{\longrightarrow} (P_0 \otimes_{\mathbf{Z}} \mathbf{Z}_p) \oplus X_p \stackrel{(\tilde{\pi},-k)}{\longrightarrow} V \longrightarrow \{0\}.$$

Here j fits into a commutative diagram of the following form:

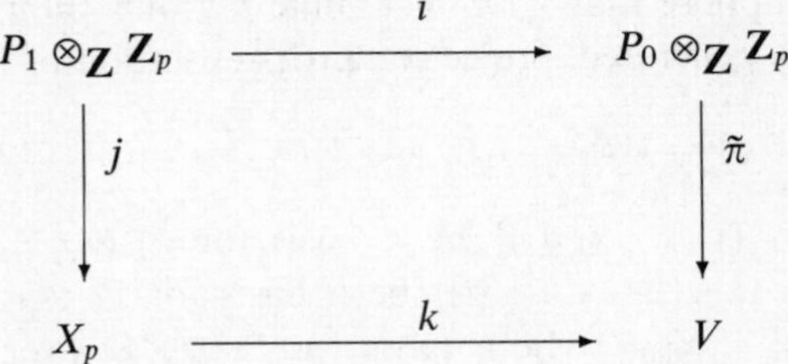

The following diagram also commutes:

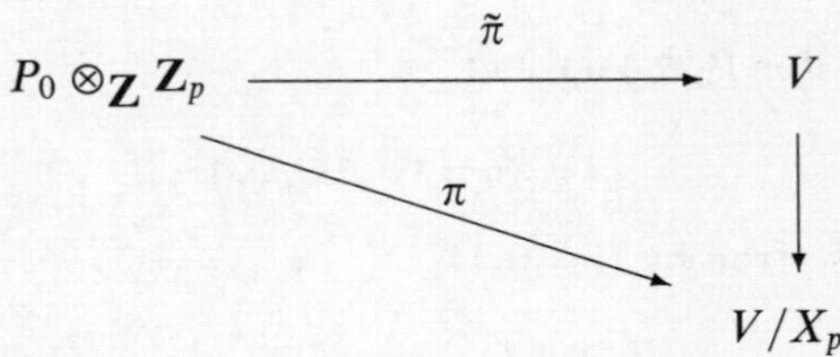

Now suppose that $X \otimes_{\mathbf{Z}} \mathbf{Q}$ has a basis, $u_0(*) = \{u_0(1), \ldots, u_0(r)\}$ over $\mathbf{Q}[G]$ and that X_p has a $\mathbf{Z}_p[G]$-basis, $u_p(*) = \{u_p(1), \ldots, u_p(r)\}$. Therefore there is a matrix, $W \in GL_r(\mathbf{Q}_p[G])$ such that

$$W u_0(*) = u_p(*).$$

Let $V \otimes_{\mathbf{Z}_p} \mathbf{Q}_p$ be endowed with the $\mathbf{Q}_p[G]$-basis

$$v_0(*) = k(u_0(*))$$

so that $v_0(a) = k(u_0(a))$ for $1 \leq a \leq r$.

Finally, let $v_p(*) = \{v_p(1), \ldots, v_p(r)\}$ be a $\mathbf{Z}_p[G]$-basis of V.

Hence we have an exact sequence

7.4.5 $\qquad \Pi: \qquad P_1 \otimes_{\mathbf{Z}} \mathbf{Z}_p \xrightarrow{(i,j)} (P_0 \otimes_{\mathbf{Z}} \mathbf{Z}_p) \oplus X_p \xrightarrow{(\tilde{\pi}, -k)} V$

to which to apply the construction of [Π] in 7.4.4.

Suppose that the matrix with entries in $\mathbf{Z}_p[G]$ which corresponds to (i, j) is

$$(V_1 V_0^{-1}, J)$$

so that J is a $t \times r$ matrix. Also let k correspond to the $r \times r$ matrix with entries in $\mathbf{Z}_p[G]$

$$K \in GL_r(\mathbf{Q}_p[G])$$

so that $k(u_p(*)) = Kv_p(*)$. Let $\tilde{\pi}$ correspond to the $t \times r$ matrix, P, with entries in $\mathbf{Z}_p[G]$.

Now we must choose ϕ in the following commutative diagram:

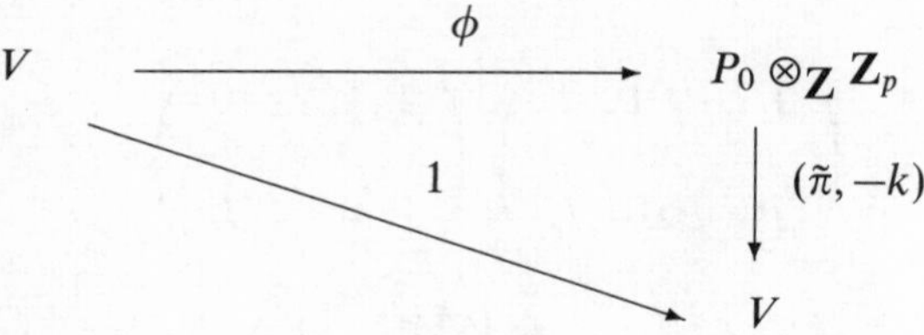

There are matrices, whose entries lie in $\mathbf{Z}_p[G]$, (D, E) such that

$$\phi(v_p(*)) = (Dx_{0,p}(*), Eu_p(*))$$

where D is $t \times r$ and $E \in GL_r(\mathbf{Q}_p[G])$. Hence the matrix of

$$(i, j) + \phi : (P_1 \otimes_{\mathbf{Z}} \mathbf{Z}_p) \oplus V \longrightarrow \begin{matrix} (P_1 \otimes_{\mathbf{Z}} \mathbf{Z}_p) \\ \oplus \\ X_p \end{matrix}$$

has the form

$$\begin{pmatrix} V_1 V_0^{-1} & D \\ J & E \end{pmatrix}.$$

In addition we have

$$I_r = PD - KE$$

and

$$PV_1V_0^{-1} = KJ.$$

Consider the following matrix computation:

$$\begin{pmatrix} I_t & 0 \\ P & I_r \end{pmatrix}\begin{pmatrix} I_t & D \\ 0 & -I_r \end{pmatrix}\begin{pmatrix} V_1V_0^{-1} & 0 \\ 0 & I_r \end{pmatrix}$$

$$= \begin{pmatrix} I_t & D \\ P & PD - I_r \end{pmatrix}\begin{pmatrix} V_1V_0^{-1} & 0 \\ 0 & I_r \end{pmatrix}$$

$$= \begin{pmatrix} I_t & D \\ P & KE \end{pmatrix}\begin{pmatrix} V_1V_0^{-1} & 0 \\ 0 & I_r \end{pmatrix}$$

$$= \begin{pmatrix} V_1V_0^{-1} & D \\ PV_1V_0^{-1} & KE \end{pmatrix}$$

$$= \begin{pmatrix} V_1V_0^{-1} & D \\ KJ & KE \end{pmatrix}$$

$$= \begin{pmatrix} I_t & 0 \\ 0 & K \end{pmatrix}\begin{pmatrix} V_1V_0^{-1} & D \\ J & E \end{pmatrix}.$$

Therefore the image of $[\Pi] \in K_1(\mathbf{Z}_p[G])$ in $K_1(\mathbf{Q}_p[G])$ is given by the following result.

Proposition 7.4.6 *In $K_1(\mathbf{Q}_p[G])$ the image of 7.4.5 is given by*

$$[\Pi] = rank(V) < -1 > +[V_1] - [V_0] - [K]$$

where $< -1 > \in K_1(\mathbf{Z}_p) \cong \mathbf{Z}_p^*$ denotes the class corresponding to $-1 \in \mathbf{Z}_p^*$.

The relation between $[K]$ and $[W]$ is given by the equation

$$Wu_0(*) = u_p(*),$$

which implies, upon applying k, that

$$K^{-1}Wv_0(*) = v_p(*).$$

Therefore 7.4.6 may be rewritten as

Proposition 7.4.7 *In* $K_1(\mathbf{Q}_p[G])$ *the image of* 7.4.5 *is given by*

$$[\Pi] = rank(V) < -1 > +[V_1] - [V_0] + [K^{-1}W] - [W],$$

where $< -1 > \in K_1(\mathbf{Z}_p) \cong \mathbf{Z}_p^*$ *denotes the class corresponding to* $-1 \in \mathbf{Z}_p^*$.

Remark 7.4.8 Note that the classes $[V_1], [V_0], [W] \in K_1(\mathbf{Q}_p[G])$ are precisely the elements which are obtained by comparing the $\mathbf{Z}_p[G]$-basis to the $\mathbf{Q}[G]$-basis for the modules P_1, P_0, X, while $[K^{-1}W]$ is the class obtained by comparing the $\mathbf{Z}_p[G]$-basis to the $\mathbf{Q}_p[G]$-basis for the module, V, in the manner of 4.2.13 and 4.6.36.

Now we shall examine the Chinburg invariant of 7.2.6 in the case of an extension, E/F, of number fields such that, for each wild prime $P \lhd \mathcal{O}_F$, the decomposition group $G(E_Q/F_P)$ is cyclic. We shall refer to such an extension as the *locally cyclic case.* The main result concerning locally cyclic extensions (Theorem 7.4.35) is rather technical. It gives the Hom-description of $\Omega(E/F, 2)$ in locally cyclic cases which satisfy a mild cohomological condition.

7.4.9 Suppose that L/K is a Galois extension of local fields with cyclic Galois group, $G(L/K)$, with a chosen generator, g, of order n.

In the 2-extension of 7.1.7 we may choose $B = \mathbf{Z}[G(L/K)]$ mapping to $\mathbf{Z}$ by the augmentation map, ϵ, given by $\epsilon(g^i) = 1$. Hence we have a short exact sequence of the form

7.4.10 $$\{1\} \longrightarrow L^* \stackrel{i}{\longrightarrow} A \stackrel{\lambda}{\longrightarrow} IG(L/K) \longrightarrow \{0\},$$

where $IG(L/K)$ is the kernel of the augmentation map,

$$\epsilon : \mathbf{Z}[G(L/K)] \longrightarrow \mathbf{Z}.$$

Let π_L denote the prime of L and let $v_L : L^* \longrightarrow \mathbf{Z}$ denote the valuation, normalised so that $v_L(\pi_L) = 1$. Let $e_{L/K}$ denote the ramification degree of L/K.

Proposition 7.4.11 *With the notation of* 7.4.9 *there exists an injective* $\mathbf{Z}[G(L/K)]$*-homomorphism*

$$\gamma : \mathbf{Z}[G(L/K)] \longrightarrow A$$

such that

(i) $\lambda(\gamma(z)) = (g-1)z$ *for all* $z \in \mathbf{Z}[G(L/K)]$.

(ii) *Let* $\sigma = \sum_{i=0}^{n-1} g^i \in \mathbf{Z}[G(L/K)]$. *There exists an element,* $x \in K^*$, *which satisfies* $v_L(x) = e_{L/K}$, *whose image generates* $K^*/(N_{L/K}(L^*)) \cong \mathbf{Z}/n$ *and such that* γ *can be chosen so that* $\gamma(\sigma) = x$.

(iii) *For* γ, x *as in* (i) *and* (ii) *there is an isomorphism of the form*

$$L^*/ < x > \cong A/(\gamma(\mathbf{Z}[G(L/K)])),$$

which is induced by i *of* 7.4.10.

(iv) *For any* $x \in K^*$ *with* $v_L(x) = e_{L/K}$, *whose image generates* $K^*/(N_{L/K}(L^*))$, $L^*/ < x >$ *is cohomologically trivial. Furthermore, if* U *is as in* 7.1.29 *and lies in* $\mathcal{O}_L^*$ *then*

$$\Omega(L/K, U) = [L^*/ < x, U >] \in \mathscr{CL}(\mathbf{Z}[G(L/K)]).$$

(v) *For any* $x \in K^*$ *as in* (iv), *there is a short exact sequence of* $\mathbf{Z}[G(L/K)]$*-modules of the form*

$$\{1\} \longrightarrow \mathcal{O}_L^* \longrightarrow L^*/ < x > \stackrel{v}{\longrightarrow} \mathbf{Z}/e_{L/K} \longrightarrow \{0\},$$

where v *is induced by* v_L.

Proof This is an elaboration of 7.3.27 and may be proved in the same manner. □

Example 7.4.12 Let p be an odd prime and let $K = \mathbf{Q}_p$, the p-adic rationals. If ξ_n denotes a primitive nth root of unity, let $L = \mathbf{Q}_p(\xi_{p^{s+1}})$ for some $0 \leq s$. In this case a suitable choice for x in 7.4.11(iv) is given by $x = (1+p)p\xi_{p-1}$.

We shall return to this example later in this section.

7.4.13 Let p be a rational prime. Suppose that E/F is a locally cyclic Galois extension of number fields and that $P \lhd \mathcal{O}_F$ is a wild prime, which lies over p. Let $Q \lhd \mathcal{O}_E$ be the chosen prime over P in 7.2.2. Hence we have $X_Q \subset \mathcal{O}_{E_Q}$ and $exp(X_Q) = 1 + X_Q \subset \mathcal{O}_{E_Q}^*$. In the construction of $\Omega(E_Q/F_P, 1 + X_Q)$ we have the fundamental 2-extension

$$\{1\} \longrightarrow E_Q^* \longrightarrow A \longrightarrow \mathbf{Z}[G(E_Q/F_P)] \stackrel{\epsilon}{\longrightarrow} \mathbf{Z} \longrightarrow \{0\},$$

which represents $inv^{-1}(1 \ (mod\ [E_Q : F_P]))$. Since

$$1 + X_Q \subset U_{E_Q}^1$$

we may apply 7.4.11(iv). Therefore we may choose any uniformiser

7.4.14 $$u_P \in F_P^*$$

whose image generates $F_P^*/(N_{E_Q/F_P}(E_Q^*))$ and then

7.4.15 $$\Omega(E_Q/F_P, 1+X_Q) = [E_Q^*/ < u_P, 1+X_Q >] \in \mathscr{CL}(\mathbf{Z}[G(E_Q/F_P)]).$$

Let $Tors(A)$ denote the torsion subgroup of A.

Lemma 7.4.16 *If $W_P = Tors(E_Q^*/ < u_P >)$ is a cohomologically trivial $\mathbf{Z}[G(E_Q/F_P)]$-module then*

$$V_P = (E_Q^*/ < u_P >)/\{Tors(E_Q^*/ < u_P >)\}$$

is a free $\mathbf{Z}_p[G(E_Q/F_P)]$-module whose rank is equal to $[F_P : Q_p]$.

Proof If W_P is cohomologically trivial then, by 7.4.11(iv) and the long exact cohomology sequence V_P is cohomologically trivial and torsion-free over $\mathbf{Z}_p$. Hence V_P is a free module, as can easily be deduced from Serre (1979, theorem 7, p. 144). □

Remark 7.4.17 From Washington (1982, lemma 13.27, p. 287) one can deduce that the cohomological triviality condition is satisfied for the finite subextensions of the cyclotomic $\mathbf{Z}_p$-extension over $F(\xi_p)$, where F is any totally real number field (cf. 7.4.39).

Under the circumstances of 7.4.13–7.4.16, choose a $\mathbf{Z}_p[G(E_Q/F_P)]$-basis for V_P ($n_P = [F_P : Q_p]$)

7.4.18 $$y_P(1), \ldots, y_P(n_P) \in V_P.$$

By 7.4.11(v), we have an exact sequence of the form

$$\{1\} \longrightarrow \mathcal{O}_{E_Q}^* \longrightarrow E_Q^*/ < u_P > \longrightarrow \mathbf{Z}/e_{E_Q/F_P} \longrightarrow \{0\}.$$

If $\mu(E_Q)$ denotes the roots of unity in E_Q then $\mathcal{O}_{E_Q}^* \cong \mu(E_Q) \oplus U_{E_Q}^1$, where $U_{E_Q}^i = 1 + \pi_{E_Q}^i \mathcal{O}_{E_Q}$, and $E_Q^* \cong \mathbf{Z} < \pi_{E_Q} > \oplus \mathcal{O}_{E_Q}^*$ so that the induced $\mathbf{Z}_p[G(E_Q/F_P)]$-homomorphism

$$U_{E_Q}^1/(Tors(U_{E_Q}^1)) \longrightarrow V_P$$

induces an isomorphism of $\mathbf{Q}_p[G(E_Q/F_P)]$-modules,

$$\{U_{E_Q}^1/(Tors(U_{E_Q}^1))\} \otimes_{\mathbf{Z}_p} \mathbf{Q}_p \longrightarrow V_P \otimes_{\mathbf{Z}_p} \mathbf{Q}_p.$$

In addition, we have a commutative diagram

$$\begin{array}{ccc} X_Q & \xrightarrow[exp]{\cong} & 1 + X_Q \\ \downarrow & & \downarrow \\ \pi_{E_Q}\mathcal{O}_{E_Q} & \xrightarrow{exp} & U^1_{E_Q} \end{array}$$

in which the vertical maps are injections with finite cokernels. Therefore the exponential is an injective map of $\mathbf{Z}_p[G(E/F_P)]$-modules

$$exp : X_Q \longrightarrow V_P,$$

which becomes an isomorphism when tensored with $\mathbf{Q}_p$.

Suppose that we have a $\mathbf{Z}[G(E_Q/F_P)]$-resolution

$$\{0\} \longrightarrow T_1 \xrightarrow{i'} T_0 \xrightarrow{\pi} V_P/(1 + X_Q) \longrightarrow \{0\}.$$

As in 7.4.5, we may construct an exact sequence of free $\mathbf{Z}_p[G(E_Q/F_P)]$-modules of the form

$$\{0\} \longrightarrow T_1 \otimes_{\mathbf{Z}} \mathbf{Z}_p \xrightarrow{(i',j')} (T_0 \otimes_{\mathbf{Z}} \mathbf{Z}_p) \oplus X_Q \xrightarrow{(\pi',-exp)} V_P \longrightarrow \{0\}$$

where π' is any lifting of π. As in 7.2.4, identify $X_P = X \otimes_{\mathcal{O}_F} \mathcal{O}_{F_P}$ with $Ind^{G(E/F)}_{G(E_Q/F_P)}(X_Q)$ and set $P_i = Ind^{G(E/F)}_{G(E_Q/F_P)}(T_i)$. Therefore we obtain a short exact sequence of free $\mathbf{Z}_p[G(E/F)]$-modules

7.4.19 $$P_1 \otimes_{\mathbf{Z}} \mathbf{Z}_p \xrightarrow{(i'',j'')} (P_0 \otimes_{\mathbf{Z}} \mathbf{Z}_p) \oplus X_P \xrightarrow{(\tilde{\pi},-Ind(exp))} Ind^{G(E/F)}_{G(E_Q/F_P)}(V_P).$$

Also, we have a short exact sequence

7.4.20

$$\Pi : \prod_{P|p} P_1 \otimes_{\mathbf{Z}} \mathbf{Z}_p \longrightarrow \left(\prod_{P|p} P_0 \otimes_{\mathbf{Z}} \mathbf{Z}_p\right) \oplus \left(\prod_{P|p} X_P\right) \longrightarrow \prod_{P|p} Ind^{G(E/F)}_{G(E_Q/F_P)}(V_P).$$

If X is the locally free $\mathcal{O}_F[G(E/F)]$-module of 7.2.3 then, considered as a $\mathbf{Z}[G(E/F)]$-module,

7.4.21

$$X \otimes_{\mathbf{Z}} \mathbf{Z}_p \cong (X \otimes_{\mathcal{O}_F} \mathcal{O}_F) \otimes_{\mathbf{Z}} \mathbf{Z}_p \cong X \otimes_{\mathcal{O}_F} \left(\prod_{P|p} \mathcal{O}_{F_P}\right) \cong \prod_{P|p} X_P.$$

By comparing a choice of $\mathbf{Z}_p[G(E/F)]$-basis for 7.4.20 with a choice of $\mathbf{Q}_p[G(E/F)]$-basis for the complex obtained by tensoring 7.4.20 with $\mathbf{Q}_p$ one obtains an element (see 7.4.4)

7.4.22 $$[\Pi] \in K_1(\mathbf{Q}_p[G(E/F)]).$$

In particular, suppose that we use a $\mathbf{Q}[G(E/F)]$-basis as the $\mathbf{Q}_p[G(E/F)]$-basis for $(P_1 \otimes_{\mathbf{Z}} \mathbf{Q}_p)$ and its image under i'' as the $\mathbf{Q}_p[G(E/F)]$-basis for $(P_0 \otimes_{\mathbf{Z}} \mathbf{Q}_p)$. Also, suppose that we use a $\mathbf{Q}[G(E/F)]$-basis for $X \otimes \mathbf{Q}$ as a $\mathbf{Q}_p[G(E/F)]$-basis for $X \otimes \mathbf{Q}_p$ and its image under $exp \otimes 1$ as a $\mathbf{Q}_p[G(E/F)]$-basis for $\prod_{P|p} Ind_{G(E_Q/F_P)}^{G(E/F)}(V_P)$. In this case the comparison of the bases for $(\prod_{P|p} P_i \otimes_{\mathbf{Z}} \mathbf{Z}_p)$ for $i = 1, 2$ yields elements

$$[V_i] \in K_1(\mathbf{Q}_p[G(E/F)]),$$

which represent the p-component of the idèle in $K_1(J^*(\mathbf{Q}[G(E/F)]))$ which corresponds to the class of $[P_i] \in \mathcal{CL}(\mathbf{Z}[G(E/F)])$ in the Hom-description of 4.2.13 and 4.6.36. Also, with these choices, the comparison of bases for $X \otimes_{\mathbf{Z}} \mathbf{Q}_p$ gives a representative of the p-component of the idèle corresponding to $N_{F/\mathbf{Q}}([X]) \in \mathcal{CL}(\mathbf{Z}[G(E/F)])$, where $[X] \in \mathcal{CL}(\mathcal{O}_F[G(E/F)])$ is the class of 7.2.6.

The class $\prod_{P|p\ wild}([P_1] - [P_0]) \in \mathcal{CL}(\mathbf{Z}[G(E/F)])$ is equal to

$$\sum_{P|p\ \ wild} Ind_{G(E_Q/F_P)}^{G(E/F)}(\Omega(E_Q/F_P, 1 + X_Q) - [W_P]),$$

where W_P is as in 7.4.16.

If l is a prime then $\Omega(E_Q/F_P, 1 + X_Q) - [W_P]$ is represented by a cohomologically trivial l-group, if $P \mid l$, and therefore has trivial p-adic idèlic representative at primes p different from l. Therefore, from 7.4.6, we have the following result (since $[\Pi] \in K_1(\mathbf{Z}_p[G(E/F)])$ its determinant lies in the Hom-description indeterminacy).

Proposition 7.4.23 *Let $v_0(*)$ be the choice of $\mathbf{Q}_p[G(E/F)]$-basis for*

$$\left(\prod_{P|p} Ind_{G(E_Q/F_P)}^{G(E/F)}(V_P)\right) \otimes_{\mathbf{Z}_p} \mathbf{Q}_p$$

described above and let $v_p(*)$ be a $\mathbf{Z}_p[G(E/F)]$-basis for

$$\prod_{P|p} Ind_{G(E_Q/F_P)}^{G(E/F)}(V_P).$$

If Φ is the invertible matrix with entries in $\mathbf{Q}_p[G(E/F)]$ given by

$$\Phi v_0(*) = v_p(*),$$

then

$$[\Phi] \in K_1(\mathbf{Q}_p[G(E/F)])$$

is a representative of the p-adic idèlic component, in the sense of the Hom-description of 4.2.13 *and* 4.6.36, *corresponding to the class of* 7.2.6,

$$\Omega(E/F,2) - \sum_{P|p \ wild} Ind_{G(E_Q/F_P)}^{G(E/F)}([W_P]) \in \mathscr{CL}(\mathbf{Z}[G(E/F)]),$$

where $W_P = Tors(E_Q^*/ < u_P >)$ is as in 7.4.16.

7.4.24 *Representing $\Omega(E/F,2)$*

From 7.4.23 and the discussion of 7.4.17–7.4.22 it is possible to describe an explicit representative for $\Omega(E/F,2) \in \mathscr{CL}(\mathbf{Z}[G(E/F)])$ in terms of the Hom-description of 4.2.13, in the locally cyclic case when all the $\mathbf{Z}[G(E_Q/F_P)]$-modules, $W_P = Tors(E_Q^*/ < u_P >)$ in 7.4.16, are cohomologically trivial. This representation is obtained merely by amalgamating 7.4.23 with the definitions of 7.2.1–7.2.6 and 4.2.13–4.2.28.

We have to produce a function lying within the group

$$Hom_{\Omega_{\mathbf{Q}}}(R(G(E/F)), J^*(N))$$

where $N/\mathbf{Q}$ is a Galois extension of the rationals containing E and all the $[E : F]$th roots of unity.

Let $a \in E^*$ be a normal basis element for E/F such that

7.4.25 $$a \in X \cap \mathcal{O}_E \subset (X \otimes_{\mathcal{O}_F} F) = E.$$

Hence $E = F[G(E/F)]a$ is the free module generated by a. Even if the identity of X is unknown in 7.2.3, one may multiply the element, a, by a large power of $[E : F]$ in order to ensure that 7.4.25 holds, without changing the class which we are going to represent in $\mathscr{CL}(\mathbf{Z}[G(E/F)])$.

For each tame prime, $P \lhd \mathcal{O}_F$, and for the chosen $Q \lhd \mathcal{O}_E$ above P we have $a_P \in \mathcal{O}_{E_Q}$ such that

7.4.26 $$\mathcal{O}_{F_P}[G(E/F)]a_P = \textstyle\prod_{R|P} \mathcal{O}_{E_R}.$$

Also

7.4.27 $$F_P[G(E/F)]a = E \otimes_F F_P \cong \textstyle\prod_{R|P} E_R$$

so that there exists $Y_P \in F_P[G(E/F)]^*$ such that

7.4.28 $$Y_P a = a_P \in \prod_{R|P} E_R.$$

Hence, since $E_R \otimes_E N \cong \prod_{S|P} N_S$,

7.4.29 $$det(Y_P) \in Hom_{\Omega_F}(R(G(E/F)), \prod_{S|P} N_S^*).$$

Write $J^*_{tame}(N)$ for the restricted idèlic product involving only those completions, N_S, for $S \lhd \mathcal{O}_N$ which lie above tame primes of F. Hence $J^*_{tame}(N)$ is an Ω_Q-module. We have a norm map (Fröhlich, 1976; Curtis & Reiner (1987, section 52.25, p. 341)

$$N_{F/\mathbf{Q}} : Hom_{\Omega_F}(R(G(E/F)), \prod_{S|P} N_S^*) \longrightarrow Hom_{\Omega_{\mathbf{Q}}}(R(G(E/F)), \prod_{S|P} N_S^*)$$

given by $N_{F/\mathbf{Q}}(f) = \prod_{g \in \Omega_{\mathbf{Q}}/\Omega_F} g(f(g^{-1} \cdot -))$.

Therefore we obtain

7.4.30 $$N_{F/\mathbf{Q}}(\{det(Y_P)\}_{P \ tame}) \in Hom_{\Omega_{\mathbf{Q}}}(R(G(E/F)), J^*_{tame}(N)).$$

Let $a \in X$ be the normal basis element of 7.4.25 and let $z(1), \ldots, z(t) \in \mathcal{O}_F$ $(t = [F : \mathbf{Q}])$ be a $\mathbf{Q}$-basis for F. Hence $z(*)a = \{z(1)a, \ldots, z(t)a\}$ is a $\mathbf{Q}$-basis for $X \otimes_{\mathbf{Z}} \mathbf{Q}$. From 7.4.21 we have an isomorphism

7.4.31 $$\lambda_p : X \otimes_{\mathbf{Z}} \mathbf{Q}_p \xrightarrow{\cong} (\prod_{P|p} X_P) \otimes_{\mathbf{Z}} \mathbf{Q}_p$$

whose P-component is characterised by

$$(\lambda_p(z(i)a))_P = a \otimes z(i) \in X \otimes_{\mathcal{O}_F} \mathcal{O}_{F_P} = X_P.$$

From 7.4.20 we also have a map of $\mathbf{Z}_p[G(E/F)]$-modules

7.4.32 $$\prod_{P|p} Ind_{G(E_Q/F_P)}^{G(E/F)}(exp) : (\prod_{P|p} X_P) \longrightarrow \prod_{P|p} Ind_{G(E_Q/F_P)}^{G(E/F)}(V_P).$$

If $y_P(1), \ldots, y_P(n_P) \in V_P$ is the $\mathbf{Z}_p[G(E_Q/F_P)]$-basis then $y_P(i) \otimes 1 \in Ind_{G(E_Q/F_P)}^{G(E/F)}(V_P)$ and

$$(y_P(*); P \mid p) = \{y_P(1) \otimes 1, \ldots, y_P(n_P); P \mid p\}$$

is a $\mathbf{Z}_p[G(E_Q/F_P)]$-basis for $\prod_{P|p} Ind_{G(E_Q/F_P)}^{G(E/F)}(V_P)$. Note that

$$\sum_{P|p} rank_{\mathbf{Z}_p}(V_P) = \sum_{P|p} [F_P : \mathbf{Q}_p] = [F : \mathbf{Q}] = t,$$

as expected. Up to multiplication by a permutation matrix, we obtain a matrix

$$B_p \in GL_t(\mathbf{Q}_p[G(E_Q/F_P)])$$

which satisfies

$$B_p \left(\prod_{P|p} Ind_{G(E_Q/F_P)}^{G(E/F)}(exp(\lambda_p(z(*)a)) \right) = (y_P(*); P \mid p).$$

By the construction of 4.2.13–4.2.28 we obtain a homomorphism

7.4.33 $$Det(B_p) \in Hom_{\Omega_{\mathbf{Q}}}(R(G(E/F)), \textstyle\prod_{S|p} N_S^*).$$

If B_p is varied by multiplication by a permutation matrix, σ, then $Det(B_p)$ is changed by multiplication by the function,

$$T \longmapsto sign(\sigma)^{dim(T)},$$

which lies in the subgroup $Det(\mathbf{Z}_p[G(E/F)]^*)$.

Let $J^*_{wild}(N)$ denote the product of the N_S^* for primes, $S \lhd \mathcal{O}_N$ lying over wild primes of F. Hence

$$J^*_{wild}(N) = \prod_p \left(\prod_{S|p} N_S^* \right)$$

where p is a rational prime lying beneath a wild prime $P \lhd \mathcal{O}_F$. Similarly, write $J^*_\infty(N)$ for the product of the Archimedean components of $J^*(N)$ so that

$$J^*(N) = J^*_{wild}(N) \times J^*_{tame}(N) \times J^*_\infty(N).$$

With this notation 7.4.33 yields a homomorphism

7.4.34 $$Det(B_p) \in Hom_{\Omega_{\mathbf{Q}}}(R(G(E/F)), J^*_{wild}(N)).$$

Theorem 7.4.35 *Let E/F be a locally cyclic Galois extension of number fields for which, in 7.4.16,*

$$W_P = Tors(E_Q^*/ < u_P >)$$

is a cohomologically trivial $\mathbf{Z}[G(E_Q/F_P)]$*-module for each wild prime,* $P \triangleleft \mathcal{O}_{F_P}$.

Define

$$s_{E/F} \in Hom_{\Omega_{\mathbf{Q}}}(R(G(E/F)), J^*(N))$$

to be the function whose coordinates are trivial in $J^*_\infty(N)$ *and are given by* 7.4.30 *in* $J^*_{tame}(N)$ *and by* 7.4.34 *in* $J^*_{wild}(N)$. *If* $[s_{E/F}]$ *is the class given by the Hom-description of* 4.2.13 *then*

$$[s_{E/F}] = \Omega(E/F,2) - \sum_{P \ wild} Ind_{G(E_Q/F_P)}^{G(E/F)}([W_P]) \in \mathcal{CL}(\mathbf{Z}[G(E/F)]).$$

7.4.36 *Cyclotomic examples*

Let p be an *odd* prime and let ξ_n denote a primitive nth root of unity. For each $s \geq 0$ let $L_{s+1} = \mathbf{Q}_p(\xi_{p^{s+1}})$ and let $x = (1+p)p\xi_{p-1} \in \mathbf{Q}_p^*$. Note that x is a uniformiser for $\mathbf{Q}_p$ with the property that $L^*_{s+1}/ < x >$ is cohomologically trivial, as in 7.4.11. Identify $G(L_{s+1}/\mathbf{Q}_p)$ with $(\mathbf{Z}/p^{s+1})^*$ in the customary manner. Set

7.4.37

$$\begin{cases} W_p(s+1) = Tors(L^*_{s+1}/ < x >) \quad \text{and} \\ V_p(s+1) = (L^*_{s+1}/ < x >)/W_p(s+1). \end{cases}$$

In this section we shall show that $W_p(s+1)$ is a cohomologically trivial $\mathbf{Z}[(\mathbf{Z}/p^{s+1})^*]$-module and that $V_p(s+1)$ is a free $\mathbf{Z}_p[(\mathbf{Z}/p^{s+1})^*]$-module of rank one. We shall also be concerned with the analogous modules for the totally real subfield

$$\begin{cases} W_p^+(s+1) = Tors((L^+_{s+1})^*/ < x >) \quad \text{and} \\ V_p^+(s+1) = ((L^+_{s+1})^*/ < x >)/W_p^+(s+1). \end{cases}$$

These $\mathbf{Z}[G(L^+_{s+1}/\mathbf{Q}_p)]$-modules are equal to the σ_{-1}-fixed points of 7.4.37, where $\sigma_a(\xi_{p^{s+1}}) = \xi^a_{p^{s+1}}$ for any $a \in (\mathbf{Z}/p^{s+1})^*$. Once again $W_p^+(s+1)$ is a cohomologically trivial $\mathbf{Z}[(\mathbf{Z}/p^{s+1})^*/\{\pm 1\}]$-module and we shall show that $V_p^+(s+1)$ is a free $\mathbf{Z}_p[(\mathbf{Z}/p^{s+1})^*/\{\pm 1\}]$-module, whose generator we shall specify (Theorem 7.4.46). We shall accomplish part of this by induction on s, using the following elementary result.

Lemma 7.4.38 *Let* p *be a prime and let* m *be a positive integer. Let* M *be a cohomologically trivial, torsion-free* $\mathbf{Z}_p[\mathbf{Z}/n]$*-module, where* $n = pm$.

If $\alpha \in M^{\mathbf{Z}/p}$ *is a free generator for* $M^{\mathbf{Z}/p}$ *as a* $\mathbf{Z}_p[\mathbf{Z}/m]$*-module of rank one then* M *is a free* $\mathbf{Z}_p[\mathbf{Z}/n]$*-module of rank one. In fact, a generator is given by any* $\beta \in M$ *such that*

$$(1 + g^m + g^{2m} + \ldots + g^{m(p-1)})\beta = \alpha,$$

where g *generates* $\mathbf{Z}/n$*.*

Proof Let N denote multiplication by $1+g^m+g^{2m}+\ldots+g^{m(p-1)} \in \mathbf{Z}_p[\mathbf{Z}/n]$. Since $H^2(\mathbf{Z}/n; M) = 0$, $M^{\mathbf{Z}/p} = N(M)$ so that there exists β such that $N(\beta) = \alpha$. Define $\psi : \mathbf{Z}_p[\mathbf{Z}/n] \longrightarrow M$ by $\psi(z) = z\beta$ then, by hypothesis, ψ is an isomorphism on $\mathbf{Z}/p$-invariants. Since M is a free $\mathbf{Z}_p[\mathbf{Z}/n]$-module of rank one, it suffices to show that ψ is surjective. For in this case $Ker(\psi)$ would be torsion free and cohomologically trivial and therefore free and necessarily trivial.

Given $x \in M$ there exists $u_1 \in \mathbf{Z}_p[\mathbf{Z}/n]$ such that $N(x) = \psi(N(u_1))$ and therefore

$$x - \psi(u_1) \in Ker(N) = (g^m - 1)M.$$

By induction there exist $u_2, u_3, \ldots \in \mathbf{Z}_p[\mathbf{Z}/n]$ such that

$$x = \psi\left(\sum_{i=1}^{\infty}(g^m - 1)^i u_i\right).$$

This element is in the image of ψ since $(g^m - 1)^p \in p\mathbf{Z}_p[\mathbf{Z}/n]$, which ensures that the series is convergent. □

Proposition 7.4.39 (i) *In* 7.4.37, $W_p(s+1)$ *is a cohomologically trivial* $\mathbf{Z}[(\mathbf{Z}/p^{s+1})^*]$*-module.*

(ii) *More generally, let* F *be a totally real number field and let* $L_1 = F(\xi_p)$*. Let* L_{s+1}/L_1 *be the completion at some p-adic prime of the subextension of the cyclotomic* $\mathbf{Z}_p$*-extension of* L_1 *such that* $G(L_{s+1}/L_1) \cong \mathbf{Z}/p^s$*. Let* L_0 *denote the corresponding completion of* F*. Suppose that* $x \in L_0^*$ *is chosen as in* 7.4.11 *then*

$$Tors(L_{s+1}^*/ < x >)$$

is a cohomologically trivial $\mathbf{Z}[G(L_{s+1}/L_0)]$*-module.*

Proof The proof of (ii) is similar to that of (i), once one knows that the p-primary roots of unity are cohomologically trivial (Washington, 1982, lemma 13.27, p. 287). In the case of (i) this fact is clear from Table 5.4 and 5.6.3 (see also 5.7.14).

From 7.4.11(v) there is a short exact sequence of the following form

7.4.40 $\{1\} \longrightarrow \mathbf{F}_p^* \times U^1_{L_{s+1}} \longrightarrow L^*_{s+1}/<x> \longrightarrow \mathbf{Z}/p^s \times \mathbf{F}_p^* \longrightarrow \{1\}$

since $e_{L_{s+1}/\mathbf{Q}_p} = p^s(p-1)$. The p-primary roots of unity, which lie in $U^1_{L_{s+1}}$, are given by the $\mathbf{Z}[(\mathbf{Z}/p^{s+1})^*]$-module, $M_{0,s+1}$, of 5.6.1, which is cohomologically trivial, by Table 5.4.

Temporarily set $Y = U^1_{L_{s+1}}/M_{0,s+1}$ so that Y is a torsion-free $\mathbf{Z}_p$-module and, as abelian groups, $U^1_{L_{s+1}} \cong M_{0,s+1} \oplus Y$.

Identify the cyclic group, $(\mathbf{Z}/p^{s+1})^*$, with $\mathbf{Z}/p^s \times \mathbf{F}_p^*$. Consider the $(p-1)$-power map, ϕ, on $L^*_{s+1}/<x>$. Since ϕ is an isomorphism on $\mathbf{Z}/p^s$ and $U^1_{L_{s+1}}$ and annihilates $\mathbf{F}_p^*$, 7.4.40 implies that there is a short exact sequence of the form

$$\{1\} \longrightarrow U^1_{L_{s+1}} \longrightarrow Im(\phi^2) \longrightarrow \mathbf{Z}/p^s \longrightarrow \{1\}.$$

Therefore $H^*(\mathbf{F}_p^*; Im(\phi^2)) = 0$. Hence $H^*(\mathbf{F}_p^*; Ker(\phi^2)) = 0$. However, there is a short exact sequence of the form

$$\{1\} \longrightarrow \mathbf{F}_p^* \longrightarrow Ker(\phi^2) \longrightarrow R \longrightarrow \{1\},$$

where $R \subset \mathbf{F}_p^*$ and $\mathbf{F}_p^*$ acts trivially on $\mathbf{F}_p^*$. The coboundary must be an isomorphism

$$\delta : H^i(\mathbf{F}_p^*; R) \cong R \longrightarrow H^{i+1}(\mathbf{F}_p^*; \mathbf{F}_p^*) \cong \mathbf{F}_p^*$$

so that we must have $R = \mathbf{F}_p^*$. Therefore $Ker(\phi^2)$ consists of all the torsion of $L^*_{s+1}/<x>$ of order prime to p and, being a cohomologically trivial $\mathbf{Z}[\mathbf{F}_p^*]$-module, it is also a cohomologically trivial $\mathbf{Z}[(\mathbf{Z}/p^{s+1})^*]$-module.

When $s = 0$ it is clear from 7.4.40 that the cohomologically trivial module, $M_{0,1} \oplus Ker(\phi^2)$ is equal to $Tors(L_1^*/<x>) = W_p(1)$, which proves the result in this case. When $1 \leq s$, if $A = M_{0,s+1} \oplus Ker(\phi^2)$, we obtain from 7.4.40 a short exact sequence of the form

$$\{1\} \longrightarrow U^1_{L_{s+1}}/M_{0,s+1} \longrightarrow (L^*_{s+1}/<x>)/A \longrightarrow \mathbf{Z}/p^s \longrightarrow \{1\}.$$

In addition the prime element, $\pi_{s+1} = 1 - \xi_{p^{s+1}} \in L_{s+1}$, maps to a generator of the right-hand group. To see that the cohomologically trivial module, $(L^*_{s+1}/<x>)/A$, is torsion-free we argue in the following manner. Let N denote the torsion subgroup. Thus N is a submodule and the action upon it must be trivial since we may embed it into the trivial module, $\mathbf{Z}/p^s$, by means of the quotient map from the exact sequence. This means that the induced map

$$H^{2i}((\mathbf{Z}/p^{s+1})^*; N) \cong N \longrightarrow H^{2i}((\mathbf{Z}/p^{s+1})^*; \mathbf{Z}/p^s) \cong \mathbf{Z}/p^s$$

factors through the trivial cohomology group,

$$H^{2i}((\mathbf{Z}/p^{s+1})^*; (L^*_{s+1}/ < x >)/A) = 0.$$

The only manner in which this can happen, since the cohomology map is the inclusion of N into $\mathbf{Z}/p^s$, is for N to be the trivial group.

Therefore we have shown that $(L^*_{s+1}/ < x >)/A$ is torsion free and the cohomologically trivial module, $M_{0,s+1} \oplus Ker(\phi^2)$, is equal to

$$Tors(L^*_{s+1}/ < x >) = W_p(s+1),$$

which completes the proof. □

In the course of the proof of 7.4.39 we also established the following result.

Corollary 7.4.41 *Let $V_p(s+1)$ be the torsion-free, cohomologically trivial $\mathbf{Z}_p[(\mathbf{Z}/p^{s+1})^*]$-module of 7.4.37. If $1 \leq s$, then there is a short exact sequence of $\mathbf{Z}_p[(\mathbf{Z}/p^{s+1})^*]$-modules of the form*

$$\{1\} \longrightarrow U^1_{L_{s+1}}/Tors(U^1_{L_{s+1}}) \longrightarrow V_p(s+1) \longrightarrow \mathbf{Z}/p^s \longrightarrow \{1\}.$$

The image of the prime, $\pi_{s+1} = 1 - \xi_{p^{s+1}}$ is a generator of the right-hand group.

Now we begin to find the $\mathbf{Z}_p[(\mathbf{Z}/p^{s+1})^*/\{\pm 1\}]$-basis for the free modules, $V^+_p(s+1)$. We will need the following result.

Proposition 7.4.42 *In the following commutative diagram, whose upper and lower rows are isomorphisms,*

$$\begin{array}{ccccc}
(U^1_{L_1})/(\mathbf{Z}/p) & \xrightarrow{\cong} & U^2_{L_1} & \xrightarrow{p} & U^{p+1}_{L_1} \\
 & & \downarrow{\scriptstyle log_p} & & \cong\downarrow{\scriptstyle log_p} \\
 & & \pi_1^2\mathcal{O}_{L_1} & \xrightarrow{\cong} & p\pi_1^2\mathcal{O}_{L_1}
\end{array}$$

the left-hand logarithm induces a Galois isomorphism of the form

$$log_p : U^1_{L_1}/Tors(U^1_{L_1}) \xrightarrow{\cong} \pi_1^2\mathcal{O}_{L_1}.$$

In addition, $\pi_1\mathcal{O}_{L_1}$ and $\pi_1^2\mathcal{O}_{L_1}$ are free $\mathbf{Z}_p[(\mathbf{Z}/p^{s+1})^*]$-modules on $\pi_1 = 1-\xi_p$ and π_1^2 respectively.

Proof In the diagram, the surjectivity of the upper pth power map follows from Hasse (1980, p. 236) since the pth power map induces isomorphisms of the form

$$\frac{U_{L_1}^{2+d}}{U_{L_1}^{3+d}} \xrightarrow{\cong} \frac{U_{L_1}^{p+1+d}}{U_{L_1}^{p+2+d}}$$

for all $d \geq 0$. The right-hand logarithm is an isomorphism whose inverse is *exp*. Hence

$$log_p : U_{L_1}^1/Tors(U_{L_1}^1) \xrightarrow{\cong} \pi_1^2\mathcal{O}_{L_1}$$

is an isomorphism of Galois modules.

From the exact sequence

$$0 \longrightarrow \pi_1\mathcal{O}_{L_1} \longrightarrow \mathcal{O}_{L_1} \longrightarrow \mathbf{Z}/p \longrightarrow 0$$

a basis for $\pi_1\mathcal{O}_{L_1}$ is given by $\{1-\xi_p, 1-\xi_p^2, \ldots, 1-\xi_p^{p-2}, p\}$. If $\sigma_j \in G(L_1/L_0)$ is given by $\sigma_j(\xi_p) = \xi_p^j$ then this basis is

$$\{\sigma_1(\pi_1), \sigma_2(\pi_1), \ldots, \sigma_{p-2}(\pi_1), (\sigma_1 + \ldots + \sigma_{p-1})(\pi_1)\}.$$

Hence $\pi_1\mathcal{O}_{L_1} = \mathbf{Z}_p[G(L_1/L_0)] < \pi_1 >$.

The argument for $\pi_1^2\mathcal{O}_{L_1}$ is similar. It suffices to show that the natural map, $\mathbf{Z}_p[(\mathbf{Z}/p)^*] < \pi_1^2 > \longrightarrow \pi_1^2\mathcal{O}_{L_1}$, is onto, since both are free modules of rank one. By p-adic approximation it suffices to show that the image of $\mathbf{Z}_p[(\mathbf{Z}/p)^*] < \pi_1^2 >$ generates

$$\pi_1^2\mathcal{O}_{L_1}/(p\pi_1^2\mathcal{O}_{L_1}) = \pi_1^2\mathcal{O}_{L_1}/(\pi_1^{p+1}\mathcal{O}_{L_1}).$$

However, this is clear since

$$\sigma_a(\pi_1^i) \equiv a^i\pi_1^i - a^i(a-1)/2\pi_1^{i+1} \quad (mod\ \pi_1^{i+2})$$

for $1 \leq a \leq p-1$ and $2 \leq i \leq p-1$ and $(\sigma_1 + \ldots + \sigma_{p-1})(\pi_1^2) = p$. □

7.4.43 Let *log* denote the usual p-adic logarithm (Curtis & Reiner, 1987, p. 356; see also 4.3.14) defined on the one-units, $U_{L_{s+1}}^1$. Let log_p denote the p-adic logarithm of Washington (1982, p. 50). Hence log_p is defined on all of L_{s+1}^* and coincides with *log* on $U_{L_s}^1$ and is normalised so that $log_p(p) = 0$.

We may define a homomorphism

$$Xog_p : V_p(s+1) \otimes \mathbf{Q}_p \longrightarrow L_{s+1} = \mathbf{Z}_p[\xi_{p^{s+1}}] \otimes \mathbf{Q}_p$$

which is an inverse to $exp \otimes 1$. In fact, define $Xog_p = log = log_p$ on

$$U^1_{L_{s+1}}/Tors(U^1_{L_{s+1}}) \subset (L^*_{s+1}/ < x >)/Tors(L^*_{s+1}/ < x >)$$

and define

7.4.44

$$\begin{aligned} Xog_p(\pi_{s+1}) &= log_p\left(\frac{\pi_{s+1}^{p^s(p-1)}}{p(1+p)}\right)/p^s(p-1) \\ &= log_p(\pi_1) - p^{-s}(p-1)^{-1}log_p(1+p). \end{aligned}$$

This formula makes sense since $\pi_{s+1}^{p^s(p-1)}/p(1+p)$ lies in $U^1_{L_{s+1}}$ and $x = p(1+p)\xi_{p-1}$ is trivial in $V_p(s+1)$.

7.4.45 Let L^+_{s+1} denote the fixed field of 'complex conjugation' (which sends $\xi_{p^{s+1}}$ to $\xi^{-1}_{p^{s+1}} = \overline{\xi}_{p^{s+1}}$) acting on L_{s+1}. Hence $(L^+_{s+1})^*/ < x >$ is isomorphic to the subgroup consisting of the elements of $L^*_{s+1}/ < x >$ which are fixed by conjugation. In addition,

$$G(L^+_{s+1}/\mathbf{Q}_p) \cong G(\mathbf{Q}(\xi_{p^{s+1}})/\mathbf{Q}) \cong (\mathbf{Z}/p^{s+1})^*/\{\pm 1\}.$$

Let $V^+_p(s+1)$ denote the subgroup of $V_p(s+1)$ consisting of the elements fixed by conjugation. Since $\overline{\pi}_{s+1} = 1 - \overline{\xi}_{p^{s+1}} = -\overline{\xi}_{p^{s+1}}(1 - \xi_{p^{s+1}})$ we see that

$$\overline{\pi}_{s+1} = \pi_{s+1} \in V_p(s+1)$$

so that $\pi_{s+1} \in V^+_p(s+1)$.

Theorem 7.4.46 *Let p be an odd, regular prime. Let $L^+_{s+1}/\mathbf{Q}_p$ and $V^+_p(s+1)$ be as in 7.4.45. Then, for $0 \leq s$, $V^+_p(s+1)$ is the free $\mathbf{Z}_p[G(L^+_{s+1}/\mathbf{Q}_p)]$-module generated by the image of $\pi_{s+1} = 1 - \xi_{p^{s+1}}$.*

Proof By 7.4.38 it will suffice to establish this result in the case of $V^+_p(1)$. Also, by 7.4.42, we have an isomorphism of the form

$$Xog_p : V^+_p(1) \xrightarrow{\cong} (\pi_1 \mathcal{O}_{L_1})^+.$$

The only non-trivial linear relations over $\mathbf{Q}_p$ between $\{log_p(1-\xi_p^j); 1 \leq j \leq (p-1)/2\}$ are (Washington, 1982, p. 74) multiples of

$$\begin{aligned}\sum_{j=1}^{(p-1)/2} log_p(1-\xi_p^j) &= (1/2)\sum_{j=1}^{p-1} log_p(1-\xi_p^j) \\ &= (1/2)log_p(p) \\ &= 0.\end{aligned}$$

Therefore, if

$$\sum_{j=1}^{(p-1)/2} \alpha_j(log_p(1-\xi_p^j) - log_p(1+p)) = 0$$

we may apply $\sigma_1, \ldots, \sigma_{(p-1)/2}$ and add the results to obtain $\sum_j \alpha_j = 0$. Hence $\alpha_1 = \ldots = \alpha_{(p-1)/2} = 0$. This means that $Xog(\pi_1)$ is a free generator of $V_p^+(1) \otimes \mathbf{Q}_p$ as a $\mathbf{Q}_p[(\mathbf{Z}/p)^*/\{\pm 1\}]$-module.

Now let us examine the circumstances under which $Xog_p(\pi_1)$ is a free generator of $V_p^+(1)$ as a $\mathbf{Z}_p[(\mathbf{Z}/p)^*\{\pm 1\}]$-module. Consider the following $(p-1)/2 \times (p-1)/2$ matrix.

7.4.47

$$\begin{pmatrix} log_p(1-\xi_p)-(p-1)^{-1}log_p(1+p) & log_p(1-\xi_p^2)-(p-1)^{-1}log_p(1+p) & \cdots \\ log_p(1-\xi_p^2)-(p-1)^{-1}log_p(1+p) & log_p(1-\xi_p^4)-(p-1)^{-1}log_p(1+p) & \cdots \\ \vdots & \vdots & \vdots \end{pmatrix}$$

The (i,j)th entry in 7.4.47 is equal to $\sigma_i(\sigma_j(Xog_p(\pi_1)))$. The determinant, R_p, of this matrix is non-zero, since $Xog_p(\pi_1)$ is a free $\mathbf{Q}_p[(\mathbf{Z}/p)^*/\{\pm 1\}]$ generator for $(\pi_1\mathcal{O}_{L_1})^+ \otimes \mathbf{Q}_p = L_1^+$. We would like to know the p-adic valuation of $R_p \in \mathbf{Q}_p^*$. By elementary row and column operations (adding all the lower rows to the top row and then subtracting the first column from each of the others) we see that

$$R_p((-1/2)log_p(1+p))^{-1}$$

is equal to

$$det\begin{pmatrix} 1 & 1 & \cdots \\ log_p(1-\xi_p^2)-(p-1)^{-1}log_p(1+p) & log_p(1-\xi_p^4)-(p-1)^{-1}log_p(1+p) & \cdots \\ \vdots & \vdots & \vdots \end{pmatrix}$$

which, in turn, is equal to

$$det\begin{pmatrix} 1 & 0 & \cdots \\ \sigma_2(Xog_p(\pi_1)) & \sigma_2(log_p((1-\xi_p^2)/(1-\xi_p))) & \\ \sigma_3(Xog_p(\pi_1)) & \sigma_3(log_p((1-\xi_p^2)/(1-\xi_p))) & \\ \vdots & & \end{pmatrix}.$$

Hence

$$R_p = (-1/2)log_p(1+p)det(a_{i,j}),$$

where $a_{i,j} = \sigma_{j+1}(log_p((1-\xi_p^{i+1})/(1-\xi_p)))$ for $1 \leq i, j \leq (p-3)/2$.

The cyclotomic units

$$\{(1-\xi_p^j)/(1-\xi_p); 2 \leq j \leq (p-1)/2\}$$

are a basis for the cyclotomic units of $\mathbf{Z}[\xi_p]^+$ modulo torsion. Hence R_p is related to the p-adic regulator, $R_p(\mathbf{Q}(\xi_p)^+)$, of Washington (1982, p. 70) by the equation

7.4.48 $$R_p = (-1/2)log_p(1+p)R_p(\mathbf{Q}(\xi_p)^+)u$$

for some p-adic unit, u.

In order to calculate the p-adic ideal generated by R_p, recall the p-adic class number formula (Washington, 1982, section 5.24, p. 71)

$$\frac{2^{(p-3)/2}h(\mathbf{Q}(\xi_p)^+)R_p(\mathbf{Q}(\xi_p)^+)}{\sqrt{d(\mathbf{Q}(\xi_p)^+/\mathbf{Q})}} = \prod_{\chi} L_p(1,\chi),$$

where the product is taken over non-trivial one-dimensional characters of $G(\mathbf{Q}(\xi_p)^+/\mathbf{Q})$. When p is a regular prime the right-hand side of the p-adic class number formula is a p-adic unit (Washington, 1982, section

5.23, p. 70). Also $2^{(p-3)/2}h(\mathbf{Q}(\xi_p)^+)$ is a p-adic unit if p is a regular prime so that

$$\frac{R_p(\mathbf{Q}(\xi_p)^+)}{\sqrt{d(\mathbf{Q}(\xi_p)^+/\mathbf{Q})}}$$

is a p-adic unit.

Therefore, by 7.4.48, the discriminant ideal of the $\mathbf{Z}_p$-lattice,

$$X_1 = \mathbf{Z}_p[(\mathbf{Z}/p)^*/\{\pm 1\}] < Xog_p(\pi_1) > \subset \mathbf{Z}_p[\xi_p]^+$$

in $\mathbf{Q}_p(\xi_p)^+$ is given by

$$\delta_{X_1} = R_p^2 = p^2 R_p(\mathbf{Q}(\xi_p)^+)^2 = p^2 \delta_{\mathbf{Z}_p[\xi_p]^+}.$$

However, $X_1 \subset X_2 = (\pi_1^2 \mathcal{O}_{L_1})^+$ so that, by Serre (1979, pp. 48–49), $X_1 = X_2$ if and only if $\delta_{X_1} = \delta_{X_2}$. On the other hand, since

$$(\mathcal{O}_{L_1}/(\pi_1^2))^+ \cong \mathbf{Z}/p,$$

the quotient of $\mathbf{Z}_p[\xi_p]^+$ by X_2 is equal to $\mathbf{Z}/p$ so that, by Serre (1979, proposition 5, p. 49)

$$\delta_{X_2} = p^2 \delta_{\mathbf{Z}_p[\xi_p]^+} = \delta_{X_1},$$

which shows that π_1 is a free generator of $V_p^+(1)$, as required. □

7.4.49 Let $E = \mathbf{Q}(\xi_{p^{s+1}})$ and $F = \mathbf{Q}$ so that $G(E/F) \cong (\mathbf{Z}/p^{s+1})^*$. We shall now consider the wild prime, p, and apply the procedure of 7.4.31–7.4.33 for obtaining the p-adic component of the Hom-description of the second Chinburg invariant for $\mathbf{Q}(\xi_{p^{s+1}}/\mathbf{Q})$.

In the notation of 7.4.31–7.4.33, $E_Q = \mathbf{Q}_p(\xi_{p^{s+1}}) = L_{s+1}$ and $F_P = \mathbf{Q}_p$ so that $G(E_Q/F_P) = G(E/F)$ in this example. In $\mathbf{Z}[1/p][G(\mathbf{Q}(\xi_{p^{s+1}}/\mathbf{Q}))]$ define idempotents by the formulae

7.4.50
$$\begin{cases} e_j = p^{-j}(\sum_{g \in G(L_{s+1}/L_{s+1-j})} g) \quad (1 \leq j \leq s) \\ \text{and} \quad e_{s+1} = p^{-s}(p-1)^{-1}(\sum_{g \in G(L_{s+1}/\mathbf{Q}_p)} g). \end{cases}$$

The integers, $\mathcal{O}_E = \mathbf{Z}[\xi_{p^{s+1}}]$, have a $\mathbf{Z}$-basis given by

$$\{\xi_{p^{s+1}}^{jp^r} \mid \ 0 \leq r \leq s, HCF(j,p) = 1, 1 \leq j \leq p^{s+1-r} - 1 \text{ and } (j,r) = (1, s+1)\}.$$

Write $\sigma_j \in G(E/F)$ for the map given by $\sigma_j(\xi_{p^{s+1}}) = \xi_{p^{s+1}}^j$. Hence, if

$$\alpha_t = \xi_{p^t} + \xi_{p^{t-1}} + \ldots + \xi_p$$

then

7.4.51 $$e_u(\alpha_{s+1}) = \alpha_{s+1-u}$$

and

7.4.52 $$\begin{cases} \sigma_j(1 - e_{r+1})e_r e_{r-1} \dots e_1(\alpha_{s+1}) = \xi_{p^{s+1}}^{jp^r} \\ \text{and} \quad e_{s+1} \dots e_1(\alpha_{s+1}) = (1-p)^{-1}. \end{cases}$$

An arbitrary element, $z \in \mathcal{O}_E[1/p]$, is of the form

$$z = \sum_{j,r} \mu_{j,r} \xi_{p^{s+1}}^{jp^r} + \mu_{1,s+1},$$

where the sum is taken over

$$\{0 \leq r \leq s, HCF(j,p) = 1, 1 \leq j \leq p^{s+1-r} - 1\}$$

and clearly $z =$

$$\sum_{j,r} \mu_{j,r}\sigma_j(1 - e_{r+1})e_r e_{r-1} \dots e_1(\alpha_{s+1}) - \mu_{1,s+1}(p-1)e_{s+1} \dots e_1(\alpha_{s+1})$$

so that α_{s+1} will serve as the normal basis element which is denoted by $a \in \mathcal{O}_E$ in 7.4.31–7.4.33. If necessary we may replace a by a suitable multiple, $p^c(p-1)^d a$, in order to guarantee that $a \in X$. This change does not alter the element represented in the Hom-description of the class-group, since it changes the representative by a function which is clearly in the image of the induction map from the class-group of the trivial group. Therefore we shall assume that $a = \alpha_{s+1}$.

Observe that if l is a prime different from p and $Q \lhd \mathcal{O}_E$ is a prime over l then α_{s+1} will also serve as a $\mathbf{Z}_l[G(E_Q/\mathbf{Q}_l]$-basis element for $\mathcal{O}_{E_Q}$ or, equivalently, as a $\mathbf{Z}_l[(\mathbf{Z}/p^{s+1})^*]$-basis element for

$$\mathbf{Z}_l[\xi_{p^{s+1}}] \cong \mathbf{Z}_l \otimes \mathcal{O}_E \cong \prod_{Q|l} \mathcal{O}_{E_Q}.$$

We are now ready to combine Theorem 7.4.46 with the method of Theorem 7.4.35 to evaluate $\Omega(\mathbf{Q}(\xi_{p^{s+1}})^+/\mathbf{Q}, 2)$. The formula will involve the value of the p-adic L-function at $s = 1$.

7.4.53 *p-adic L-functions*

Let p be an odd prime. Choose a generator, $g \in (\mathbf{Z}/p^{s+1})^*$ and let $\xi_n = exp(2i\pi/n)$. Let

$$\chi_j : (\mathbf{Z}/p^{s+1})^* \longrightarrow \mathbf{C}^*$$

denote the Dirichlet character given by

$$\chi_j(g) = \xi^{\alpha}_{p^s(p-1)}$$

if $j = g^{\alpha} \in (\mathbf{Z}/p^{s+1})^*$. Denote by f_χ the *conductor of* χ (Washington, 1982, p. 19) so that $f_{\chi_j} = p^{\epsilon}$ is the least positive integer such that χ_j factors through $(\mathbf{Z}/p^{s+1})^* \longrightarrow (\mathbf{Z}/p^{\epsilon})^*$.

Now let us recall Leopoldt's formula (7.4.55), which connects the p-adic logarithms and p-adic L-functions (Washington, 1982, p. 63). Let $\chi : (\mathbf{Z}/p^{s+1})^* \longrightarrow \mathbf{C}^*$ be a Dirichlet character with conductor, $f_\chi = p^{s+1}$. Write $\tau(\chi)$ for the Gauss sum

7.4.54 $$\tau(\chi) = \sum_{a \in (\mathbf{Z}/p^{s+1})^*} \chi(a)\xi^a_{p^{s+1}}.$$

If $\chi = 1$, the trivial character, then set $\tau(1) = 1$.

In this situation the p-adic L-function of χ at $s = 1$, $L_p(1, \chi)$, satisfies

7.4.55 $$L_p(1, \chi) = -\tau(\chi) \sum_{a \in (\mathbf{Z}/p^{s+1})^*} \overline{\chi}(a)p^{-(s+1)} log_p(1 - \xi^a_{p^{s+1}}).$$

We would like to write $L_p(1, \chi_j)$ in the form of 7.4.55 for all χ_j. In Washington (1982, p. 63) 7.4.55 is proved for non-trivial characters, χ, with $\chi(-1) = 1$ (χ is *even*) and $f_\chi = p^{s+1}$. However, if $\chi(-1) = -1$ (χ is *odd*), then both sides of 7.4.55 are zero. If $\chi = 1$ then the right-hand side of 7.4.55 equals

$$\sum_{a \in (\mathbf{Z}/p^{s+1})^*} log_p(1 - \xi^a_{p^{s+1}}) = log_p(p) = 0,$$

so that we shall define $L_p(1, 1)$ to be zero. On the other hand, if χ is even and $f_\chi = p^t > 1$ then $L_p(1, \chi)\tau(\chi)^{-1}$

$$= - \sum_{a \in (\mathbf{Z}/p^t)^*} \overline{\chi}(a)p^{-t} log_p(1 - \xi^{ap^{s-t+1}}_{p^{s+1}})$$

$$= - \sum_{a \in (\mathbf{Z}/p^t)^*} \overline{\chi}(a)p^{-t} \sum_{b \in (\mathbf{Z}/p^{s+1})^*, b \equiv 1 (p^t)} log_p(1 - \xi^{ab}_{p^{s+1}})$$

$$= - \sum_{a \in (\mathbf{Z}/p^{s+1})^*} \overline{\chi}(a)p^{-t} log_p(1 - \xi^a_{p^{s+1}}).$$

Thus 7.4.55 holds in general and we may write

7.4.56 $$L_p(1,\chi_j) = -\sum_{a\in(\mathbf{Z}/p^{s+1})^*} \overline{\chi_j}(a) f_{\chi_j}^{-1}\tau(\chi_j) log_p(1-\xi^a_{p^{s+1}})$$

for all $j \in (\mathbf{Z}/p^{s+1})^*$.

7.4.57 $\Omega(\mathbf{Q}(\xi_{p^{s+1}})/\mathbf{Q}, 2)$

Following the recipe of 7.4.35, there exists

$$B^+_{p,s+1} \in \mathbf{Q}_p[(\mathbf{Z}/p^{s+1})^*/\{\pm 1\}]^*$$

such that

7.4.58 $$B^+_{p,s+1}(\alpha^+_{s+1}) = Xog_p(\pi_{s+1}) = log_p\left(\frac{\pi_{s+1}^{p^s(p-1)}}{p(1+p)}\right)/p^s(p-1)$$

where $\overline{\alpha} = \sigma_{-1}(\alpha)$ is the complex conjugate of α and $\alpha^+_{s+1} = \alpha_{s+1} + \overline{\alpha_{s+1}}$.

For $\mathbf{Q}(\xi_{p^{s+1}})^+/\mathbf{Q}$ we have decomposed the Chinburg invariant

$$\Omega(\mathbf{Q}(\xi_{p^{s+1}})^+/\mathbf{Q}, 2) \in \mathscr{CL}(\mathbf{Z}[(\mathbf{Z}/p^{s+1})^*/\{\pm 1\}])$$

into the sum of a finite module, $W^+_p(s+1) = W_p(s+1)^{<\sigma_{-1}>}$ (the invariants of $W_p(s+1)$ under complex conjugation), and a module whose Hom-description representative (for tame primes, see Fröhlich, 1976, theorem 1, p. 388) is

$$\chi \longmapsto \begin{cases} det(B^+_{p,s+1})(\chi) & \text{at } p \\ (\alpha^+_{s+1} \mid \chi)^{-1}(a^+_l \mid \chi) & \text{at other } l \end{cases}$$

where (a_l) is as in 7.2.1 and 7.2.2 and α^+_{s+1} is a normal basis element for $\mathbf{Q}(\xi_{p^{s+1}})^+/\mathbf{Q}$. Here $(z \mid \chi)$ denotes a *resolvent* defined by the formula

$$(z \mid \chi) = \sum_g g(z)\chi(g^{-1}).$$

In 7.4.50–7.4.52 we observe that α_{s+1} would serve as a $\mathbf{Z}_l$ normal basis element if l is different from p. Hence a_l may be taken to equal α_{s+1} (and similarly $\alpha^+_{s+1} = a^+_l$) so that

$$(\alpha^+_{s+1} \mid \chi)^{-1}(a^+_l \mid \chi) = 1$$

at primes, l, different from p in 7.4.57.

Next we remark that

$$B^+_{p,s+1}(\alpha^+_{s+1}) = Xog_p(\pi_{s+1})$$

so that for all χ (Fröhlich, 1976, proposition 1.3, p. 386)

$$(Xog_p(\pi_{s+1}) \mid \chi) = det(B^+_{p,s+1})(\chi)(\alpha^+_{s+1} \mid \chi)$$

in $\mathbf{Q}_p(\xi_{p^{s+1}})^*$. In addition, if χ is one-dimensional (Fröhlich, 1976, (1.4), p. 385),

$$(Xog_p(\pi_{s+1}) \mid \chi) = \sum_{\gamma} \gamma(Xog_p(\pi_{s+1}))\overline{\chi}(\gamma)$$

(summed over $(\mathbf{Z}/p^{s+1})^*/\{\pm 1\}$)

$$= \sum_{\gamma} [\gamma(log_p(1 - \xi_{p^{s+1}})) - p^{-s}(p-1)^{-1}log_p(1+p)]\overline{\chi}(\gamma)$$

(from 7.4.44)

$$= (1/2) \sum_{a \in (\mathbf{Z}/p^{s+1})^*} [log_p(1 - \xi^a_{p^{s+1}}) - p^{-s}(p-1)^{-1}log_p(1+p)]\overline{\chi}(a)$$

$$= -(f_\chi/2\tau(\chi))L_p(1, \chi)$$

if χ is non-trivial, by 7.4.56. Also,

$$(Xog_p(\pi_{s+1}) \mid 1) = -(1/2)log_p(1+p).$$

Lemma 7.4.59 *Let χ be a one-dimensional complex representation of $G(\mathbf{Q}(\xi_{p^{s+1}})/\mathbf{Q})$ with conductor $1 < f_\chi = p^{j+1}$. Then*

(i) $(\alpha_{s+1} \mid \chi) = p^{s+1}f_\chi^{-1}\tau(\overline{\chi})$,

(ii) $(\alpha^+_{s+1} \mid \chi) = (\alpha_{s+1} \mid \chi) = p^{s+1}f_\chi^{-1}\tau(\overline{\chi}), if\ \chi(-1) = 1$,

(iii) $(\alpha^+_{s+1} \mid 1) = (\alpha_{s+1} \mid 1) = -p^s$.

In the above formula the Gauss sum is the purely local *Gauss sum of* 7.4.54.

Proof The proof is left as an exercise (see 7.5.20). □

This discussion, together with 7.4.35, yields the following result.

Theorem 7.4.60 *Let p be a regular, odd prime. For $\mathbf{Q}(\xi_{p^{s+1}})^+/\mathbf{Q}$ the Chinburg invariant*

$$\Omega(\mathbf{Q}(\xi_{p^{s+1}})^+/\mathbf{Q}, 2) \in \mathcal{CL}(\mathbf{Z}[(\mathbf{Z}/p^{s+1})^*/\{\pm 1\}])$$

decomposes into the sum of the finite module, $W_p^+(s+1)$, and a module whose Hom-description representative is

$$\chi \longmapsto \begin{cases} f_\chi L_p(1,\chi) & \text{at } p, \chi \text{ non-trivial,} \\ -p\log_p(1+p) & \text{at } p, \chi = 1, \\ 1 & \text{at other } l. \end{cases}$$

Proof By 7.4.35, $\Omega(\mathbf{Q}(\xi_{p^{s+1}})^+/\mathbf{Q}, 2)$ is the sum of $W_p^+(s+1)$ and an element whose Hom-description is trivial at primes different from p. At p a one-dimensional character, χ, is sent to

$$(Xog_p(\pi_{s+1}) \mid \chi)(\alpha_{s+1}^+ \mid \chi)^{-1}$$

$$= \begin{cases} -(f_\chi/2\tau(\chi))L_p(1,\chi)(p^{s+1}f_\chi^{-1}\tau(\overline{\chi}))^{-1} & \chi \text{ non-trivial,} \\ -(1/2)\log_p(1+p)(-p^s)^{-1} & \chi = 1, \end{cases}$$

$$= \begin{cases} -(1/2)f_\chi p^{-s-1}L_p(1,\chi) & \chi \text{ non-trivial,} \\ (1/2)p\log_p(1+p)p^{-s-1} & \chi = 1, \end{cases}$$

since $\tau(\overline{\chi})\tau(\chi) = \chi(\sigma_{-1})f_\chi = f_\chi$.

However, the function sending χ to $(-1/2)^{dim(\chi)}$ equals $Det(-1/2)$, which lies in $Det(\mathbf{Z}_p[((\mathbf{Z}/p^{s+1})^*)/\{\pm 1\}]^*)$. Similarly the function sending χ to $(p^{-s-1})^{dim(\chi)}$ lies in $Det(\mathbf{Z}_l[((\mathbf{Z}/p^{s+1})^*)/\{\pm 1\}]^*)$ for all primes, l, different from p. Hence, dividing by the global function

$$(\chi \longmapsto (p^{-s-1})^{dim(\chi)}) \in Hom_{\Omega_{\mathbf{Q}}}(R(((\mathbf{Z}/p^{s+1})^*)/\{\pm 1\}), \mathbf{Q}^*)$$

and multiplying by these l-adic determinants alters the Hom-description to the required form. □

7.4.61 $\Omega(F_{s+1}/\mathbf{Q}, 2)$

Let $F_{s+1}/\mathbf{Q}$ be the subextension of $\mathbf{Q}(\xi_{p^{s+1}})/\mathbf{Q}$ given by $F_{s+1} = \mathbf{Q}(\xi_{p^{s+1}})^{(\mathbf{Z}/p)^*}$. Hence $G(F_{s+1}/\mathbf{Q}) \cong \mathbf{Z}/p^s$ with a generator, γ, say, which may be taken to be the image of $1+p \in (\mathbf{Z}/p^{s+1})^*$.

In this situation we shall prove the following result.

Theorem 7.4.62 *Let p be an odd, regular prime. Then, as predicted by the conjecture of* 7.2.14,

$$\Omega(F_{s+1}/\mathbf{Q}, 2) = 0 \in D(\mathbf{Z}[\mathbf{Z}/p^s]).$$

This result will be proved in the course of the subsequent discussion.

Lemma 7.4.63 *In* $\mathscr{CL}(\mathbf{Z}[\mathbf{Z}/p^s])$ *the function given on one-dimensional representations,* χ, *by*

$$\chi \longmapsto \begin{cases} f_\chi & \text{at } p, \chi \text{ non-trivial} \\ p^2 & \text{at } p, \chi = 1, \\ 1 & \text{at other } l \end{cases}$$

represents the trivial element.

Proof In $Hom_{\Omega_{\mathbf{Q}}}(R(\mathbf{Z}/p^s), \mathbf{Q}^*)$ we have the global function given by

$$\chi \longmapsto \begin{cases} f_\chi & \chi \text{ non-trivial} \\ p^2 & \chi = 1. \end{cases}$$

Dividing by this we may change the Hom-representative of this element so that it is represented by the function which is trivial on all characters at p and on all non-trivial characters away from p. At all other primes the function sends the trivial one-dimensional character to p. However, each component of this function is a local determinant since, by Fröhlich (1983, proposition 2.2, pp. 23–24),

$$Det(\mathbf{R}[\mathbf{Z}/p^s]^*) = Hom^+_{\Omega_{\mathbf{Q}}}(R(\mathbf{Z}/p^s), (\mathbf{Q}(\xi_{p^s}) \otimes_{\mathbf{Q}} \mathbf{R})^*)$$

and (if l is different from p)

$$Det(\mathbf{Z}_l[\mathbf{Z}/p^s]^*) = Hom_{\Omega_{\mathbf{Q}}}(R(\mathbf{Z}/p^s), (\mathbf{Z}(\xi_{p^s}) \otimes_{\mathbf{Z}} \mathbf{Z}_l)^*).$$

This completes the proof, since these functions represent the trivial element in the class-group. □

Next, we observe that the contribution to $\Omega(F_{s+1}/\mathbf{Q}, 2)$ from $W_p(s+1)$ is trivial since this is represented by $W_p(s+1)^{(\mathbf{Z}/p)^*}$. By 7.5.21, this has trivial action and no elements of order p. Therefore, by 5.3.9, this module represents a Swan module, which is trivial in the class-group in the case of a cyclic group (4.2.48; Curtis & Reiner, 1987, p. 344). Now let us examine the other part of the Chinburg invariant.

From Washington (1982, pp. 118–123), if ψ and θ are complex (one-dimensional) characters of $\mathbf{Z}/p^s$ and $(\mathbf{Z}/p)^*$, respectively, then $\chi = \theta\psi$ is a character of

$$(\mathbf{Z}/p)^* \times \mathbf{Z}/p^s \cong (\mathbf{Z}/p^{s+1})^* \cong G(\mathbf{Q}(\xi_{p^{s+1}})/\mathbf{Q}).$$

Furthermore, if $\chi(-1) = 1$ then

7.4.64 $$L_p(1, \chi) = f((p+1)\overline{\psi}(1+p) - 1, \theta),$$

where

$$f(T, \theta) = g(T, \theta)(T-p)^{-1}(1+p)$$

and $g(T, \theta) \in \mathbf{Z}_p[[T]]$. Since we chose the generator, γ, to correspond to $1 + p \in (\mathbf{Z}/p^{s+1})^*$ we see that

$$f((p+1)\overline{\psi}(1+p) - 1, \theta) = \overline{\psi}(f((p+1)\gamma - 1, \theta))$$

since $\gamma = 1+p$ corresponds to $1+T$ under the isomorphism (Washington, 1982, p. 113)

$$\mathbf{Z}_p[[T]] \cong \lim_{\overleftarrow{s}} \mathbf{Z}_p[\mathbf{Z}/p^s].$$

Note that this isomorphism ensures that $g((p+1)\gamma - 1, \theta)$ makes sense in $\mathbf{Z}_p[\mathbf{Z}/p^s]$.

The character, χ, factors through $G(F_{s+1}/\mathbf{Q}) \cong \mathbf{Z}/p^s$ if and only if $\theta = 1$ and $\psi = \chi$. In this case

7.4.65 $$L_p(1, \chi) = \overline{\chi}(g((p+1)\gamma - 1, 1))(\overline{\chi}(\gamma - 1))^{-1}.$$

However, $g(T, 1) \in \mathbf{Z}_p[[T]]^*$ (Washington, 1982, p. 125, lemma 7.12) so that

$$g((p+1)\gamma - 1, 1) \in \mathbf{Z}_p[\mathbf{Z}/p^s]^*$$

and the function

7.4.66 $$\chi \longmapsto \overline{\chi}(g((p+1)\gamma - 1, 1)) = Det(g((p+1)\gamma^{-1} - 1, 1))(\chi)$$

lies in $Det(\mathbf{Z}_p[\mathbf{Z}/p^s]^*) \subset Hom_{\Omega_{\mathbf{Q}}}(R(\mathbf{Z}/p^s), \mathbf{Z}_p[\xi_{p^s}]^*)$. Therefore the element of $Hom_{\Omega_{\mathbf{Q}}}(R(\mathbf{Z}/p^s), J(\mathbf{Q}(\xi_{p^s}))^*)$ given by the function of 7.4.60

7.4.67
$$\chi \longmapsto \begin{cases} f_\chi L_p(1,\chi) & \text{at } p, \chi \text{ non-trivial,} \\ -plog_p(1+p) & \text{at } p, \chi = 1, \\ 1 & \text{at other } l. \end{cases}$$

represents the same class in $\mathscr{CL}(\mathbf{Z}[\mathbf{Z}/p^s])$ as does

$$\chi \longmapsto \begin{cases} (\overline{\chi}(\gamma) - 1)^{-1} & \text{at } p, \chi \text{ non-trivial,} \\ -p^{-1}log_p(1+p)(g(0,1))^{-1} & \text{at } p, \chi = 1, \\ 1 & \text{at other } l. \end{cases}$$

Lemma 7.4.68 *The function on one-dimensional representations, χ, given by*

7.4.69
$$\chi \longmapsto \begin{cases} 1 & \text{at } p, \chi \text{ non-trivial,} \\ -p^{-1}log_p(1+p)(g(0,1))^{-1} & \text{at } p, \chi = 1, \\ 1 & \text{at other } l \end{cases}$$

represents the trivial element in $D(\mathbf{Z}[\mathbf{Z}/p^s])$.

Proof Observe that $-p^{-1}log_p(1+p)(g(0,1))^{-1}$ is a p-adic unit, which is congruent to 1 modulo p. We may write this unit as $u_N v_N$ where v_N is an integer prime to p and u_N is a unit which is congruent to 1 modulo p^N. Since u_N is a p^{N-1}th power, the element of $D(\mathbf{Z}[\mathbf{Z}/p^s])$ represented by

$$\chi \longmapsto \begin{cases} 1 & \text{at } p, \chi \text{ non-trivial,} \\ u_N & \text{at } p, \chi = 1, \\ 1 & \text{at other } l \end{cases}$$

is divisible by p^{N-1} while the element represented by

$$\chi \longmapsto \begin{cases} 1 & \text{at } p, \chi \text{ non-trivial,} \\ v_N & \text{at } p, \chi = 1, \\ 1 & \text{at other } l \end{cases}$$

is trivial, being a Swan module. Hence the class represented by 7.4.69 is divisible by p^m for all m in the (finite) class-group and hence is zero.

Finally, therefore, we see that 7.4.67 represents the same class as

$$\chi \longmapsto \begin{cases} (\overline{\chi}(\gamma) - 1)^{-1} & \text{at } p, \chi \text{ non-trivial}, \\ 1 & \text{at } p, \chi = 1, \\ 1 & \text{at other } l. \end{cases}$$

However, the p-adic values of this representative lie in $\mathbf{Z}_l[\xi_{p^s}]^*$ for all primes, l, different from p. Therefore the argument which was used in the proof of 7.4.63 shows that this formula gives a representative for the trivial element, as required. □

The Fröhlich–Chinburg conjecture of 7.2.14 (sometimes referred to as the Second Chinburg Conjecture (Cassou-Nogués *et al.*, 1991, section 4.2; Chinburg, 1985, section 3.1; Holland, 1992, section 1.1) anticipates that the invariant, $\Omega(E/F, 2)$, will be equal to a class of order two which has a Hom-description in terms of the Artin root numbers of the irreducible symplectic representations of $G(E/F)$. In particular, this conjecture would predict that $\Omega(E/F, 2)$ would vanish when $G(E/F)$ were abelian. Theorem 7.4.62 adds some new, wildly ramified examples to the list of cases in which the conjecture is true.

7.5 Exercises

7.5.1 (*Serre,* 1979, p. 202) Prove Theorem 7.1.20.

(*Hint:* Show that the cohomology classes of 7.1.20 are *fundamental classes* for a local class formation in the sense of Serre (1979, chapter XI, section 3) and use this fact to reduce the calculation to the unramified case.)

7.5.2 Let F_q denote the finite field with q elements and let F_q^c denote its algebraic closure. Let F_0 denote the Frobenius map given by the q^dth power.

(i) Prove that $H^1(G(F_{q^{ds}}/F_{q^d}); F_{q^{ds}}) = 0$. (That is, prove that

$$F_{q^{ds}} \xrightarrow{F_0 - 1} F_{q^{ds}} \xrightarrow{1+F_0+\ldots+F_0^{s-1}} F_{q^{ds}}$$

is exact in the middle.)

(ii) Prove, for any $y \in F_{q^{de}}$, that there exists $w \in F_q^c$ such that $F_0(w) - w = y$.

7.5.3 Verify that the 2-cocycle, $f : G(L/K) \times G(L/K) \longrightarrow L^*$, of 7.1.23 is *normalised*. (That is, show that

$$f(x,1) = f(1,x) = 1.)$$

7.5.4 From the description of the cohomology generators in terms of 2-extensions of 7.1.8 and 7.1.20, prove that the following diagram commutes:

$$\begin{array}{ccc} H^2(G(L/K);L^*) & \xrightarrow{inv} & \mathbf{Q}/\mathbf{Z} \\ \downarrow & & \downarrow (-\cdot[F:K]) \\ H^2(G(L/F);L^*) & \xrightarrow{inv} & \mathbf{Q}/\mathbf{Z} \end{array}$$

7.5.5 Let F be an intermediate extension of L/K. Show that

$$Res^{G(L/K)}_{G(L/F)}(\Omega(L/K,2)) = \Omega(L/F,2) \in \mathscr{C}\mathscr{L}(\mathbf{Z}[G(L/F)]).$$

7.5.6 Let $N \lhd G$ be finite groups. In terms of the Hom-description for $K_0T(\mathbf{Z}[G])$, verify that $Inf^G_{G/N} : R(G/N) \longrightarrow R(G)$ induces the homomorphism which sends the module, T, to its N-fixed points, T^N.

7.5.7 Let L/K be a Galois extension of number fields and let

$$G(E/N) \lhd G(E/K)$$

be a normal subgroup. Let

$$inf_{E/N} : \mathscr{C}\mathscr{L}(\mathbf{Z}[G(E/K)]) \longrightarrow \mathscr{C}\mathscr{L}(\mathbf{Z}[G(N/K)])$$

denote the canonical map (induced by inflation on representations or by taking $G(E/N)$-fixed points on modules).

Use 7.1.57 to show that

$$inf_{E/N}(\Omega(E/K,2)) = \Omega(N/K,2) \in \mathscr{C}\mathscr{L}(\mathbf{Z}[G(N/K)]).$$

7.5.8 Let $G(L/F) \leq G(L/K)$ be as in 7.3.43 with coset representatives $x_1, \ldots, x_d$.

(i) Verify that the homomorphism

$$u_1 : \oplus_{i=1}^d \mathbf{Z}[G(L/F)] \longrightarrow \mathbf{Z}[G(L/K)]$$

given by $u_1(b_1, \ldots, b_d) = \sum_{i=1}^d b_i x_i$ is an isomorphism of left $\mathbf{Z}[G(L/F)]$-modules.

(ii) Verify that

$$u_1(\sigma(L/F), \ldots, \sigma(L/F)) = \sigma(L/K).$$

(iii) Define

$$u_2 : \oplus_{i=1}^d \mathbf{Z}[G(L/F)] \longrightarrow \oplus_{i=1}^d \mathbf{Z}[G(L/F)]$$

by $u_2(b_1, \ldots, b_d) = (b_1, b_2 + b_1, \ldots, b_d + b_1)$. Verify that

$$u_2(\sigma(L/F), 0, \ldots, 0) = (\sigma(L/F), \ldots, \sigma(L/F)).$$

(iv) Verify that the homomorphism

$$v : IG(L/K) \longrightarrow IG(L/F) \oplus (\oplus_{i=2}^d \mathbf{Z}[G(L/F)])$$

given by ($a_i \in \mathbf{Z}[G(L/F)]$)

$$v\left(\sum_{i=1}^d a_i x_i\right) = \left(\sum_{i=1}^d a_i, a_2, \ldots, a_d\right)$$

is an isomorphism of $\mathbf{Z}[G(L/F)]$-modules.

(v) With the identifications of (i)–(iv), verify that 7.3.46 may be identified with 7.3.47 as a diagram of $\mathbf{Z}[G(L/F)]$-modules.

7.5.9 Let $G(L/M) \lhd G(L/F)$ be as in 7.3.50.

(i) Prove that there are $\mathbf{Z}[G(M/F)]$-module isomorphisms

$$\left(\frac{L^*}{1+X}\right)^{G(L/M)} \cong \frac{(L^*)^{G(L/M)}}{(1+X)^{G(L/M)}} \cong \frac{M^*}{1+X^{G(L/M)}}.$$

(ii) Let $y_1, \ldots, y_m$ be a set of coset representatives for

$$G(L/F)/G(L/M)$$

and let $\sigma(L/M)$ be as in 7.3.46. Show that the map which sends $y_s G(L/M)$ to $y_s \sigma(L/M)$ induces $\mathbf{Z}[G(M/F)]$-module isomorphisms of the form

$$a : \mathbf{Z}[G(M/F)] \xrightarrow{\cong} (\mathbf{Z}[G(L/F)])^{G(L/M)}$$

and

$$a : IG(M/F) \xrightarrow{\cong} (IG(L/F))^{G(L/M)}.$$

(iii) Show that the homomorphism, a, of (ii) induces an isomorphism

$$c(m - \sigma(M/F))\mathbf{Z}[G(M/F)] \cong ((r - \sigma(L/F))\mathbf{Z}[G(L/F)])^{G(L/M)}.$$

(iv) With the identifications of (i)–(iii) verify that the $G(L/M)$-invariants of 7.3.47 may be identified with 7.3.51.

7.5.10 Prove that the $G(L/M)$-invariants of 7.3.47 may be identified with 7.3.53.

7.5.11 Let G be a finite group. Let v be a non-zero integer and also multiplication by v, $v : \mathbf{Z} \longrightarrow \mathbf{Z}$. In the notation of 7.3.11, show that f_v^* has a canonical factorisation $g \in Hom_{\Omega_{\mathbf{Q}}}(R(G), \mathscr{I}(E))$ where

$$g(\chi) = \begin{cases} v & \text{if } 1 = \chi, \\ 1 & \text{if } 1 \neq \chi, \text{ irreducible.} \end{cases}$$

7.5.12 Let G be a finite group. Let w be a non-zero integer and also let multiplication by w, be denoted by $w : \mathbf{Z}[G] \longrightarrow \mathbf{Z}[G]$. In the notation of 7.3.11, show that $f^*_{w\mathbf{Z}[G],\mathbf{Z}[G]}$ has a canonical factorisation $h \in Hom_{\Omega_{\mathbf{Q}}}(R(G), \mathscr{I}(E))$ where

$$h(\chi) = w^{dim_{\mathbf{C}}(\chi)}$$

for all $\chi \in R(G)$.

7.5.13 Let G be a finite, cyclic group. Let j be an injective $\mathbf{Z}[G]$-module homomorphism in a short exact sequence of the following form:

$$0 \longrightarrow IG \xrightarrow{j} \mathbf{Z}[G]/(\sigma_G\mathbf{Z}[G]) \longrightarrow \mathbf{Z}/\#(G) \longrightarrow 0.$$

Show that f_j^* has a canonical factorisation $s \in Hom_{\Omega_{\mathbf{Q}}}(R(G), \mathscr{I}(E))$, where

$$s(\chi) = \begin{cases} 1 & \text{if } 1 = \chi, \\ \text{ideal} < \chi(x) - 1) > & \text{if } 1 \neq \chi, \text{ irreducible.} \end{cases}$$

Here x is a generator of G, $\sigma_G = \sum_{g \in G} g$ and *ideal* $< \chi(x) - 1) >$ lies in $\mathscr{I}(\mathbf{Q}(\chi))$.

7.5.14 Let $H \leq G$ be groups and suppose that $\#(G) = n$. Let σ_G be as in 7.5.12. Let $\mathscr{R}$ be a set of coset representatives for G/H and assume that $1 \in \mathscr{R}$.

(i) Show that

$$A = (IG^H)/((n - \sigma_G)\mathbf{Z}[G])^H)$$

is isomorphic to the abelian group on generators

$$\{\lambda_g \mid g \in \mathscr{R} - \{1\}\}$$

subject to the relations

$$n\lambda_g = 0 \text{ for } g \in \mathscr{R} - \{1\}$$

and

$$\#(H)\left(\sum_{g \in \mathscr{R} - \{1\}} \lambda_g\right) = 0.$$

(ii) Prove that

$$\#(A) = \#(H)n^{(|G:H|)-2)}.$$

(*Hint:* Calculate a determinant, as in Theorem 5.2.33(proof).)

7.5.15 Let L/K be a Galois extension of number fields with group, $G(L/K)$. Let S be a finite, $G(L/K)$-stable set of places of L which includes the infinite places, the ramified places and such that, for $K \leq F \leq L$, the S-class number of F is one. Set

$$J_{S,L} = \prod_{v \in S} L_v^* \times \prod_{v \notin S} \mathcal{O}_{L_v}^*$$

and

$$U_{S,L} = \{x \in L^* \mid x \in \mathcal{O}_{L_v}^* \ if \ v \notin S\}.$$

We have a diagonal embedding of $U_{S,L}$ into $J_{S,L}$ and a resulting short exact sequence of abelian groups

$$0 \longrightarrow U_{S,L} \longrightarrow J_{S,L} \longrightarrow C_{S,L} \longrightarrow 0.$$

Global class field theory shows that $\{H^2(G(L/K); C_{S,L})\}$ is a *class formation* (Serre, 1979, pp. 166 and 221). Consequently there are universal cohomology classes which represent the *Weil group extension*

$$C_{S,L} \longrightarrow W_{S,L} \longrightarrow G(L/K).$$

(i) Use the Weil group extension to construct a class

$$\Omega(L/K,S) \in \mathscr{CL}(\mathbf{Z}[G(L/K)])$$

(cf. 7.1.34).

(ii) How does $\Omega(L/K,S)$ vary with S ?

(iii) Let $G(L_w/K_v) \leq G(L/K)$ be a decomposition group. What is the relationship, if any, between $\Omega(L/K,S)$ and the local Chinburg invariant of 7.1.29.

7.5.16 Embed $\mathbf{Q}_2(\sqrt{(-5)})$ as a maximal subfield into a 2-adic division algebra whose Hasse invariant is equal to 1/2. (See Reiner 1975, p. 148, for example, for models of division algebras and consider the matrix

$$X = \begin{pmatrix} \xi_3 - \overline{\xi}_3 & 1 \\ -2 & \overline{\xi}_3 - \xi_3 \end{pmatrix}$$

where $\xi_3 = exp(2\pi i/3)$ and $\overline{z}$ denotes the complex conjugate of z.)

7.5.17 Embed $\mathbf{Q}_2(\sqrt{(-2)}, \sqrt{(-5)})$ as a maximal subfield into a 2-adic division algebra, D, of index four, whose Hasse invariant is equal to 1/4. (If $\alpha = \xi_3 - \overline{\xi}_3$ and $\pi^2 = -2$ in D consider $z = \alpha(1 + \pi^2 + \pi^3) + 2\pi$ in the model for D given in Reiner (1975, p. 148).)

7.5.18 If $\alpha, \beta \in \mathbf{Q}_2^*$ then the *Hilbert symbol*

$$(\alpha, \beta) \in \{\pm 1\}$$

takes the value one if there exist $x, y \in \mathbf{Q}_2$ such that $\alpha + \beta x^2 = y^2$ and is equal to minus one otherwise.

(i) Suppose that $u, v \in \mathbf{Q}_2^*$ are non-squares. Show that there exists a quadratic extension $L/\mathbf{Q}_2(\sqrt{(u)}, \sqrt{(v)})$ such that $L/\mathbf{Q}_2$ is Galois with $G(L/\mathbf{Q}_2) \cong Q_8$ if and only if

$$(u, v)(-1, u)(-1, v) = 1.$$

(Cf. Snaith, 1989b, p. 102, Example (2.31).)

(ii) Construct a Q_8-extension containing $\mathbf{Q}(\sqrt{(-2)}, \sqrt{(-5)})$.

7.5.19 Complete the details of the proof of 7.4.11, following the argument which was used to establish 7.3.27.

7.5.20 Use 3.3.3 to prove the formulae of 7.4.59.

7.5.21 Let $W_p(s+1)$ and $W_p^+(s+1)$ be as in 7.4.37, where p is an odd prime.

(i) Determine the Galois action of $(\mathbf{Z}/p^{s+1})^*$ on $W_p(s+1)$.

(ii) Show that $W_p^+(s+1)$ is a finite group of order prime to p on which the p-Sylow subgroup of $(\mathbf{Z}/p^{s+1})^*/\{\pm 1\}$ acts trivially.

Bibliography

Adams, J.F. (1963), On the groups J(X) I, *Topology*, **2**, 181–195.

Adams, J.F. (1965a), On the groups J(X) II, *Topology*, **3**, 137–171.

Adams, J.F. (1965b), On the groups J(X) III, *Topology*, **3**, 193–222.

Adams, J.F. (1966), On the groups J(X) IV, *Topology*, **5**, 21–71.

Bentzen, S. & Madsen, I. (1983), On the Swan subgroup of certain periodic groups, *Math. Ann.*, **264**, 447–474.

Boltje, R. (1989), Canonical and explicit Brauer induction in the character ring of a finite group and a generalisation for Mackey functors, Augsburg Univ. thesis.

Boltje, R. (1990), A canonical Brauer induction formula, *Représentations linéaires des groupes finis*, Astérisque #181–182, Société Mathématique de France, 31–59.

Boltje, R., Cram, G-M. & Snaith, V.P. (1993) Conductors in the non-separable residue field case, NATO ASI Series Vol. 407 Proc. *Algebraic Topology and Algebraic K-theory*, Lake Louise, Kluwer, 1–34.

Boltje, R., Snaith, V. & Symonds, P. (1992), Algebraicisation of Explicit Brauer Induction, *J. Alg.*, **148**, 504–527.

Brauer, R. (1951), Bezeihungen zwischen Klassenzahlen von Teilkörpern eines Galoischen Körpers, *Math. Nachr.*, **4**, 158–174.

Brylinski, J-L. (1983), Théorie du corps de classes de Kato et revètements abéliens de surfaces, *Ann. Inst. Fourier*, **33**, 23–38.

Burns, D.J. (1991), Canonical factorisability and a variant of Martinet's conjecture, *J. London Math. Soc.*, **142**, 24-46.

Cassou-Noguès, Ph. (1978), Quelques théorèmes de base normal d'entiers, *Ann. Inst. Fourier*, **28**, 1–33.

Cassou-Noguès, Ph., Chinburg, T., Fröhlich, A. & Taylor, M.J. (1991), L-functions and Galois modules, London Math. Soc. 1989 Durham Symposium, *L-functions and Arithmetic*, Cambridge University Press, 75–139.

Chinburg, T. (1983), On the Galois structure of algebraic integers and S-units, *Inventiones Math.*, **74**, 321–349.

Chinburg, T. (1985), Exact sequences and Galois module structure, *Annals of Math.*, **121**, 351–376.

Chinburg, T. (1989), *The Analytic Theory of Multiplicative Galois Structure*, Mem. A.M. Soc. No. 395.

Coombes, K. (1982), Algebraic K-theory and the abelianised fundamental groups of curves, University of Chicago thesis.

Coombes, K. (1986), Local class field theory for curves, *Contemp. Math.*, **55**, 117–134.

Curtis, C.W. & Reiner, I. (1981, 1987), *Methods of Representation Theory*, vols. I & II, Wiley.

Dickson, L.E. (1958), *Linear Groups*, Dover.

Fröhlich, A. (1976), Arithmetic and Galois module structure for tame extensions, *J. Reine. Angew. Math.*, **286/7**, 380–440.

Fröhlich, A., ed., (1977), *Algebraic Number Fields*, London Math. Soc. 1975 Durham Symposium, Academic Press.

Fröhlich, A. (1983), *Galois Module Structure of Algebraic Integers*, Ergeb. Math. 3 Folge-Band No. 1, Springer-Verlag.

Fröhlich, A. (1984), *Classgroups and Hermitian Modules*, Progress in Math. No. 48, Birkhäuser.

Fröhlich, A. (1988), Module defect and factorisability, *Ill. J. Math.*, **32**, 407–421.

Gérardin, P. & Labesse, J-P. (1979), The solution of a base change problem for $GL(2)$ (following Langlands, Saito, Shintani), Amer. Math. Soc. Proc. Symp. Pure Math., *Automorphic Forms, Representations and L-Functions*, 115–134.

Gorenstein, D. (1968), *Finite Groups*, Harper and Row.

Green, J.A. (1955), The characters of the finite general linear groups, *Trans. Amer. Math. Soc.*, **80**, 402–447.

Greenberg, M. & Harper, J. (1971), *Lectures on Algebraic Topology*, W.A. Benjamin.

Hasse, H. (1980), *Number Theory*, Springer-Verlag Grund. Math. Wiss. No. 229.

Hilton, P.J. & Stammbach, U. (1971) *A Course in Homological Algebra*; Grad. Texts in Math. No. 4, Springer-Verlag.

Holland, D. (1992), Additive Galois module structure and Chinburg's invariant, *J. reine angew. Math*, **425**, 193–218.

Holland, D. & Wilson, S.M.J. (1992), Fröhlich's and Chinburg's conjectures in the factorisability defect class-group, preprint.

Hunter, J. (1964), *Number Theory*; Oliver and Boyd.

Huppert, B. (1967), *Endliche Gruppen I*, Springer-Verlag.

Hyodo, O. (1987), Wild ramification in the imperfect residue field case, in *Galois Representations and Arithmetic Algebraic Geometry* (ed. Y. Ihara), Adv. Studies in Pure Maths. No. 12, Kinikuniya Co., 287–314.

Illusie, L. (1979), Complexe de de Rham-Witt et cohomologie cristalline, *Ann. Sci. Ec. Norm. Sup.*, **12**, 501–661.

Isaacs, I.M. (1976), *Character Theory of Finite Groups* Academic Press.

Iwasawa, K. (1986), *Local Class Field Theory*, Oxford Math. Monographs, Oxford University Press.

Kato, K. (1987), Swan conductors with differential values, *Advanced Studies in Pure Math.*, **12**, 315–342.

Kato, K. (1989), Swan conductors for characters of degree one in the imperfect residue field case, *Contemp. Math.*, **83**, 101–131.

Kim, S. (1991), A generalization of Fröhlich's theorem to wildly ramified quaternion extensions of **Q**, *Ill. J. Math.*, **35**, 158–189.

Kim, S. (1992), The root number class and Chinburg's second invariant, *J. Alg.*, **153**, 133–202.

Lang, S. (1970), *Algebraic Number Theory*, Addison-Wesley.

Lang, S. (1984) *Algebra*, 2nd ed., Addison-Wesley.

Macdonald, I.G. (1979), *Symmetric Functions and Hall Polynomials*, Oxford Math. Monographs, Oxford University Press.

Martinet, J. (1977a), H_8, London Math. Soc. 1975 Durham Symposium, *Algebraic Number Fields*, ed. A. Frölich, Academic Press, 525–538.

Martinet, J. (1977b), Character theory and Artin L-functions, London Math. Soc. 1975 Durham Symposium, *Algebraic Number Fields*, ed. A. Frölich, Academic Press, 1–87.

Matsumura, H. (1990), *Commutative ring theory*, Cambridge Studies in Advanced Mathematics No. 8, Cambridge University Press.

Maunder, C.R.F. (1970), *Algebraic Topology*, Van Nostrand Reinhold.

Milne, J.S. (1980), *Étale cohomology*, Princeton Mathematical Series No. 33, Princeton University Press.

Oliver, O. (1978), Subgroups generating $D(\mathbf{Z}[G])$, *J. Algebra*, **55**, 43–57.

Oliver, O. (1980), SK_1 for finite group rings II; *Math. Scand.*, **47**, 195–231.

Quillen, D.G. (1971), The Adams conjecture, *Topology*, **10**, 67–80.

Quillen, D.G. (1972), On the cohomology and K-theory of the general linear groups over a finite field, *Annals of Math.*, **96**, 552–586.

Rota, G-C. (1964), On the foundations of combinatorial theory I. Theory of Möbius functions, *Z. Wahrsch verw. Gebiete*, **2**, 340–368.

Reiner, I. (1975), *Maximal Orders* , L.M.Soc. Monographs No. 5, Academic Press.

Rogers, M (1990), On a multiplicative-additive Galois invariant and wildly ramified extensions, Columbia University thesis.

Sehgal, S.K. (1978), *Topics in Group Rings*, Mono. in Pure and Applied Math. No. 50, Marcel Dekker.

Sen, S. (1969), On automorphisms of local fields, *Annals of Math.*, **90**, 33–46.

Serre, J-P. (1960), Sur la rationalité des représentations d'Artin, *Annals of Math.*, **72**, 405–420.

Serre, J-P. (1977), *Linear Representations of Finite Groups*, Grad. Texts in Math. No. 42, Springer-Verlag.

Serre, J-P. (1979), *Local Fields*, Grad. Texts in Math. No. 67, Springer-Verlag.

Shintani, T. (1976), Two remarks on irreducible characters of finite general linear groups, *J. Math. Soc. Japan*, **28**, 396–414.

Snaith, V.P. (1988a), A construction of the Deligne–Langlands local root numbers of orthogonal Galois representations, *Topology*, **27**, 119–127.

Snaith, V.P. (1988b), Explicit Brauer Induction, *Inventiones Math.*, **94**, 455–478.

Snaith, V.P. (1989a), Invariants of representations, NATO ASI Series No. 279, *Algebraic K-theory: Connections with Geometry and Topology*, Kluwer, 445–508.

Snaith, V.P. (1989b), *Topological Methods in Galois Representation Theory*, C.M.Soc Monographs, Wiley.

Snaith, V.P. (1990a), On the class group and the Swan subgroup of an integral group-ring, McMaster University preprint No. 3.

Snaith, V.P. (1990b), Restricted determinantal homomorphisms and locally free class groups, *Can. J. Math.*, **XLII**, 646–658.

Snaith, V.P. (1990c), *The Yukiad*, a novel, The Book Guild (ISBN 0-86332-478-9).

Sullivan, D. (1974), Genetics of homotopy theory and the Adams conjecture, *Annals of Math.*, **100**, 1–79.

Swan, R.G. (1960), Periodic resolutions for finite groups, *Annals of Math*, **72**, 267–291.

Symonds, P. (1991), A splitting principle for group representations, *Comm. Math. Helv.*, **66**, 169–184.

Tate, J.T. (1977), Local constants, London Math. Soc. 1975 Durham Symposium, *Algebraic Number Fields* (ed. A. Fröhlich), Academic Press, 89–131.

Tate, J.T. (1984), *Les conjectures de Stark sur les fonctions L d'Artin en* $s = 0$, Progress in Mathematics No. 47, Birkhauser.

Taylor, M.J. (1978), Locally free classgroups of groups of prime power order, *J. Algebra*, **50**, 463–487.

Taylor, M.J. (1980), A logarithmic approach to class groups of integral group rings, *J. Algebra*, **66**, 321–353.

Taylor, M.J. (1981), On Fröhlich's conjecture for rings of integers of tame extensions, *Inventiones Math.*, **63**, 41–79.

Taylor, M.J. (1984), *Class Groups of Group Rings*, L.M.Soc. Lecture Notes Series No. 91, Cambridge University Press.

Thomas, A.D. & Wood, G.V. (1980), *Group Tables*, Shiva Math. Series 2, Shiva Publishing.

Wall, C.T.C. (1966), Finiteness conditions for CW complexes, *Proc. Royal Soc. A.*, **295**, 129–139.

Wall, C.T.C. (1974), Norms of units in group rings, *Proc. London Math. Soc.* (3), **29**, 593–632.

Wall, C.T.C. (1979), Periodic projective resolutions, *Proc. London Math. Soc.* (3), **39**, 509–553.

Washington, L. (1982), *Introduction to Cyclotomic Fields*, Springer-Verlag Grad. Texts in Math. No. 83.

Wilson, S.M.J. (1990), A projective invariant comparing integers in wildly ramified extensions *J. reine angew. Math.*, **412**, 35–47.

Index

UNIVERSITY
LIBRARY